Forests of Central Africa
Nature and Man

Table of contents

	Preface	7
	Foreword	9
1	The mineral world	13
2	Climate and forests	39
3	Ecological gradients	69
4	Temporal gradients	99
5	The life in the forest	147
6	The fauna	181
7	Man and the forest	239
8	Conservation	281
9	Great challenges	305
	Epilogue	349
	Appendix: the main protected areas of forested Central Africa	352
	Bibliography	354
	Index	359
	Acknowledgements and credits	367

1 (PAGE 1) Papilio zalmoxys is a large blue butterfly, endemic to the dense humid forests of the guineo-congolian region.

2 (PAGE 2) The Mambili, a small tributary of the Congo River, remains the most important way to travel inside the Odzala-Kokoua National Park in Congo. For many centuries rivers were the only safe way to travel over long distances in the Congo Basin, and in many places they are still unavoidable.

3. The African elephant Loxodonta africana and its immediate ancestors have profoundly "managed" the guineo-congolian forest. Without these animals these forests would probably have been very different from what they are today.

Preface

Successfully writing about tropical forests is a challenge almost impossible to meet as this would imply both exposing an elusive world (due to its multifarious nature) and conveying the feeling of enormous power and mysterious danger it generates while at the same time letting its tremendous frailty and almost total dependence on the will of a few decision-makers filter through. Just like the sea, the forest mesmerizes man as it helps him live, think and dream. This applies to all men, including those who have always lived off it, the foresters harvesting the forest's fibre resources, naturalists and scientists monitoring and studying them as well as the artists attempting to render the inspiration it spawns.

Its strange beauty, the melting pot of living organisms silently competing against each other, the contrasting dark and light tangle made of sundry botanical forms that come in all shades of green, arouse wonder, fear or the desire to dominate. Hopefully, as fascination affects each and every one of us, no one can claim to possess it exclusively.

The extreme wealth of the tropical forest's biological diversity has been amply demonstrated and further confirmed by the constant discovery of new species. The intimate knowledge of the forest peculiar to the people who made the forest their abode often inspires respect and humility. As the forest is such a vast repository of species and its ecology is such a dynamic process, shouldn't it serve as the prime example of the principles of homeostasis, claiming that the more complex the biotope, the more stable it is? A great number of external factors have disrupted the theoretical system and shaken up the equation: clearing of major species, indiscriminate and unchecked hunting, transforming of the environment for farming and breeding purposes, etc. Against the host and scope of such aggressions the "dark and hostile" forest as described by explorers can no longer defend itself alone. The ecological network that developed throughout the centuries is silently eroding away, and in an increasing number of tree clumps the canopy is protecting but a small fraction of the original biodiversity.

Yet we have plenty of reasons to wisely manage the asset represented by the many resources of Central African forests and to ensure that as many social, economic and environmental interests benefit from the product of such resources. In other terms, a consensus is required as each party has different desires and focuses efforts on extracting just a few specific resources, a few suits if you will in this house of cards…

Mankind's forest is neither an infinite resource nor a burning library. Its content is naked and alive and provides Central Africa with a foundation on which it can build its development. Conversely, wasting the forest's resources causes further risks of under-development. The tropical forest is the pre-eminent sanctuary of the large primates as well as the cradle of humanity. If this book had only one moral it would probably be that that very same humanity, whether tangled up in abject poverty or anxious to become too rich too quickly, is destroying its own house, its own attic and legacy. We can only try to imagine what degree of insanity, arrogance and despondency it has reached. Luckily for the forest an increasing number of voices are currently speaking in favour of preserving it and ensuring its sustainable management.

Against such a complex backdrop, where the dynamic force between man and his vital resources is so complicated, it is risky to venture without a proper guide. Jean-Pierre Vande weghe shares his immense knowledge of African nature with us as well as his wide experience of the various players called upon in this book, which serves as a comprehensive survey and a particularly relevant tool. Let us hope that his message will be heard: to ensure the sustainability of the forest we will need both the political will and willingness of all people to reverse the destructive process, as well as ample funds and swiftness in order to overcome the race against time.

Jean-Pierre d'Huart

4. Not unlike our ancestors a few million years ago, a group of western lowland gorillas, led by its "silverback" goes out to look for food in a grassy clearing. Odzala-Kokoua National Park, Congo.

Foreword

Until hardly 30 or 40 years ago, tropical forests represented for most Westerners no more than an infernal setting. They were, of course, a fabulous world of tree ferns and palms, a world of giant trees supported by immense buttresses and loaded with strange orchids, a world of monkeys swinging from liana to liana, of simmering butterflies and birds of paradise with psychedelic colours and shapes. But it was also a world of dangers. Each leaf hid a snake, a scorpion or an enormous hairy spider. The torrential rains, the clouds of insects, the toxic or stinging plants and the innumerable parasites rendered life there unbearable. At night, the traveller, overcome by the suffocating heat and strange noises, tossed and turned on his bed dreading the approach of beasts. In Africa, the Elephant, Buffalo and Forest Leopard were much more vicious than their grassland congeners and the Gorilla was the mythical King-Kong that only the most intrepid hunters would confront at the risk of their life. Like the swamps, synonyms of pestilence and symbols of anti-humanity, the tropical forests were no more than a heresy of nature and an obstacle to all development. They produced wood, of course, but sooner or later they would have to give way to croplands or let themselves be tamed and managed. The idea of protecting them was completely preposterous. Besides, they were inexhaustible and no one would ever get to the end of them. Unfortunately, this vision was not based only on fantasies. Until a few years ago older missionaries remember the time – it was well before the Second World War – when nearly half their colleagues succumbed to fevers before even reaching their posts in the Congo basin. As for the coasts of Liberia and Sierra Leone, weren't they called "the White Man's Grave"!

In the second half of the 20th century, this way of seeing things changed. Of course, it was found that tropical forests harboured some fearsome viruses – the recent epidemics of Ebola are there to remind us – but at the same time it was realised that they were in general not as hostile as previously believed. It was even possible to venture there without leather boots or pith helmet and the Gorilla was often no more than a peaceful, fearful giant that tourists ended up approaching and observing, armed only with their cameras. It was also discovered that tropical humid forests constitute the richest habitat on earth. Even if butterflies, orchids, snakes, scorpions and hairy spiders were sometimes much rarer than had been imagined, these forests still harboured more than half of all living species of the planet. Little by little, attention and funds thus turned away from the African savannas which, until the 1970s, symbolised the last bastions of wild nature, monopolised the attention of conservationists and inspired generations of writers, poets and film-makers. Their large fauna, although enormously spectacular, was abandoned to the monotony of the plains, to misty horizons and the weapons of hunters.

However, as the world became conscious of the immense value of these forests, it measured the extent to which they were threatened, not only by the demographic explosion and profound changes of the local populations, but also – perhaps above all – by the economic growth of the industrialised world and the exponential increase of its needs. Curiously, the forests of Central Africa seemed better able to resist than others.

5. Deep into the forest of the Kivu Region, a liana bridge thrown over a small tributary of the Luhoho River shows that Man is everywhere.

FOREWORD

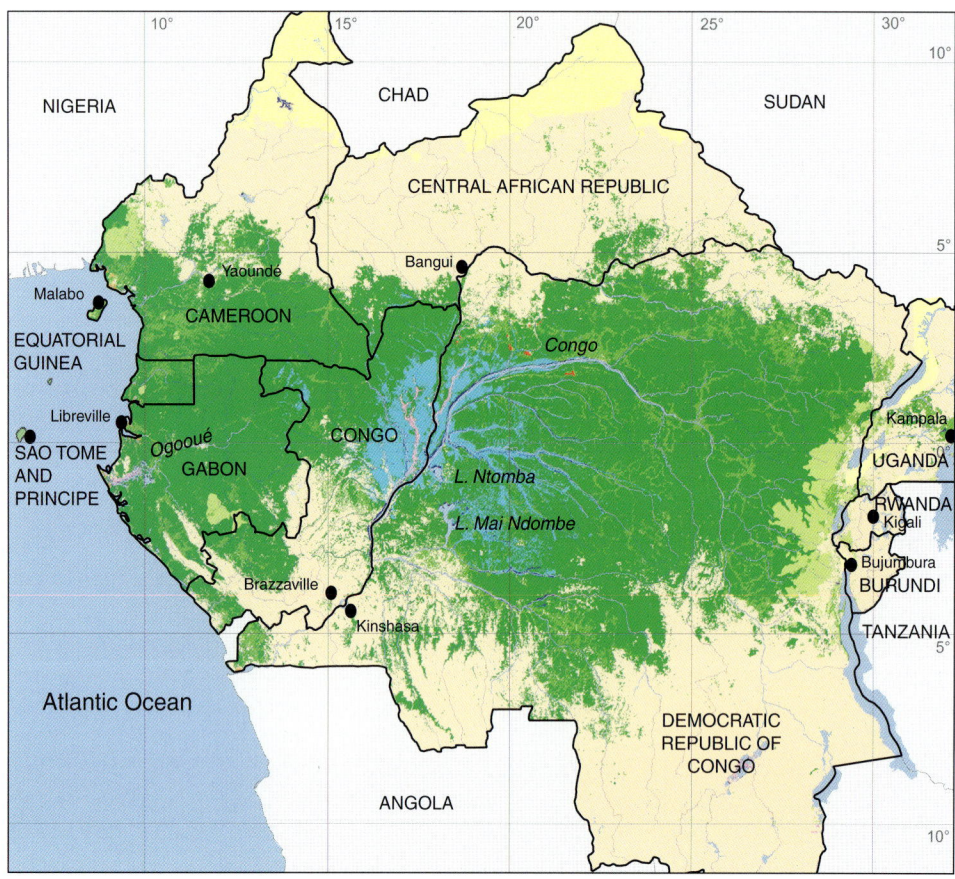

6. This book covers the forests of Cameroon, Gabon, Equatorial Guinea, Central African Republic, Congo, Democratic Republic of the Congo, western Uganda, Rwanda, Burundi and Sao Tomé and Principe. It covers mainly the dense humid forests, the lowland forests and the mountain forests, but it has also an eye for the many open habitats included in the forest belt. Finally it does not ignore the fact that dry forests and moist riparian forests extend far into the grassland regions.

Of course, forest exploitation, often anything but sustainable, and hunting turned into carnage have irremediably degraded Central African forests, but they have not yet undergone massive conversion. A context of unsettled social and economic conditions, of population explosion, of planned and unplanned immigration, of cultural imbalances, of recurrent unrest and of relentless global economic pressures, however, threatens soon to plunge Central Africa's forests into the same upheavals as those that have engulfed the forests of West Africa and Asia.

Since the 1980s, much effort has been expended on the protection of the forests of Central Africa: from the Gulf of Guinea to Lake Victoria, more than 14 million square kilometres of forest have been included in some sixty faunal reserves and national parks, not to mention forest reserves and hunting domains. Among these sites, there are Biosphere Reserves, World Heritage sites and the largest national park in Africa, with an area of 3.6 million hectares for it alone. Finally, in the border region of the Democratic Republic of the Congo, Uganda, Rwanda and Burundi, exists the most extensive and nearly continuous trans-border network of protected areas. Existing reserves are regularly enlarged and new reserves proclaimed. In 2001, for example, Odzala National Park in the Congo was increased from 128,000 to 1,350,000 hectares. In 2002, Gabon gazetted 13 national parks with a total area of 3,007,445 hectares. For some, this is excessive and only represents a disguised despoliation of the African populations. But, whatever is done, no more than 10% of the area of existing forests will ever end up being protected and the 90% remaining are condemned to degradation or conversion.

The struggle to preserve the forests of Central Africa therefore can no longer limit itself only to protected areas. It requires human and financial means that the states concerned do not have. The only recourse left is international aid. Unfortunately, funds that can be mobilised are largely insufficient and increasingly absorbed by the development of strategies and action plans or by the organisation of seminars, colloquia and workshops whose effects are not always very visible. While the industrial world has never been as rich and powerful as it is today – no one is unaware of it, not even in the deepest part of the forest – the sums allotted to action in the field are becoming more and more derisory. In addition, they are often still diverted to development programmes, under the pretext that without development there is no conservation.

More than ever, available funds have to be used in an optimal way. But here one runs up against a clear lack of knowledge: vast regions of Central Africa remain virtually unexplored, and knowledge of the distribution of many plants and animals is only fragmentary. How, then, can realistic programmes be conceived? How can our rare resources be best used? How can conservation, development and sustainable exploitation be reconciled? These problems have become all the more complex now that everybody has something to say about conservation, and that human sciences experts or economists tend to replace biologists, sweeping away a hundred years of biological research and forestry experience in the name of a fashion which will probably fade away like all the others. And as if all that were not enough, activists, although sometimes very useful, come along and throw oil on the fire by placing tropical forests in the middle of controversies highly publicised by the media where emotions win over good sense.

This situation is all the harder given that the sociopolitical context is ever more constraining and that authoritarian management of protected areas and forestry concessions have had to give way to forms of management more attentive to the needs and constraints of the human populations who are among the most destitute on the planet.

It was therefore urgent to re-situate the forests of Central Africa in a more serene context, to give back to biology the place it deserves and to bring the peoples of the forest into a more realistic and perhaps more respectful light. Only the pretext was missing. The European Commission's ECOFAC programme, financed by the European Fund for Development within the framework of the Lomé conventions, provided one. Since 1993, this ambitious regional programme, active in Cameroon, the Congo, Gabon, Equatorial Guinea, the Central African Republic and São Tomé and Principe, has tackled in the field several problems raised by the conservation of forest ecosystems, on a biological, as well as human, plane. Wherever the opportunity arises, this work draws upon examples from results and experiences acquired during the first ten years of the ECOFAC programme.

Among the innumerable difficulties it had to, and still must, face, is the fact that the protected forested areas of Central Africa remain very little known both from lack of communication and from lack of visitors. Literature dealing with nature and its conservation in sub-Saharan Africa is not lacking, however, but it is either hard to find for the non-professional naturalist or superficial. Or else, the problem of the Central African forests is drowned in a world context where western Africa, South-east Asia and Amazonia are mingled with it. Central Africa has its own particularities, however, and amply deserves separate treatment.

The lack of visitors in protected areas is obviously explained by the rather sombre – if not catastrophic – regional context, but it is also largely linked to the closed nature of the forest habitat, very disappointing for any unprepared or non-supported visitor. Indeed, the trees hide the landscape and look hopelessly alike. Birds can be heard, but are observed only very furtively. The dark environment is dominated by amphibians and insects that one never sees, and monkeys often betray their presence only by the fruits that they drop. Although tropical forests harbour more than half of the living species of the planet, this diversity only rarely constitutes an attraction in itself. The richest and most complex nature must imperatively be placed within its context and interpreted.

This work attempts to do this, without pretending to be complete. Some subjects are only lightly touched upon and others benefit perhaps from a too detailed treatment. A choice had to be made, and it inevitably reflects the author's personal experience.

The first six chapters outline the physical context of Central Africa and provide a biological introduction to its forests, including the innumerable open habitats they contain and which largely contribute to their biological diversity. Central Africa has been understood in a fairly broad sense: in this book it extends from the Cross River to Lake Victoria and includes the heights of Cameroon and the Albertine Rift. The scope is not limited to large, apparently still mostly undisturbed forest tracts, but also includes a number of fragmented and marginal habitats which unfortunately foreshadow the forests of the future.

The three last chapters illustrate the human context, both historical and present. Indeed, the historical context is too often forgotten and African nature, the forest especially, then becomes a sort of vestige of the Garden of Eden which escaped from Man's grip. Of course, there are intact habitats in Africa and old forests – whatever some foresters might think – survive very well without human intervention. But in many places Man's impact is so complex that it is difficult to distinguish the natural from the human: forests with an absolutely virgin appearance are thus no more than the result of millennia of interaction between man and nature, others appearing just as wild, are hardly a century old and owe their existence to the tragic events that marked the opening of Africa to the world. Finally, this human impact varies considerably from one region to another, especially when montane forests are compared to lowland forests.

As for the present human context, no one denies its importance. Its demographic and socio-economic factors cannot, however, be dissociated from the growing pressures of the industrialised countries that always want more while always paying less. Generally, this human context is at the root of the problems experienced today by the national parks and reserves. It forces us to accept more and more compromises in the desire to reconcile conservation and the needs of human populations. Finally, it questions the very existence of protected areas. In each continent or subcontinent this human context is differently shaped, however, and experiences acquired in Amazonia, Asia or even West Africa cannot immediately be transferred to Central Africa without some adaptation.

Finally it should be mentioned that this book, completed at the end of 2002, will be published just at the moment when a new partnership initiative, including the Central African countries and the Western World, is being launched with the aim to save these Central African forests.

I | The mineral world

Covering an area of a little over two million square kilometres – nearly four times the size of France – the forests of Central Africa are the second largest block of tropical forest in the world but their fauna and flora comprise far fewer species than those of Amazonia or south-east Asia. Biodiversity is not expressed in figures, however. It is described. Like the forests of the other continents, with which they share many similarities, the Central African forests have their own originality and thus contribute significantly to the biological diversity of our planet. For example, the large mammals are still much better represented than elsewhere and many of the species are unique. They include the Okapi, the numerous species of forest duikers, the anomalures or flying mice and three great apes: Gorilla, Chimpanzee and Bonobo, to mention just a few.

This biological treasure came down to us, to the dawn of this 21st century, relatively intact, thanks perhaps to African *Homo sapiens*, who seems to have been less destructive than his counterparts on other continents for the past tens of thousands of years. The uniqueness of the fauna and flora has its origin, however, in a very much older array of phenomena that involve the living and the mineral kingdoms. The present-day faunas and floras of the Central African forests are in fact the outcome of a very long history in which progress has in large part been oriented and punctuated by major geological events. On several occasions these phenomena nearly wiped out everything but, after each cataclysm, life got the upper hand and started up again on a new track.

The foremost factor in the shaping of the characteristics of the Central African forests is that the majority of the continent is made up of very old rocks and that its relief is hopelessly flat, unlike tropical Asia, for example. Later, after the break-up of Gondwanaland, which occurred mainly in the Cretaceous period, 65 to 130 million years ago, Africa was completely isolated from the other continental masses for nearly 80 million years, a period during which most of the present-day families and species, both plant and animal, came into being. Finally, during the last 20, 30 or 40 million years, a number of violent geological events profoundly changed the eastern and western margins of Central Africa, creating immense rifts, thrusting up volcanoes, lifting new mountain ranges, changing stream courses and creating immense lakes. These upheavals put in place Central Africa's present-day relief, breaking the sad monotony of the plains and plateaux, modulating the climate and creating favourable conditions for the diversification of faunas and floras. Therefore it is impossible to apprehend the characteristics of Central African forests and their problems without taking into account the geological history of the region.

1.1. Since the Tertiary, the movements of the earth's crust lifted the western margins of Central Africa and rejuvenated the hydrographic networks. In creating spectacular rapids and falls, like those of Koungou on the Ivindo in Gabon, they built the barriers that would, for centuries, prevent foreign penetration into the interior of the continent.

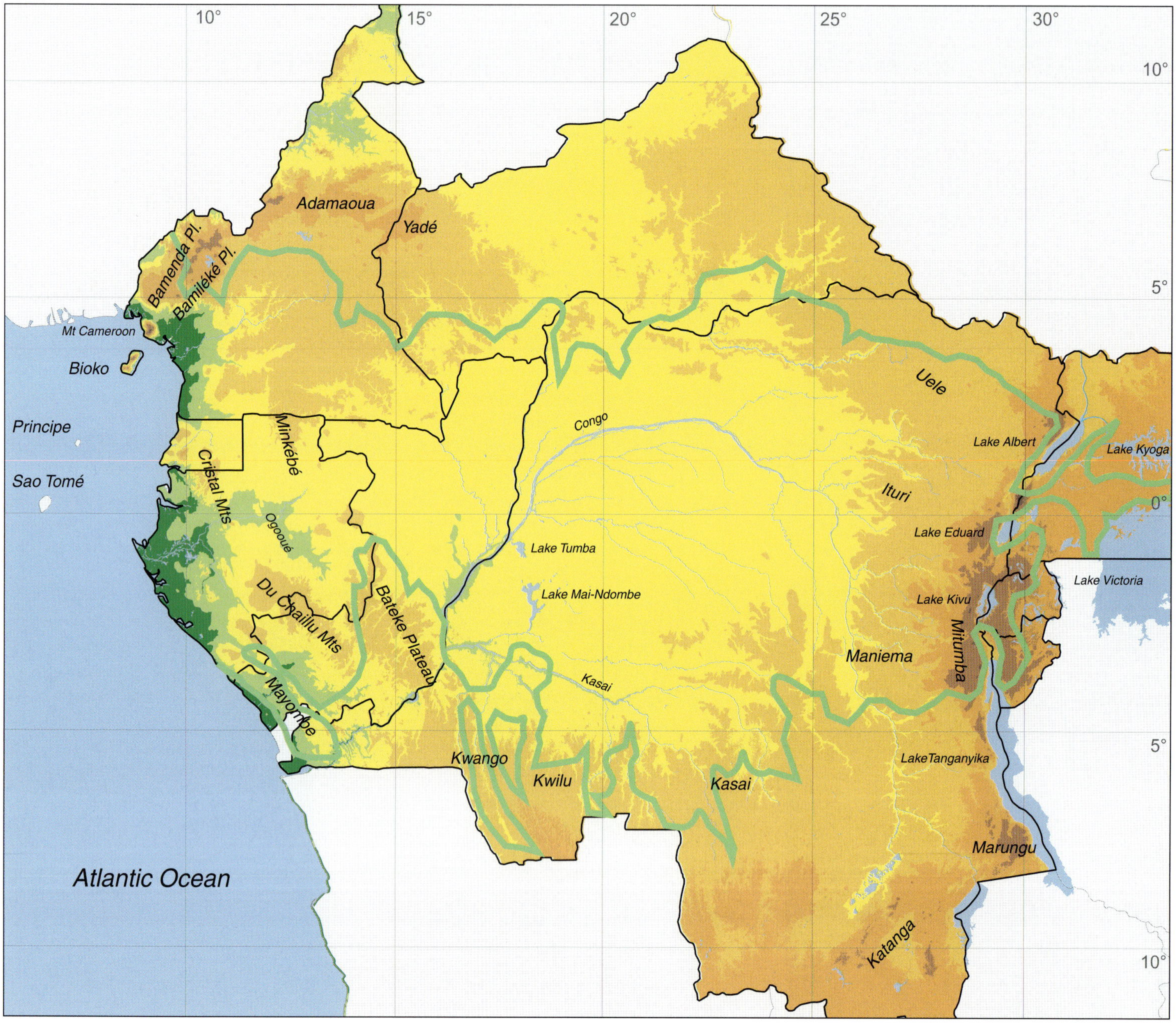

Figure 1.2. Central Africa : relief and river network

0 - 100 m
100 - 300 m
300 - 600 m
600 - 1000 m
1000 - 1650 m
1650 - 3000 m
> 3000 m

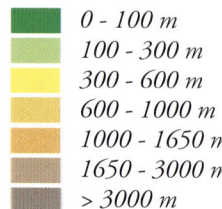

1.2. Central Africa: relief and hydrography.

The origin of the rocks

Africa is the largest continent after Asia, but its relief is very flat and often very monotonous. Forested Central Africa, in particular, is mainly an immense basin where the Congo River and its innumerable tributaries tirelessly drain their alluvia. In the centre, at around 300 metres altitude, lies one of the largest marshes in the world. All around, plateaux and more or less mountainous massifs rise, with altitudes culminating at 600 to 1,500, even 2,000, metres (FIGURE 1.2). They separate the Congo basin from the Chari basin in the north, from the Nile in the north-east, from the Zambezi in the south, and the Ogooué in the west. In spite of being very old and often very eroded, these features exist. The eastern edge of the basin lies up against the Albertine Rift, whose two nearly parallel mountain ranges culminate at between 3,000 and 5,100 metres and, with the string of great lakes that they frame, form an immense barrier stretching some 1,000 kilometres from north to south. In the north-west, the high volcanic mountains of the Cameroon Range extend for

I THE MINERAL WORLD

more than 800 kilometres in an arc from Mount Cameroon at the edge of the ocean to the Yade plateau in the Central African Republic and include the mountains and plateaux of the Bamenda highlands, the Bamiléké region and the Adamaoua (FIGURE 8).

Spectacular as they might be, on a continental scale these elevated formations cover only relatively small surfaces. In spite of this, they have played, and still play, a very important role biologically. The rest of the continent remains flat and monotonous. This very characteristic appearance is linked to its history, well known from mining research and, more recently, petroleum prospecting.

We now know that the Earth was born around 4.5 billion years ago. Of the Hadean, the very first epoch of its geological history, there remains no trace in Central Africa, where the oldest rocks are 3.5 billion years old. They belong to the Archaean epoch, the second in the Earth's history, which spanned the period 3.9 to 2.6 billion years BP (before present). This epoch was characterised by the birth of cratons, thick shields of eruptive rocks – principally granites, diorites, gabbros, dolerites and migmatites – several hundred kilometres in diameter and 15 to 20 kilometres thick, that prefigured the continents. For a long time it was thought that these cratons were perfectly rigid and unchanging masses, but today we know that they have undergone important reshaping during the pan-African cycle of rejuvenation, some 600 million years ago.

Four such cratons probably joined to form the African continent (FIGURE 1.4). Central Africa, in particular, lies mostly on the Congo craton, nowadays largely buried under an enormous coating of sediments. However, its rocks show on the surface in a vast discontinuous crown around the perimeter of the Congo basin (FIGURE 1.3). They are found in southern Gabon and the Congo, mainly on the western boundary of the Mayombe Range and the du Chaillu massif. They are encountered again throughout northern Gabon – from north-east of the Lopé National Park and the Crystal Mountains to the Minkébé massif and the Bélinga Mountains – a large part of Cameroon – including the region of the Dja – and the western part of the Central African Republic. To the north-east of the basin, these Archaean rocks show on the surface in the eastern part of the Central African Republic, in the north-east of the Democratic Republic of the Congo and in all of northern Uganda. South of the basin, in the Kasai, they show at the surface in a more discontinuous way.

Immense mobile zones between the rigid cratons were invaded by shallow seas that accumulated sediments eroded from emergent surfaces and filled them

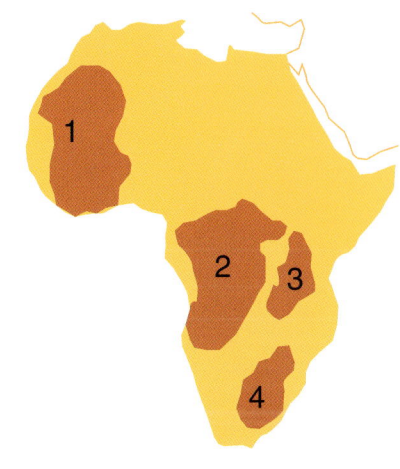

1.3. By 3.5 billion years ago, the Earth had cooled sufficiently and its Archaean cratons began to form. The welding together of the cratons of West Africa (1), of the Congo (2), of Tanzania (3) and of the Kalahari (4) created the African continent. This process began 2.6 billion years ago and ended around 600 million years ago. Beginning in this epoch, the African continent formed a rigid unit. Until 150 to 200 million years ago it was, however, part of a much larger block: Gondwanaland (after the Geological atlas of UNESCO).

1.4. Simplified geology of Central Africa (after the Geological atlas of UNESCO).

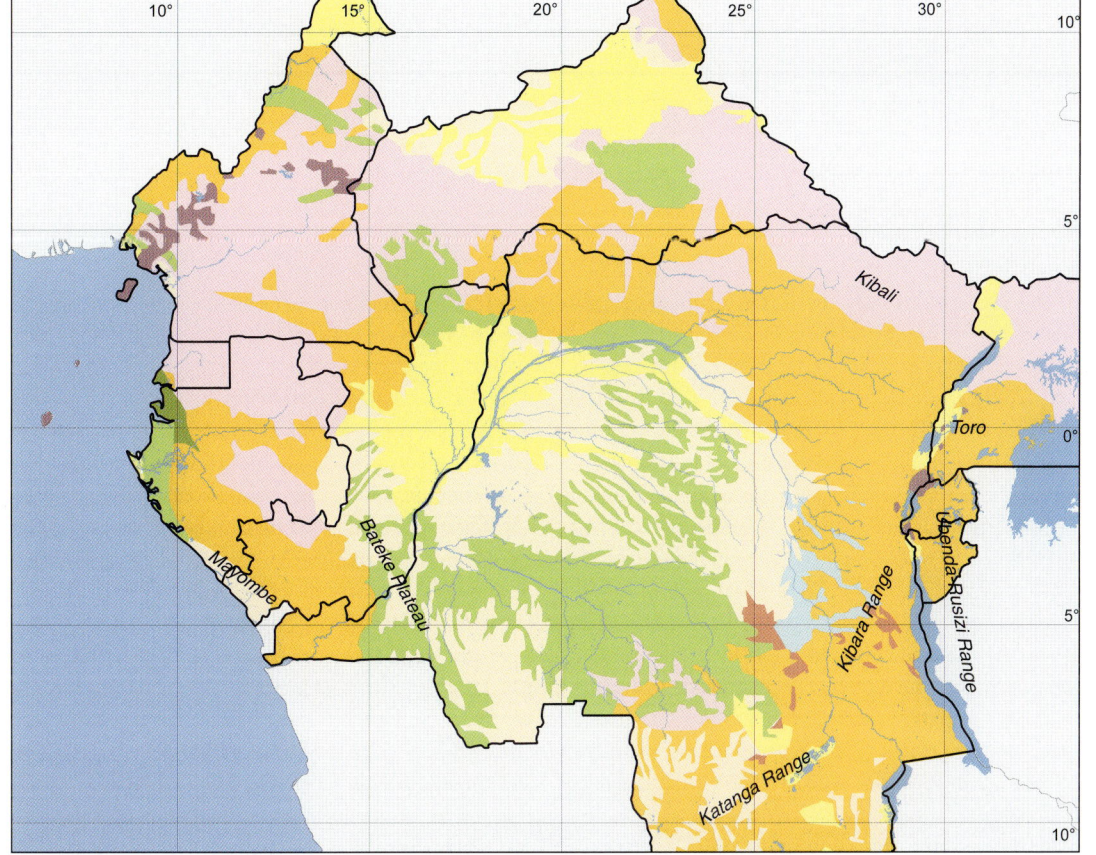

Figure 1.4. Simplified geology of central Africa (from the UNESCO geographical atlas)

- Neogenous and quaternary volcanic layers
- Sedimentary layers:
- quaternuary (<2My)
- neogenous (23-2 My)
- cretaceous (142-65 My)
- jurassic (206-142 My)
- triassic (248-206 My)
- permo-carboniferous (570-225 My)
- proterozoic (2600-570 My)
- archean (3500-2600 My)

(My : millions of years before present)

1 THE MINERAL WORLD

1.5. The high bare crests of Mount Brazza, that dominate the savannas of the Lopé, are formed of profoundly altered metamorphic rocks that are 2.3 billion years old, like the mountains that we perceive in the distance.

selves in. Throughout the Proterozoic, the third phase of the Earth's history, that began 2.6 billion years ago and ended 545 million years ago, the various cratons began to weld together in this way. Animated by the convection of the magma, they continued to be subjected to constant motion. In approaching each other they compressed, lifted and folded the masses of sediments that would become immense mountain ranges. In the south, there was the Kasai, in the west, the Mayombe. This massif sprung up between the Congo and the Brazil cratons which were then nearly side by side. In the north-east, the Kibali range was thrust up and extended to Uganda, while in the east, between the Congo and the Tanzania cratons, the Ubendian-Rusizian range formed (FIGURE 1.4).

Some 1,600 million years ago, the African continent formed in this way a single continental mass. At the same time that this binding-together was achieved, there also appeared the first fissures. As they increased in size they created new marine expanses that, in their turn, accumulated sediments engendering new foldings. In the second part of the Proterozoic, 1,600 to 600 million years ago, most of the mobile zones were thus reshaped or rejuvenated. The vast Kibara range whose rocks are found today from Uganda to Katanga, was formed 1,250 to 850 million years ago. The Katanga range, in the south-east of the present-day Democratic Republic of the Congo, developed between 1,100 and 600 million years ago. In the west, a range running north-south, parallel to the present coast of the continent, took form 740 to 620 million years ago and profoundly rejuvenated the Mayombe which, once again, found itself wedged between the Congo and Brazil cratons.

From the beginning of the Proterozoic, depressions also formed at the interior of the Congolese craton. Just like the mobile zones, these would accumulate deposits that would undergo an alternation of sedimentation phases and folding phases, of which the most important – the pan-African orogenesis – would take place around 600 million years ago and coincide with the Katangan foldings. Of these intra-cratonic covers there remain, among others, the Gabonese Francevillian deposits which we see around Booué (FIGURE 1.7).

These fold chains followed each other throughout the Proterozoic. They would not only affect the relief, but also profoundly transform the very nature of the rock involved. Under immense mountain ranges, in the depths of the terrestrial crust, sedimentary masses were in effect remelted and recrystallised by intense heat and pressure. In peripheral areas, the transformation was often incomplete and engendered alternating layers of hard quartzites and softer schists. But at the heart of the process, metamorphosis was total. It created gneiss, metamorphic granites and many new rocks. Throughout the terminal period of the Proterozoic, intrusive granitic rocks also continued to be created: masses of magma found their way through superficial rock layers but never reached the surface. Much later, when the softer superficial layers eroded away, they would give rise to the inselberge (PAGES 92-93).

A large part of Central Africa is thus made up of very diverse Precambrian rocks. Because of their age, most of these rocks are extremely altered everywhere where they show on the surface. The soils that they generate are often very leached and poor. But the richness of this immense region is not to be found on the surface. It is hidden in the depths of the Earth. In fact, the Precambrian holds important deposits of minerals such as copper, cobalt, tin, manganese, gold, uranium and diamonds. Some, like chrome and asbestos, are found in Archean rocks. Others are concentrated in the metamorphic formations, especially in zones of contact between ancient mobile areas and craton rocks.

Considering the extent of these kinds of formations, it is not surprising that Central Africa appears to be a veritable "geological scandal". For biological diversity, however, this geological history represents a constant and inescapable threat. In the past, the underground treasure has been a determinant element in colonial history. Today it conditions mining activities – legal and illegal – with a complete cortège of social and environmental perturbations. It is unlikely that such important mineral deposits will remain unexploited forever just because of nature conservation concerns. Even in Europe and the United States that does not work. In many cases, this wealth also engenders and maintains violent conflicts that annihilate all conservation and development efforts. The war in the Democratic Republic of the Congo which lasted since 1996 would probably never have involved as many actors without the mineral richness of the Congo Basin.

1.6 The Akrum inselberg in Equatorial Guinea, a mass of intrusive rocks born in the second phase of the Proterozoic.

1.7. Near the small town of Booué, just east of the Lopé National Park, the Ogooué River flows on a bed of rocks belonging to the Francevillian. This group of sedimentary rocks, accumulated in the very interior of the Congolese craton, underwent a strong metamorphosis around 1.8 billion years ago.

1.8. This essentially mineral realm, made of water and rocks, is the habitat of the Podostemaceae. Attached to exposed rocks and bathed by the current, these angiosperms with minute pink flowers are reminiscent of algae and mosses.

1 THE MINERAL WORLD

The emergence of the continent

The last foldings took place in the Ordovician, 480 million years ago. Since then, and unlike most other continents, the major part of sub-Saharan Africa would no longer be affected by major tectonic movements. Except for some episodes of volcanic activity, the continent was subjected only to an unflagging erosion that got the better of the Precambrian ranges and transformed it into a vast peneplain. Its surface was, however, animated by the internal convection of the magma. In places it lifted up and gave rise to the great plateau that today forms eastern and southern Africa. Elsewhere, it lowered and created the vast depressions of the Sudd, the Chad, the Congo and the Okavango. This is how, for 620 million years, the Congo craton was carved out in its centre and little by little filled with sediments carried from its peripheral areas.

Independently from this strictly geological story, there was also a profound transformation of the Earth's atmosphere by cyanobacteria and the first oxygen-producing plants. About 450 million years ago, there thus appeared a layer of ozone that put up an effective barrier to ultraviolet rays. The first plants, related to mosses, as well as the first arthropods, could then colonise the continents.

But the lands that would become Africa formed at that time the central nucleus of Gondwanaland, an enormous continental mass that "drifted" in the southern hemisphere. In the Carboniferous, which began 354 million years ago, this mass came in contact with the continental masses of the northern hemisphere to form a single super-continent, Pangaea. At the beginning of this epoch, the climate was still hot and humid, and the seas covered a large part of the continents. Subsequently, the climate cooled and became more seasonal. Enormous ice caps accumulated on the poles and the seas retreated. While the regions that were going to be Europe and North America bathed in a tropical climate, the southern half of what would become Africa was covered with immense glaciers. They were centred on the north of present-day South Africa and Zimbabwe but extended to the Congo basin. This immense shell of ice finished the levelling of the relief. On the eastern edge of the basin, by Lake Albert, the glaciers dug deep valleys still used today by the rivers that flow from the mountains of the Albertine Rift to the Congo River.

Throughout the Permian, that began 290 million years ago, Pangaea drifted northward (FIGURE 1.9). Finally, it collided with Siberia, the last continent still to be independent. Meanwhile, its margins began to crumble away. In Central Africa, the glaciers retreated and the global climate changed once more. Northern Pangaea became hot and arid, the south remained cool and humid.

In the course of the Triassic, which began 248 million years ago, Pangaea continued to drift northwards. Its breaking-up intensified and Laurasia was again separated from Gondwanaland. From the onset of the Jurassic, 206 million years ago, Gondwanaland in its turn started to come apart. A rift announced the birth of the South Atlantic Ocean, but the separation of the Antarctic and Africa was accompanied by immense eruptions of basalts from which the table mountains of the Cape region, in South Africa, remain to this day. This fragmentation of continental masses was accompanied by a significant warming of the climate. The polar ice caps disappeared for a long period and enormous portions of the continents were again submerged by marine transgressions.

In the Cretaceous, that began 142 million years ago, north-south faults would open first the South Atlantic and, later, the North Atlantic (FIGURE 1.10). In the process of enlarging, they freed the coasts of Gabon and the Lower Congo. Finally, they separated South America from Africa. In the eastern part of Gondwanaland, similar rifts would give birth to the Indian Ocean. This vast continent thus shattered into a multitude of fragments that would go off in all directions in the southern hemisphere, forming South America, the

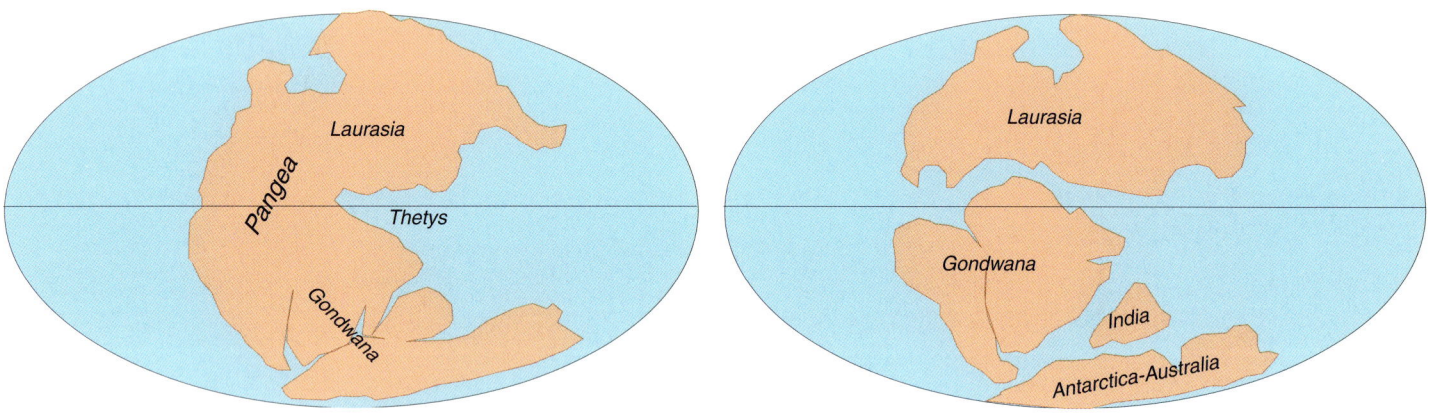

1.9. Some 250 million years ago, at the end of the Permian, all the continents were stuck together into a single enormous continental mass – Pangea – which had already happened between 750 million and 570 million years ago. The level of the oceans was at the time very low, the global climate cool and dry and the emerged surface of the lands was at its maximum. Tree ferns still dominated the "forests" of the epoch, but were soon going to be struck by a massive wave of extinctions that occurred at the end of this era, 248 million years ago.

1.10. At the beginning of the Cretaceous, 135 million years ago, Laurasia was separated from Gondwanaland by the Tethys Sea and Gondwanaland itself began splitting up.

Antarctic, Madagascar, Indonesia and Australia. In the centre, an immense bloc that remained nearly immobile was completely isolated by ocean expanses (FIGURE 1.11). This mass comprised Africa and Arabia. The climate was hot and humid. The ocean rose more than 200 metres above its present level and a shallow sea invaded the Sahara from the Gulf of Guinea to Libya, isolating West Africa from the rest of the continent.

But this splitting-up continued. In this way, 65 million years ago, the first phenomena that were to give rise to the Arabo-African Rift took place. Around the middle of the Tertiary, in the Miocene, India went and "butted" Asia, lifting up the Himalayas and Tibet, while the Red Sea and the Gulf of Aden appeared. Like India, Arabia went and abutted Asia, pushing up the high mountain ranges that stretch from Turkey to Afghanistan.

Unlike the other continents, the surface of Africa was hardly ever affected by the drift of plates because of its central position and its relative immobility. The important changes in the arrangement of the continental masses was accompanied, however, by changes in the ocean currents and the distribution of energy in the atmosphere. They probably also affected the parameters of the earth's rotation. All these phenomena finally brought about changes of climate and sea-level, precipitated by enormous volcanic eruptions that accompanied the formation of the rifts. They thus had a decisive impact on the history of life and were very probably at the origin of the great waves of extinction that marked the end of the Permian, the Triassic and the Cretaceous.

The freeing of Africa, which took place mainly from the Cretaceous to the Miocene, had equally enormous regional consequences. Not only did the continent move northward and pivot on itself, but while South America, the Antarctic and Australia would still keep some contact, Africa very soon became isolated from the other continental masses, an isolation lasting 70 to 80 million years. It is precisely during this long period that the tropical forest plant species we know today appeared and that the mammals, birds, snakes and lizards diversified. For the mammals, in particular, this precocious isolation would have important consequences. The first representatives of their lineage, the mammal-like reptiles, appeared in Gondwanaland during the Carboniferous and, just before the break-up of the super-continent during the Mesozoic, they had engendered seven large lineages: the Insectivores, the Primates, the Bats, the Scaly Ant-eaters or Pangolins, the Rodents, the Hares and the Ungulates. During the Tertiary, these diverse groups would give rise to the families, genera and species that we know today. Each continent would thus see the appearance of its own orders and families.

From the Miocene, this isolation would nevertheless be broken with the formation of the Arabo-African Rift and the separation of Arabia 20 million years ago. Unlike India, this mass conserved links with its continent of origin and became a real bridge between Africa and Asia. The Primates and hippopotamuses dispersed

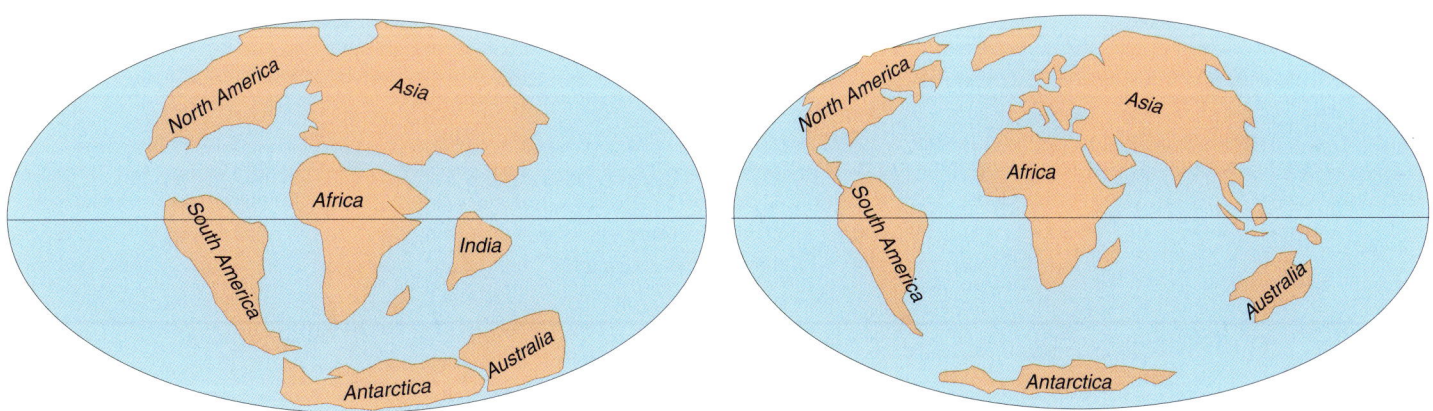

over Eurasia while numerous Asiatic families colonised Africa. This is how the African fauna became enriched with bovids, equids, rhinocerotids, tragulids, viverrids, mustelids, canids, leporids, sciurids, hystricids and murids. It is possible that even the cats did not appear in Africa before this epoch. Very rapidly, all the families developed typically African forms, such as the zebras and the numerous species of antelopes of the savannas. Much later, there were also exchanges with Europe, but the Cervidae never got south of the Sahara. These last exchanges were thus of little importance in the history of the forests.

1.11. At the Cretaceous-Tertiary transition, 65 million years ago, Africa was completely isolated, but there were still contacts between Australia, Antarctica and South America.

1.12. It is not until the Miocene, 18 million years ago, that Africa would again come into contact with other continental masses through the formation of the Red Sea and the separation of Arabia.

The origin of the relief

Although Africa was completely isolated from other continents and its surface hardly disturbed for nearly 400 million years except for the unflagging erosion and the sea-level fluctuations, by the end of the Cretaceous, local tensions in the earth's crust would provoke a slow lifting of the entire western margin of Central Africa. This phenomenon was accompanied by a general flexing of the continental plate, probably at the break-up of Gondwanaland and the formation of the Atlantic rift. This provoked cracks and disrupted the hydrographic network. It reactivated the erosion that washed away the older surfaces, renewing soils and slowly exposing the present-day features.

Around the Gulf of Guinea, these tectonic movements lifted up and caused masses of sediments accumulated in the Atlantic rift to emerge. They thus formed the coastal plain that extends from Cameroon to Angola (FIGURE 1.13). In Gabon the plain is around 100 kilometres wide, but is cut in two by a narrow north-south range formed by the Lambaréné horst, a band of continental material, lifted and detached from the continent itself during the creation of the rift. The eastern half of the plain was deposited at the end of the Jurassic and the beginning of the Cretaceous in a rift that was still continental, probably comparable to the East African rift that we know today. Its more recent western half was formed after the marine incursions coming from the south had invaded the trench. It has marine sediments, and notably important salt-bearing layers, covered over with calcareous and, later, sandstone layers. At the delta of the Ogooué, one also finds clay or calcareous sediments dating from the Palaeocene, the Eocene and the Miocene. At Kouilou and Cabinda, the plain, much narrower there, was not constituted until the Pliocene, the final phase of the Tertiary. Recent sediments are unflaggingly added to these old formations. They trace the abrupt, sandy, rectilinear coast with its successive littoral cordons isolating vast expanses of lagoons. They also invade the estuaries and contribute to their slow filling in.

During the formation of Cretaceous deposits, enormous masses of decomposing organic material were buried under the marine sediments. Compressed and cut off from any source of oxygen, they are the origin of petroleum deposits, exploited all along the Gulf of Guinea, on land and at sea.

The flexing of the continent also accentuated the depression of the Congolese basin and slowed its drainage. For hundreds of millions of years, the basin accumulated immense quantities of sediments carried from the surrounding mountains and plateaux. The Permo-Carboniferous, Triassic and Jurassic deposits bear witness (FIGURE 1.4). In the Cretaceous, the erosion was even more active. It deposited enormous coatings of fluvio-lacustrine sediments that show at the surface nowadays in the entire southern part of the basin (FIGURE 1.4). In the Palaeogene, under more arid climate conditions, these sediments were partially covered over by sandstone deposits incorporating aeolian sands from the Kalahari. In the Neogene, again under humid conditions, these latest deposits were in their turn covered over by thick layers of ochre sands. Together, these Paleogenian and Neogenian deposits form the bedrock of the present Batéké plateau (FIGURE 1.2, 1.15) that juts out from Angola like a gigantic bare tongue separating the Congo basin in the east from that of the Ogooué in the west, and fades away in the region of Odzala-Kokoua National Park. Finally, in the Pliocene and the Quaternary, the entire centre of the basin was covered over with new fluvio-lacustrine sediments (FIGURE 1.4). The alluvial part of the basin, covering a surface of around 700,000 square kilometres, is thus very homogeneous structurally, which explains the monotony of its landscapes.

The basin's margins, on the contrary, have been carved out of Archaean and Proterozoic rocks of various degrees of hardness. They were submitted to differing degrees of erosion in different places, which freed the hardest rocks and cleared away the softer materials. In this way a great diversity of landscapes was created, whose structural complexity was further accentuated by the alternation of long humid periods with a forested mantle and of drier periods with an herbaceous plant cover. This alternation in fact favoured the development of significant ferruginous cuirasses or iron-pans that, in their turn, strongly influenced erosion. Landscapes of the highest parts of Central Africa are

1.13. The coastal plain bordering the Gulf of Guinea was deposited mainly in the Cretaceous and emerged from the upper Eocene, 45 million years ago, after the flexing of the African continent and the lifting of its western margin, phenomena that probably accompanied the opening of the Atlantic. Since then, its sandstones and limestones have been submitted to an intense erosion which created innumerable cirques, such as those of the Wonga-Wongué Presidential Reserve in Gabon. Most of the cirques, like the one in the photograph, are old and covered by vegetation, but some are still in full activity and resemble the Lékoni cirque on the Batéké Plateau (Figure 20).

1 THE MINERAL WORLD

thus very diverse and highly complex. They are divided into three main types.

In the north-west, from the Congo to Cameroon, the Archaean formations give rise to landscapes of worn hills interspersed in places with inselberge. Among the most striking of these formations are the Crystal Mountains (FIGURE 1.14) and the Chaillu Mountains in Gabon, and Mount Alén in Equatorial Guinea. In areas where there has been recent tectonic movement or in zones of contact between very different types of rocks, ridges and domes with a more striking relief have formed. In places, the flexing of the continent even provoked cracks that would give rise to long escarpments, like the one that more or less follows the border of Gabon in the Odzala-Kokoua National Park.

In the Mayombe region, between Angola and the Ogooué, completely smoothed out in previous epochs, the upliftings of the Tertiary brought with them the resumption of erosion. This activity freed a new range of the Appalachian type, composed of an alternation of hard quartzite rocks and soft schisteous rocks.

Elsewhere, as in a great part of the Central African Republic, the sandstone and calcareous formations of the Proterozoic age or even the Cenozoic give rise to massive flat-topped features with steep sides or to more or less undulating plains with indefinite or discontinuous drainage systems.

1.14. The lifting of the western margins of Central Africa, beginning in the Eocene, created new features, such as the Crystal Mountains in the north-west of Gabon. Their highest summits are hardly more than 900 metres high, but they are not very far from the Atlantic and, because of this, the westernmost peaks are very often shrouded in masses of clouds that come from the Gulf of Guinea. Their vegetation is thus very distinctive, notably rich in species of Begonia.

1.15. The Batéké Plateau reaches from the Angolan border to the region of Odzala-Kokoua National Park in the Congo. In Gabon vast herbaceous expanses are found around Franceville. As in the coastal plain, its sandstones lend themselves to cirque-forming erosion. The Lékoni cirque (opposite) is still in full activity.

1.16. The volcanic highlands of Cameroon extend from Mount Cameroon (4,070 metres) to the Adamaoua, where Tchabal Mbabo, the highest summit, culminates at 2,480 metres. To the southeast this range extends into the Gulf of Guinea where it forms the islands of Bioko, Principe, São Tomé and Annobon.

The volcanism of Cameroon

The opening of the Atlantic brought about the formation of innumerable faults all along the western coast of the African continent. In places it also provoked more violent events. Thus, from the end of the Cretaceous, there appeared a vast area of subsidence, a trough extending from the Gulf of Guinea to the Tibesti (FIGURE 1.16). In its Cameroonian part, this depression was initially limited by two great, nearly vertical, faults whose sides were nearly 1,000 metres high. Today, this relief is softened, but to the east lies the Nlonake fault, lined by little volcanoes, and in the west, the Rumpi Hills, extending over 280 kilometres from the coast to the Nta-Ali Mountains. At the Cross River, faults also appear, cutting the main fault at right angles.

A range of volcanoes thrusts up in the axis of this great depression. Today the continental branch comprises Mount Cameroon, the Manengouba and Bambouto massifs. Mount Cameroon is the highest volcano in the range (figure 1.17). It is a typical central volcano, with a summit culminating at 4,000 metres and whose base is 50 kilometres wide. The summit itself, Fako, is constituted by the edge of an ancient, very degraded, crater in the process of being filled in. Immediately to the west, towards the highest point, at 4,060 metres, there is a small sulphur-covered chimney

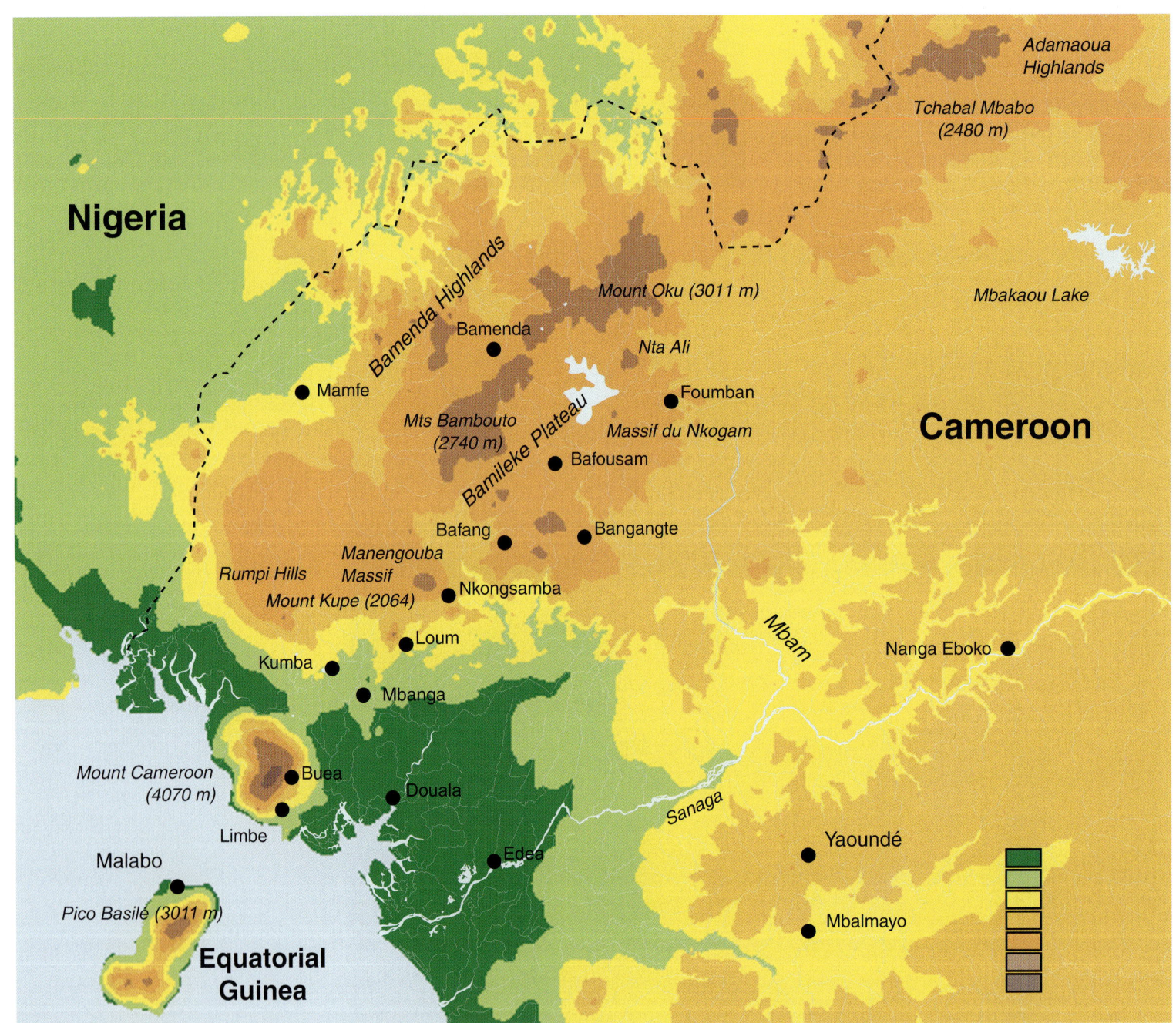

1 THE MINERAL WORLD

that gives off sulphurous gases. On the south side gapes an active crater 100 meters wide and 50 metres deep. Its walls are vertical and hot vapours escape from its burning fissures. At about one kilometre to the south-west, there is the very old cone of the Mateer.

Mount Cameroon's volcanic activity began in the Cretaceous. In the Eocene, 55 to 34 million years ago, it was considerable and produced enormous quantities of basaltic lavas. Activity started up again in the Miocene, 20 million years ago, with the emission of lighter lavas. In the Pliocene and the Pleistocene, the volcano produced new sheets of basaltic lavas that now form a layer more than 1,000 metres thick. More recently, in 1909, a great eruption again created small volcanic cones at about 2,400 metres altitude from which sprang a river of lava 800 metres wide, 20 metres thick and 6 kilometres long. In 1922, came an eruption that formed, at around 3,600 metres, the Mateer crater whose lavas reach the sea. For the rest, there was still a period of activity in 1959 and the very most recent eruption happened in 1999.

North and north-east of Mount Cameroon, in a western fault and just to the east of the Rumpi Hills, there are some lava flows of very different ages that cover the Precambrian rocks. In the depression itself, 90 kilometres from the summit of Mount Cameroon, rises Mount Kupe whose summit is some 2,050 metres high. Farther north-east, there is the Manengouba massif. It is 25 kilometres wide and 2,450 metres high; and its mass incorporates the remains of two calderas fit together, the largest of which is eight kilometres in diameter. All volcanic activity has ceased in this area. Finally,

there are the Bambouto Mountains, part of the Bamenda range, 180 kilometres long and 2,679 metres high.

Independently from the great volcanic formations, the western region of Cameroon also includes several explosion craters (Figure 1.18). They are formed without much outpouring of lava by the contact of infiltrating water with the magma. The Barombi Mbo crater, some 160 kilometres to the north-east of Mount Cameroon, is a nice example.

1.17. From sea level to over 4,000 metre, the slopes of Mount Cameroon harbour a unique continuum of vegetation types. Forests are strongly disturbed, however, by the human and volcanic activities.

1.18. One of the numerous explosion craters in the northern part of the Cameroon Highlands.

23

1 THE MINERAL WORLD

1.19. The southern coast of the island of São Tomé.

1.20. On the Gabonese coast, not far from Libreville, a lighthouse rises from Pointe Ngombé. It is set on a mass of basalt which represents the only witness in Gabon to volcanic activity that accompanied the opening of the Atlantic.

The islands of the Gulf of Guinea

On the Gabon coast, far to the south of Cameroon, a small isolated volcanic eruption occurred probably in the second half of the Cretaceous (FIGURE 1.20). However, some very much more important volcanic events took place out at sea. The great Cameroon fault was prolonged into the Atlantic in the direction of Saint Helena and engendered the islands of the Gulf of Guinea: Bioko, Principe, São Tomé and Annobon (FIGURE 1.16).

Bioko, formerly Fernando Po, is the largest and covers 2,020 square kilometres. It still harbours some active volcanos. In the north, Pico de Basilé, 3,011 metres high, is the second highest mountain in the region after Mount Cameroon. It is extinct. In the south, its relief is dominated by the Gran Caldeira de Luba, reaching 2,261 metres, and Pico Biao at 2,010 metres. Situated on the continental plate some 32 kilometres from the Cameroon coast, this island was linked to the continent each time that the waters of the Atlantic subsided. This was still the case during the last glaciation and its present-day isolation does not date back more than 12,000 years. This explains why the fauna and flora of this island remain very close to those of Mount Cameroon.

The three other islands are all situated far beyond the continental plate. Principe, 220 kilometres from the coast, has an area of 128 square kilometres. Being strictly volcanic, this island is constituted mainly of basalts. Its summit culminates at 948 metres. It has two "satellites": Ilhéu Caroço or Boné do Joquei, an islet 300 metres high, three kilometres to the south-east, and the Tinhosas Islands, 22 kilometres away. Tinhosa Grande covers 20 hectares, Tinhosa Pequeña around three. Only Boné do Joquei is covered with some vegetation. The other islets are absolutely bare but harbour large seabird colonies: White-tailed Tropicbird *Phaethon lepturus*, Brown Booby *Sula leucogaster*, Brown Noddy *Anous stolidus*, White-capped Noddy *Anous minutus*, and Sooty Tern *Sterna fuscata*.

São Tomé, located 280 kilometres from the continent and covering a surface of 836 square kilometres, was considered until a few years ago as having thrust up from the bottom of the ocean. Since then, it has been discovered that São Tomé represents a fragment of Gondwanaland, abandoned out in the ocean during the opening of the Atlantic, a little like the Seychelles were left out in the Indian Ocean during the moving away of India. Its bedrock comprises sandstone rocks identical to some formations that are found on the Cameroon coast, notably near Douala, and in Nigeria. Long after the first volcanic activity of Principe, mainly in the Pliocene, some basaltic eruptions were grafted onto this island that gave it its present-day configuration. It has a tortured relief and its jagged coasts are very steep. Its highest point, São Tomé peak, reaches 2,024 metres above sea level and is surrounded by a dozen peaks with altitudes of around 1,000 metres. In the northern part of the island, there are craters and very visible lava flows. The dominant rocks are compact basalts, generally covered with a layer of clay.

Annobon, also known by the name of Pagalu, is 340 kilometers from the mainland, the farthest of the islands. With a an area of 1,700 square kilometres, it is strictly volcanic, like Principe. The island is mountainous with a summit which is 645 metres high and has a crater in its centre.

The Albertine Rift

Contrary to the fairly localised events in Cameroon and the Gulf of Guinea, the Arabo-African Rift very profoundly upset all of East Africa. This enormous complex of faults split the earth's crust for a distance of some 6,000 kilometres, from Palestine to Mozambique. In the first phase, it detached Arabia and created the Red Sea. Next, it extended from Ethiopia to Malawi.

Its formation was preceded by the uplifting of domes several hundred kilometres in diameter centred on Ethiopia, Kenya and the Kivu. Pushed upwards by the convection of the magma, the earth's crust was raised some 700 to 1,000 metres above the average altitude of the plateaux, which were already situated at 1,000 metres above sea level.

The successive up-thrusting of the relief set off an intense erosion that carried off the Cretaceous surfaces and deeply transformed hydrologic networks. Below the earth's surface, the convection of the magma reheated and remelted the deepest layers of the crust, which thus became thinner and lost its resistance. At the hottest places, the thrusting of the magma opened up passages towards the surface and made volcanoes spurt forth. But at the summit of the domes, the thinned crust was unable to resist the tensions. It finally cracked and founded long parallel faults. The land situated between the two faults sank to form an enormous depression, while the sides continued to rise up. These movements started at the very beginning of the Miocene, 25 or even 30 million years ago.

The first great faults appeared in the region of the Arabo-Ethiopian dome. They gave rise to the Gulf of Aden, the Red Sea, the Danakil Rift and the Afars depression. A southern branch cut the Ethiopian part of the dome in two, extending into the Kenyan one and reaching northern Tanzania to form the eastern rift or Gregory Rift.

The western branch of the rift formed on the western edge of the East African plateau. It developed a bayonet trace nearly 2,000 kilometres long, from the Assoua lineament to the Zambezi lineament, two very old, deep fractures of the terrestrial crust (FIGURE 26). At the deepest part of the trench, the crust's thinning was never more than 25%, however. Unlike the Red Sea, the Albertine Rift always kept the appearance of a continental rift.

Although, in the north, these faults extended all the way into the Archaean rocks, it took form essentially in the zone of Precambrian orogeneses, between the Congolese and Tanzanian cratons. In places, it even superimposed the ancient Palaeozoic and Mesozoic fractures without following them.

Today, these faults form a highly complex network with innumerable parallel fractures intersected again by lateral fractures sometimes at right angles with the main trough. This trough did not form all at once, but rather by linking up fairly distinctive basins averaging about 100 kilometres long. The trough of Lake Tanganyika for example, 650 kilometres long, was constituted by the juxtaposition of seven individual basins, not counting that of the Rusizi River. The first faults appeared at the beginning of the Miocene, 25 to 30 million years ago. They formed the Lake Albert and Lake Kivu basins. Later, they extended southwards. They reached Malawi two to three million years ago.

While some of the land sank, the horsts rose. These uplifts mainly involved the edge of the trough but sometimes the structures situated in the middle of the

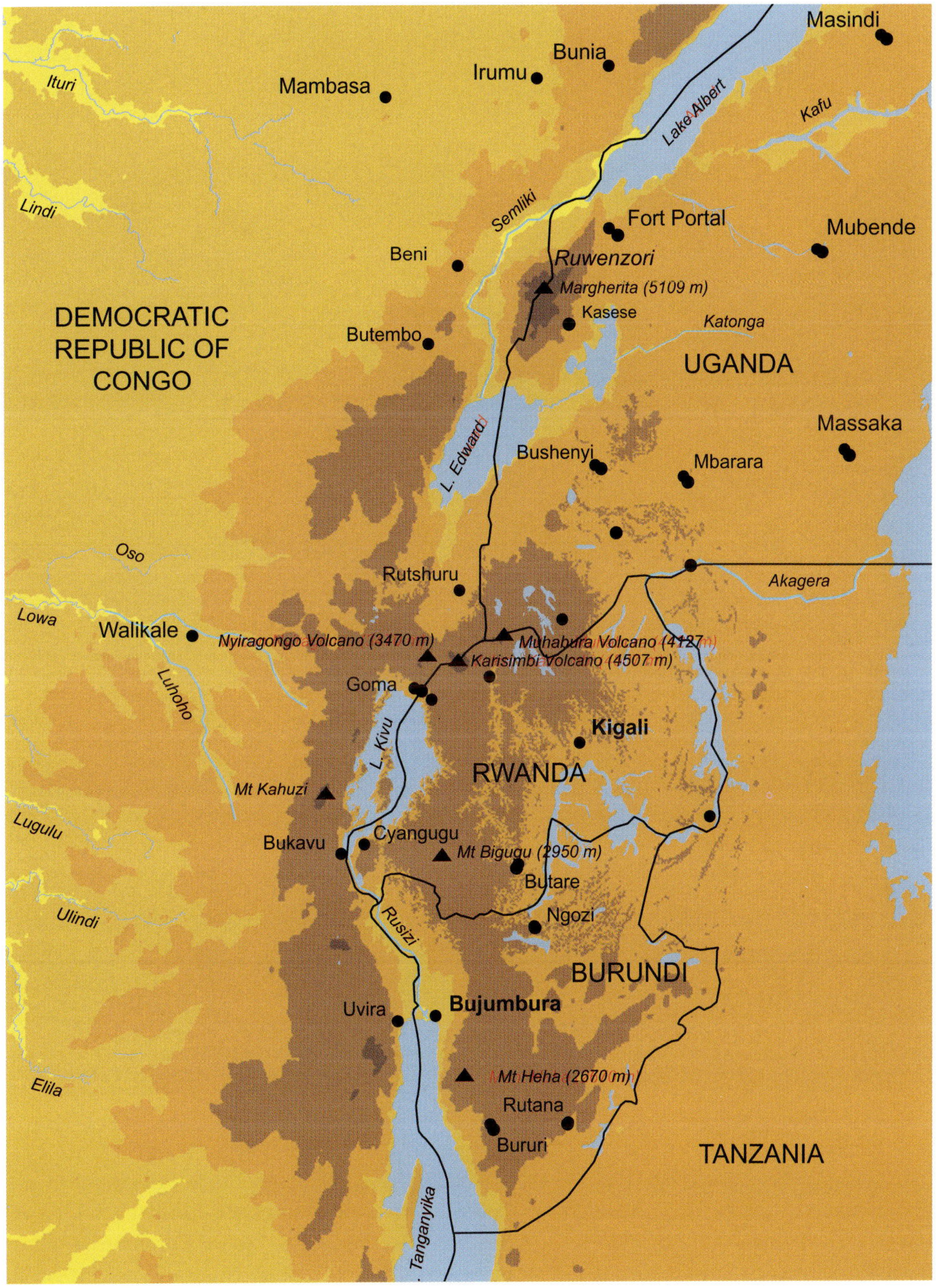

Figure 1.21. The western branch of the Great Rift Valley, known as the Albertine Rift, with its string of great lakes lying between two almost parallel mountain chains.

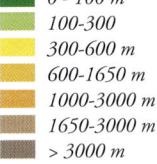

0 - 100 m
100-300
300-600
600-1650 m
1000-3000 m
1650-3000 m
> 3000 m

1 THE MINERAL WORLD

1.22. Mount Rwenzori, with its numerous glaciers at its summit – the Margherita – 5,110 metres high, is one of the last places in equatorial Africa where it is possible to study the continuous and nearly intact transition between the lowland forest and the Afro-alpine vegetation.

trough floor, as well. Although the initial dome was real, most of the uplifting was done well after the fault formed. West of the trough, the Mitumba range was formed. At the level of Rwanda and Burundi, the uplifting gave rise to the Congo-Nile ridge and, right in between the two faults of the Kivu, provoked the emergence of Idjwi Island.

Farther north, the Rwenzori, the most spectacular horst, emerged (FIGURE 1.22). Just like Idjwi Island, it was formed in the middle of the subsidence trough, separating the Lake George basin from the Semliki River basin some two million years ago. It is formed by an enormous block of more or less metamorphosed Precambrian rocks – mainly granites, gneiss and amphibolites – that were not only lifted up but also toppled over. Because of its relatively young age, the Rwenzori has kept a very steep relief. Its summit is over 5,000 metres high and it has remained covered with immense glaciers for most of its existence (FIGURE 25), which have created its long U-shaped valleys, its moraines and its numerous marshes and small lakes.

The formation of the Albertine Rift was accompanied by volcanism, although it was much less widespread and later than in the Eastern Rift. In Kenya, the first eruptions occurred 22 to 23 million years ago and gave rise to Mount Elgon, Mount Kenya and Mount Kilimanjaro, three great central volcanoes situated at a distance from the axis of the trough. In the Albertine Rift, the first volcanic eruptions did not take place until about 12.6 million years ago, thus 11 million years after those of the Eastern Rift. They have not stopped since. Contrary to those of the Eastern Rift, they took place in the axis of the trough, particularly in the zones of accommodation located at the junction of two basins and crossed by perpendicular slide faults.

The very first volcanic activity thus appeared between the basins of Lake Edward and Lake Kivu and gave rise to the Virunga, an ensemble of eight large central volcanoes and several hundred adventive cones. They spewed out enormous quantities of lava and because of this would profoundly affect the entire hydrographic network of the Albertine Rift region. In barring the course of several rivers, they engendered the lakes Kivu, Bunyoni, Mutanda, Bulera and Luhondo, to mention a few of the main ones.

The three eastern volcanoes are the oldest. They include Muhabura (4,127 metres), Mgahinga or Gahinga (3,474 metres) and Sabyinyo (3,634 metres). The first two have kept well formed cones and Gahinga has a vast marshy crater at its summit. Sabyinyo, on the contrary, is mostly collapsed. Only its north-western crater wall remains, giving it a jagged appearance and a very steep relief (FIGURE 1.25).

Next was born the central group. Mikeno (4,100 metres) somewhat resembles Sabyinyo with its single very pointed tooth (FIGURE 1.24). Bisoke or Bushokoro (3,711 metres) has a lake some hundred metres in diameter in its crater (FIGURE 1.26). Karisimbi (4,507 metres) comprises a main cone often covered with hail, and an adventive crater that has formed a vast "shoulder". Its youngest lavas are no more than 10,000 years old. Apparently, this group is extinct, but in 1,957, there was another small eruption on the north flank of Bisoke.

The western group is the most recent and is still in full activity. Most of its lavas are no more than 15,000 years old. Nyamulagira (3,058 metres) is a vast caldera three kilometres in diameter. It erupted violently from 1938 to 1940, then was active again in 1951 and 1952. The most recent eruptions took place in 1956 and 1957. To this day it gives off smoke and gas through small fissures in the bottom of its caldera and, around its central cone, adventive eruptions burst forth from adjacent cones nearly every five to seven years. Eight similar cones were thus formed between 1951 and 1994. These

1.23. *The Virunga seen from the Rwindi Plain in the Virunga National Park. From the left to the right: Muhabura, Gahinga and Sabyinyo. In the foreground, the sandy cliffs of the Rutshuru River.*

small eruptions are usually preceded by earthquakes. They start with gas explosions and continue with the eruption of lava. In a few weeks or months they can form a cone 100 to 150 metres high and emit thick viscous flows that spread over distances of 10 to 20 kilometres. The lava of Nyamulagira is basic and has the consistency of solid porridge.

Nyiragongo (3,470 metres), located 14 kilometres south-east of Nyamulagira, also has a caldera. Since 1928 it contained a vast lake of molten lava that raised and lowered with the internal pressures of the magma. At night, this mass of lava lit up the clouds gathered above the volcano with a red glow that could be seen from the extreme south of Kivu and reminded the city of Goma of the danger that slept at its door. In December 1976, this lake reached an unprecedented level and on 10 January 1977, it emptied brusquely through three fissures in the wall of the main cone. In a few hours, an ultrabasic lava, extremely fluid, spread to a distance of 10 kilometres. On leaving the fissures, its speed was of the order of 60 kilometres per hour, but because of its fluidity, the flows were not more than a metre thick. The lava stopped around Goma, which escaped destruction that time. After it had emptied, the Nyiragongo crater was more than 1,000 metres deep with a diameter of 1,200 meters at its summit. Its terraces collapsed. In the years that followed, the lava lake reformed and at times the clouds were again illuminated at night. Finally, in mid-January 2002, a new eruption, again accompanied by earthquakes, surprised the city. Its inhabitants fled – mostly towards Rwanda – and a flow of lava, over 10 metres thick in some places, cut the city nearly in two.

A little over ten million years ago, the eruptions of South-Kivu started between the Kivu and the Rusizi basins. In Rwanda, they created the basalt plateau of Gisakura, on the edge of the Nyungwe Forest. In the Democratic Republic of Congo, they founded the volcanoes Kahuzi (3,308 metres) and Biega (2,790 metres) as well as the flows of Ngweshe-Karehe, the Lugulu west of the Biega volcano, Mwenga-Kamituga and Moeso. Today this region seems calm but it still experienced a small eruption at the beginning of the century.

Finally, 450,000 years ago, volcanic activity began in Toro and Ankole, in Uganda, just east and south-east of the Rwenzori. It did not cease until 4,000 years ago, after having created a multitude of explosion craters, nowadays very often occupied by small lakes. Lava flows were rare.

The thirty-odd craters of Queen Elizabeth National Park are part of this system. The most picturesque ensemble just north of the Kasinga Channel, was probably formed some 5,000 to 10,000 years ago. Some of the craters are occupied by lakes, like the Kitagata (FIG-

1 THE MINERAL WORLD

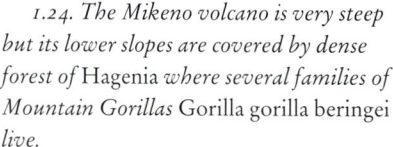

1.24. *The Mikeno volcano is very steep but its lower slopes are covered by dense forest of* Hagenia *where several families of Mountain Gorillas* Gorilla gorilla beringei *live.*

1.25. *Sabyinyo's cone has mostly collapsed and only its very steep north-eastern wall remains.*

1.26. *The vast crater lake of Bisoke or Bushokoro with the summit of Mount Mikeno in the distance.*

URE 1.27). Their waters are frankly alkaline and several gave birth to traditional salt exploitation. Another ensemble of craters spirals out in the middle of Maramagambo Forest, at the foot of the Rift escarpment. It includes Lake Nyamusingeri and Lake Kasianduku. All those explosion craters buried the Lake Edward region under cinders. A great many of the national park's soils are thus made up of volcanic dusts.

Apart from explosion craters and lava eruptions, the volcanic activities of the Rift also engendered hot springs. In Uganda, they can be found at Mongiro, in the north-eastern part of the Semuliki National Park and in a few craters situated north of the Kazinga Channel. In Kivu, they occur in several places the Virunga National Park. In Rwanda, there are some near Ruhengeri, near the Virunga and at Mashyuza, in the south-west of the country; in Burundi, in the Musongati and Bururi. In some springs the water temperature does not exceed 40°C, but, at the outlet of the Mongiro springs, it reaches 106°C. Most of that water comes from rain. It infiltrates through the faults, comes into contact with the very hot rocks, turns into vapour and rises back to the surface where it heats more superficial water tables. The hot springs indicate the presence of lava near the surface that lacks the strength to forge itself a passage to the surface. They are thus a sign of the abatement of volcanic activity. Apart from their geological interest, hot springs form very unusual habitats with endemic plant and animal species.

Altogether, the Albertine Rift thus constitutes a major element in equatorial Africa, even only taking into account the mountains and lakes that it engendered. Although the Rift barely touches the Congolese forest region, it considerably influences its climate in many ways. On a regional scale, it accentuates the contrast between humid Central Africa and dry East Africa. However, it does not actually constitute a barrier, but rather an enormous transition and contact zone between the humid habitats of the west and the arid habitats of the east. On a more local scale, the amplitude and complexity of its relief creates very abrupt ecological gradients. Not only does it thus engender the montane and Afro-alpine habitats that, without the Rift, would never have existed in the region, but it also brings nearly into contact extremely contrasted habitats such as the arid plains of Lake George and the dense forests of Maramagambo, Kalinzu and Kasyoha-Kitomi. Through the cyclic variations of the climate during the last two million years – we will come back to this – the Albertine Rift has always been able to keep very diverse habitats. And so it has been a refuge for many species and a vast natural laboratory.

In human terms, also, the Albertine Rift has played a very important role. Of course, the highlands, located above 2,000 metres, were not colonised until in the very last centuries. But the middle altitudes, between 1,200 and 2,000 metres, have attracted and fixed Man for several thousand years. Thanks to a rather "temperate" climate, the absence of some diseases and the fertility of the land, of volcanic origin, the perimeter of Lake Kivu and the plateaux lying just east of the Rift in Uganda, Rwanda and Burundi, have seen the formation of very high densities of population and have been transformed into genuine human nurseries. Not only have these populations profoundly influenced the natural habitats and pushed back the eastern margin of the dense forests, but they have regularly unloaded their population excesses on the neighbouring regions. Very recently, the impact of these populations extended even beyond the region of the Rift and reached most of the Congolese basin – we will come back to this in chapter 7. In creating a great diversity of ecological conditions, the formation of the Rift has thus put in place the conditions for a real explosion of biodiversity, but paradoxically it has also engendered the mechanisms for its destruction.

1.27. On the steep slopes that dominate the small salt lake of the Kitagata crater, one of many explosion craters of the Queen Elizabeth National Park, remnants of dry forest grow. The dominant tree is an African variety of the European Olive Olea europaea. *In the distance Mount Rwenzori, wrapped in its clouds as usual, is barely visible.*

1.28. *The hydrographic network of Central Africa belongs largely to the Congo basin (1), but the extreme north-east belongs to the Nile basin (2), the extreme north to the Chad basin (3), and the north-west to the Benoue and Niger basin (4). The coastal regions bordering the Gulf of Guinea are drained by a series of coastal rivers. From north to south, the main ones are the Cross (5), the Wouri (6), the Sanaga (7), the Nyong (8) and Ntem (9) in Cameroon; the Wele (10) and the Muni (11) in Equatorial Guinea; the Komo (12), the Ogooué (13) and the Nyanga (14) in Gabon; the Kouilou (15) in the Congo.*

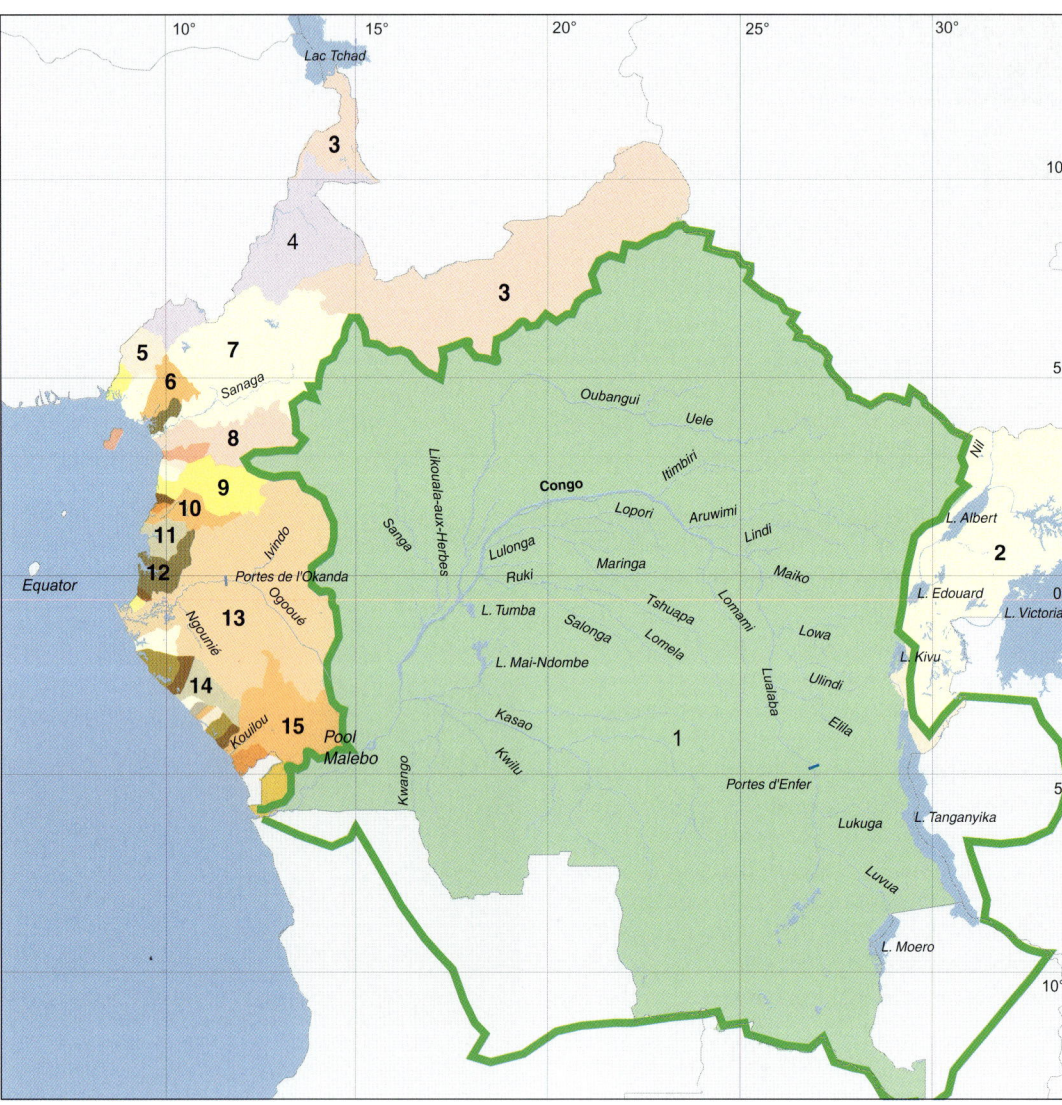

Rivers of Central Africa

	length in km	watershed in km²	flow in m³/sec
Congo	4,374	3,600,000	410,000
Wouri	-	8,250	317
Sanaga	1,043	131,500	2,072
Nyong	-	26,400	440
Ntem	-	31,000	290
Ogooué	1,200	203,000	4,400
Nyanga	-	25,000	450
Kouilou	-	56,000	916

After Bricquet, 1990.

Rivers, streams and lakes

Forested Central Africa is drained into the Atlantic mainly by the Congo River whose basin extends into Tanzania, Zambia and Angola. In the east, a narrow fringe of montane forests in the Democratic Republic of the Congo, Uganda, Rwanda and Burundi belong to the Nile basin, while a major part of Cameroon, Gabon, Equatorial Guinea and south-western Congo are drained by innumerable coastal rivers (FIGURE 1.28).

Among them, many are hardly worth naming and their flow is not even sufficient to keep the ocean from blocking their mouth by a sand bar (PAGE...). Others are more important and the Sanaga and the Ogooué are among the great rivers of Africa. From Cameroon to the Congo, all the coastal rivers have a profile that appears to be very young: their middle and lower courses are invariably broken by rapids or water falls (FIGURE 1.1). Yet these rivers originated in the Cretaceous and are therefore as old as the Atlantic, but their flow has been perturbed by the widespread lifting which affected the western margins of Central Africa for more than 45 million years. As for the mouths of these rivers, notably of the Wele, the Muni, the Mondah and the Komo, they often seem disproportionate. In fact, they have been drowned and transformed into rias by the marine transgression that occurred since the end of the last glacial epoch, around 12,000 years ago. The largest is the Komo, better known under the name of the Gabon Estuary.

The Sanaga is the most important river in the forested region of Cameroon. Among its main tributaries, the Djerem drains the south slope of the Adamaoua plateau and the Lom the drains the highlands on the border between Cameroon and the Central African Republic. Downstream from Nachtigal falls, 80 kilometres north of Yaoundé, the Sanaga also receives the waters of the Mbam, which drains the mountains of north-western Cameroon and the Bamiléké country. Then, it goes over

1.29. Below Makokou, in north-eastern Gabon, the black waters of the Ivindo pour into the rapids of the Loa-Loa. In July, when the waters are at their lowest, innumerable white amaryllis Crinum purpurascens *appear between the exposed rocks and in the pockets of calm water.*

1.30. The Collared Pratincole Glareola nuchalis *is specific to torrential rivers where it hunts sometimes at the surface of the churning water, sometimes high in the sky. It breeds during low water and lays its eggs on the bare rock.*

1.31. The Ogooué River clears a passage through the Portes de l'Okanda by the Lopé National Park.

falls and rapids, tumbling downward 350 metres in 140 kilometres. After Edéa, at last, the Sanaga flows peacefully towards the ocean. Its floods are rather moderate, all the more so that nowadays its course has been particularly regulated by dams.

The Ogooué River has its source on the Batéké plateau of the Congo, at an altitude of about 600 metres. Near Franceville, in south-eastern Gabon, it goes over the Poubara Falls and collects the waters of the Mpassa, its first major tributary. It then flows over two series of rapids and collects the Lolo and, mostly, the Ivindo (FIGURE 1.38). The latter stream is its main tributary and drains essentially all of north-eastern Gabon. Downstream from the town of Makokou, its black waters are thrown into a succession of rapids (FIGURE 1.29) – the most important being those of Mingouli – and over the falls of Koungou (FIGURE 1.1). Below the Boué region, the Ogooué, sometimes wide, sometimes narrower, enters a new succession of rapids and small falls (FIGURE 1.7). Just before reaching the Portes de l'Okanda (FIGURE 36), it also receives the waters of the Offoué, a small watercourse that forms part of the eastern limit of the Lopé National Park. Down to Ndjolé its course remains fairly rough and strewn with innumerable pointed rocks, but even in the dry season experienced boatmen navigate their dugouts without too much difficulty. Beyond, the river enters the coastal plain where it receives its second largest tributary, the Ngounié, which partially drains the Chaillu Mountains. Its current slows and its bed spreads over two to three kilometres. During low water, it uncovers immense sand banks (FIGURE 1.32) which shift from year to year and seriously impede navigation but offer essential breeding sites for several species of birds, including a near-endemic

I | THE MINERAL WORLD

tern. Downstream from Lambaréné, the river splits into a multitude of channels enveloping islands covered with swamp and riverine forests. Away from the main channels and the masses of alluvium they carry, innumerable lakes with capricious contours stretch northward and southward, (FIGURE 1.34). The largest and the most picturesque are Onanqué and Oguémoué. From Lambaréné to the shores of the ocean, the Ogooué ploughs up a passage through a valley some 120 kilometers long and 2 to 60 kilometres wide, whose whole axial part has been filled in by its alluvial deposits. Only

1.32. From its confluence with the Ngounié, just above Lambaréné, the Ogooué widens. Its current slows down and during low water, from July to September, immense sand banks fall dry. Scattered with craters that are the nests of the fish Heterotis niloticus, *a species of Sudanian origin introduced in the Ogooué basin and the Fernan Vaz lagoon in 1959, they also harbour many nesting birds, notably the Little Tern* Sterna albifrons, *the last viable population of the African form of the Common Tern* Sterna hirundo *and even some pairs of the Caspian Tern* Sterna caspia.

1.33. The commonest bird is, however, the Grey Pratincole Glareola cinerea, *a species that occurs from Mali and the Niger to the Congo basin.*

1.34. Lake Evaro is two or three hours away from Lambaréné by dugout. With the Ezanga, the Onangué and the Oguémoué, it is part of a vast complex of lakes surrounded by dense forests and marshy expanses where multitudes of aquatic birds, the Hippopotamus and Manatees live side by side.

32

the most distant bays have remained open and harbour lakes that very gradually fill in with the progression of little interior deltas.

Very near the coast, the river comes up against Mandji Island on which is built the town of Port Gentil. It then explodes into a veritable fan of more tenuous channels. The main one continues its course due west and empties into the Gulf of Guinea just south of Port-Gentil. Others turn northward and empty through 20 to 30 kilometres of mangroves into Lopez Bay, east of Port Gentil. Further south, still others reach the lagoon of Nkomi Fernan Vaz, which form the northern extremity of the vast lagoon complex of southern coastal Gabon.

Without taking into account all the lagoons that extend to the Congolese border, the delta itself covers a total surface of more than 500,000 hectares. It thus is the largest delta of sub-Saharan Africa after the delta of the Niger. Although it extends inland for nearly 150 kilometres, from the ocean to Lambaréné, the river's flow is sufficient to keep the salt water from reaching upstream more than 20 to 25 kilometres. Most of the Ogooué delta is thus under a fresh water regime.

The Ogooué is at its lowest from July to September and at its highest around November and April. Given that its entire drainage basin lies south of the climatic equator, the difference between its maximum and minimum flow is enormous and can reach a factor of five. At Ndjolé, the flood crests can reach a height of five to six metres and some years they inundate the low-lying parts of the town. In the delta, however, they never exceed three to four metres.

Most of the interior of Central Africa belongs to the basin of the Congo, the second largest river in the world after the Amazon. Its source is located at around 1,400 metres above sea level, on the high plateaux of the Katanga. Its upper course, called the Lualaba down to Kisangani, is a torrent at first, but calms down when it reaches the vast marshy depression of the Upemba. Then it collects the waters of the Luvua, which drains Lake Moero, and the Lukuga, which drains Lake Tanganyika, Lake Kivu and most of Rwanda, Burundi and north-western Tanzania. Beyond the *Portes d'Enfer*, it encounters its tributaries, Luama, Elila, Ulindi, Lowa and Maïko – to name only the most important ones – which drain the mountains of the Albertine Rift (FIGURE 1.35).

Just before Kisangani, the river passes a second series of spectacular rapids and becomes the Congo. In its middle course, it widens and turns towards the west, then to the south-west. Its banks become lower and are finally reduced to simple alluvial accumulations separating the waters of the river from those of the surrounding swamp forests. The Congo does not drop more than 100 metres in altitude over a distance of nearly 2,000 kilometres. It thus flows very slowly and its bed, strewn with islands and sand banks, is 35 kilometres wide at its confluence with the Oubangui. From the north, it collects the waters of the Aruwimi, the Itimbiri and the Mongala, then the Oubangui, the Sangha and the Likouala. From the south, it receives the Lomami, whose source flows from the Katanga, and numerous streams that drain the centre of the basin, the two main ones being the Lulonga and the Ruki. Still farther south, the Congo collects the waters of the Kwa which unites the Kwango, the Kwilu and the Kasai, three large watercourses whose sources are located on the Angolan plateau. After this last great confluence, the bed of the Congo narrows as it cuts across the Batéké plateau before widening again in the Malebo Pool – formerly Stanley Pool. Twenty-five kilometres wide and 35 long, it occupies the bottom of a vast basin surrounded by

1.35. The Irangi scientific reserve – just north of the lower part of Kahuzi-Biega National Park – is crossed by a small tributary of the Luhoho River that flows into the Lowa. Their cold, clear waters come down from the mountains of the Albertine Rift.

1.36. *The Congo River is the second largest river in the world, and is the vital water way of the interior of Central Africa. Its flow is interrupted in several places by rapids or waterfalls. Nevertheless, this river is essential for transport in the Democratic Republic of the Congo. On its banks have been erected, facing each other, Brazzaville and Kinshasa, whose tall buildings are visible on the horizon. Currently, these two towns are inhabited by probably about 10 million people. In the foreground are Water Hyacinths* Eichhornia crassipes. *Imported from South-America for its beautiful flowers, this plant has invaded a large portion of the Congo basin, where it disturbs the ecosystems and the navigation.*

hills. Brazaville sits on its fairly steep northern bank. Kinshasa is on its south bank, a vast marshy plain. At its centre, the Pool holds Mbamou Island and a multitude of smaller periodically flooded banks similar to those of the Ogooué.

Upon leaving the Pool, the Congo reaches its lower course, marked by more than 30 falls – including the famous Inga Falls – that extend for 300 kilometres and lower the course by 265 metres. Below Matadi, it spreads out in its estuary, some ten kilometres wide at its mouth.

Because of the size of its basin, the Congo River has a complex regime. It drains the regions on both sides of the equator and hence is subject to very different patterns of precipitation in terms of both quantity and of seasonal distribution. The flow of the tributaries coming from the north, such as the Oubangui and the Mbomou, is at its maximum in October and November, at its minimum in March and April. Inversely, the flow of the Kwango and the Kasai is maximum in April and May, minimum around October. And the rivers of the central basin generally have two high-water seasons and two low-water seasons. Thus, the various parts of the basin partly compensate each other, while the length and very weak slope of the tributaries slow flooding. Lastly, at the Pool, the maximum flow is recorded in December with two low water periods centred on March and July-August, broken by a secondary increase in May. In March and April, the waters of the Congo flow back several hundred kilometres into the Ubangi. The monthly variations of the flow are largely buffered by the immense marshy expanses in the centre of the basin. At the Pool, they are no more than of the order of one to two, while those of the Kwango or the Kwilu vary from one to six. In comparison with the Ogooué, the Congo is a very regular river.

During different geological periods, the central basin of the Congo incorporated vast lacustrine and fluvio-lacustrine expanses. The enormous masses of sediments, accumulated since the Mesozoic and, especially, the Cretaceous, bear witness to their existence. To this day the Congo forms, with the Likouala, the Sangha and the Ubangi, one of the largest marshes in the world, covering over 200,000 square kilometres, and of the same magnitude as the Pantanal in South America. At the heart of this marsh lie Lake Ntomba and Lake Mai Ndombe. The first is relatively recent. It was probably formed after the damming of the mouth of a small stream by alluvial accumulations of the Congo. It remains in contact with the river, however, via a narrow channel. Its banks are sharply cut and often very steep. It has some islands and small rivers that heavily supply the little deltas. Lake Mai Ndombe extends from north to south over a distance of 140 kilometres and its greatest width is 60 kilometres. Its banks are rather low. It is fed by numerous rivers that drain the surrounding humid forests and swamp forests. Its outlet empties into the Lukenie with which it forms the Fimi, which in its turn empties into the Kasai and forms the Kwa. Like Lake Ntomba, its waters are black, heavily loaded with humus material and very acid. They contrast with the waters of the Fimi, which carry muddy alluvia. The black and the brown waters do not mix immediately and flow for a great distance in parallel ribbons. In the same way, the waters of the Fimi form, in their turn, a long blackish ribbon flowing along side the brown-red waters of the Kasai.

The estuary has been the object of much speculation. Some have proposed that the Congo long ago belonged to the basin of the Chad. Others have suggested that it emptied into the Atlantic via the Sanaga, later via the Ogooué and only very recently – for at most 10,000 years – via its present estuary. The sediments deposited in the Atlantic offshore from this estuary show, however, that it has functioned at least since the Eocene-Oligocene transition, 35 million years ago. The apparent youth of the lower course of the coastal rivers located farther north is only the result of the relatively recent tectonic movements that have lifted up the Mayombe range, among other things.

These tectonic movements also affected the middle course of the Congo. To the north, this river may have captured the Ubangi River, previously belonging to the basin of the Chad. This capture would have been the work of a small tributary of the Mpoko and would have taken place near Fort-Possel, where the river turns suddenly southwards. Tectonic movements continue today in the central part of the basin. They generate a westward drift of the confluence of the Sangha and the Likouala-aux-Herbes.

Finally, the Congo basin seems to have been disrupted by the birth of the Albertine Rift. It may have captured, probably at the level of the Portes d'Enfer, the southernmost branches of the Nile that would thus have become the Congo's upper course. How and at what moment the capture of the Lualaba might have taken place remains a mystery because in this region the raising of the Rift obliterated all traces of any old hydrographic network. To this day, however, the Lualaba harbours species of fish identical, or nearly so, to species of the Nile and absent in the rest of the Congo basin.

Farther north, the Congo would have, on the other hand, abandoned to the Nile several small rivers that drained Uganda and western Kenya, notably the Mara and the Kafu. This history can still be followed in detail and without any doubt. It began in the Miocene, with the collapse of the Rift fault, the blockage of the east-west drainage and the flooding of the region of Lake Albert. Later, this inundation continued towards the south to form the vast "Lake Semliki". Later still, the the fault's movements lowered the level of this lake and the lifting of the Rwenzori cut it in two. In the south, a great lake was thus formed that prefigured Lake George and Lake Edward. It collected the waters of the Kivu, south-west of Uganda and Rwanda. It emptied into Lake Albert via the Semliki River. The Rift, however, continued its splitting and, as the bottom of the trough sank, its sides rose. Around 750,000 years ago, the Kafu and the Katonga were thus blocked and their waters gave rise to Lake Kyoga and Lake Victoria. Since then, Lake Victoria has had a fairly turbulent existence and its level fluctuated greatly in function of the oscillations of the climate. During the last glaciation, it was even reduced to a few small residual lakes with no outlet and very salty. For very long periods, perhaps tens of thousands of years, the Semliki was thus the only headwater of the Nile. But the Albert Nile itself did not exist except intermittently and its flow was interrupted during the driest periods.

South of Lake Edward, the birth of the Virunga would profoundly disturb the landscapes and return to the Congo River a small part of its basin that had been taken over by the Nile. In obstructing the flow of the waters in the axis of the Rift, the Virunga's lavas created in effect an enormous natural dam, above which was born Lake Kivu. Like Lake Edward and Lake Albert, this lake experienced substantial variations in its water level, but got an outlet towards Lake Tanganyika (FIGURE 1.37). Secondarily, the eruptions of the Virunga also obstructed the small lateral valleys of the Rift and created a multitude of smaller lakes, presently at altitudes of 1,600 to 1,850 metres. Some are without any outlet, such as Lake Mutanda in Uganda, the Mokoto lakes in the Democratic Republic of Congo, or Lake Karago and Lake Bikinga in Rwanda. Lake Bunyoni empties, however, into Lake Edward, while Lake Bulera and Lake Luhondo empty into the Akagera via the Mukungwa River. In blocking the waters of the Mukungwa, the lavas of the Virunga also modified the entire hydrographic system of Rwanda that was diverted largely towards the Akagera and Lake Victoria by the tilting of the relief. This tilting affected the entire Rift region and hindered the drainage of many high valleys also that were transformed into bogs. The two most extensive ones are located in Rwanda: the Rugezi, in the north of the country at altitudes of 2,000 to 2,100 metres, covers more than 8,000 hectares; the Kamiranzovu, at an altitude of about 1,950 metres in the Nyungwe Forest, covers some 1,300 hectares. But there are smaller ones in the Kahuzi-Biega National Park in Kivu, in the Bwindi Impenetrable National Park in Uganda and in the Kibira National Park in Burundi.

This abundance of lakes and wetlands is thus a char-

The Lakes of the Central Congo Basin

	depth in m	area in km²	altitude in m
M. Ndombe	3	2,300	300
Ntomba	3-8	765	325

The Lakes of the Albertine Rift

	depth in m	area in km²	altitude in m
Victoria	80	69,500	1,133
Albert	56	6800	620
Edouard	112	2325	914
George	3	270	914
Kivu	485	2055	1,460
Tanganyika	1,475	60,300	800

1 THE MINERAL WORLD

1.37. The Rusizi River, seen from the Kamaniola escarpment around 1990, drains the waters of Lake Kivu into Lake Tanganyika. Until the 1970s, its deeply set valley harboured beautiful riverine forests.

1.38. In addition to the main rivers an infinite number of nameless small streams and channels constitute an incredible, natural communication network which deeply influenced the human history of Central Africa.

acteristic of the Albertine Rift region. Lake Albert and Lake Tanganyika are situated well outside of the forested region and are thus of no regard to it except for their very probable influence on the climate. Lake Tanganyika, 650 kilometres long, constitutes, in effect, an interior sea and is the second deepest lake on earth after Lake Baikal. Lake George and Lake Edward, even though more modest in size and surrounded by grasslands, are more important for the forested region. Lake Victoria comes in contact with it along its northern and north-western shores. In rising to their present level, some 10,000 years ago, the very fresh waters of the lake isolated the Sese Islands, covered mostly with very humid forests, and created a vast zone of contact between the dense forests and the most varied aquatic habitats. Finally, there are only Lake Kivu, numerous smaller volcanic lakes and high altitude peat bogs of the central part of the Rift that have been truly an integral part of the forest zone, at least until Man cleared the forests that surrounded them. The very steep shores of Lake Kivu do not leave much room for the development of humid habitats.

Thus, from the Atlantic coast to Lake Victoria, the forests of Central Africa are scattered and surrounded by vast wetlands. Variable in size and type, they greatly contribute to the biodiversity of the region. Not only do they introduce into the heart of the forest a great number of distinctive habitats, often very dynamic, but, because of their extent, they must also have a significant influence on the climate. Lake Victoria, for example, lying astride the equator, forms a gigantic natural kettle that maintains a high rainfall zone, covered with dense forests, on the Sese Islands and the north-western shore of the lake. For the fauna, the large watercourses, such as the Sanaga, the Ogooué and the Congo, constitute indisputable barriers that have strongly conditioned distribution and oriented speciation, perhaps most of all of the mammals (PAGE 59).

One could imagine that, for humans, the immense marshes of the centre of the basin constituted an inhospitable region and that they considerably stimulated the imagination even of local populations. Thus, people periodically speak of the *Mukilibembe*, the Lake Télé Monster, a competitor of the Loch Ness Monster. These marshes deflected human migration (PAGES 244-245), but the thousands of kilometres of wide, calm watercourses also formed an enormous natural network of waterways that greatly facilitated the penetration of present-day peoples into the forested region and, later, the entire process of colonial seizure. The rapids and falls that, just inland from the coasts, almost systematically blocked access to the watercourses of the interior, contributed substantially to the isolation of the interior of Africa, particularly the Congo basin. The human history of the Congo basin is thus very different from that of Amazonia where large boats went without difficulty upriver into the interior of the continent for more than 1,000 kilometres. From the 16th century, the Spanish were able to reach the heart of the Amazonian forest, while the forests of the central basin of the Congo would have to wait until the mid-19th century to see the appearance of the first Europeans.

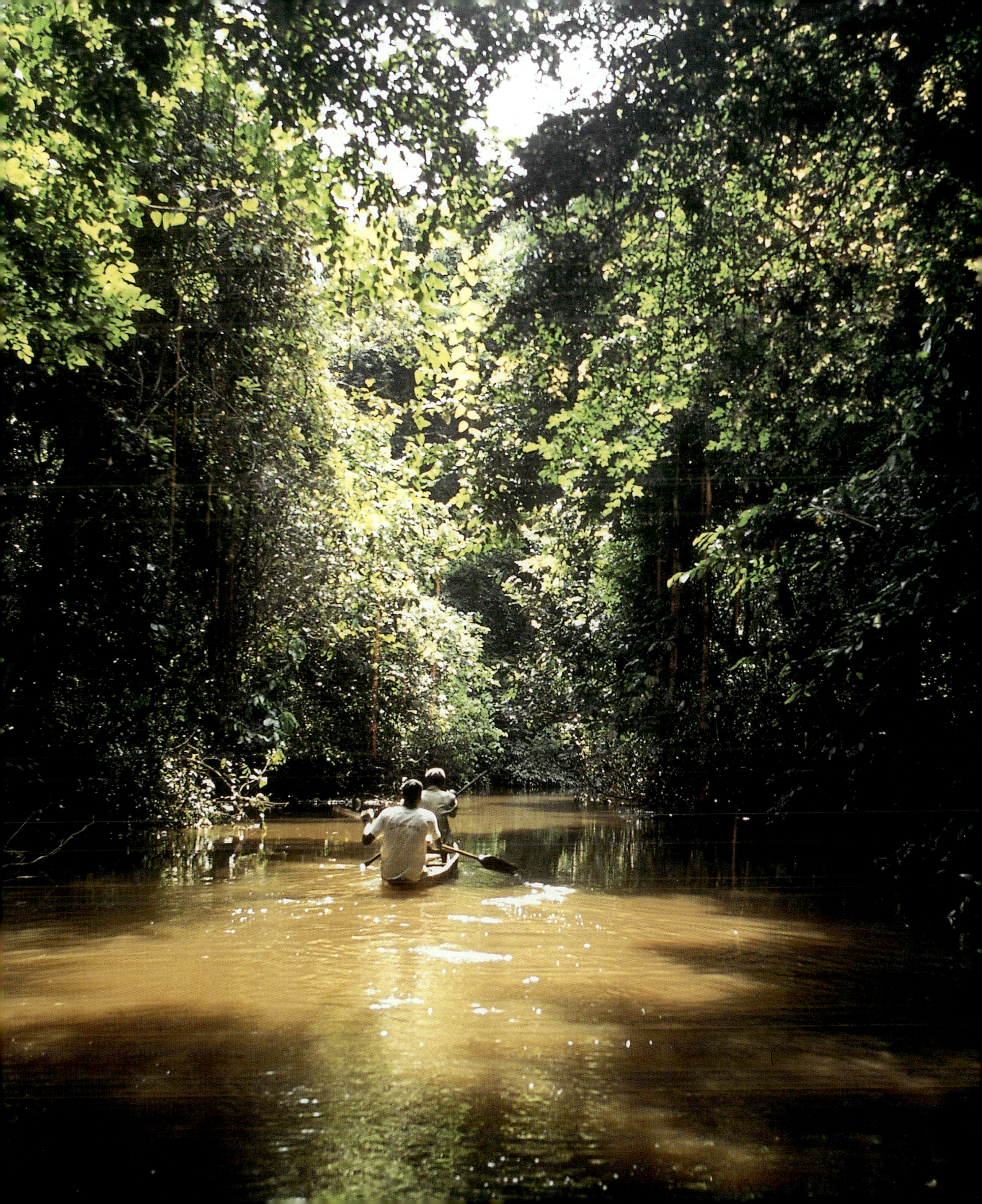

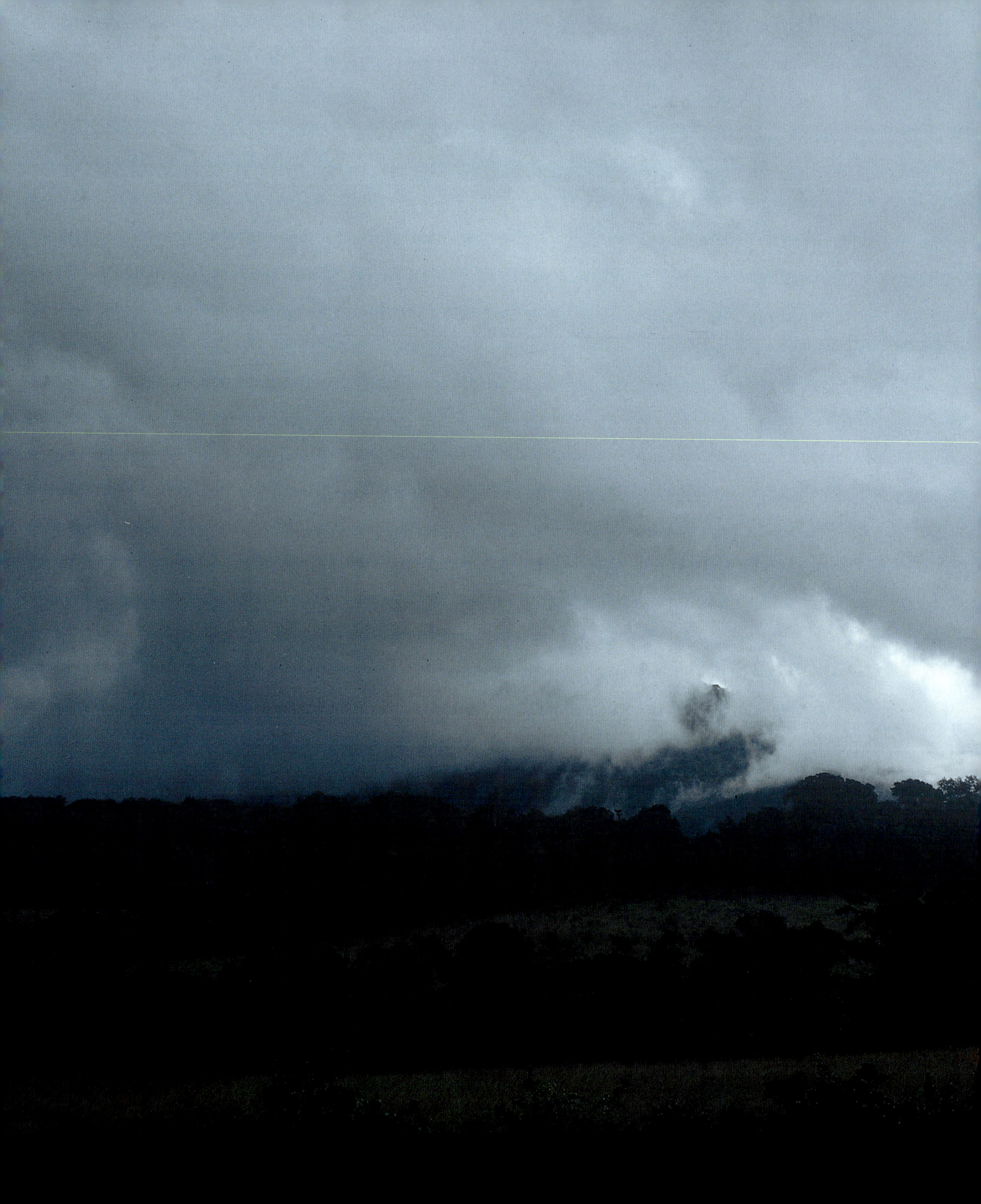

2 | Climate and forests

Life has existed on earth for nearly four billion years, but the continents were colonised only 400 million years ago. Forest-like plant formations developed very soon, but the flowering plants appeared only as recently as one hundred million years ago and it is not until the Miocene, 15 to 20 million years ago, that most of the present-day species were established. Since then, their communities and associations have been constantly fashioned by variations in climate, especially in the last two million years. Forests as we know them today took form at about the same time as the human species. Some of the formations that we are now in the process of destroying are even only 2,000 years old, and some date back no further than the end of the 19th century. Although they include some ancient species, in existence since the time of the last dinosaurs, dense and humid tropical forests are not static, anachronic relicts from an obscure past but an ensemble of contemporary, constantly evolving, habitats. They also constitute the richest and most complex habitats on the planet, so complex that we are not able to grasp them or to use them in a really sustainable way. Most disconcerting to us is the fact that the history of the forests unfolds on a time scale that eludes our understanding, and that most effects of environmental changes only become visible after a long period of time.

Although relatively young, these forests are, like all present-day species and the communities they form, the product of a very long story whose roots plunge into the very origins of life and whose successive chapters were written and rewritten by continual transformations of the environment. The genesis and shape of the continents have played an important role, often a decisive one, but the climate has always been one of the principal actors. Through the effects of temperature and its fluctuations, moisture availability and seasonal contrast, it is the climate that has always determined which species would survive and where they would be able to thrive. The history of the forests and that of the climate are just two aspects of one and the same story. And it is not just an old story. Inexorably, it continues under our very eyes, even though, on the time scale of a human life, its effects may not be clearly perceptible. So if the immediate future of the forests is mainly in the hands of those who exploit them, their future in the long term also depends on the climate. Until about a hundred years ago this power was out of our reach, but we have since acquired the ability to warm the Earth. In the long term, the future of the forests does not depend only on those who exploit their resources, but on the behaviour and the choices of all of humanity.

2-1. Storm over the Lopé National Park. Rains that fall on dense and humid tropical forests can be very violent and are accompanied by veritable tornadoes. Unlike regions near the tropics of Cancer and Capricorn, equatorial regions never experience hurricanes. This is a very important fact for the dynamics of the forests.

2.2. *On a biogeographical level, the Mediterranean regions of the African continent are part of the Palaearctic region, with Europe and most of Asia. Sub-Saharan Africa and Madagascar constitute the Afrotropical region, even though Madagascar is so different from the continent that many scientists consider that it should be made into a separate region of its own. Whatever it is, sub-Saharan Africa is divided into four main terrestrial biomes: dense humid forests (1), savannas (2), deserts and arid formations (3) and Afro-alpine habitats (in black). At the extreme southern end of the continent, centred on the Cape region, the Fynbos (5) survives, an ensemble of plant formations recalling Mediterranean habitats. In spite of its restricted area, it represents unquestionably the richest biome of the continent botanically, with an enormous number of endemic species, to such an extent that some scientists would want to make a separate region of it, like Madagascar.*

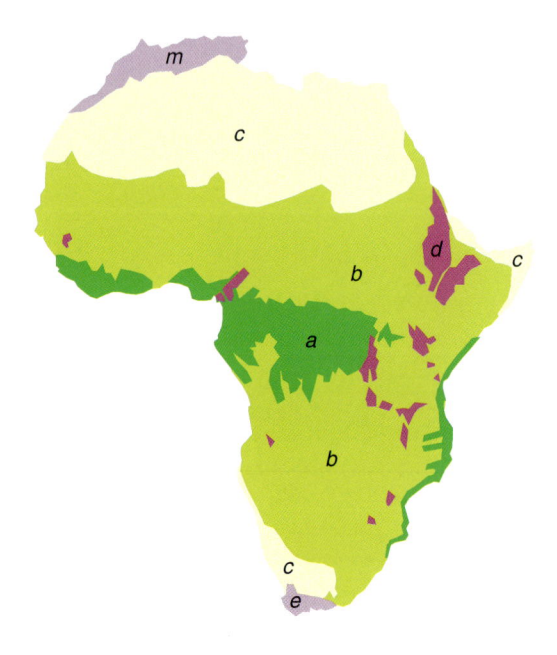

The great biomes

In French-speaking Africa, the word *"forêt"* often designates all that lies beyond the inhabited or cultivated world, whether or not there are trees. In Rwanda, one speaks of the *"forêt du Mutara"*, a vast grassy landscape, in places completely devoid of all shrubby or arborescent vegetation. In usual French, the word *"forêt"* evokes a natural or semi-natural habitat composed essentially of trees. Technicians and scientists also agree on the fact that without trees there is no *"forêt"*. Beyond this point their opinions can be quite divergent. Some take into account only morphological or structural criteria. They thus attach much importance to the form and height of the trees. Others pay more attention to ecological criteria. The FAO considers as forest all habitats in which trees cover at least 10% of the surface. On this basis, most of the savannas are included in the category of forest and the Central African Republic is credited with forests covering over 60% of its territory, while most biologists evaluate its forest cover at no more than 20 %. These divergences are not just simple differences between schools. They are the source of a good many misunderstandings and sterile discussions. For example, how can the reduction of forested areas be evaluated if there is not even agreement on the definition of forests? When speaking of forests, one must first state the concept used.

In this book, we adhere to an ecological vision that divides the extraordinarily varied vegetation of sub-Saharan Africa into four main biomes. By this we mean four ensembles of habitats, often diverse, but each regulated by the same eco-climatic constraints. The forest biome is centred on the Congo basin and the coasts of the Gulf of Guinea. Far to the north and far to the south, extend the Saharan and the Somalian deserts and the Namib and Kalahari deserts. Between these two extremes, is interposed the savanna biome, an immense multifaceted zone of transition. Finally, scattered like an archipelago on the ocean, is the Afro-alpine vegetation, taking refuge on the highest mountains (FIGURE 2.2).

This very schematic division of the continent's vegetation is conditioned mainly by the climate in spite of the fact that in some places the substrate may attenuate or reinforce its influence. The Afro-alpine biome, adapted to frequent nocturnal frosts that can occur at any season, is the only one to depend mainly on temperature. The three others all depend on the availability of water. The vegetation of the sub-desert and arid regions resists periods of drought lasting several years and manages to exploit rare and unpredictable precipitation to the full. This vegetation includes stunted, ligneous or succulent forms and many species that survive drought only in seed form. Savannas or tropical grasslands and woodlands are adapted to more abundant and more frequent rains interspersed by generally predictable annual dry seasons. During these periods, which can last from three to seven months, many trees lose their leaves and many herbaceous plants eliminate their aerial parts, surviving underground as rhizomes, tubers, corms or bulbs. Dense humid forests develop only where there is no frost and where periods of water shortage last for no more than one to two, exceptionally three, months. In more arid regions, with less than 1,000 millimetres of rainfall and two dry seasons of four to five months, savannas usually occur but dry forests can nevertheless develop, just as savannas can survive in very humid regions.

The altitudinal limit of Central African dense forests is determined by frost. On their periphery, the limit is conditioned rather by rainfall. But many other factors intervene and this limit is not always found where one could expect it. But wherever it lies, the edge between the forest and the savanna is generally well marked, is often even very abrupt, and the distinction between forest and savanna does not pose any great problem. However, in some places, there exist densely wooded savannas and woodlands, notably in the Central African Republic, in the south of the Democratic Republic of the Congo, in Uganda, and elsewhere in eastern Africa (FIGURE 2.3), which are reminiscent of forests. In French they are called *"forêts claires"*. Except for ecologists who know the problem well, this unhappy nomenclature brings about much serious confusion, for this type of habitat belongs, in fact, to the savanna biome and not to the forest one.

In tropical regions, the difference between forest and savanna is based more on the herbaceous vegetation than the ligneous vegetation. Bushed or wooded savannas and woodlands have a fairly developed shrub or tree stratum, open grass savannas do not. Nevertheless, all have a well developed herbaceous layer where gramineous species form the dominant element. Invariably, savannas, including woodlands, are subjected to frequent fires. And the regular occurrence of fire constitutes perhaps the most fundamental difference between forest and savanna.

Nowadays, fire is almost always of human origin, but in other times it came naturally. It existed well before the appearance of Man. In the sediments of the Niger delta, for example, traces of bush fires dating back more than 10 million years have been found and we know that the fern forests of the Carboniferous, 300 million years ago, burned. One can then suppose that savannas experienced fire from the beginning of their existence – 20 to 25 million years ago – and this long history inevitably engendered a series of adaptations. Unlike forests, the savanna vegetation is more or less resistant to fire, the sensitive species having been eliminated. However, many savanna plants need fire to propagate. Some bulb plants, for example, do not flower except immediately after its passage, other species have seeds that cannot germinate until after they have been somewhat "grilled".

Forests and savannas form, then, two very different worlds having only a few species in common. In one case, there is the world of trees, constant humidity and shade, in the other, the world of grasses, marked seasons and fire. They divide up a large part of the continent in function of climate, essentially by the intermediary of fire. In the absence of fire, a continuum of forest, going from the most humid formations to the driest ones, is able to extend from the equator to the approaches of the desert. But since the earth's atmosphere contains sufficient oxygen, fire is inevitable. In the dry regions it has reduced the forests to minute relicts which have not been able to survive except in sufficiently humid areas. During some periods, and in some places, forests were able to regain some territory, as they do today in Cameroon, Gabon and the Congo. Forests and savannas are engaged in a harsh competition and their limits come and go according to the vagaries of the climate which determine the extent of the fires.

We now know that the earth's climate has always varied and that the vegetation has always adapted, or tried to do so, for better or worse. During several periods these climatic variations were fatal for many species, but it is very probable that they were also a strong stimulus for the diversification of flora and fauna. The history of the species and their associations, in the case of Central African forests, is therefore inseparable from that of the climate. But first, let us look at present-day relations between the forests and the various parameters of the climate.

2.3. *In Tanzania, the plains of the Moyowozi, a tributary of the Malagarazi which flows into Lake Tanganyika and thus belongs to the Congo basin, are covered with wooded savannas or open forests with* Isoberlinia *and* Brachystegia. *This is the miombo, a very rich habitat, which extends over a vast part of Tanzania, Mozambique, the Democratic Republic of the Congo, Angola and Zimbabwe. Under the large trees that form a light but nearly continuous canopy is a grassy carpet which burns each year and keeps this habitat from becoming a true dry forest.*

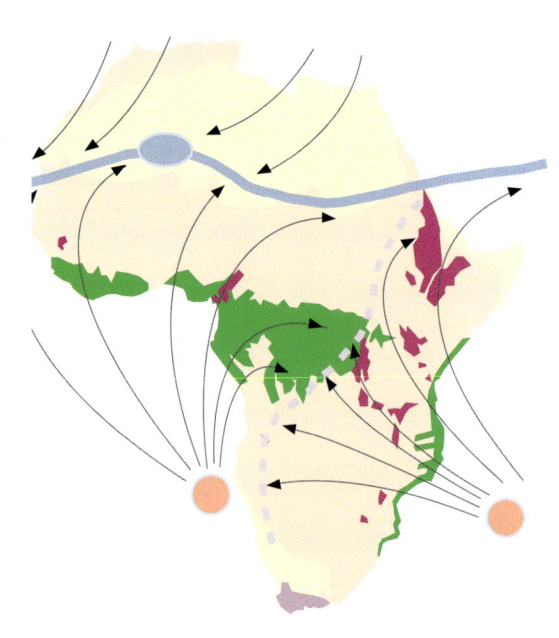

2.4. In July, the Intertropical Zone of Convergence is centred on the Sahara between 15 and 20°N, while the anticyclones of South Africa, the Indian Ocean and Saint Helena dilate. Cold water upwellings then form in front of the coasts of the Congo and Gabon and the cold Benguela current pushes cold waters from Namibia to the equatorial latitudes (after Wauthy in Maley et al. 2000).

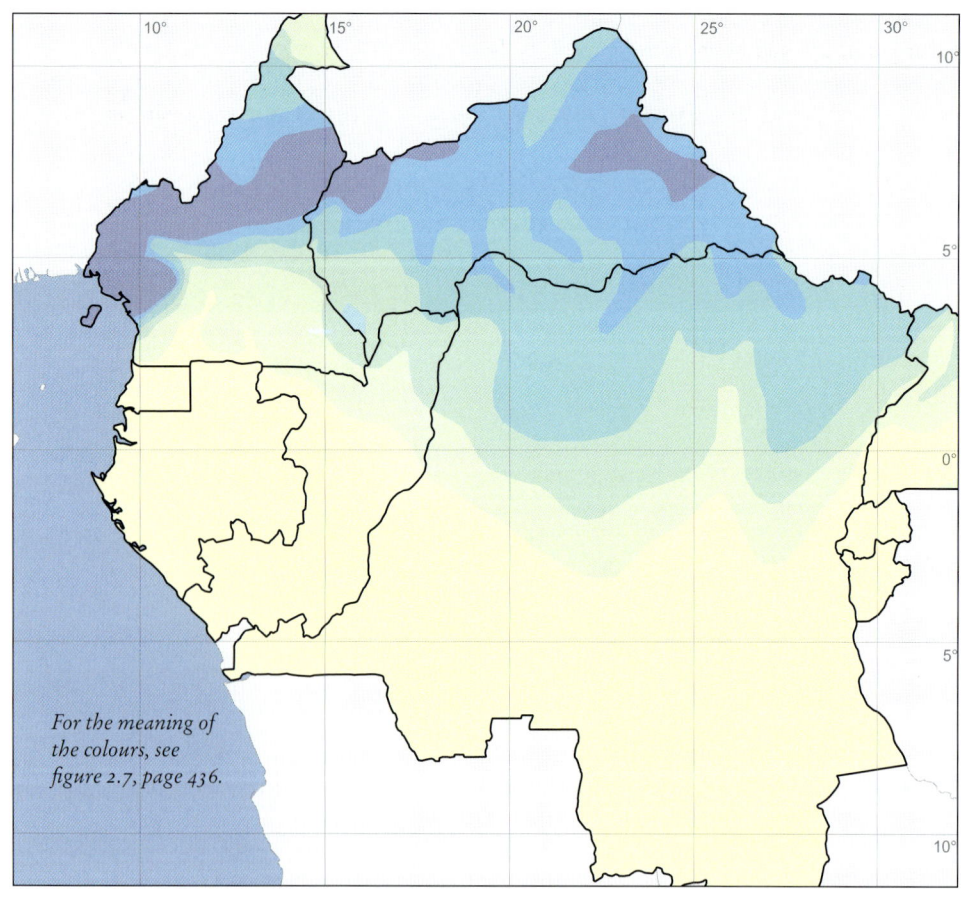

2.5. During the boreal summer, the entire northern part of the Congolese forest massif experiences heavy rainfall (after Leroux, 1983).

For the meaning of the colours, see figure 2.7, page 436.

The climate today

The two main climatic variables that determine the vegetation everywhere in the world are temperature and precipitation. In temperate and cold regions the seasons are marked by the alternation of summer and winter and temperatures play an essential role. In tropical regions, on the contrary, precipitation governs the annual climate cycle through an alternation of wet and dry seasons, while temperature remains relatively constant throughout the year.

Temperature

Everywhere in the Central African forest zone, whether 500 kilometres north or south of the equator, average daily temperatures vary little and seasonal fluctuations do not exceed 3°C. On the islands of the Gulf of Guinea, variations can even be less than 1°C. Variations recorded between the daily minimum, just before dawn, and the noon maximum are greater. Inside the forest, the difference reaches 6 to 7°C, and in large clearings, 10 to 12°C. Differences are less than in savannas because the high ambient humidity within the forest prevents very high temperatures.

Variations in temperature related to altitude are also important. Temperature decreases by about 0.6°C per 100 metres of elevation. At around 2,000 or 2,500 metres above sea level, the climate becomes cool. Not only does this temperature gradient condition a gradient of the vegetation, but above 3,000 to 3,300 metres freezing occurs. Since seasonal variations in temperature are very slight, freezing can occur at any time of the year, generally in the form of nocturnal frosts. The upper forest limit is determined in this way.

It is not only from an altitudinal perspective that the direct influence of temperature on forest vegetation is expressed. Its effects on the vegetation are more indirect but still very important through evaporation and mainly evapotranspiration – the water that the vegetation restores to the atmosphere by means of its transpiration.

Precipitation

Most of the rain that falls on the forests of Central Africa is the result of local evaporation and evapotranspiration. Only a third is brought by the winds, mainly from the South Atlantic. These winds are conditioned by the relative importance of the four high-pressure zones centred on the Azores, the Sahara, Saint Helena, South Africa and the Mascarene Islands. Wind determines the position of the Intertropical Convergence, a low-pressure zone that constitutes the climatic equator and forms, in theory, in regions where the sun reaches the zenith. During the year, this zone moves from one tropic to the other (Figures 2.4 and 2.6).

Evaporation and the cloud formation above the South Atlantic depend mostly on surface water temperatures, themselves influenced by cold currents and the resurgence of the eastern part of the tropical Atlantic.

During the boreal summer – from June to August – the anticyclones of the southern hemisphere dilate and push the Intertropical Convergence back above the

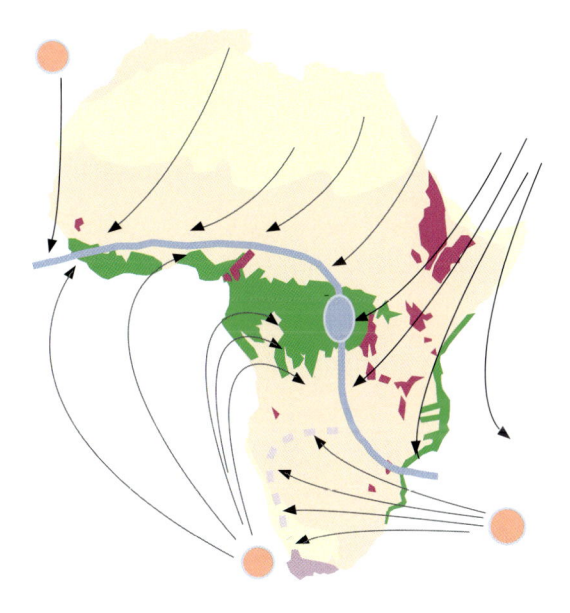

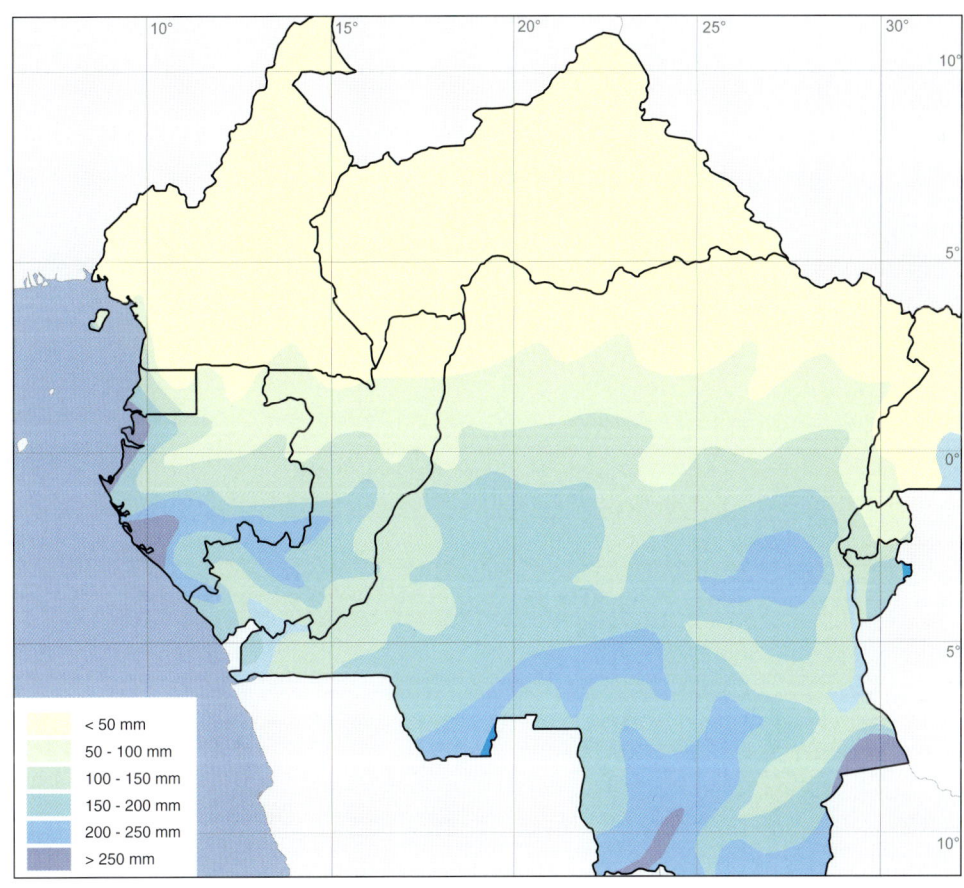

Sahara between 15° and 20°N (FIGURE 2.4). The Saint Helena anticyclone spreads out as far as the continent. The trade winds that it generates head towards the Intertropical Convergence, but turn eastwards north of the geographic equator. They then rain down on all of western Africa, Cameroon, the Central African Republic and southern Chad (FIGURE 2.5). The upwellings of cold water that they generate in front of the Gabonese and Congolese coasts, and the Benguela Current, that they push into the area around the Gulf of Guinea, provoke an atmospheric stabilisation that is marked by the formation of a thick stratiform cloud ceiling. It does not rain in Gabon, in most of the Congo nor in the western part of the Democratic Republic of the Congo, but the sky remains hopelessly grey in these regions and the average daily temperature drops about 3°C. The absence of the sun compensates for the lack of rain and plays a fundamental ecological role. By greatly reducing evaporation, it allows the maintenance of the forest cover, for which three months of dry season would otherwise be unbearable. The south-eastern and eastern parts of the Democratic Republic of the Congo and the Great Lakes region are, on the other hand, subjected to trade winds from the south-east, coming from the Indian Ocean. South of the equator, it is a dry wind but, in northern Uganda and north-eastern regions of the Democratic Republic of the Congo, there is some rainfall.

During the austral summer – from December to February – the Saharan and the Azores anticyclones dilate. The Saint Helena anticyclone retracts (FIGURE 2.6) and the Zone of Intertropical Convergence moves southwards along an S-shaped trajectory which is located around 5°N in West Africa, and extends to 15°S at the level of Zambia. North of the front, the continental trade wind, or harmattan, blows winds from the Sahara, hot, dry and often loaded with dust. In the south, the Benguela Current weakens and gives way to the hot waters of the Gulf of Guinea which block the upwellings of cold waters in front of the Gabon coast. Pushed by the Saint Helena anticyclone, the trade wind is then heavily loaded with moisture, and the air masses, heated at the base, that it sends onto the Atlantic coast of Central Africa become unstable. In January and February, the continental trade wind from the north-east is able to push back the winds coming from the Atlantic, and a small dry season – or rather one of lesser rainfall – can then settle in on Gabon and the western Congo (FIGURE 2.7).

Between these two extreme periods, the distribution of the rains more or less follows the displacement of the Intertropical Convergence. The extreme eastern part of the Democratic Republic of the Congo and the region of the Great Lakes, like all of eastern Africa, thus have a maximum of rainfall in November-December and April-May. These two periods are interspersed with more or less pronounced dry periods when the Intertropical Convergence recedes far to the north or to the south. These dry periods, that nothing attenuates, are too long to allow the development of humid forests, except at higher elevations. Rains occur nearly throughout the year in western Central Africa because the Intertropical Convergence is never very far away.

2.6. In January, the Saint Helena anticyclone retracts and moves away from the coast of Africa. The Sahara and the Azores anticyclones dilate and push the Intertropical Convergence Zone southward along an S-shaped line that lies near 5°N in West Africa and 15°S in the Zambian region.

2.7. During this austral summer, heavy rains fall on southern Gabon and on the southern and south-eastern parts of the Democratic Republic of the Congo. Only the northern fringes of the Congolese forest massif experience drought at that time (after Leroux, 1983).

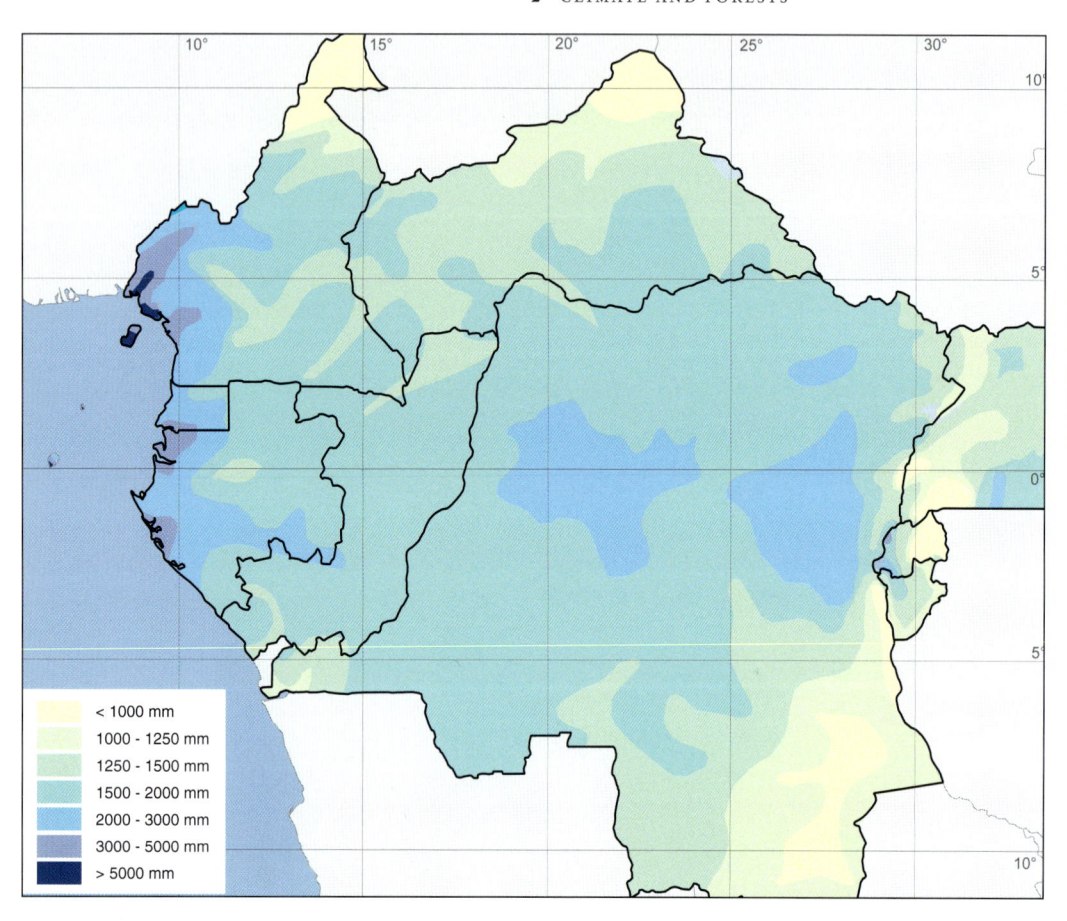

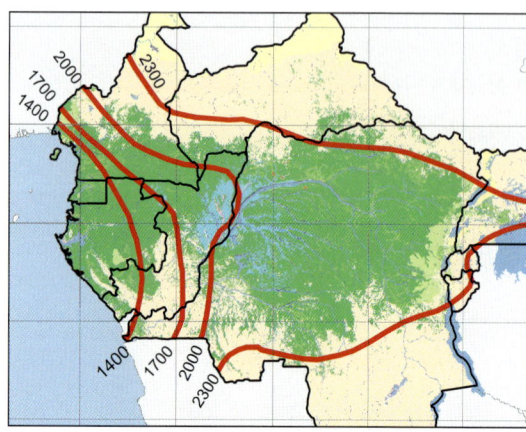

2.8. *Average annual distribution of rainfall in Central Africa (after Leroux, 1983).*

2.9. *Insolation varies from 1,300 hours per year on average along the Gulf of Guinea to 2,300 hours on the eastern margins of the Congo basin and to more than 2,700 hours in northern Cameroon and the Central African Republic.*

The relative size of the two subtropical anticyclones and the displacement of the Intertropical Convergence that this balance induces depend essentially on the distribution of energy between the two hemispheres. Since the Antarctic is on average more than 11° colder than the Arctic, the meteorological equator is located much farther north than the geographical equator, on average.

The average annual rainfall in most of the forested region of Central Africa is between 1,800 and 2,200 millimetres (FIGURE 2.8). The shores of the Gulf of Guinea between Libreville and Nigeria are, however, much wetter. Thus, it rains an average of 2,900 millimetres per year in Libreville, 3,200 millimetres at Bata and 3,400 millimetres at Douala. Debundscha, a small settlement south-west of Mount Cameroon, holds the record with nearly 12 metres. On the islands of the Gulf of Guinea, rainfall varies. While the north coast of Bioko receives no more than 1,000 or 1,200 millimetres of rain, the south-western part of the island receives nearly ten metres. In a similar way, rainfall in the south-western part of São Tomé is around 6,000 millimetres, while the plains in the north of the island receive only 1200 millimetres.

On the continent, as well, rainfall can reach 2,200 to 2,500 millimetres, notably on the Chaillu Mountains in Gabon, the central and the eastern Congo basin, the saddle between the Karisimbi and Bisoke volcanoes, the eastern part of the Nyungwe Forest in Rwanda, the north-western shore of Lake Victoria and the Sese Islands in Uganda. North and south of the equator, rainfall diminishes progressively and the limit of dense and humid forest formations is located roughly between the 1,500 and 1,700 millimetre isohyets.

Lastly, in the region of the Albertine Rift, the rainfall gradient is sometimes very abrupt. Areas of very low rainfall, centred on the plains around Lake Albert, Lake Edward and the Rusizi, are deeply interwoven into the forested zone.

Insolation and solar radiation

As in all equatorial regions, the length of days and nights varies only slightly in the course of a year and the angle of incidence of the sun's rays at noon is never less than 43°. In theory, the conditions of light and radiation should then be very uniform over Central Africa. The duration of insolation varies, however, in function of cloud cover. Not only does it increase as one moves away from the equator, but it also varies according to an east-west gradient. Insolation reaches 2,200 to 2,500 hours in the eastern part of the central basin, but only 1,300 hours on the edge of the gulf of Guinea (FIGURE 2.9). Solar radiation reaches 500 calories per square centimetre per day in the east and only 350 in the west (FIGURE 2.10).

With very similar temperatures and rainfall, the forested regions in the eastern part of the Congo basin are thus subjected to climatic conditions very different from those that prevail in Gabon, for example.

Mists and fogs

Alongside measurable rainfall, there is also a moisture supply that cannot be directly measured. It comes from the condensation of dew or mists and fogs at the level of the vegetation. Significant along water courses, on the periphery of marshes or in steep-sided valleys, it is especially abundant in mountainous regions where mists and fogs are common and tend to linger (FIGURE 2.11). The quantity of water captured by the vegetation, often loaded with epiphytes – mainly lichens, mosses, ferns, orchids – can represent 30 to 50%, sometimes even 100%, of the measurable precipitation. Although taking it into account presents problems of measurement, this hidden precipitation is extremely important for the vegetation. We shall come back to this further along (PAGE 70-71).

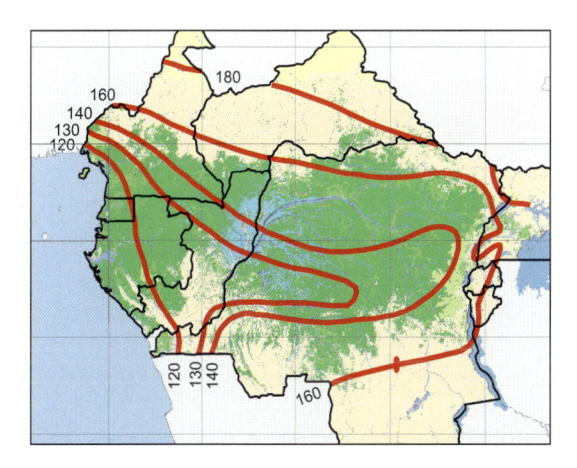

Evaporation and evapotranspiration

The amount of water available for the vegetation does not depend only on the quantity of rainfall. It also depends on evaporation. This is influenced by an ensemble of factors often difficult to measure, beginning with temperature and relative air humidity. Relative humidity is itself narrowly linked to the temperature and thus undergoes important daily fluctuations, mainly in the forest canopy. Inside the forest this relative humidity rarely goes below 95%, but it can drop to 65% in the dry season.

In general, evapotranspiration reaches 1,000 to 1,500 millimetres per year in forested regions. The balance between rainfall and evapotranspiration is positive and excess water returns to the ocean via the rivers. It is only for very brief periods that this balance can be negative and the more water that evaporates, the more it falls.

The wettest forests – called evergreen forests (PAGES 72 and 73) – generally need 2,000 millimetres of rain and do not tolerate more than two months of dry season. In Central Africa, however, there are evergreen forests with a rainfall of 1,700 millimetres and nearly three months of dry season. This is possible only if the soil holds important reserves of water or if the temperature is sufficiently lowered, either by altitude or by the presence of a heavy cloud cover, as is often the case on the shores of the Gulf of Guinea. Other regions, on the other hand, are covered with savanna with a rainfall of 2,000 millimetres. This situation is usually related to the soil or to fire.

The climatic regions

Forested Central Africa is divided into five climatic regions according to the length and distribution of dry seasons. In regions subjected to equatorial climate conditions, situated on the climatic equator, the rains are clearly bimodal. As can be observed at Odzala National Park (FIGURE 2.12.3) and in the Dja Faunal Reserve (FIGURE 2.12.2), they thus are separated into two seasons of heavy rain, centred respectively on April and November, interspersed with two seasons of lesser rainfall, centred on January and July. Given the amount of rainfall, no month can be considered as dry. Away from the climatic equator, there extends a zone of transitional equatorial climate where one of the two dry seasons tends to be reduced to a simple season of lesser rainfall. In the south, for instance in the Lopé National Park in Gabon, the July-centred dry season is extended while the January dry season is hardly perceptible (FIGURE 2.12.4). In the north, such as at Ngotto in the Central African Republic, the reverse happens: it is the January season that is extended while the July season

2.10. Radiation varies somewhat in the same way: from 300 Kcal along the Gulf of Guinea to 500 Kcal on the eastern edge of the basin.

2.11. Fog, which forms every night in the valleys of the Nyungwe Forest in Rwanda, greatly increases the effect of the rains and maintains a lush flora of epiphytes, especially of ferns and orchids. After dawn, when the sun rises and the temperature increases, this fog slowly retreats but we have often to wait for hours before it is totally gone.

2 CLIMATE AND FORESTS

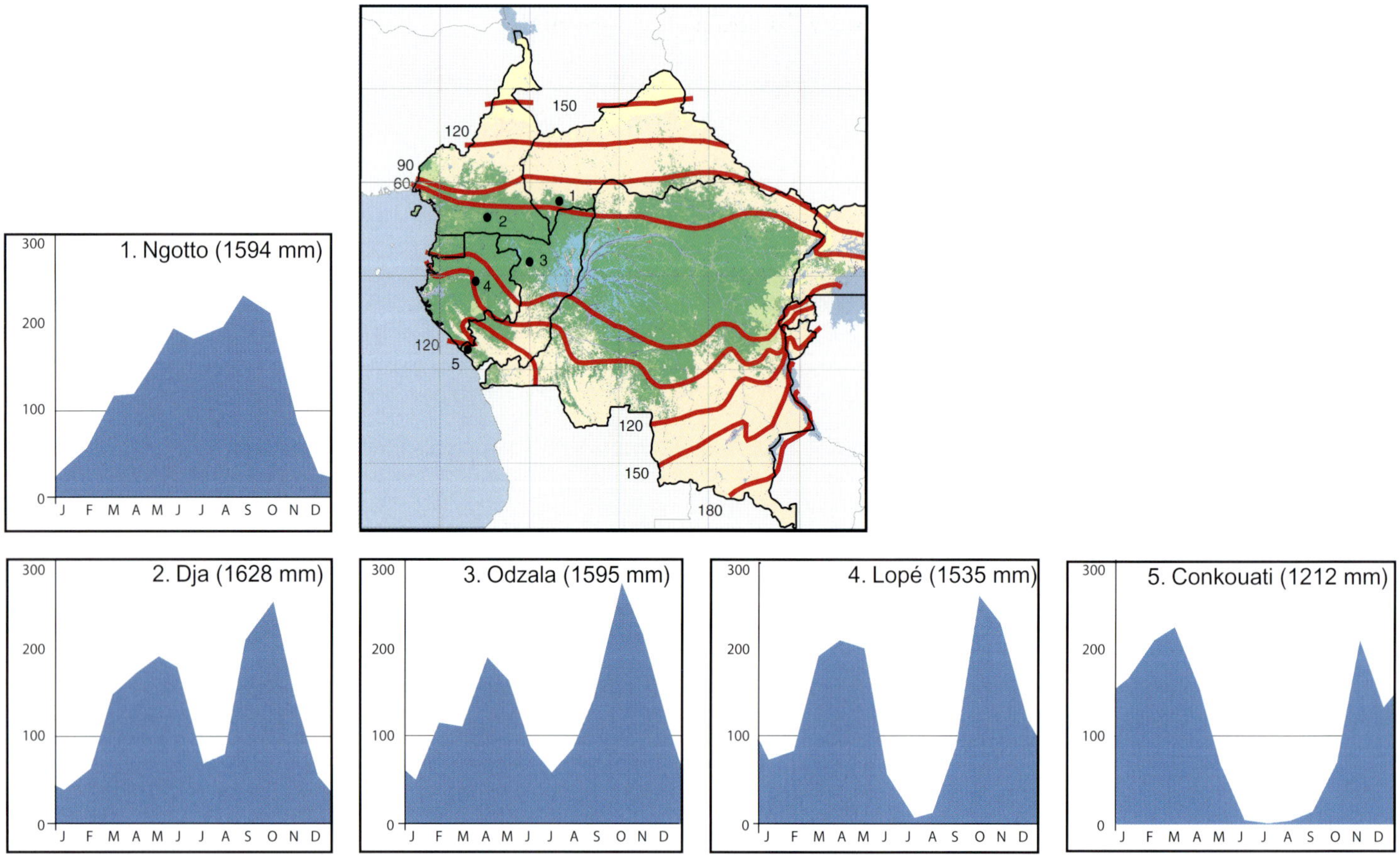

2.12. Central Africa is divided into five climatic zones according to the distribution and duration of the dry seasons. The map shows the duration of the dry seasons in number of days. Rainfall diagrams illustrate the differences in distribution.

In the forests of Ngotto in the Central African Republic (1), the tropical transition climate has only a single dry season of three months centred on January.

In the faunal reserve of the Dja in Cameroon (2), the climate is bimodal with a main dry season around January and a period of lesser rainfall in July.

In Odzala National Park in Congo (3), the climate is also bimodal and the main dry season is centred on January.

In the Lopé National Park in Gabon (4), situated to the south of the climatic equator, the climate is bimodal with a main dry season centred on July and a period of lesser rainfall in January.

In the Conkouati National Park in south-western Congo (5), the climate becomes unimodal with a long dry season lasting from May to October.

contracts (FIGURE 2.l2.1). Still farther from the climatic equator lie tropical climate zones characterised by a single long dry season that alternates with a single rainy season. In the south, this dry season is centred on July (FIGURE 2.12.5); in the north, on January.

Whatever their rainfall might otherwise be, the regions located on the periphery of the Gulf of Guinea are under a distinct Atlantic influence characterised by the frequent presence of stratiform clouds. In mountainous regions the climate is subjected to the cooling influence of altitude, but rainfall can be quite variable depending on the relief and orientation of the slopes in relation to the winds.

Inter-annual variation

As is the case everywhere, forested Central Africa's climate varies from year to year. The annual volume of rainfall can thus vary around 50% in relation to the average, just as the rainiest months or the driest ones can shift. These irregularities perturb the life of farmers and cause every generation to say that "the climate is not what it used to be". These irregularities have always existed, however, and their ecological impact is not to be underestimated. In fact, extreme climate conditions constitute true bottlenecks for the vegetation. The wettest years eliminate species sensitive to an excess of water; the exceptionally dry years, on the contrary,

eliminate those with the greatest water requirements.

These variations are in part local, in part the expression of more extensive phenomena, in time as well as in space. Thus, over periods of several years or tens of years, there appear clear-cut trends of an increase or decrease of rainfall in relation to the average at long term. A detailed study of the Congo-Gabon region shows that between 1951 and 1970 rainfall was above average, while from 1971 to 1990 there was rather a deficit.

Studies on a continental scale show that all of sub-Saharan Africa does not fluctuate in phase, however, and Bigot and his collaborators showed that for the period 1951-1988 the continent divides into five sectors (FIGURE 2.13). Atlantic Central Africa and forested western Africa form one. Eastern Africa, including the eastern parts of the Democratic Republic of the Congo, forms another. The centre of the basin, as well as the north-eastern and south-western parts of the Congo basin, form rather a transition zone. The Congo basin is divided then between two more or less independent climatic zones. Altogether, its rainfall has shown, however, a clear-cut regression since 1960, which is translated by a reduction in the annual flow of the Congo River at the Malebo Pool (FIGURE 2.14).

Over the course of the last dozen years, we have discovered that these climatic variations are largely related to interactions between the oceans and the atmosphere, in particular, the exchange of humidity linked to surface water temperatures. In the tropics these can vary considerably depending on the upwelling of deep water. Upwellings are, themselves, activated by the trade winds that blow in the direction of the Zone of Intertropical Convergence and their appearance is thus more or less seasonal. They are also conditioned by the configuration of the coasts and the proportion of submarine waves channelled along the equator by the cancelling of the Coriolis force – the force that also causes water to eddy as sket it is swallowed up into the drain of a sink.

As we have already seen, the lowering of the temperature of the ocean's surface waters lowers the water-vapour content of the monsoon that passes over it, stabilises convections, favours the formation of vast stratiform cloud covers and pushes the Zone of Intertropical Convergence northward. This phenomenon, observed from the end of May to September in the Gulf of Guinea, brings to Gabon and Equatorial Guinea the "long dry season". But some years the Saint Helena anticyclone moves westward or reduces. Then the trade winds slow, upwellings diminish and the hot tropical waters of the Gulf of Guinea extend farther south along the west coast of Africa. All of that brings abnormally heavy rains to Angola and Namibia, generally arid countries, and suppresses the "short dry season" north of the forested block. Sometimes, notably in 1984, even the "long dry season" is practically quashed at the level of Gabon. This phenomenon, known by the name of *El Niño* in the Pacific, also happens in the tropical Atlantic.

The study of innumerable data accumulated since 1950 shows that the North Atlantic, the South Atlantic and the Pacific do not evolve in phase. On the one hand, there are oscillations between the eastern Pacific and the western Pacific. On the other hand, there are also oscillations in the North Atlantic and the South Atlantic. Altogether, there appear in this way four types of configuration of ocean surface temperatures. Each one influences the rainfall of South America and of Africa in its own way. Rainfall on western and central Africa is influenced by the equilibrium between the North Atlantic and the South Atlantic. The Pacific oscillations have only a reduced importance. As for the rainfall on eastern Africa – including the region of the Albertine Rift – it is rather influenced by the surface temperatures of the Indian Ocean.

Some spectacular events have recently shown that climatic oscillations can have grave consequences for the forests. It suffices to recall the disastrous fires that ravaged the forests of Borneo during the powerful *El Niño* of 1983 or those that hit the whole of South-east Asia in 1997. Apart from these exceptional and happily rare events, however, it appears that tropical forests also react very rapidly to much more discreet fluctuations. Tree mortality as well as recruitment varies from year to year according to subtle fluctuations. Large-scale studies, launched by S. Hubbell and R. Foster of the Smithsonian Institute's Tropical Research Institute of the Barro Colorado Island Reserve in Panama, were able to demonstrate this objectively. No comparable study has been made in Central Africa, but there is every reason to believe that the observations in Panama can be extrapolated to it.

Ocean surface temperatures thus have certainly had a capital influence throughout the history of the forests. They are likely to continue to have one in the years to come. Global warming, which no one doubts any longer in spite of all the efforts of the powerful industrial lobbies to minimise or deny it, seems in fact to be accompanied by a greater frequency and a greater intensity of phenomena of the *El Niño* type.

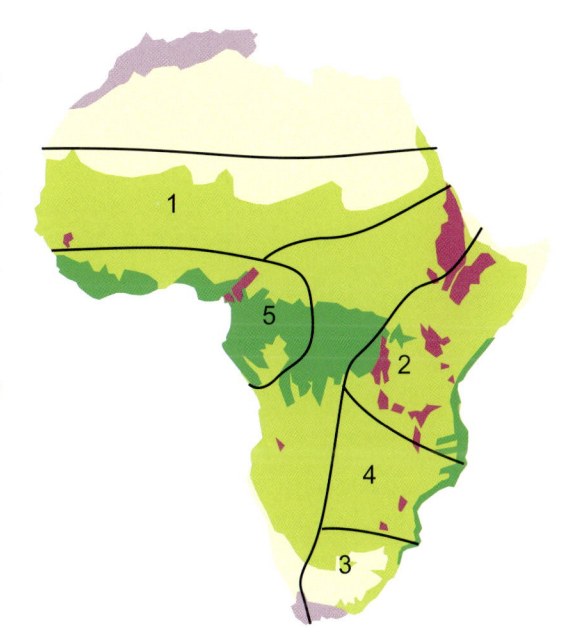

2.13. In studying rainfall data from 466 stations between 1951 and 1988, Bigot and his collaborators discovered that sub-Saharan Africa had five large sectors of coherent inter-annual rainfall variability. The western half of the Central African forests belong to a sector extending around the Gulf of Guinea (5). The eastern margins of these forests belong to the East African sector (2). The central parts of these forests are a transition zone.

2.14. Evolution of rainfall between 1950 and 1990 in Gabon (a), southern Cameroon (b) and the Congo (c). Parallel evolution of the flow of the Ogooué River (d) and Congo River (e) (after Olivry et al., 1995).

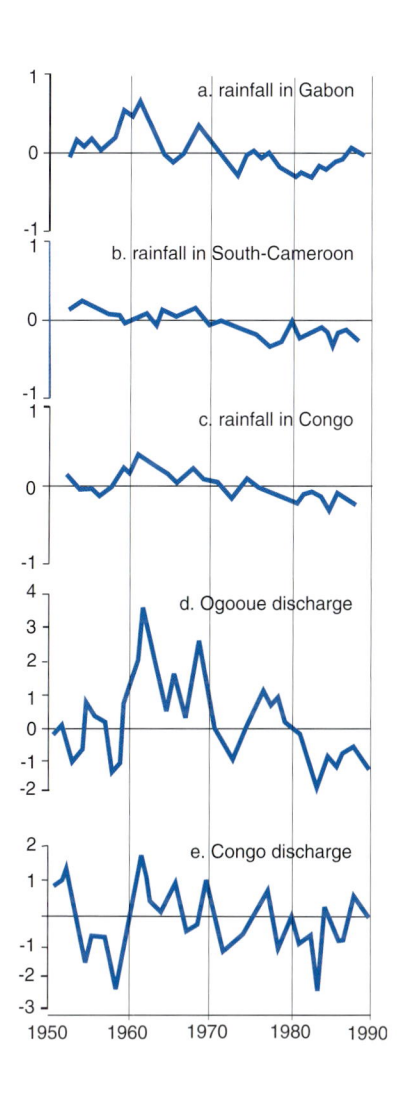

2.15. *The main continental sites which have furnished palynological data on the history of the environment in Central Africa:*
(1) Lake Barombi-Mbo (Maley & Brenac, 1998); (2) Lake Ossa (Reynaud-Farrera et al., 1996); (3) Lopé National Park (Oslisly & White, ...); (4) Kitina, (5) Sinnda and (6) Ngamakala (Vincens et al., 1999); (7) Kakamoéka (Maley & Giresse, 1998); (8) Bilanko (Elenga & Vicens, 1990); (9) Kouyi (Maley, Caballe & Sita, 1990); (10) Nouabale-Ndoki (Maley); (11) Bokuma and (12) Imbonga (Preuss, 1990); (13) lake Mahoma (Hamilton, 1983); (14) Mubwindi, (15) Akakagyezi and (16) Muchoya (Jolly et al., 1997); (17) Kamiranzovu and (18) Rwasenkoko (Hamilton, 1982); (19) Kashiru (Bonnefille & Riollet, 1988); (20) Kuruyange (Jolly et al., 1997).

The history of the forests

Since the end of the 19th century, we know that in the past the climate was subjected to considerably more variation than we have been able to observe in the course of the last century. The glaciers of eastern Africa have been much more extensive than they are today. On the Rwenzori, for example, there are ancient glacial moraines lying at about 2,100 metres altitude while the lower limit of present-day glaciers is at altitudes of around 4,500 or 4,600 metres. Some high mountains of Kenya, Ethiopia and the Sahara even bear traces of glaciers that have completely disappeared. Also, the water level of the sub-Saharan great lakes has been much higher than it is at this point in time and the dunes of the Sahara extended as far as 400 or 600 kilometres south of the present limit of the desert. These observations cannot be explained except by climatic changes with fluctuations that must have gone far beyond those of the *El Niño* or *La Niña* type oscillations. They must rather be closer to those that affected the temperate regions during the great glaciations.

Just like the temperate forests, the tropical forests must have inevitably experienced enormous fluctuations in area. During some periods they must have been much vaster. Fossil stumps, found in regions where there is no trace of forest nowadays, attest to this. At other epochs, they must have been, on the other hand, strongly reduced. The innumerable stone lines that are found under present-day forests in the Congo basin prove it. These more or less continuous layers of buried stones are in fact witness to ancient and very strong erosion that could not have happened under a forest cover (PAGE 163).

But the history of the forests is not a chronicle of just their distribution with all the consequences that would go with it. It also takes into account the origin of species. These two chronicles are situated on very different time scales, but they are finally orchestrated by the same ensemble of forces that manage the energetic distribution of our biosphere, ocean surface temperatures and distribution or rainfall, in other words, solar activity, the cyclic variations of the orbital parameters of our planet, the dividing up of the continental masses and variations in the level of the oceans. The history of the forests and that of the climate can thus not be separated.

Without reliable means of dating it was impossible to establish relationships between all the observations that could have helped to retrace this history. Until 1950, it was thus believed that during the glacial periods in temperate regions there were corresponding pluvials – cold wet periods – in the tropical regions. Carbon 14 dating shows, however, that the highest lacustrine levels coincided with warm periods and that the glaciations of the north were accompanied by a cooling and drying of the climate in tropical regions. The refinement of this technique was thus a major event and permitted at last the start of a detailed reconstruction of the history of the climate and of the forests. It became possible to establish a coherent link between old lacustrine water levels, moraines and firn lines of east African mountains, dunes and fossil valleys of arid regions, fossilised plant remains (leaf prints, fruits, stumps, charcoal, etc.) and, above all, pollen conserved in sediments.

It is palynology, or the study of pollen, that has delivered most of the elements of the history of vegetation and climate. The first sources of fossilised pollen to be studied during the 1970s were the peat bogs of the mountainous regions of eastern Africa and the Albertine Rift. For forested Central Africa, it was necessary to wait until the 1980s and 1990s. Data were then obtained from Cameroon, Gabon and Congo (FIGURE 2.15). Research is still in progress in Gabon, notably in the Lopé National Park.

Since the 1980s, continental data have been completed by numerous marine drillings undertaken within the framework of petroleum exploration. Formerly, pollens from ancient sediments, especially Cretaceous and Tertiary, were simply described as shapes having no more than a stratigraphic significance. Some petroleum companies are still at this stage. But the accumulated

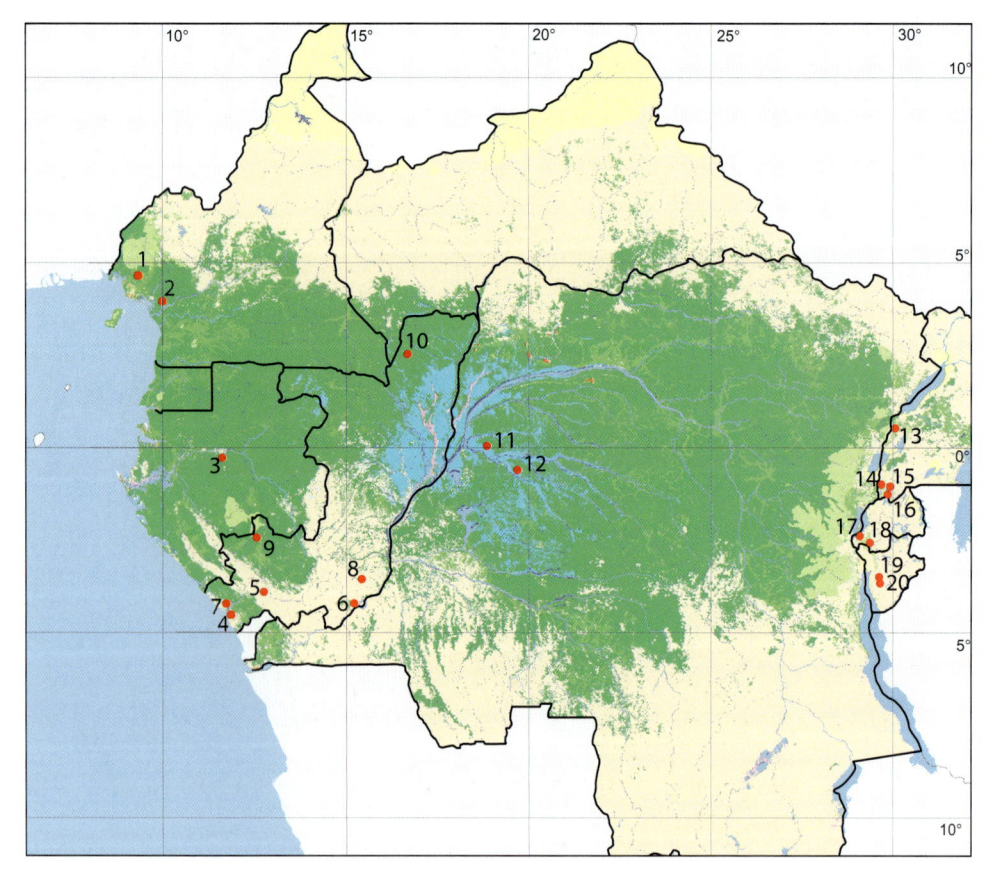

2 CLIMATE AND FORESTS

experience of the past 20 years, the much better knowledge of present-day pollens and the constitution of enormous reference collections – the Laboratory of Palynology of Montpellier University possesses more than 40,000 sample slides – permit identification of ancient pollens, or at least their detailed comparison with modern ones.

After the perfection of carbon 14 dating methods, the development of palynology scored a second great advance of science. And even if data are still lacking for many places, the two techniques, used in association and completed by the study of sediments, have led to the unveiling of some chapters in the history of the Congolese forests in a very detailed way.

The history of the Angiosperms

Except for a few rare exceptions – notably the small family Gnetaceae – the present-day tropical forests are made up exclusively of flowering plants or Angiosperms. Yet, these plants, so familiar today and dominant in all the tropical and temperate regions of the world, did not appear until the last 140 million years and did not acquire their dominance until the last 65 million years.

And yet, life on Earth has existed for nearly four billion years. The first terrestrial plants had already colonised the continents in the Devonian, 390 million years ago – at the same time as the first arthropods – and the first formations that could be qualified as forests took shape around 375 million years ago.

This first terrestrial vegetation was composed of plants that reproduced by spores. Some resembled mosses, others were ferns, lycopods and horsetails. Towards the end of the Devonian, seed plants or Gymnosperms appeared: Pteridospermales or seed ferns, Cycads (FIGURE 2.16), Conifers, and Ginkgophytes. During the transition from the Permian to the Triassic, 249 million years ago, the living world was hard-hit by a wave of extinctions that carried off at least 70% of the species of the time. The Pteridophytes never recovered. And when, after some millions of years, life diversified anew, it was the turn of the Gymnosperms to dominate the world. They must have kept this first place for most of the Mesozoic, only to be supplanted by the Angiosperms during the course of the Cretaceous (FIGURE 2.17). From the beginning of this period, Pteridospermales totally disappeared. The Cycads would just barely survive the Cretaceous-Tertiary and were reduced to relicts. Of the Ginkgophytes, only a single species has survived, *Ginkgo biloba* of China. Only the Conifers still made advances during the Tertiary and were even able to keep a clear supremacy in boreal and mountainous regions.

The first Angiosperms made their appearance at the very beginning of the Cretaceous and the oldest fossil, *Archaefructus*, is 140 million years old. It comes from China. While their predecessors produced spores or seeds whose dispersal or pollination was still largely dependant on the wind, the newcomers were distinguished by their flowers, delicate and ephemeral organs especially adapted for pollination by insects. The diver-

2.16. The Cycads dominated the plant world alongside the dinosaurs for millions of years. Today the few remaining species survive often as ornamental plants.

2.17. The Pteridophytes (ferns, horsetails, club mosses and selaginellas) were dominant in the Devonian and the Carboniferous. After the great wave of extinction that marked the end of the Permian and that coincided with a dramatic drop in the level of the seas, the formidable volcanic eruptions in Siberia and a profound change in the world climate, the Gymnosperms (Pteridosperms, Cycads, Gynkophytes and Conifers) took over. They may not have been very perturbed by the extinctions of the Triassic-Jurassic and the Jurassic-Cretaceous transitions, but from the end of the Cretaceous the Angiosperms (mono- and dicotyledons) took over. These latter were hard-hit by the wave of extinction of the Cretaceous-Tertiary transition, probably provoked by the fall of an enormous meteorite on Central America and the strong volcanic eruptions on the Deccan plateau, in India, provoking another lowering of the level of the seas and global climate changes (after WCMC, 2000; Palmer, 2000).

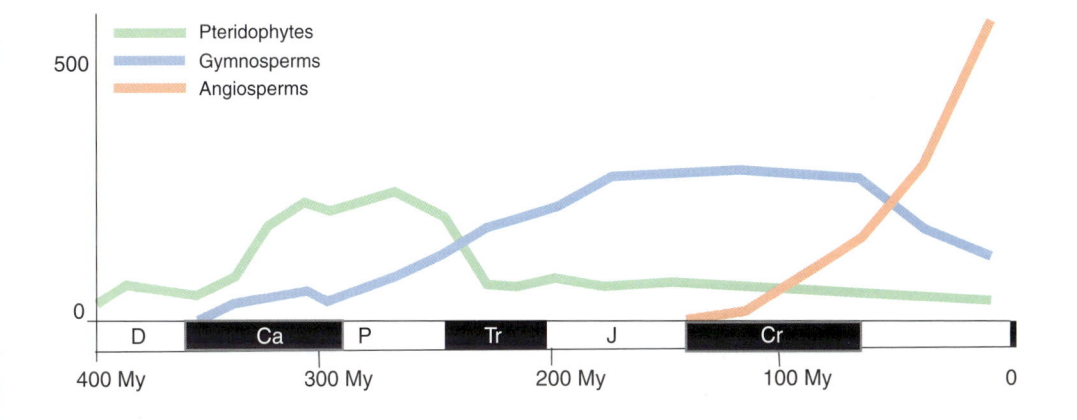

49

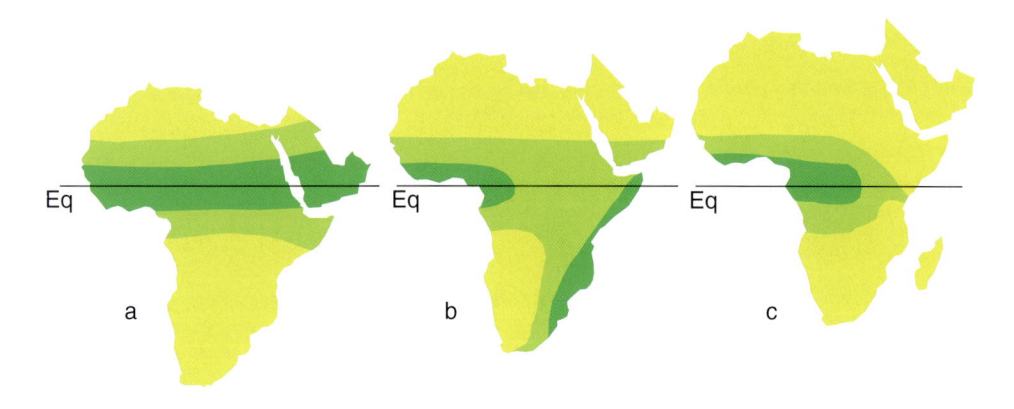

2.18. During the Cretaceous and the Tertiary, Africa drifted northward and pivoted on itself, isolating itself from the other fragments of Gondwanaland. The distribution of bauxites (dark green) and iron pans (pale green), allow a reconstitution of the paleoenvironment and, more particularly, the distribution of humid and arid regions (after J. Maley, following the works of Tardy et al., 1991; Guiraus & Maurin, 1991; Parrish et al., 1982).
(a) at the beginning of the Cretaceous, 70 million years ago.
(b) at the middle Eocene, 50 million years ago.
(c) at the Quaternary and the present.

sification of the Angiosperms therefore paralleled that of the insects. It took place during the Cretaceous and, just before the Cretaceous-Tertiary transition, the Angiosperms finally supplanted the Gymnosperms. Their history, at least for the part of the world we are looking at, was reconstituted by Jean Maley. We will pick out just the main points.

The story begins 130 to 140 million years ago. Africa was then still included in the enormous continental mass of Gondwanaland that was beginning to break up (PAGES 18 and 19). The part of Gondwanaland that went on to become Africa found itself clearly farther south than it is nowadays. It was inclined to the east (FIGURE 2.18) and was bordered in the north by the Tethys Sea. The region around the Gulf of Guinea was subjected to a dry continental climate. With its nascent rift, studded with salt lakes, that stretched from Namibia to Biafra, it must have resembled present-day eastern Africa. Its vegetation was made up of many cycads, well adapted to hot dry regions, and conifers. The conifers were represented not only by Podocarpaceae, probably limited to high altitude regions, but also Araucariaceae, nowadays gone from Africa.

It is in this landscape that the first, still very primitive, Angiosperms developed. They appeared somewhere in the northern part of Gondwanaland which at the time occupied equatorial latitudes, probably on the shores of the Tethys Sea and the nascent Atlantic Ocean. As much as can be judged by the appearance of their pollen, they probably resembled some herbaceous Magnolids, like the ones found nowadays in the undergrowth of dense and humid forests of Southeast Asia, and Nympheales on river banks and lake shores. They were therefore very probably pioneers, adapted to disturbed areas like the banks of watercourses or openings in the Gymnosperm forest.

Very soon, however, more complex pollens would appear, similar to present-day dicotyledons. They are reminiscent of the Winteraceae, a family that has disappeared from Africa, but which survives, like the Podocarpaceae, in the austral forests of South America, Madagascar and Australia, as well as at high altitudes in some tropical regions.

After some hesitation, the diversification of the Angiosperms continues with the appearance of species that announce the Didymelaceae (today a family endemic to Madagascar) the Euphorbiaceae, the Proteaceae, the Caryophyllaceae and the Amaranthaceae. Initially, the Gymnosperms remained dominant and it is not until around the middle of the Cretaceous that the Podocarpaceae start to regress. At the same time, the plants with pollen resembling those of the Winteraceae disappear. They apparently would survive, however, until the beginning of the Miocene, 25 million years ago, in the Cape region of South Africa.

Near the end of the Cretaceous, 70 million years ago, the Gymnosperms regressed very strongly, while the Angiosperms continued their diversification and became the dominant element of the flora. Among the new families, there were the Poaceae or grasses, the Proteaceae, the Ulmaceae, the Caesalpiniaceae and the precursors of the Apocynaceae. The world climate was then hot and humid. The oceans, in full transgression, inundated vast portions of the continents. The equator passed through Liberia, the Sudan and Arabia, while the desert expanses of southern Africa reached the Congo basin. They probably persisted until the end of the Neogene, five million years ago. Northern Africa, still partially covered by shallow seas, benefited, on the other hand, from a stable, hot and humid equatorial climate. The emerged land was covered with a luxuriant vegetation including many dicotoledons and palms. Besides the Arecaceae, present-day palms of Africa, there were even the Nypaceae, palms that no longer exist except in Asia. In the west, this band of palm-rich vegetation extended to the entire northern part of South America where it still survives.

The "take-over" of the Angiosperms in the tropics thus took 35 million years. One would be tempted to make it coincide with the extinction wave of the Cretaceous-Tertiary transition, 65 million years ago. In reality, the take-over happened progressively throughout the Cretaceous and could not in any case have been the result of a sudden phenomenon. On the contrary, this wave of extinction, that carried off the last dinosaurs and many Gymnosperms – except for some Gnetaceae they would completely disappear from the low-altitude tropical flora – caused the disappearance of a great number of Angiosperms that had "hardly" seen the light of day (FIGURE 2.19).

J. J. Midgley and W. J. Bond think that the flowering plants would have been favoured in their competition with the Gymnosperms by their capacity to regenerate much more rapidly. According to Maley, the diversification and domination by the Angiosperms was rather linked to profound changes in climatic conditions following the splitting up of Gondwanaland. Before this major geological event, the immense arid expanses of this super continent were in effect occupied by the Gymnosperms which excelled in the trapping of atmospheric humidity brought by dew, nocturnal fogs and low clouds. With the progressive opening of the Atlantic during the Cretaceous, the low clouds gave way to stormy formations accompanied by much heavier rainfall. While the tropical Gymnosperms had developed in dry hot continental conditions, the Angiosperms were better adapted to hot humid conditions. This adaptation was achieved only very progressively, however. The first Angiosperms, those that probably resemble the Winteraceae, still had climatic requirements very close to those of the Podocarpaceae and disappeared from the hot, humid tropical regions at the same time as they did. The fact that the Conifers survived very well in cold temperate regions underscores the importance of climatic factors in the emergence of the flowering plants.

The establishment of the present-day flora

During the Tertiary, Africa, completely isolated from the other continents, continued its northward drift and pivoted on itself around an axis located offshore from Senegal. The eastern part thus went up further than the western part. Towards the middle of the Eocene, 50 million years ago, the continent had nearly reached its present position.

By changing the distribution of the continental masses, the splitting-up of Gondwanaland probably changed the rotational parameters of the Earth. After 200 million years of rather hot, stable climate, a global cooling settled in. Antarctica – one of the pieces of Gondwanaland – was covered by an ice cap that reached sea level 37 million years ago, during the Eocene-Oligocene transition. The enormous accumulation of ice then provoked a profound energetic imbalance between the northern hemisphere and the southern hemisphere and the powerful Saint Helena anticyclone was born (PAGES 42-43). This situation subjects equatorial Africa to a monsoon regime – trade winds loaded with rain –, and pushes the meteorological equator well to the north of the geographical equator. It thus counteracts the effects of continental drift. During the Oligocene and Miocene, the Antarctic ice underwent periods of extreme extension that must have accentuated the climatic contrast between the two hemispheres even more.

Two other phenomena, moreover, would influence the climate of Central Africa. First of all, there was the uplifting of the Himalayas and Tibet following the encounter of the Indian and Asiatic continental plates. This phenomenon engendered the powerful jet-stream that still blows today above India and Arabia in the direction of the Sahara. It accentuates the aridity of northern Africa and increases the thermal contrast with regions located around the Gulf of Guinea. Next, near the end of the Eocene, the cold Benguela current was born, and its effects must have been accentuated in the Miocene by the appearance of cold water upwellings off the coasts of the Gulf of Guinea.

From the end of the Cretaceous to the Miocene, all of North Africa, from the centre of Cameroon to the Sudan and up to the Mediterranean shores of Libya and Egypt, experienced a humid tropical climate. This area was in large part covered with vegetation that must have somewhat resembled that which is found nowadays in Cameroon, at the limit between the forest and the Sudanese savannas. Among the species, some are today typical of humid forests, others of semi-deciduous forests. Many, however, have disappeared from Africa.

The composition of the flora evolved in this way and continued its diversification throughout the Paleogene. Progressively, the dense forests of Africa began to

2.19. Apparition and extinction of plant families in Cameroon and Gabon (after Maley, 1996; adapted from Salard-Cheboldaev & Dejax, 1991).

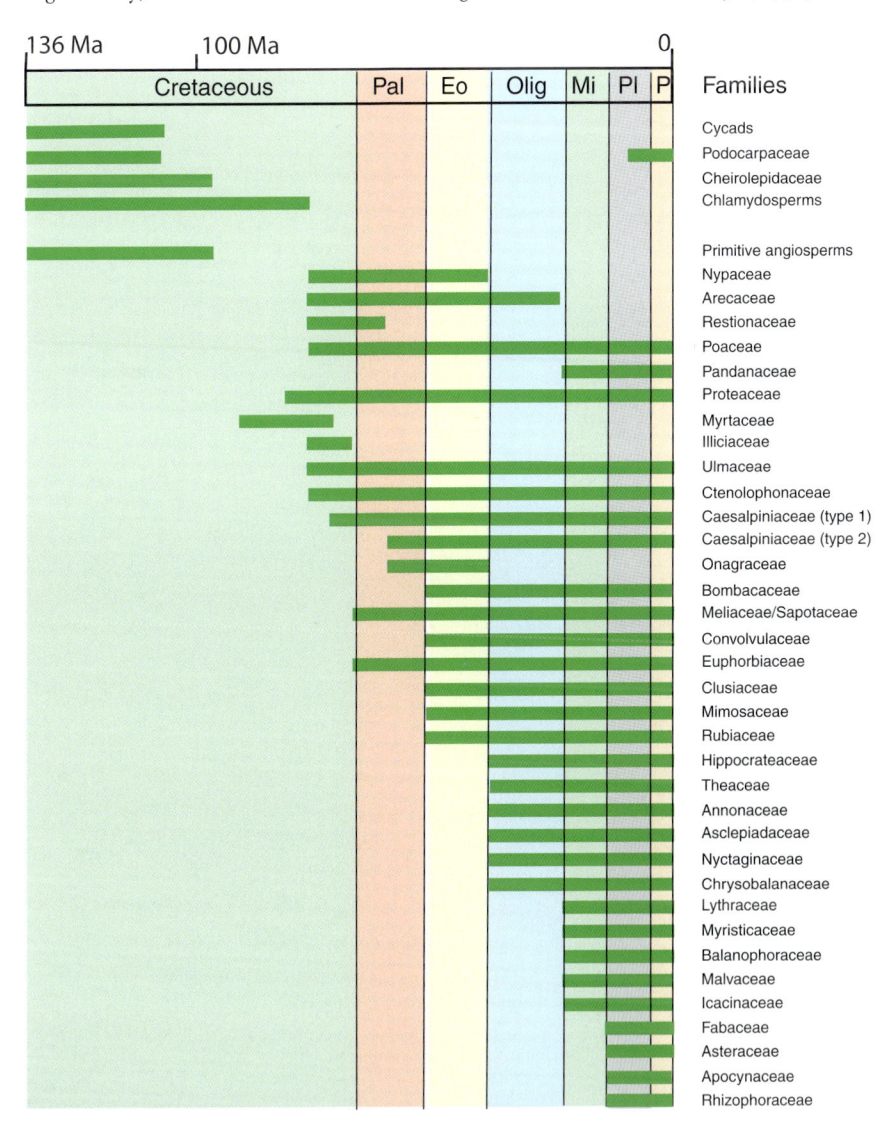

resemble present-day forests. In the Miocene, nine new families appeared, including the Rhizophoraceae. The first mangroves appeared on the coasts of the Gulf of Guinea in this epoch and the flora of the Central African forests must have been very near to that of the present.

In parallel with the diversification and appearance of new genera and species, this African forest flora was, however, impoverished by dry periods that marked Africa much more than other continents. From the Oligocene, palms start to regress and, from the beginning of the Tertiary, there were episodic proliferations of grasses. At least 25 million years ago, there thus appeared savannas comparable to those that we know today and, during the Miocene, the desert flora appeared also.

This impoverishment can be seen in the present-day flora, and some families, such as the Bromeliaceae and the Cactaceae, are thus virtually absent from Africa, while the Lauraceae, Arecaceae, Araceae, Piperaceae, Gesneriaceae, Melastomataceae and Urticaceae, so abundant and so diversified in the humid undergrowth of South America and Asia, are not represented except by very few genera and species. The general impoverishment of the African flora linked to the Cretaceous-Tertiary transition was thus accentuated by the notorious impoverishment of the humid forests, mostly towards the end of the Eocene and the end of the Miocene.

From the beginning of the Pliocene, four to five million years ago, the Sahara appeared and a climatic gradient settled in on sub-Saharan Africa that was fairly similar to what we know today. The humid tropical forests contracted. Palynological data from the Niger delta, covering a period from 10.5 to two million years, show, however, that their extension varied considerably. During the dry periods, the savannas expanded and the presence of grass cuticles in sediments shows even that they burned.

The Niger delta pollen samples show, on the other hand, that *Podocarpus* appeared or reappeared in abundance near the end of the Tertiary (Figures 2.21-2.22). This conifer had been rare or absent in preceding periods but today it again populates the montane forests of Cameroon, the Albertine Rift region and eastern Africa. In fact, the Podocarpaceae had been abundant until the beginning of the Cretaceous. Then they were lost track of. Perhaps they survived in small number in marginal mountainous habitats or perhaps they again colonised the tropical regions of the Gulf of Guinea as a result of the climatic changes. Whatever its origin, this expansion of the *Podocarpus* happened on two occasions, between 10.5 and 9.2 million years and between 2.9 and 2 million years BP (FIGURE 2.23). Each time, it coincided with a progression of the grasses and a very marked marine regression. The first was correlated with an expansion of the Antarctic ice. The second with an expansion of Antarctic and Arctic ice. The development of montane conditions thus paralleled those of the savannas. In other words, the drying of tropical low altitude regions accompanied a global cooling of temperatures.

It is in this way that at the end of the Tertiary there appeared in Central Africa elements of an ancient flora, adapted to a cool climate and a high atmospheric humidity. It had been supplanted with the splitting up of Gondwanaland and the formation of the Atlantic Ocean, but had probably survived in southern Africa. Its reappearance in the tropical regions was the joint result of new climatic changes and of the rebirth of mountainous formations.

2.20. The Melastomataceae are very widespread in sub-Saharan Africa. They are found at all altitude levels of the forest, often on the forest edge, like this beautiful Dissotis *that borders marsh edges in Kibira National Park in Burundi. The family is, however, much less diversified in Africa than in South America.*

Cyclic glaciations

For four to five million years everything seems to indicate that Africa experienced a climatic distribution somewhat similar to that which we know today and that the great biomes, such as we described in the beginning of the chapter, took on the appearance that we know. Their distribution has constantly varied, however.

Depending on the variations of its rotation parameters around the sun, the earth has, for 2.5 million years, experienced cyclic periods during which the polar ice caps of the two hemispheres expand. According to marine sediments, these glacial periods initially had a moderate amplitude and a periodicity of 41,000 years, but for 800,000 years they are more contrasted and their periodicity is around 100,000 years.

During cold periods, the average global temperatures lower and great quantities of water accumulate in the polar ice caps. Consequently, the level of the oceans lowers substantially and rainfall diminishes in the tropical regions. In Central Africa, this phenomenon is accompanied by a lowering of the surface water temperature of the Gulf of Guinea – 4 to 9° C during the last glaciation – which stabilises the trade winds and favours the formation of stratiform clouds producing only very little rain. To the general lowering of temperature and the reduction of measurable rainfall, is added a reduction of solar radiation and an augmentation of hidden precipitation, favourable to the montane vegetation. Being indeed very low, the stratiform clouds often envelop the summits of hills and mountains with fog. They favour in this way the formation of cloud forests in which *Juniperus* and *Podocarpus* conifers flourish, as well as do some Angiosperms, notably of the genera *Olea*, *Ilex* and *Erica*.

According to the Niger delta sediments (FIGURE 2.23), the beginning of the Pleistocene seems to have been a very dry climatic period with a pronounced advance of savanna and montane vegetation, and therefore a fragmentation of lowland forests. Sediments from the Congo estuary show that the period situated between 800,000 and 300,000 BP was itself also very dry and allows the supposition that vast portions of the Congo basin had not been covered by forests. For 250,000 years the climate again seems to have been altogether more humid.

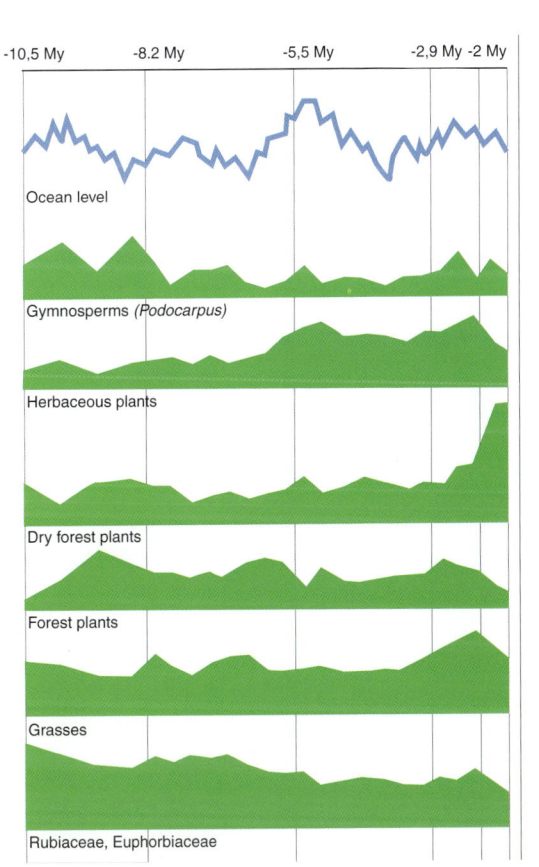

2.21-22. Genus Podocarpus *is represented in Africa today by two main species or groups of species: the group* milanjianus, *with relatively large leaves, is confined to damp mountainous regions; the group* gracilior, *with much narrower leaves, is also found in dryer regions. On the island of São Tomé a third species exists:* Podocarpus mannii *(above) with definitely larger fruits.*

2.23. Sediments from the Niger delta show us that changes of the vegetation at the end of the Neogene – two to ten million years age – followed the great oscillations of the level of the oceans. In fact, these are only a reflection of the extension of the Arctic and Antarctic ice caps. These changes in the vegetation were therefore clearly linked to global climate variations (after Poumot, 1989).

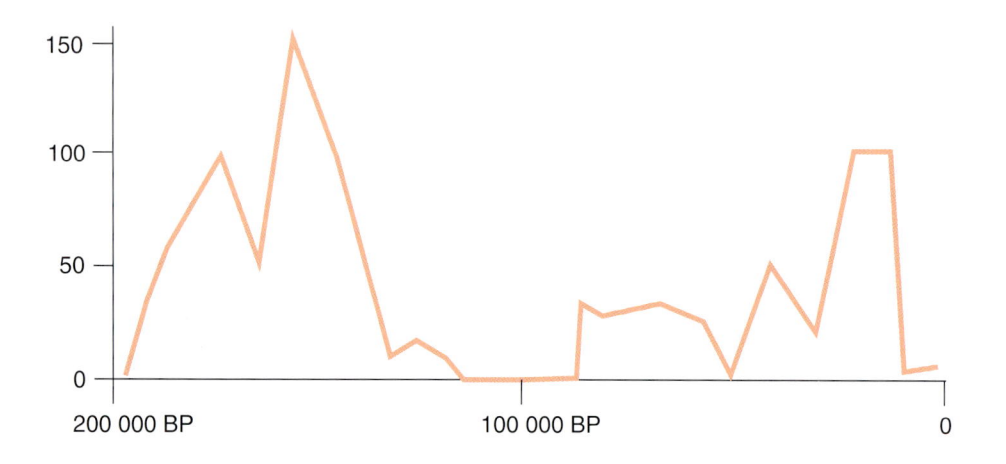

2.24. *Studying the marine sediments deposited by the Congo River, Paul Giresse and his colleagues calculated an aridity index for the last 200,000 years.*

2.25. *According to the study of pollen preserved in the sediments of Lake Barombi Mbo in Cameroon (a), Jean Maley and Patrice Brenac were able to follow the advances and retreats of the forests and savannas during the last 28,000 years: after a drastic reduction around 15,000 BP, the forests had a maximal expansion between 10,000 and 4,000 bp. The brief extension of the grasslands between 25,00 and 2,000 BP is clearly visible. A similar study, made by Raymonde Bonnefille and Guy Riolet (b) at Kashiru marsh in Burundi, covers the last 40,000 years. In this region, at 2,100 metres above sea level, situated on the edge of the Central African forests, and covered with more extensive grasslands, the last 20,000 years show the same type of changes in vegetation as in Cameroon. For older periods, dates are less precise, and results more hypothetic. Several authors working in the Albertine Rift region, have recently put together their observations regarding the last 18,000 years. The data of Muchoya swamp alone in Uganda (c) give a good idea of the evolution of the vegetation:* Stoebe, Artemisia, Anthospermum *and* Hagenia *were dominant during the last glaciation;* Podocarpus, Olea, Macaranga, Ilex *and* Celtis *became dominant during the maximal extension of the forests (Jolly* et al.*, 1997).*

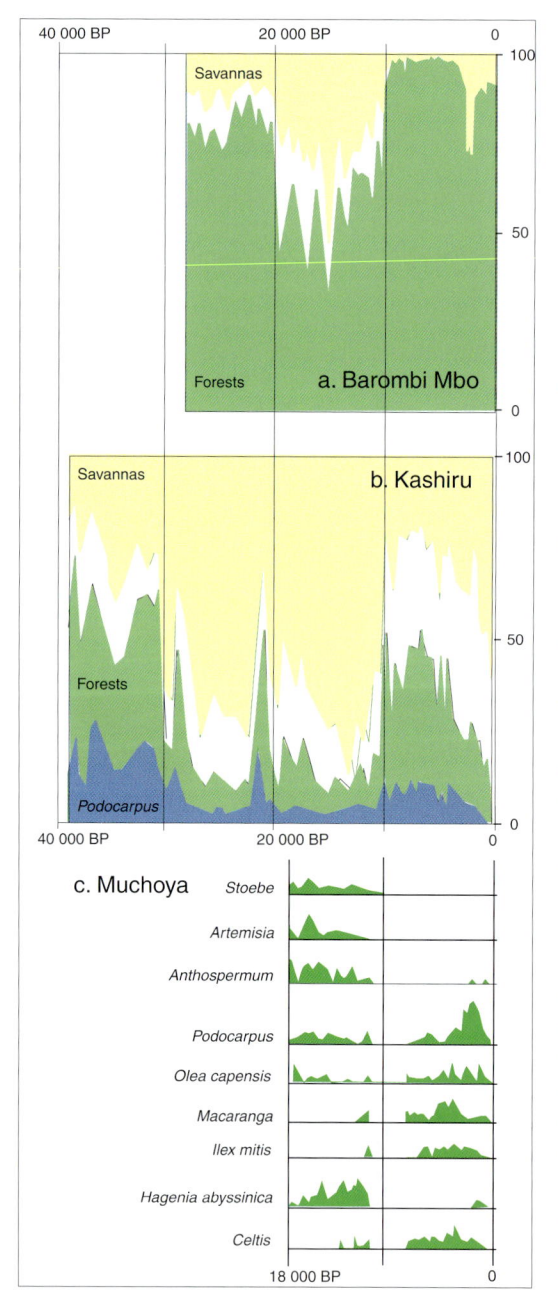

The last 200,000 years

The study of marine sediments of the Gulf of Guinea, especially in front of the estuary of the Congo River, as well as of some lacustrine sediments, allows a fairly detailed retracing of the principal variations of landscapes and climate over the past 200,000 years (FIGURE 2.24).

From 140,000 to 128,000 BP the climate was very cool and this period corresponded to the end of a glacial period that had begun around 160,000 BP. From 128,000 to 118,000 BP, the climate was hot and humid, a little like it has been during the last 10,000 years. The dense and humid forest attained then a maximum expansion, probably comparable to that reached 6,000 to 7,000 years ago. From 118,000 to 75,000 BP, the climate was still altogether fairly hot, except for the cooler intermediaries of 105,000 and 85,000 BP.

Beginning in 75,000 BP, there was a new cool period that would last until around 10,000 BP. Within this long period, the climate nevertheless underwent variations. Between 70,000 and 40,000 BP, it was cool and dry and the forests were less extensive than they are today. Vast portions of Gabon, Cameroon, the Congo and the centre of the Democratic Republic of the Congo were covered with savannas interspersed with forested galleries, probably a little like those we can still observe nowadays at Kwilu, Kwango and Kasai or, on a smaller scale, in the northern part of the Lopé National Park. At times there was very heavy erosion. This action left the stone lines that can be found in the soil of a good many forests. Between 40,000 and 30,000 BP, the climate was a little more humid and probably less cool. Savannas survived in the Congo basin, but the Bateke Plateau was covered with forests, at least in part. In the coastal regions, the mangroves underwent a great expansion.

In the Albertine Rift region, the peats of the Kamiranzovu swamp in Rwanda (FIGURE 2.26) and those of Kashiru marsh in Burundi conserve traces of the evolution of these landscapes over the last 40,000 years (FIGURE 2.25). They show that, before 30,000 BP, *Podocarpus* forests and forests of *Olea* and *Ilex* were well represented but interspersed with *Cliffortia* thickets and Ericaceae heaths. Until around 35,000 BP there were also important forests of *Macaranga kilimanscharica* and *Syzygium guineense*. The presence of *Begonia*, *Impatiens* and *Anthocleista* in the beginning of this period around the Kamiranzovu suggests that there were then very humid forests, like those that are still found today in this marshy basin. After 38,000 BP, the cooling of the climate causes these species to disappear in favour of *Hagenia* and *Rapanea* formations. From 30,000 BP the retreat of the forests was very clear-cut while herbaceous formations expanded.

Abundant data allow the retracing of the climate and the history of the forests over the last 28,000 years with even more precision (FIGURE 2.25). Between 29,000 and 20,000 BP, Lake Barombi Mbo, located at about 300 metres altitude in an explosion crater some 80 kilometres from the sea and 60 kilometres north-east of Mount Cameroon, was surrounded by forests with abundant Cesalpinaceae but also harbouring montane forest species such as *Olea capensis*. These were probably linked to the presence of cloud forests on the crests and summits, but this phenomenon was not isolated. On the Bateke plateau there were forests of montane affinities with *Podocarpus*, *Ilex mitis* and *Olea capensis*. Farther east, on the mountains of the rift, there were always *Podocarpus* forests with *Olea*, *Macaranga*, *Hagenia* and *Rapanea*. But beginning in 22,000 BP there also appeared thickets of *Cliffortia* and heaths or high altitude meadows of *Artemisia* and *Stoebe*. The latter genus is now gone from Central Africa but survives on the dryer mountains of eastern Africa. The landscapes were more open than nowadays and the climate dryer and colder.

Beginning in 20,000 BP, drought accelerated. It was accompanied by a general and very marked regression of the lowland forests that once more gave way to herbaceous formations. This situation lasted until around 14,000 BP and saw its apogee between 18,000 and 15,000 BP. It corresponded to the last maximum expansion of the glaciers in the northern hemisphere. Africa experienced then its most arid period and the Sahara reached as far as 600 kilometres south of its present limit. The level of the Atlantic lowered 110 to 120 metres and the Gabon coast moved more than 60 kilometres to the west.

In the rift region, temperatures lowered 4 to 6°C on average and rainfall reduced probably by 50%. The glaciers of the Rwenzori descended to around 3,500 metres and all the vegetation zones were displaced downward by 1,000 meters. The giant lobelias *Lobelia spp.* and arborescent senecios *Senecio spp.* descended to around 2,500 metres above sea level and the *Philippia* thickets to 2,000 metres. The eastern slope of the mountains, swept by dry winds, was covered by low ericaceous vegetation, meadows with *Stoebe* and *Artemisia* and thickets of *Hagenia* (FIGURE 2.27) or *Cliffortia*. Patches of humid montane forests survived, however, on the western slope, probably at the head of valleys situated between 1,000 and 2,000 meters altitude. The most

2.26. Kamiranzovu marsh is situated at around 1,950 metres altitude in the western part of the Nyungwe forest in Rwanda. It covers an area of 1,300 hectares and its central part is occupied by *Cyperus* beds. Its periphery supports a swamp forest with *Syzygium rowlandi*, *Anthocleista zambesiaca* and *Carapa grandiflora*.

2.27. *Hagenia abyssinica*, of the Rosaceae family, often associated with *Cliffortia*, also a Rosaceae species, forms large thickets around Rwasenkoko marsh in the eastern part of Nyungwe forest. The landscapes that surround this marsh very probably resemble those that covered the Congo-Nile crest around 14,000-15,000 BP and represent in this respect a true natural monument.

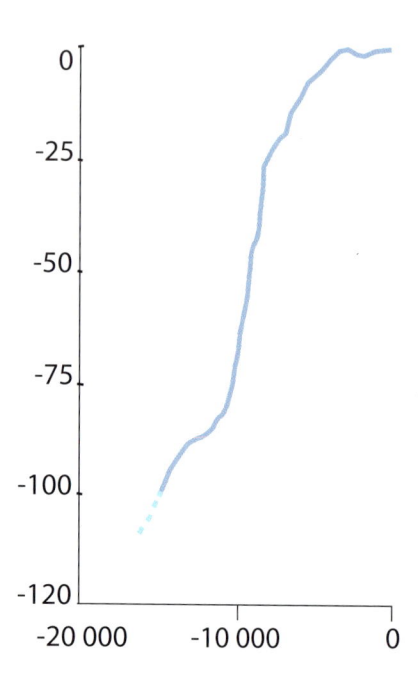

2.28. Around 15,000 to 18,000 BP the level of the Atlantic Ocean in Gabon was 120 metres lower than today. In front of Libreville, the coast line was 60 kilometres farther west. Between 11,000 and 8,000 BP the rise was very rapid and reached two metres per century. The invasion of the Gabon estuary began 7,500 years ago and the islands of Corisco and Elobey (Equatorial Guinea) were separated from the continent about 5,000 years ago. For 4,000 years the level has been near the present one.

important mountain refuge must have been south-west of present-day Bukavu, in the Itombwe region, but smaller refuges must also have existed at the level of Mount Kahuzi and on the western flank of the eastern fold of the rift, notably in the basin of the Kamiranzovu.

Still other refuges were located in the region of the Bwindi forest in south-western Uganda, perhaps even on the western flank of the Rwenzori. In fact, there must have been a constellation of small refuges of very varying size, perhaps somewhat like the ones that can be seen today on the Nyika plateau in Malawi or that could still be seen hardly 25 years ago in the south of Burundi. By that as it may, these mountain refuges were probably well separated from the lowland forest refuges located much farther to the west. Perhaps both forest types were linked by narrow forest galleries, but altogether, the western piedmont of the rift mountains formed an open landscape. The thick stone lines of Ossokari, a small village located along the Bukavu-Kisangani road, testify to this. In Cameroon, as well, isolated montane forest massifs might have survived on the high features of the landscape since tree pollen never disappears completely from the sediments of Lake Barombi Mbo.

Between 14,000 and 9,500 BP, the climate rapidly re-became more humid. Between 4,000 and 3,000 BP, the ocean rose and reached a level comparable with the present (FIGURE 2.28). The Sahara retracted. It was invaded by powerful water courses and by the immense expansion of Lake Mega-Chad and Lake Sudd. In Central Africa, these changes were accompanied in their initial phase by very strong erosion. New stone lines were formed in this way, notably in Cameroon and in the eastern part of the Democratic Republic of the Congo. They were accompanied by a great increase in alluvial deposits in the submarine delta of the Congo River which Paul Giresse and his collaborators could precisely date to 11,500 BP. From 9,500 BP, reforestation was rapid and the forests reached their maximum expansion around 7,000 to 6,000 BP. They then expanded well beyond their present limits. In the Congo, fossil tree stumps dating from this epoch were found near Pointe-Noire, a region presently covered by coastal savannas.

In the Albertine Rift region, the Rwenzori glaciers retreated beginning in 14,500 BP and, from 12,500 BP, in Uganda, Rwanda and Burundi, forests dominated by *Hagenia* developed which seemed to reach their maximum expansion around 11,000 to 11,500 BP. Starting in 11,000 BP, lowland forest species appeared that were mainly colonists, such as *Alchornea*, *Macaranga kilimanscharica* and *Polyscias fulva*. In southern Burundi, forests of *Podocarpus* also appeared, but they regressed from 10,000 BP, without disappearing completely.

Between 3,800 and 3,700 BP, while the Sahara was undergoing a very humid phase, the regions bordering the Gulf of Guinea were subjected to a reduction in rainfall. After a brief respite, this drought continued starting in 3,000 BP. It hit even the montane forests of south-western Uganda where there was an increase in *Celtis* – a genus of trees typical of semi-deciduous forests – while at higher altitudes *Olea* forests progressed and, probably higher still, *Podocarpus* forests (FIGURE 2.25c). The drought was at its maximum between 2,500 and 2,000 BP. It brought with it very violent forest fires of which traces are found in the form of layers of charcoal, mainly throughout the entire north-eastern Congo basin. Farther west, in Cameroon and Gabon, these fires were apparently more limited and probably even largely of human origin. Lacustrine sediments of Lake Barombi Mbo dating from this epoch contain hardly any micro-carbons. These fires were followed by heavy erosion (they once more produced stone lines), a transient progression of savanna vegetation and finally a real explosion of typical pioneer forest vegetation.

This last great retreat of the forests brought with it a reduction of some species that afterward were never again able to attain their previous distribution. Okoumé *Aucoumea klaineana* – the main species presently exploited in Gabon – is a good example. At the maximum of forest expansion, 3,000 to 7,000 years ago, it reached southern Cameroon, 170 kilometres north of its present limit. One can thus imagine that some species disappeared for ever.

In the region of the Great Lakes, this retreat did not escape the attention of researchers, but it was attributed to human influences and put in relation with the immigration of the Bantu peoples. Yet, despite a much higher population density and despite very frequent fires, the forests are today in full expansion. Moreover, the retreat that took place 2,000 to 2,500 years ago happened simultaneously throughout sub-Saharan Africa, it therefore corresponds well to a world climatological phenomenon.

Contrary to the dry periods of the great glaciations, which were cold periods, these recent periods of drought were accompanied by a warming. They probably resulted more from an increase in evaporation and seasonal contrast than from a real reduction of rainfall. For this reason, Maley qualifies them as climatic "worsenings" rather than dry phases.

After 2,000 BP, the forests again invaded vast areas of savanna. On several occasions their expansion was, however, slowed by new climatic worsening, notably between the 9th and 12th centuries. Altogether, the forests have not yet been able to reach their limits of 6,000 to 7,000 years ago, but their expansion continues.

Thus, little by little, the islets of savanna of the coastal region and of the Lopé National Park in Gabon, or in the Odzala National Park in the Congo, are shrinking. Even in more densely inhabited southern Cameroon, the forest progresses. It has gained several hundreds of thousands of hectares since the beginning of the 20th century. The town of Yaoundé, for example, created in 1888 on the edge of the forest, now finds itself 60 kilometres inside the forest.

On a world scale it is interesting to note that the last extreme reduction of the forests during the glacial maximum, 15,000 to 18,000 years ago, struck Central Africa and South America at the same time. Subsequently, there were, however, important differences between the two continents and, for the last 8,000 years, the Amazonian forest often retreated while the African forest progressed or vice-versa. This discrepancy is observed not only between Congolese forests and Amazonian forests. It also exists between Congolese forests and Guinean forests. The history of the forests thus seems to be influenced by an entanglement of several types of climatic cycles.

There are obviously great global cycles of about 100,000 years, linked to variations in the Earth's orbital parameters, such as those brought to light by Milankovitsch in 1941. They determine an alternation of cold, dry periods and warm, humid periods, of which the extreme phases do not last altogether more than 10% of historical time and for 80% of the time the climate seems a little cooler and a little dryer than it is today.

Upon this "long" cycle, is superimposed a shorter cycle. Detailed analyses of palynological data from western Cameroon reveal the existence of 2,200- to 2,500-year cycles in the relative abundance of pollen of the main forest species. They betray perhaps the phenomena of succession and could thus be the expression of the dynamism of the forests. Cycles of similar duration were found in Greenland, the Antarctic, North America, the Indian and Pacific oceans, as well as in the fluctuations of lake levels and glaciers in Europe and North America. Some scientists have put them in relation with the main cycle of solar activity which is around 2,300 years.

Lastly, there are still shorter and more irregular oscillations. They are thought to be linked to variations in the surface water temperatures of the oceans and induce phenomena of the *El Niño* type. They have for an effect that the globally hot periods are not always the periods of maximum expansion of the forests. Sometimes, they are accompanied even by great destruction, notably by fire, and thus a fragmentation of the forested domain. This latter type of oscillations is interesting to follow in so far as it may permit the anticipation of effects of global warming that Man is presently facing.

Lessons from Carbon 13 and Carbon 12

In addition to carbon 14, which we use for dating, carbon is also represented in nature by carbon 12 and carbon 13. Starting from the carbon dioxide of the atmosphere, these two isotopes are assimilated very differently by savanna grasses and forest vegetation. The ratio between carbon 12 and carbon 13 in the organic matter of the soil is thus less than 27 per thousand under forest cover, while it reaches 13 to 15 per thousand in the savanna. In measuring this ratio at different places and at different depths, and in comparing the results with carbon 14 dating, it is possible to trace a soil profile that reflects the history of plant cover rather well.

This technique was applied in the Lopé National Park in Gabon by Lee White and his collaborators (FIGURE 2.29). There the main vegetation types are grasslands (ocre), pioneer forests and Marantaceae forests (pale green) and mature forests (dark green). Profile 1, located in grassland, shows than these grasslands are at least 10,000 years old. Profile 2, located in a mature forest at 650 metres above sea level, shows that these forests are several thousands of years old. Profile 3, located in a young *Sacoglottis-Aucoumea* forest, suggests that here the forest were converted to grasslands between 3,000 and 2,000 BP, but later reinvaded the area. Profile 4 was situated in a mature forest not far from a place where iron was smelted 170 to 250 years ago. The forest there is very ancient, and after the departure of the human population it regenerated rapidly. An intermediate grassland stadium never developed.

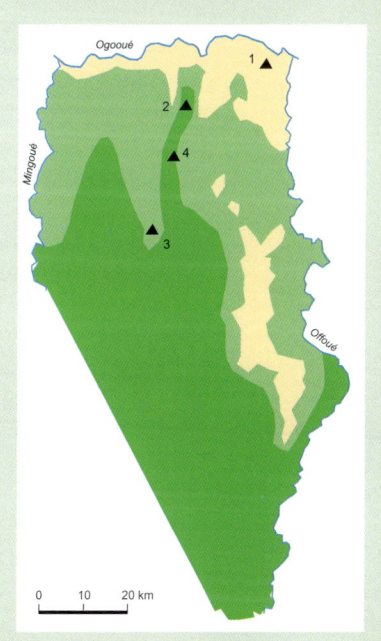

These data where compared to the current distribution of the main vegetation types and the archeological data collected by Richard Oslisly. They show that even at the maximum of the forest transgression, about 7,000 years ago, savannas always persisted along the Ogooué River. However, during the period of climatic worsening, 2,000 to 2,500 years ago, these savannas covered a much large area than today. Then they probably included most of the belt of young forests, which apparently started to grow some 1,400 years ago at a time when human populations disappeared in the area (PAGE 246).

All these climatic cycles seem to have appeared at the end of the Tertiary, but they are probably much older. As long as the world climate was distinctly warmer and more humid, as in the Cretaceous, they had very little effect on the vegetation, however. It is only with the general cooling of the climate throughout the Tertiary that the coolest phases start to show up.

Biogeography or the mark of history

Given its length and complexity, it would be very surprising if this history of the forests of Central Africa were not reflected in the distribution of the species that inhabit them. It is, moreover, problems of distribution, apparently not explicable by ecology, that have driven scientists to discover it. However, the fascinating diversity of species does not facilitate the solving of these problems and it is often very difficult to distinguish the role of history from that of ecology. On a much more fundamental level, they incite two questions that are themselves intimately linked to the history of these forests: how have all these species come into existence and how have they been able to persist over the course of time; in other words, how did they manage to coexist?

The answers are all the more complex in that, beyond the misleading homogeneity of tropical forest habitats, the numerous species and species groups have at times very different patterns of distribution. It is therefore necessary first of all to know the distribution of the species in detail. For the flora this information is not available. Botanical knowledge leaves a lot to be desired in Central Africa and vast territories are still nearly unexplored scientifically. In Gabon, for example, the total number of botanical specimens collected there represents only one specimen per 500 hectares. In all, around 5,500 species of plants have actually been collected, but, according to Marc Sosef and Frans Breteler of the University of Wageningen, there could be at least 7,000.

In fact, since 1960, very few systematic investigations have been conducted. Floristic inventories are rare, have not been done except on very small surfaces or centred on very specific groups of species. The great majority of inventories conducted in Central African forests take into account only ligneous species with a known or potential commercial interest, and then only above a certain diameter. Some attractive families, notably the Orchidaceae and the Begoniaceae, have been the object of more important collections. Even for these families, it is enough to visit a still little-known region to be able to find undescribed forms.

To the insufficiency of collections, is added the uncertainty of classification, which renders any interpretation dangerous. Except for genera and species which have been the object of recent, very detailed systematic revisions, many species of plants remain in effect known under different names in different regions, while other species, really different, remain confused under the same nomenclature.

All together, the existing botanical data are obviously not negligible, but they are not enough to be able to understand their biogeography. Therefore zoological data are a welcome complement. The distribution of vertebrates, especially mammals and birds, as well as of some privileged groups of invertebrates, such as some families of butterflies, is in fact much better known. Even though it is not directly superimposeable on that of plants and reflects different time scales, the biogeography of animals constitutes a good indicator of the history of the forests and often permits a clarification of problems that botany alone does not manage to achieve.

The origin of species

Let us come back to our two main questions. The coexistence and thus the survival of numerous species can be explained by competition, the distribution of species by niches, the relationships between prey and predator, as well as the structure and dynamics of habitats. None of these strictly ecological processes can, however, explain how these species appeared. According to the theory of allopatric speciation by Ernst Mayr, recognised today by a great majority of biologists, all specific differentiation requires, in effect, that, populations belonging to the same ancestral species at a given moment in their history be separated in space, therefore isolated from each other.

Naturally, some biologists have searched for the mechanisms leading to the isolation of populations and their speciation in the history of the forests with their alternation of contractions and transgressions. So was born the theory of refuges, based on the distribution of birds, developed initially for Amazonia by Jurgen Haffer, a German geologist and ornithologist. First published in 1974, it was very quickly adapted to Africa, notably by Diamond and Hamilton, Crowe and Crowe, Mayr and O'Hara, and Grubb.

According to this theory, African forest species would have come into existence in the refuges, the last islets of forest during the most unfavourable climatic phases. Populations, isolated for long periods in different islets, inevitably undergo a genetic drift, all the more rapid, for that matter, since they comprise small numbers of individuals. During the forest transgressions, these populations expand and again enter into contact. If their genetic differences are sufficiently expressed, they will no longer "recognise" each other and will behave as two different species. In the opposite case, they will again mix and melt into a single population. At most, genetic diversity will increase.

This theory had much success and incited a great number of publications. Very soon, however, it also had its detractors. At the time of its publication, paleopalynology was not yet as advanced as nowadays and great uncertainty persisted as to the reality and extension of

---

What is the origin of the enormous diversity of the tropical forests?

That question has been harassing biologists for years, and until now no completely satisfying answer has been given. The absence of marked seasonality, and thus of periods of strong environmental stress, can partly explain the phenomenon. Spatial heterogeneity at different scales, horizontally and vertically as well, can add to its understanding.

This heterogeneity means that the forest can be seen as a complex mosaic of sub-habitats. However, in itself it depends largely on the great number of plant species and their many different life forms. In some ways this can be compared to a vicious circle, and we can ask the question if diversity is not enhancing itself.

refuges. Many scientists put their existence frankly in doubt. Some field observations even seemed to contradict the theory.

The most classic example is that of the Dahomey gap. The Guineo-Congolese forests are in effect cut into two very distinct massifs at the level of Togo and Benin by a band of savannas. Some 600 kilometres wide, it should constitute a substantial barrier to the distribution of forest species. Now this seems not to be the case, neither in the distribution of plants nor in that of animal species. Robbins and Grubb showed it for small mammals; Moreau, Crow and Crow for birds; Amiet for amphibians; Carcasson for butterflies and Brenan for plants. The species of western Nigeria are thus sometimes closer to those of Ghana and to the Ivory Coast than to those of Cameroon. In many cases, the separation between Guinean forests and Congolese forests is located, in fact, much more to the east, more precisely at the level of the Cross River, where the continuity of the forest massif is not broken except for the narrow ribbon of the river. Now, if present-day fragmentation of the forests is not reflected in the distribution of species, how does one allow that ancient fragmentation was able to play a role.

Various authors have proposed alternative models of speciation. Most were developed for the Amazon, but can as well be applied to Africa. Haffer himself commented on them. They repose on tectonic movements, the barrier effect of the great watercourses, variations of the vegetation resulting from climatic oscillations of the Quaternary, and environmental gradients.

For Central Africa, Marc Colyn also proposed a theory of transgressions. According to him, the Congolese forests should be seen particularly in their more or less fragmented form, as they are 80% of the time (PAGE 57). The brief transgression phases would engender new species by hybridisation of populations already more or less differentiated and put back into contact secondarily.

Among these models, some attach more importance to one or the other of these phenomena; others accept a combination of factors. The objective evaluation of proposed models, including ones based on refuges, runs up against the fact that we do not yet know the exact configuration of the supposed refuges, the real age of species, the detailed distribution of the species and present ecological conditions on a sufficiently fine scale. It is thus very probable that biogeographers fall back on history a little too often in order to explain the distribution of species. Without a consistent supply of new data, as much on present-day conditions as on conditions in the past, a solution cannot be found for this problem.

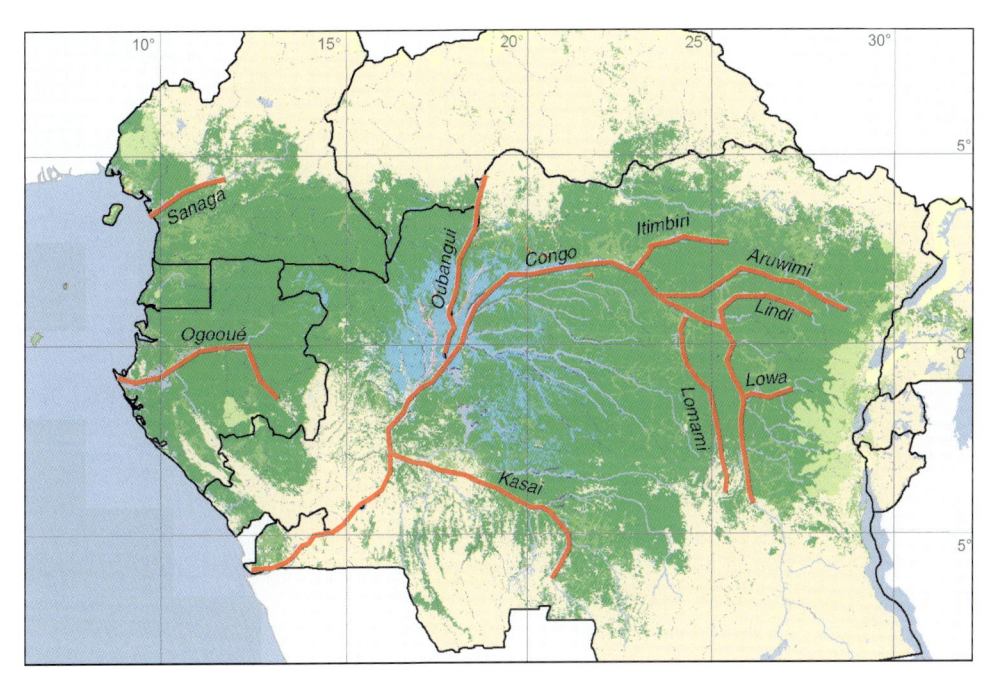

In the light of recent discoveries, it would seem, nevertheless, that the only unlikely model would be one based on ecological gradients. It supposes that speciation could occur in a parapatric manner, thus without the geographical isolation of implicated populations. Studies on bird distribution in sub-Saharan Africa, analyses of the distribution of Amazonian forest birds and biochemical studies on the rodents of the Andes reject it. The other models can all be "proved" by irrefutable observations. The distribution of the Cercopithecidae in the Congo basin, studied notably by Colyn, give substance to the model based on river barriers (FIGURE 2.30). No model can, however, explain the totality of speciation phenomena, neither in Amazonia nor in Central Africa. Until proof to the contrary, the refuge theory explains it best, and, so far, no palaeoenvironmental data has been able to demonstrate that it is false.

Independently from the role that they have played in extinction and speciation phenomena, the continual "pulsations" of the forest massif have also had an influence on plant associations. At each new transgression, and as they were able to re-invade lost spaces, the surviving species are in effect split up as a result of the fact that they do not all "migrate" at the same speed. Among plants, species dispersed by wind or by birds are clearly more efficient colonists than those having large heavy fruits that fall on the spot. Birds themselves – especially those of the canopy – and bats and insects with a strong flight are favoured compared to reptiles, amphibians, micro-mammals or crawling arthropods. What constitutes a barrier for some, such as a large river, for example, can be perfectly well crossed by others.

2.30. In Central Africa fluvial barriers play a fundamental role in the speciation of the Cercopithecid monkeys, but are of no importance for birds and plants.

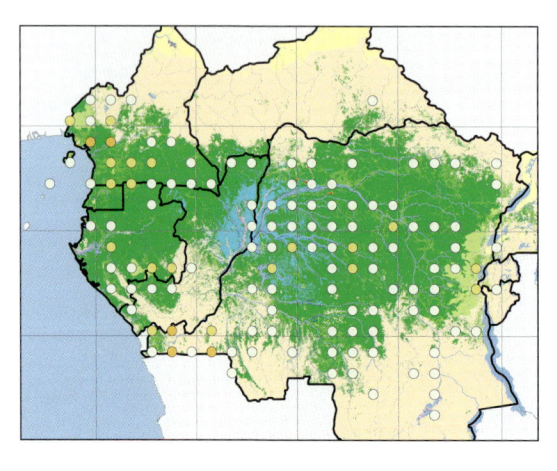

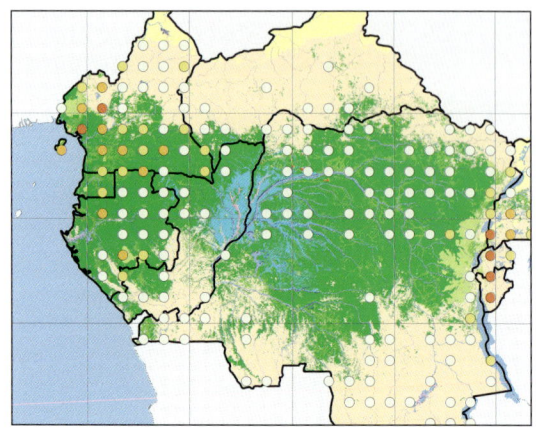

2.31-34. Fjeldsa and Lovett compared the distribution of ancient and recent species. Among plants, they took as ancient species (2.31) 23 species of trees present in both the Guineo-Congolian and the eastern forests of genera Allanblackia, Mammea, Zenkerella, Angylocalyx *and* Scorodophloeus. *For recent species (2.32), they considered the 92 species of genus* Impatiens, *herbaceous plants of the humid forest understorey. Among birds, they chose for ancient species (2.33) representatives of genera* Tigrornis, Agelastes, Afropavo, Pteronetta, Himantornis, Canirallus, Corythaeola, Pseudocalyptomena, Picathartes, Pseudochelidon, Ixonotus, Nicator, Phyllanthus *and* Hemitesia. *For recent species (2.34), they took forest species of genera* Campephaga, Lobotos *and* Coracina, Lamprotornis, Poeoptera, Onychognatus *and* Cinnyricinclus, Turdus *and* Zoothera, Hirundo, Petrochelidon *and* Ptyonoprogne, Zosterops *and* Speirops, *as well as* Nectarinia.

The number of species per one-degree square is indicated on each of the maps as follows:

○ 1-2
○ 3-5
○ 6-10
○ 11-15
● > 15

This phenomenon explains in part why the forest zones that correspond to ancient refuges are nowadays the richest and why biological diversity diminishes as one moves away from them. It also shows that the absence of a species in one place is not necessarily linked to ecological factors. The species in question perhaps has simply not had the time to reach this particular place. That the rich regions are all in high rainfall areas is explained by the fact that during the cold dry periods of the glaciations the amount of rainfall was strongly reduced, but the relative distribution of precipitation seems to have remained nearly unchanged.

Globally, the turbulent history of the forests seems very largely responsible for the distribution of species. Many present-day species – the pollen proves it – have existed, however, for much longer than the cyclic variations of the climate became evident. Very instructive in this way is the comparison between the Guineo-Congolese forests and those that border the Indian Ocean. To this day, numerous genera of Leguminosae, and even some species, exist in the two forest massifs. Now these were separated in the Oligocene or at the very beginning of the Miocene, but some of them already existed in the Paleogene and could thus be as much as 25 to 40 million years old. As far as the plant world is concerned, where speciation seems to be in general clearly slower than in the animal world, it would be insane to want to explain patterns of speciation through the climatic phenomena of only the very last glaciations.

Within this perspective, the studies of Jon Fjeldsa and John Lovett are very interesting. In comparing the distribution of ancient and recent species, among both plants and birds, these researchers were able to show that regions corresponding to ancient refuges are especially rich in old species, dating back to well before the Pleistocene and often having a very wide distribution (Figures 2.31 and 2.33). Those with a very reduced distribution more than probably represent relicts of groups for the most part now gone and are concentrated in regions of high rainfall, which must have been very stable throughout time – in other words, refuges. Recent species are especially abundant in the peripheral zones of the forest massif and in mountainous zones where there is a mosaic of very contrasted habitats and sources of high humidity (Figures 2.32 and 2.34).

According to this work, the richness of contemporary African forests rests on at least two complementary mechanisms. Species and ancient genera, dating from the Paleogene, maybe even from the upper Cretaceous, would have survived the "difficult" epochs of the Tertiary and the Quaternary in refuges which, because of this, functioned as veritable conservatories. The most recent species, having appeared in the last two million years, came into being in the periphery of the forest massif: the forest galleries within the zone of contact between the forest and the savanna, the zone of contact between the forest and the wetland habitats and the mountainous regions. All these regions are characterised by the coexistence of very different habitats and the presence of permanent humid habitats.

At first glance this hypothesis seems to contradict the refuge theory and could even render some credibility to parapatric speciation. This is not at all certain, for it is in effect more than probable that during the most difficult climatic phases these marginal regions of the forest massif harboured micro-refuges centred on their humid zones. In fact, one could fall again into the refuge theory, but the most important refuges in terms of speciation appear to be the small marginal refuges and not the large ones.

Whatever the real mechanisms might be, it is almost certain that the cyclic variations of the vegetation linked to the Milankovitch cycles during the Tertiary and the Quaternary constituted a very important "motor" of speciation, a true "species pump" as John Terborgh wrote.

These considerations on the origin of forest species are not just controversial subjects for scientists in need of publications. They also have important implications for conservation. As such, innumerable strategies and action plans relative to nature conservation and biodiversity could no longer avoid devoting a passage to the refuges of the Pleistocene. It is important, indeed, that ancient refuges be included in networks of protected areas, but this approach is not enough. Although allowing the conservation of many species, including ancient species with high genetic, aesthetic or ethical value, they may not encompass the true centres of speciation. These centres would be more difficult to circumscribe as they are dispersed over complex and ecologically very dynamic regions. Without prestigious endemic taxa, their conservation would also be more difficult to defend at a political level. But they are probably just as important as the refuges for the future of biodiversity. In some ways they represent "laboratories", while the former are no more than "museums".

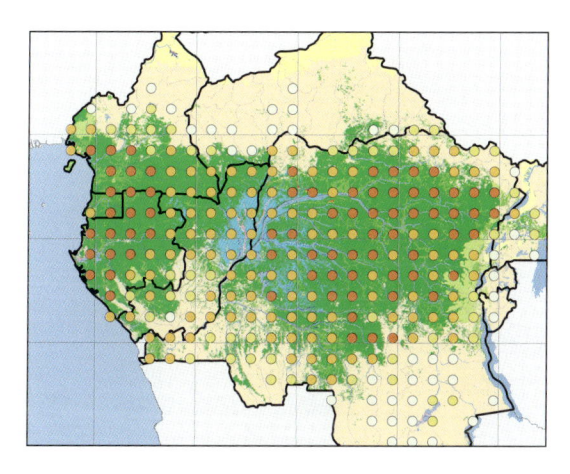

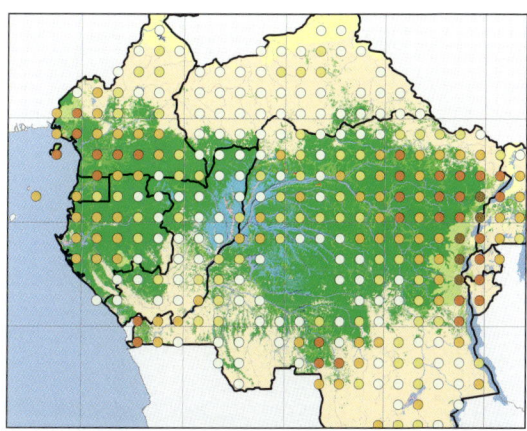

2.35. The many species of genus Impatiens *of the Balsaminaceae family typically represent a recent radiation. They are abundant in the cool humid forests of the Albertine Rift.*

The appearance of genera and species

In the **Eocene** 40-45 million years ago, the genera *Pentaclethra*, *Calpocalyx* and *Acacia* (Mimosaceae), *Alchornea* (Euphorbiaceae), *Combretum* and *Terminalia* (Combretaceae); the species *Bombax buonopozense* (Bombacaceae), *Mitragyna inermis* (Rubiaceae) and *Symphonia globulifera* (Clusiaceae).

In the **Oligocene**, 24 to 37 million years ago, there appeared the genera *Sindora*, *Fillaeopsis*, *Amblygonocarpus*, *Tetrapleura*, *Pseudoprosopsis*, *Xylia*, *Adenanthera* and *Leucena* (Mimosaceae), *Afzelia* and *Brachystegia* (Caesalpiniaceae), *Klaineanthus* (Euphorbiaceae), *Merremia* (Convolvulaceae), *Pentadesma* (Clusiaceae), *Randia* and *Macrosphyra* (Rubiaceae), *Parinari* and *Hirtella* (Chrysobalanaceae), *Scaevola* (Goodeniaceae), *Tacazzea* (Asclepiadaceae), *Annona* (Annonaceae), *Lovoa* and *Trichilia* (Meliaceae) and *Hippocratea* and *Campylostemon* (Hippocrateaceae).

In the **Miocene**, 5 to 25 million years ago, there appeared the genera *Gardenia*, *Oligodon* and *Morelia* (Rubiaceae), *Mimusops* and *Manilkara* (Sapotaceaea), *Melia* (Meliaceae); the species *Pycnanthus angolensis* (Myristicaceae), *Iodes africana* (Icacinaceae) and *Rhodognaphalon brevicuspe* (Bombacaceae) (after Maley, 1996)

2.36. *The seven main centres of endemism of sub-Saharan Africa according to White:*
1. *The Guineo-Congolese region,*
2. *The Zambesian region,*
3. *The Sudanian region,*
4. *The Somali-Masai region,*
5. *The Karoo-Namib region,*
6. *The Cape region,*
7. *The Afromontane region.*
Regions of transition:
8. *Guineo-Congolian-Sudanian,*
9. *Guineo-Congolian-Zambesian,*
10. *Lake Victoria regional mosaic,*
11. *Zanzibar-Inhambane regional mosaic,*
12. *Kalahari-Highveld regional transition zone,*
13. *Tongaland-Pondoland regional mosaic,*
14. *Sahel transition zone,*
15. *Sahara regional transition zone.*
The Afro-alpine region, impossible to indicate on the map, is no more than an extremely impoverished part of the Afromontane region.

Specificity of the tropical forest flora

The flora of tropical forests has only very few families which are not found in other regions of the world. Moreover, the few families endemic to tropical forest are represented by a small number of genera and species, the relationship of which is often obscure. The most diversified families such as the Annonaceae, Arecaceae, Caesalpiniaceae, Ebenaceae, Euphorbiaceae, Lauraceae, Moraceae, Myrtaceae, Sapotaceae and Zingiberaceae have all many species outside the tropics.

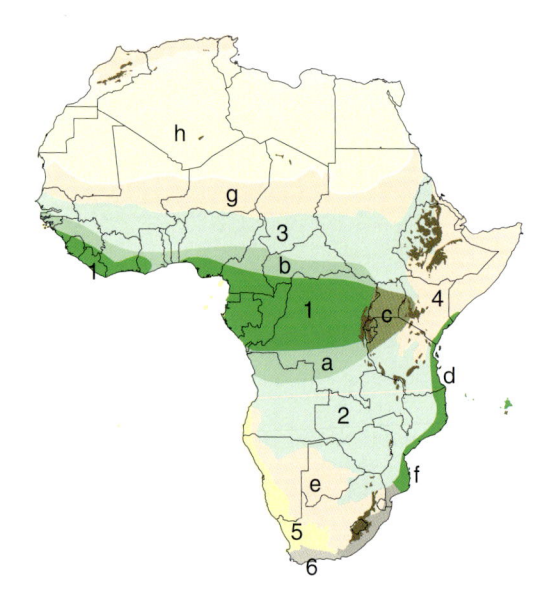

Biogeographic regions

Within the three great biomes, principally defined by eco-climatic conditions (PAGE 40), vegetation varies from one region to the other in function of its history. In the savanna biome, for example, the Sudanese and Zambesian regions largely share the same ecological conditions and have a number of species in common. But they are distinguished by the fact that each one also possess its own endemic species. On the basis of species distribution, Frank White divided the Afrotropical region into seven centres of endemism, separated by regions of transition (FIGURE 2.36).

The dense humid forests of Central Africa are thus split up into the Guineo-Congolese region, from Liberia to Uganda, and the Afromontane region. According to the species we consider, these two regions can be split into a number of smaller units, however. A good example are the studies by Marc Sosef on the distribution of members of Begoniaceae belonging to the subgenera *Loasibegonia* and *Scutobegonia*. This ensemble includes some 40 species incapable of surviving outside of the undergrowth of very humid forests and endowed with very reduced dispersal faculties. Combining their respective distributions reveals the existence of 21 centres of endemism in Central Africa: 4 montane centres and 17 lowland ones (FIGURE 2.37). All probably correspond to refuges of the last glaciation.

Lowland forests

Lowland forests are part of the Guineo-Congolese region which extends from eastern Nigeria to Uganda. They contain some 8,000 species of plants, of which 80% are endemic. This number is impressive, certainly, but there are much richer regions in the world. Venezuela, for instance, has more than 21,000 species, Brazil more than 56,000 and Indonesia nearly 30,000. Even South Africa has more than 23,000 species, of which 8,580 just for the small Cape floristic region. Instead of speaking of the "richness" of the Central African forests, it would thus perhaps be better to speak of their "poverty". This situation is found again in the number of tree species more than ten centimetres in diameter that are found on a single hectare. In Central Africa this number is in general situated between 60 and 90, while in Asia it is around 108 and 240 and in South America between 56 and 285.

As for the composition of the flora, some families, such as the Euphorbiaceae, Meliaceae, Moraceae and Sapotaceae – probably very ancient families – are nearly as well represented as in South America. One species of tree, *Symphonia globulifera*, of the Clusiaceae family, occurs on the two continents apparently without having given rise to distinct new geographical forms. On the other hand, the Bromeliaceae, Cactaceae and Luxembourgiaceae are only represented by a single species.

With the Asian forests, those of Central Africa share the Irvingiaceae, represented locally, among others, by the genera *Irvingia*, *Klainedoxa* and *Desbordesia*. Like the South American forests, the African forests are distinguished by the near absence of the Dipterocarpaceae. This family, so abundant and diversified in Asia, is represented by only one species, the Ntana, *Marquesia excelsa*, practically limited to the humid forests of Gabon.

Other families, such as Araceae, Arecaceae (palms), Gesneriaceae, Melastomataceae, Orchidaceae, Piperaceae and Urticaceae, are fairly well represented but still much less diversified than in South America or tropical Asia. The Orchidaceae number altogether more than 20,000 species, but only 1,200 in Africa. It is interesting to note that most of these poorly represented families are very rich in epiphytes.

Nine families are endemic: the Dioncophyllaceae, Hoplestigmataceae, Huaceae, Lepidobotryaceae, Medusandraceae, Octoknemaceae, Pandaceae, Pentadiplandraceae and Scytopetalaceae. Most of these families, however, are represented by a very small number of genera and species.

The Congolese region is divided into two distinct biogeographical regions. The Cameroono-Gabonese, or Atlantic coastal or Lower Guinean region, stretches from the Cross River to the Ubangi and Congo rivers.

The Congolese region strictly speaking, or Congolia, extends farther eastward. The two regions are separated by the immense marshes of the Congo, the Ubangi and the Sangha.

Depending on the species studied, these two regions in their turn divide up into a finer mosaic. On the scale of the Begoniaceae – as we have already seen – they divide up into some 16 zones of endemism, of which 13 are in the Atlantic coastal region and 3 are in the central Congo basin (FIGURE 2.37).

Altogether, the Atlantic coastal region represents the richest part of the Congolese forests. It is very rich in Euphorbiaceae, mainly in the Ogooué basin, and many relict species of very humid forests has been preserved, not only many Begoniaceae, but also several species of palms (FIGURE 2.38) and species belonging to essentially American families such as the Ozouga, *Sacoglottis gabonensis*, of the Humiriaceae, or the Izombé, *Testulea gabonensis*, of the Luxembourgiaceae. These forests also harbour species which do not occur farther east in the Congo basin, but are found again in the hyper-humid forests of Liberia.

The regions south of the Sanaga have kept the Okoumé. A study of the genetics of this species, conducted by Nicole Muloko of the Centre international de recherches médicales de Franceville (CIRMF), shows that it has two genetically distinct populations with centres of dispersal that coincide with the Crystal Mountains, to the north, and the Chaillu massif, to the

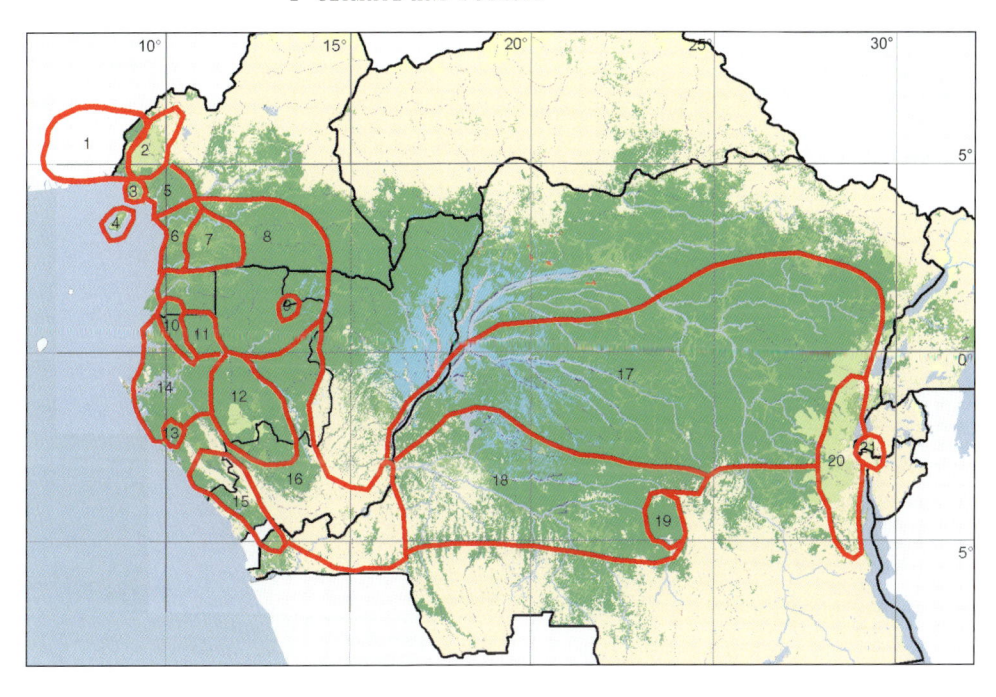

2.37. On the basis of the distribution of begonias of subgenera Loasibegonia *and* Scutobegonia, *Marc Sosef distinguishes 21 centres of endemism for Central Africa: (1) South-eastern Nigeria, (2) The mountains of western Cameroon, (3) Mount Cameroon, (4) The island of Bioko, (5) The northern part of coastal Cameroon, (6) The southern part of coastal Cameroon, (7) The Cameroon plateau, (8) The Cameroono-Gabonese plateau, (9) The Belinga Mountains, (10) The western Monts de Cristal, (11) The eastern Monts de Cristal, (12) the massif du Chaillu, (13) The Doudou Mountains, (14) The coastal zone of Gabon, (15) The Mayombe, (16) The western part of the Congo and the Democratic Republic of the Congo, (17) The centre-north of the Congo basin, (18) The centre-south of the Congo basin, (19) The Sankuru zone, (20) The eastern part of the Democratic Republic of the Congo, (21) The mountains of Rwanda and Burundi.*

2.38. The very humid forests of the Atlantic coastal region, in places subjected to a mean annual rainfall of 3,000 millimetres or more, harbour three relict species of palm: Raphia regalis, Podococcus barteri *and* Sclerosperma mannii *(opposite). The latter is a small palm of the understorey, very abundant in the swamp forests of the Gabon estuary. It also exists in some very humid valleys of the Lope Reserve.*

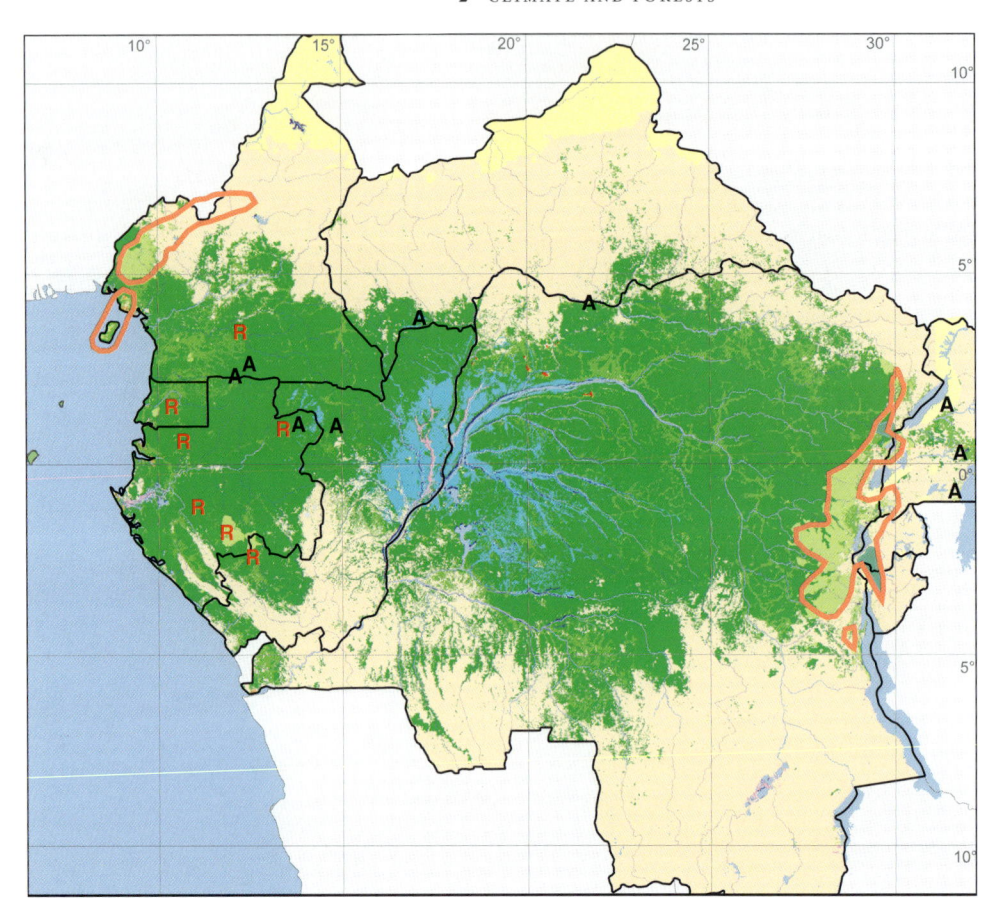

south. This allows the supposition that, at an epoch of extreme reduction of the forest, probably 2,000 to 2,500 years ago, the Okoumé forests were reduced to only these two mountainous areas.

Montane forests

Montane forests, and to a lesser extent, submontane forests, represent a somewhat archaic world, without palms, but with tree ferns, two genera of conifers (*Podocarpus* and *Juniperus*) and more mosses, liverworts and lichens than the lowland forests. Until some thousands of years ago the forests of the Albertine Rift even harboured Restionaceae, a family of primitive Angiosperms today absent from Central Africa.

The two islets of montane forest of Central Africa, that of Cameroon and that of the Albertine Rift, are part of a much vaster "archipelago" equally encompassing the mountains of Ethiopia, eastern Africa and Angola. Today this discontinuous ensemble harbours around 4,000 species of plants of which at least 75% are endemic. It includes two endemic families, the Barbeyaceae and the Oliniaceae.

Although each islet has its particularities and its endemic species, many species, not only of plants but also of animals, occur in several islets, often very distant from one another. In spite of the 2,000 kilometres that separate them, the montane forests of the Albertine Rift and those of Cameroon have many species in common or are inhabited by very closely related species whose separation is apparently not very old – sometimes they have been considered as representing no more than geographical forms of one and the same species. Obviously, the various Afromontane islets have been in communication and their separation must not be very old. It is even probable that connections existed at several epochs.

An observation related by Dwight Lawson and Michael Klemens seems to support the existence of ancient connections. Among the Amphibians, which are not outstanding for their aptitude in dispersal, the "viviparous" toads of genus *Nectophrynoides* – nowadays split into three very close genera – are not in fact known except from Mount Nimba of Liberia, Ivory Coast and Guinea, and the mountains of Tanzania and Ethiopia, with no intermediate populations.

Other observations indicate, on the other hand, "recent" connections. Numerous Afromontane species, notably of trees and birds, have, in fact, small isolated populations that suggest connections via the mountains of Gabon, the Congo and Angola, as well as the plateaux that separate the basins of the Congo and the Zambezi (FIGURE 2.39). Pollen from *Podocarpus* and other montane species on the Batéké highlands in the Congo support this theory.

2.39. *The Afromontane forests are divided into two ensembles: those of the Cameroon-Nigeria region and those of the Albertine Rift. Between these two regions, however, some Afromontane relicts survive (R) in Equatorial Guinea, Gabon, the Congo, Angola, Zambia and the southern part of the Democratic Republic of the Congo, which leads one to suppose that there could have been an east-west connection passing south of the Congo basin. The existence of unexpected populations of* Apalis jacksoni *(A), a species of the montane and submontane forests of Cameroon and the Albertine Rift, offers, however, the possibility of an ancient connection passing north of the basin.*

2.40. *The small* Impatiens bururiensis *is endemic to Bururi Forest in Burundi. Even there, its distribution is limited to the valley of the Sikuvyaye River.*

Françoise Dowsett-Lemaire and Robert Dowsett show that there could also have been a communication passing to the north of the Congo basin. The Black-throated Apalis, *Apalis jacksoni*, a small, well-known warbler of the canopy of montane and submontane forests, notably of Cameroon and the Albertine Rift, also occurs in southern Cameroon, north-eastern Gabon, northern Congo, the south-western part of the Central African Republic and still farther east on the border between the Central African Republic and the Democratic Republic of the Congo. In fact, all these populations, even the one in southern Cameroon, are different from those of the mountains of Cameroon but identical to those of the Rift.

Although all the forests are not yet completely explored, the Albertine Rift is the richest Afromontane islet in Africa. Though they extend over 800 kilometres from north to south and are nowadays very fragmented by human activity, the Rift's forests constitute a very homogeneous ensemble. Many species are distributed throughout this mountainous region without having engendered well-marked subspecies. The forests surrounding Lake Kivu and the north end of Lake Tanganyika – the forests of Itombwe, Kahuzi-Biega, Virunga, Nyungwe and Kibira – include the most humid formations and are the richest, the richest individual massif being the Itombwe, located south-west of Bukavu (FIGURE 2.41). The more distant forests of Lendu, Rwenzori, Mount Kabobo and Mount Mahale (in Tanzania), often isolated by low dry regions, are markedly poorer as one moves away from the Itombwe.

The montane forests of Cameroon-Nigeria form a distant element of the Afromontane archipelago and are poorer in species because of this, but they are proportionally richer in endemic species. They divide up, in their turn, into several isolated massifs.

The Mount Cameroon zone forms a well individualised ensemble, with at least 50 species and three genera of endemic plants. It shares a good number of species with the forests of Bioko and, because of its altitude, it harbours Afroalpine species that do not occur in other parts of this region.

Farther north lie the Rumpi Hills with their very humid submontane forests. To the north-east, lie those of the Kupe, Nlonako and Manengouba mountains. The first are very humid, the latter dryer and more frankly montane. Farther yet are the forests on Mount Oku, dominating the Bamiléké plateau, of the Obudu plateau in Nigeria, and, lastly, of the Tchabal Mbabo, the highest summit of the Adamaoua. All the forests are isolated from one another by lowland forests or by non-forested habitats and each has its own character with its cortège of endemic subspecies and species. The richest are the forests of the Rumpi Hills and those of the Kupe-Manengouba chain, but all of them are

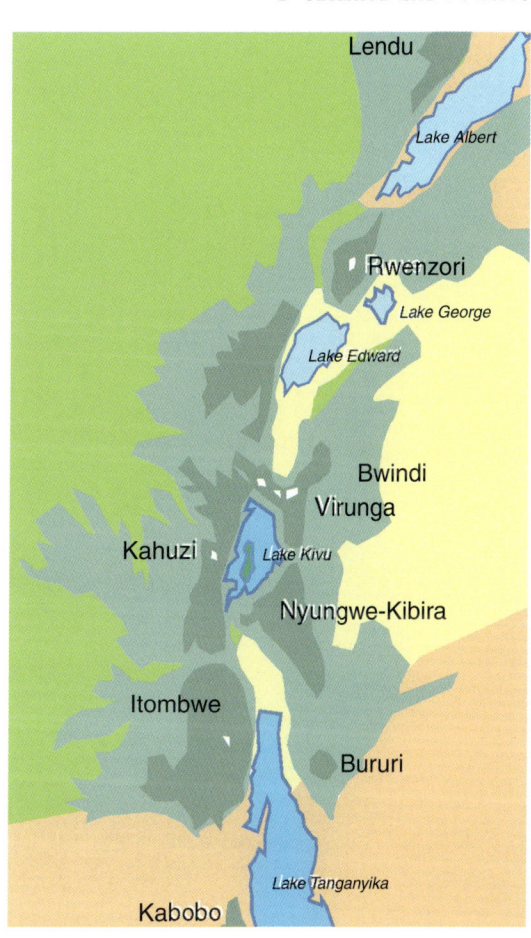

2.41. The montane forests of the Albertine Rift belong to nine more or less isolated blocks: the Lendu massif, the forests west of Lake Edward, the Kahuzi-Biega massif, the Itombwe, Mount Kabobo, the Rwenzori, the Virunga-Bwindi-Gishwati massifs, the Nyungwe-Kibira massif and Bururi Forest.

2.42. The montane forests of Cameroon are scattered between Mount Cameroon and Tchabal Mbabo. Most important are Mount Cameroon, Mount Kupe, Mount Nlonako, Mount Manegouba, Mount Oku and Tchabal Mbabo. On the Rumpi Hills, Bakossi and Nta Ali mountains grow submontane forests. On most of the Bamileke Plateau and the Adamaoua Plateau these forest have been replaced by grasslands since centuries.

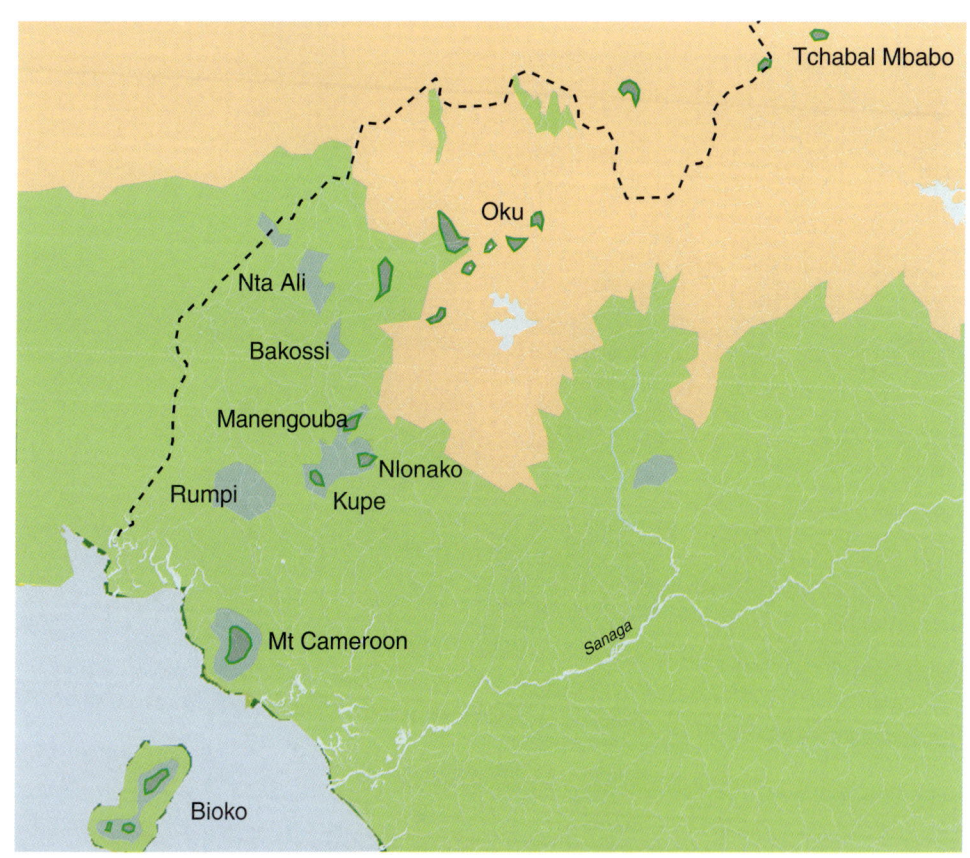

2 CLIMATE AND FORESTS

important for conservation and all are equally threatened with disappearance. Those of Tchabal Mbabo were not well known, but recent exploration has shown that they harbour several species considered as endemic of the Cameroon Highlands.

The islands of the Gulf of Guinea

Bioko, the largest of the islands, 40 kilometres from the coast of Cameroon, has, according to Figueiredo, 1,105 species of plants. Isolated for only 1,000 to 1,2000 years, its flora is very close to that of Mount Cameroon and shows at most a beginning of differentiation. Its fauna, on the contrary, includes endemic species and subspecies, which shows once more that speciation in animals is more rapid than in plants.

Principe and São Tomé are true oceanic islands, located respectively 220 and 280 kilometres from the coast, and their flora has come mainly from Gabon and Equatorial Guinea. Some species have not changed. Others have given rise to endemic subspecies and species. Altogether, it is a completely original flora, which is becoming better known, notably through the work of Exell. Work accomplished in the past few years by the ECOFAC team shows, however, that much remains to be discovered, especially on Principe where 49 species of orchids and at least 29 other plants were added to the flora in 1998 and 1999.

According to the IUCN, 601 species of plants were recorded on São Tomé in 1990, of which 108 are endemic; 314 were recorded for Principe, including 35 endemics. Additions in recent years have increased these figures, but have not modified the overall picture: some species are common to the two islands, but many occur on only one of them. Of orchids, for example, Tarig Stevart found a total of 129 species, 36 of which are present on both islands, 65 uniquely on São Tomé and 28 only on Principe. However, during recent investigations in unexplored regions of the Monts de Cristal, in Gabon, some of these apparently endemic island orchids were found on the continent. The level of plant endemism may be perhaps a little less high than the first investigations suggested.

The flora of these islands is also very rich in Pteridophytes, which is not surprising since these plants are excellent long-distance colonists, are very apt at colonising new lavas (PAGE 104) and love humidity above all.

While it is generally admitted that the flora and fauna of Principe are older than those of São Tomé, it is possible that the forests of São Tomé have kept at least one species of Gondwanaland origin (PAGE 24). *Podocarpus manni*, long considered as a derivative of *Podocarpus latifolius*, still present in Cameroon, Equatorial Guinea and the Congo, is dioecious – with separate male and female trees – and produces much larger fruits than those of *P. latifolius* (PAGE 53). It could be a Gondwanaland relict.

As for Annobón, this small island, difficult of access, still has primary forests in its central and southern parts, but these have been very little explored and almost nothing has been published about them.

2.43. Bulbophyllum sandersoni is one of the Central African orchids which was able to colonise São Tomé. Orchids have extremely small seeds, easily transported by the wind.

2.44. Costus giganteus, of the Zingiberaceae family, is endemic to the island.

2.45. The submontane forests on São Tomé and Principe are the result of a constantly very humid climate.

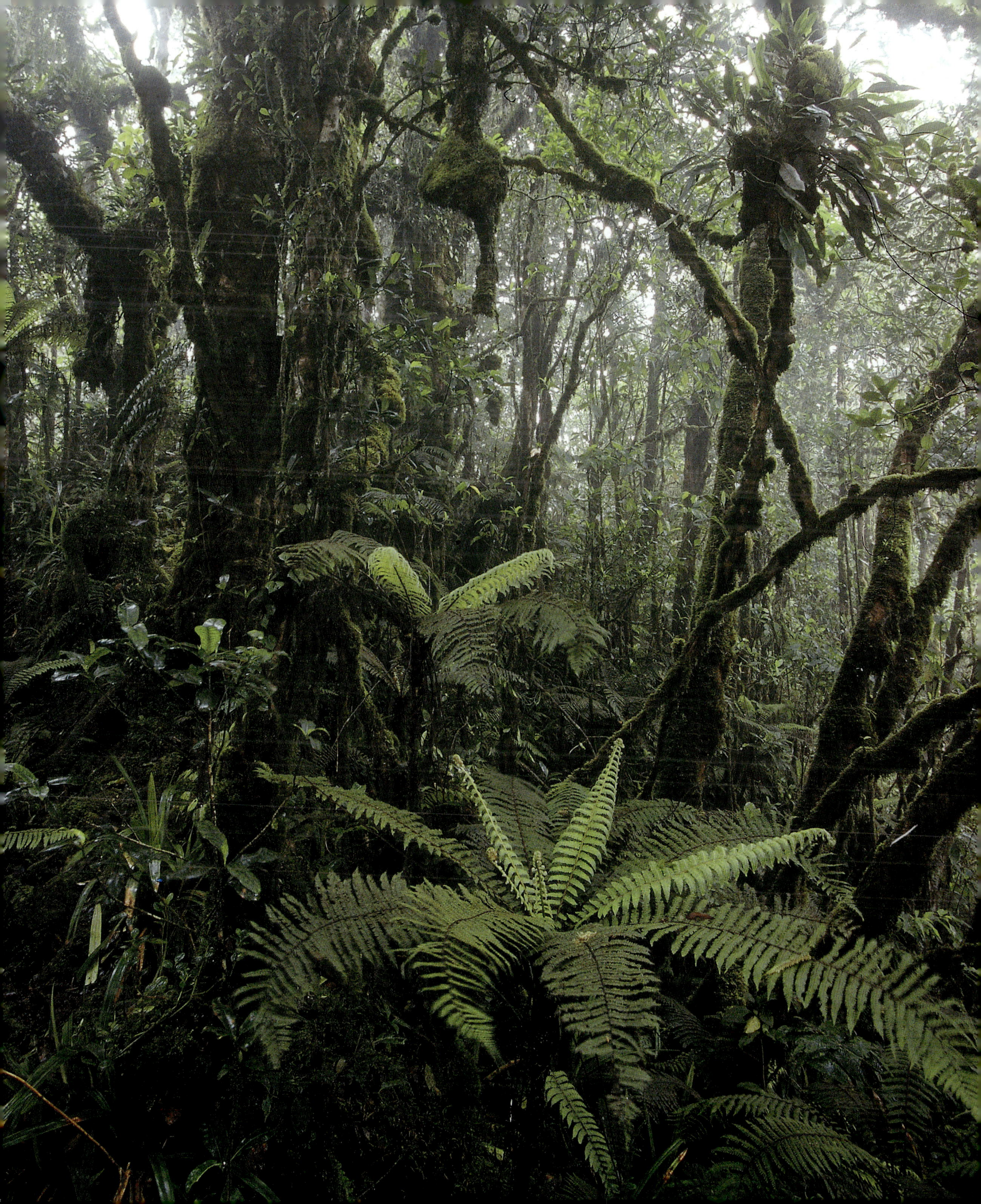

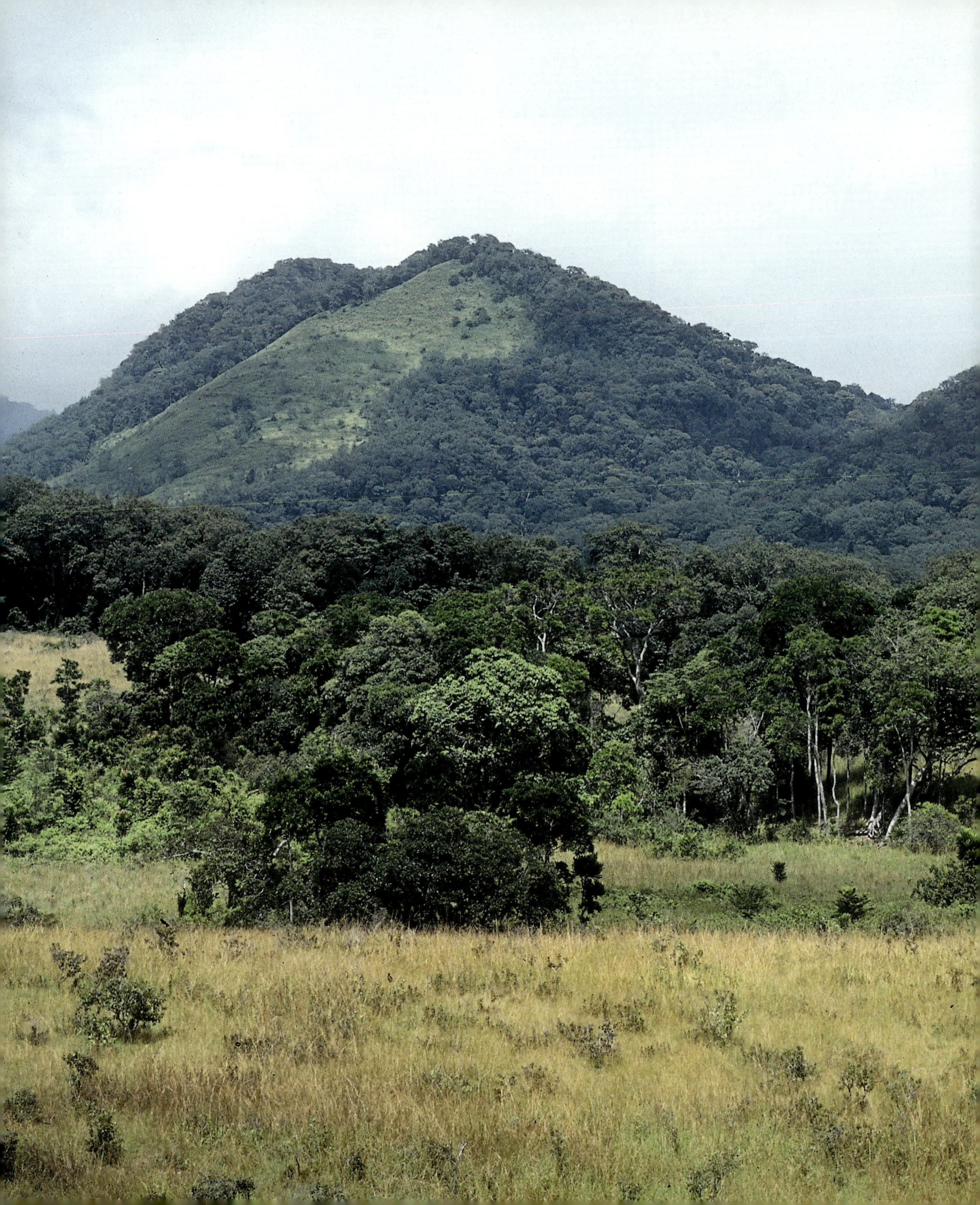

3 Ecological gradients: altitude, climate and soil

The forests of Central Africa, highly fragmented 2,000 to 2,500 years ago, were reconstituted and, in spite of growing human presence, invaded most of the area between the Gulf of Guinea and Lake Victoria. Until very recently, only water courses, a few large lakes, marshes, savannas and some high mountain summits broke this ocean of green. Seen from the sky, the forest seemed to have an exasperating uniformity and more than one aeroplane pilot lost above this wilderness has searched in vain for a landmark before being swallowed up in it. Nowadays new cuts appear in the densely populated areas and along roadways, but, for the most part, this vast forest massif remains nearly continuous. Once inside the forest, the observer sees, however, that it is far from homogeneous. The distribution of plant species and their associations constitutes a complex multi-dimensional mosaic with zones of transition, sometimes abrupt, sometimes imperceptible, that can be arranged along five main axes. The first expresses the biogeographical gradient, already mentioned in the previous chapter. The next three axes express ecological gradients: The second, based on altitude, divides montane and submontane forests from lowland forests. The third reflects climatic conditions, mainly the quantity and distribution of rainfall. It separates the evergreen forests from the semideciduous and sclerophyllous forests. The fourth reflects edaphic conditions. It integrates variables of the soil, mainly its composition, texture and resources in water. It distinguishes terra firme forests from inundatable and swamp forests, as well as crest forests from valley forests, and all the formations linked to particular soils. The fifth and last axis is temporal, which we will deal with in the next chapter. These gradients are in practice indissociable and influence each other mutually. Thus, each parcel of forest is located in the multidimensional mosaic by its position along the five main axes or gradients. Except for some values, each environmental parameter can become a limiting factor. If rainfall is insufficient, the dry season too long, the soil unable to retain moisture or the temperature too low, the forest gives way to other plant communities: savannas, meadows and thickets on inselberge, marshy clearings or high-altitude moorlands. Narrowly intertwined with the forest massif, these open formations play an important role ecologically and, in spite of their appearance, they are an integral part of the forest ecosystem of Central Africa.

3.1. In the north of the Lopé National Park, in the heart of the Gabonese forest, rainfall is marginal for the development of the forest and the soil is poor. In places, forest formations are limited to valley bottoms, while the ridges are covered by grasslands. A similar landscape must have covered immense parts of Central Africa during the cool dry climate phases of the past.

3.2. In regions bordering the Gulf of Guinea, low clouds from the Atlantic come up against the mountains, linger in the valleys and enshroud the high crests. Not only do they locally increase the atmospheric humidity, but they also reduce radiation and temperature.

Ecological gradients

The altitudinal gradient

As we have seen in the previous chapter, lowland forests and montane forests represent two very different worlds. Not only do they not react in the same way to climatic variations, but neither their structure nor their floristic composition are the same. In addition to two endemic families of trees (PAGE 64), montane forests also possess several endemic genera, notably *Afrocrania*, *Balthasaria*, *Ficalhoa*, *Hagenia* and *Xymalos*. They are the only forests to harbour members of the Podocarpaceae and Ericaceae in equatorial regions and have more "archaic" plants – lichens, bryophytes and pteridophytes – than the lowland forests do.

Where does the limit between lowland forests and montane forests lie? In many places it is easy to answer this question because of the fact that nowadays the two types of forest are separated by areas of cultivation. In places where there is still a continuity, the situation is much more insidious. Some scientists have attempted to distinguish these limits. Lebrun, Gilbert and Pierlot proposed limits for the eastern part of the Democratic Republic of the Congo, Lewalle for western Burundi, Letouzey for the Mount Cameroon region and Exell for the islands of the Gulf of Guinea. Detailed studies by Hamilton on altitudinal gradients in Uganda show that there is no discontinuity. Just as temperature decreases very regularly about 0.5 - 0.6°C per 100 metres of elevation at upper altitudes, the number of species and families diminishes as well as the height of the trees and the complexity of the canopy, the average size of the leaves, the percentage of species having compound leaves, the percentage of deciduous species and the frequency of buttresses. All these factors vary in a gradual way without a well marked threshold at any point. It is not until near the upper limit of the gradient, above 2,300 metres altitude, that "stages" seem to appear because only relatively few species remain and the dominant species impose a characteristic physiognomy on the forest.

Thus two different worlds but nothing to allow an objective separation! Some scientists have concluded that lowland forests and montane forests are part of a single biome. This view is, however, contradicted by biogeographical data, and the absence of a clear-cut limit is perhaps just a consequence of history. First of all, one can imagine that, during certain periods, when the forests were much more fragmented than they are now, the montane forests were separated from the lowland forests or, at most, linked by narrow forest galleries (PAGE 56). The gradual transition would thus be largely the result of the present forest transgression. We must also remember that the distribution of montane forests has varied considerably in time and that their inferior limit must have oscillated within an altitudinal zone of the order of 1,000 metres. There probably are two different biomes which, at times, come into contact, but their zone of overlap displaces in altitude according to climate variations. Lastly, the enormous latency that characterises the response of forest formations to climatic changes must also contribute to the suppression of clear-cut limits.

In practice, we recognise three distinct zones within the altitudinal gradient: lowland forests constituted uniquely of low-altitude elements, mountain forests constituted almost uniquely of montane elements and transitional or submontane forests comprising a mixture of low-altitude elements, montane elements and some species strictly limited to intermediate altitudes. These transitional forests thus represent a sort of "battlefield" where the montane forests and the lowland forests meet, penetrating each other and mixing.

In the west, in the region of the Gulf of Guinea, lowland forests reach 600 or 800 metres altitude and montane forests begin at 1,500 or 1,800 metres, even at 1,100 metres on São Tomé. Their upper limit is situated at about 2,500 metres. In the east, in the region of the Albertine Rift, lowland forests extend up to 1000 or even 1,200 metres above sea level and montane forests begin at 1,650 or 1,800 metres. Their upper limit is situated between 2,700 and 3,400 metres.

This difference between east and west is probably based on several independent phenomena. First of all, numerous observations show that the higher the annual rainfall is, the lower the altitudinal limits are and vice versa. This phenomenon is observable within the same forest massif – Nyungwe Forest in Rwanda, for example – where the altitudinal limits of distribution of many species are clearly higher on the dry eastern slopes of the Congo-Nile divide than on its more humid western slopes. This could be linked to the fact that the more it rains, the more cloud cover there is, therefore less insolation and lower average temperature. Nearly all the high altitude regions bordering the Gulf of Guinea receive at least 2,500 millimetres of rain annually, but no more than 1,200 to 1,300 hours of sunlight in the same period, while the forested regions in the east receive only 1,300 to 2,200 millimetres rain annually but more than 2000 hours of insolation (PAGE 44).

Secondly, measurable rainfall also varies with altitude: the drier the piedmont of a mountain, the higher the zone of maximum rainfall is situated. In the very humid region of Mount Cameroon, rainfall decreases rapidly with altitude, while in the forested regions of the rift maximum rainfall is not observed before around 2,000, 2,500 or even 3,000 metres (FIGURE 3.3).

Lastly, clouds and non-measurable precipitation must be taken into account. In the Gulf of Guinea region, ridges and summits are very frequently shrouded for long periods at altitudes of 500 or 600 metres and above (FIGURE 3.2). Even in the Lopé National Park, it is not unusual for the cloud ceiling to be at around 350 or 500 metres and to engulf summits culminating at 400 to 800 metres.

In the Albertine Rift region, the effects of clouds are very different. When this region is crossed by large masses of clouds, most often cumuliform and coming from the east, they envelop summits and ridges above 2,000-2,500 metres, but pass fairly rapidly. They do not linger except just on the west side of the summits as a result of turbulence – it is very visible on Mount Bururi in Burundi – or keep from being trapped except in deep valleys of the interior of the mountain massifs or on their west flank, as in the Nyungwe Forest in Rwanda or the Kalinzu Forest in Uganda. Their effect is reinforced there by nocturnal fogs that fill the valleys after a nocturnal temperature inversion, frequent in these regions with imposing features.

At the risk of simplifying things a little, it could be said that montane forests of the Albertine Rift are mainly high altitude forests while those bordering the Gulf of Guinea are rather cloud forests.

3.3. Average annual rainfall varies with altitude. The more humid the piedmont of a mountain is, the lower is its zone of maximum rainfall. On very high mountains there is apparently a second maximum at high altitudes.
a. South-western slope of Mount Cameroon,
b. South-eastern slope of Mount Cameroon,
c. Eastern slope of Mount Cameroon,
d. Southern slope of the central group of the Virunga (Karisimbi, Bisoke)
After Lauer and personal observations.

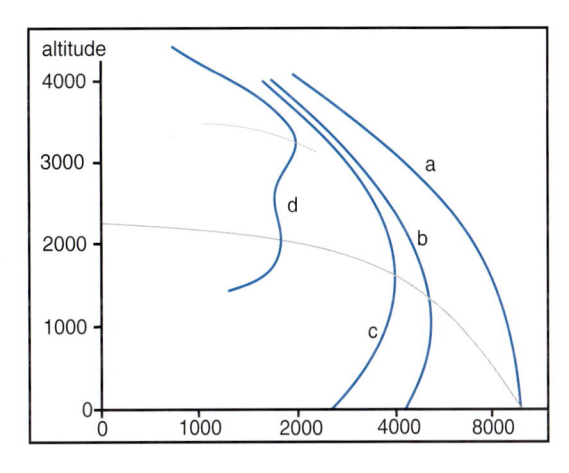

3 ECOLOGICAL GRADIENTS

The climatic gradient

Since distribution of tropical forests depends essentially on available humidity, mostly rainfall, it is natural that rainfall would have a capital influence on their appearance, structure and composition. The effect of the rains is, however, largely modulated by temperature. The higher the temperature, the greater the evaporation and the more rainfall is required to compensate it. Inversely, the cooler the climate, the greater the availability of humidity. The monthly quantity of rain necessary to compensate evaporation is therefore a function of altitude: at sea level, it is of the order of 100 millimetres; at 1,500 or 2,000 metres, it is reduced to about 50 millimetres. The climatic gradient is therefore greatly influenced by the altitudinal gradient. It is also influenced by the nature of the soil, but less by its chemical composition than by its capacity to retain water.

Apart from this influence of altitude and temperature, Central African forests divide up according to a gradient of available humidity going from hyperhumid forests to the driest forests. Once more, the transitions are nevertheless very gradual, sometimes hardly perceptible and often in mosaic.

The effect of rainfall upon the vegetation depends essentially on two factors: the annual total and the seasonal distribution, in other words, the length of the dry season. Which is more important? Some studies show that the length of the dry season is determinant; others that it is the annual total. Often correlated, these two parameters probably act more or less independently. In regions of high rainfall, the length of the dry season, usually very short, matters little. In regions where the total annual rainfall is marginal for the survival of a forest, the length of the dry season could, on the contrary, have a decisive effect.

In the most humid regions, with more than 2,000 millimetres of rain and less than two dry months per year, evergreen forests develop. In principle, the trees never lose their leaves simultaneously, but renew them continuously. In Gabon, there are evergreen forests in regions where the dry season lasts three months. This would not be possible except for the Atlantic influence on the climate, which covers the entire country with a uniform cloud layer that blocks the sun's rays and lowers the temperature during the dry season (PAGES 42 and 43).

In regions where the average annual rainfall is less than 2,000 millimetres, or the dry season too long, some large trees lose all their leaves during periods of hydric stress (Figures 3.4 and 3.5). The drier the climate or the longer the dry season the greater the proportion of deciduous trees in the forest. There is, therefore, an entire gradation of more or less deciduous formations. Apart from evergreen forests, most authors recognise two categories: semi-evergreen forests and semi-deciduous forests. Aubréville, for example, speaks of semideciduous dense humid forests and semi-deciduous dense dry forests. Lebrun and Gilbert, on the contrary, recognise only two categories in all: evergreen umbrophilous forests and semi-deciduous mesophile forests. This apparent confusion in terminology simply shows that no objective delimitation exists between evergreen forests and those in which some of the trees lose their leaves during some seasons. In most of the Congo basin, there are always at least some trees that lose all their leaves at times.

In practice, however, three main types can be perceived. The mainly evergreen forests include patches of hyperhumid forests surrounded by humid forests, and are strictly limited to the islands of the Gulf of Guinea (Bioko, São Tomé and Principe) and to the coastal areas of Gabon, Equatorial Guinea and Cameroon. The mainly deciduous forests constitute the entire periphery of the forest massif from Cameroon and the Mayombe region to the eastern part of the Democratic Republic of the Congo and Uganda. In between the two are interspersed the extensive transitional forests.

Dropping leaves in order to resist water shortage is a good strategy when the dry periods are long and when the rains return at predictable times of the year. In regions near the equator, around the Albertine Rift and

3.4. Some species lose all their leaves simultaneously. Others never do. However, all the individuals of a given species do not necessarily lose their leaves at the same time and defoliation never happens at the height of the dry season. It can also be of a very brief duration, sometimes only a week or two. In most of the Congolese forests, there is always a tree that loses it leaves at any time of the year, even in forests said to be evergreen bordering the Gulf of Guinea.

Lake Victoria, average annual rainfall is no more than about 800 to 1,200 millimetres in places. The climate is characterised by the occurrence of two relatively short annual dry seasons, but the return of the rains cannot be predicted to within a month. In these regions dropping every leaf could be dangerous. Also forest trees, instead of losing their leaves in the dry season as trees of the semi-deciduous forests do, have small shiny coriaceous leaves called sclerophyllus. Endowed with a thick cuticle, these leaves are very well protected against water loss, but they have only to open their pores to get back into action when the climate permits. In fact, these forests somewhat resemble the laurisylves of Mediterranean regions, all the more so that the olive tree *Olea europaea* is often an important element of it.

The edaphic gradient

Independently from the influence of climate through temperature or rainfall, tropical forests are also strongly influenced by some properties of the soil or edaphic factors: chemical composition, depth and texture, permeability, declivity and drainage. Correlations between the nature of the soil, its texture, the floristic composition of forest formations and their structure are not, however, always very clear. At times one even runs up against obvious discordances. In some places, this phenomenon can rest on the history of forest formations that have not yet had the time to adjust to present conditions (PAGES 56-60). Studies conducted outside of Central Africa, however, show that the chemical composition of the soil mainly has an influence on the floristic composition of the forests, while the texture of the soil and its hydric resources – its capacity to conserve water – have an influence rather on the structure of the forest cover.

High dense forests dominated by a single species of the Caesalpiniaceae family, frequent throughout the lowland forest zone of the Congo basin, form on extremely varied but poor soils. The trees of this family manage to get around the rarity of nutrients in the soil by establishing a symbiosis with fungi, the ectomycorrhizas (PAGE 166). The much more varied but rather low forests that develop on the white sands of coastal alluvial plains or on exposed rock pavements, such as those that can be seen in a good part of the Mondah forest near Libreville, are other examples of formations linked to the texture of the soil.

The most remarkable examples of the influence of soil can be found in steep hilly regions. The shallow, well drained soils of steep slopes and ridges support forests very different from those of the deeper, less well drained soils of the piedmont. Nevertheless, the distribution of the various formations depends on rainfall.

3.5. A Kapok tree Bombax buonopozense, *of the Bombacaceae family, in flower at Makokou, in Gabon. This tree has a wide distribution, geographically and ecologically as well, but is more common in drier forests.*

In regions of low rainfall – 1,200 to 1,500 millimetres per year on average – dry forests of ridges and summits are often disturbed by fire or transformed into grasslands. In the most unfavourable conditions, they are reduced to simple galleries along streams (FIGURE 3.6). This is what can be seen in the northern part of the Lopé National Park (Figures 3.1), on the Bateke plateau, in southern Burundi (FIGURE 3.7) and on the eastern flank of the Itombwe south of Bukavu.

In regions of higher rainfall – 1,800 to 2,500 millimetres per year on average –, the summits and upper slopes are often covered with dense forests, while the lower slopes and steep slopes are often covered with very sparse forests (FIGURE 3.8). In these areas, deep soils are often gorged with water during heavy rains, becoming unstable and preventing large trees from remaining in place. Characteristic species of these very open formations, *Entandrophragma excelsum* and some species of genus *Cassipourea*, all have pivot roots that allow a deep anchorage. The distribution of these forest formations can be seen in Nyungwe forest in Rwanda, in the northern part of Kibira National Park in Burundi (FIGURE 3.9) and on the western flank of the Itombwe and Kahuzi-Biega massifs.

Not only does the arrangement of forest formations differ from one climatic region to another, but some trees can also have variable ecological preferences. In drier regions, *Newtonia buchananii*, for example, grows in galleries along streams (Figures 3.6 and 3.7), but in well watered montane forests it grows on ridges. In the case of other species, notably *Symphonia globulifera*, *Podocarpus latifolius*, *Ilex mitis* and *Xymalos monospora*, this difference in behaviour is even more marked: at high altitude they often grow on well drained slopes; at low altitude, they are found in hollows or even in flooded areas.

In the absence of well marked features, drainage can become the most important edaphic factor. When it is insufficient it provokes flooding. In function of the amplitude and duration of these floodings we separate inundatable forests from swamp forests.

Periodically inundated forests can undergo strong fluctuations of the water level: at times one needs a dugout to move about in them, at times they are dry (PAGES 124-125). In swamp forests, on the contrary, variations in the water level are very slight. The water is always just above or just below ground level and the soil is continually waterlogged. These forests are richer than periodically inundated forests, but their trees are generally of a lesser size. In both types of forest there are species with stilt roots, such as *Pandanus* and some

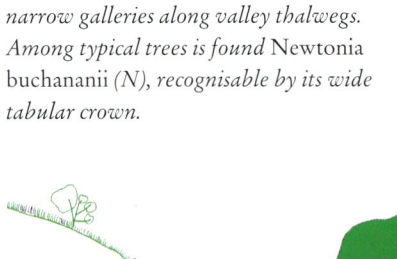

3.6. In low-rainfall regions of the Albertine Rift, the forest is often reduced to narrow galleries along valley thalwegs. Among typical trees is found Newtonia buchananii *(N), recognisable by its wide tabular crown.*

3.7. In this valley, at around 1,800 metres altitude in southern Burundi, dense forest galleries are limited to the thalweg. The dry forests of the hills have long ago been replaced by grass lands and fern heaths. Some think that this landscape was created by Man. The great botanical richness of the herbaceous formations, and notably the fact that there are numerous plant species that are found again on the Nyika plateau in Malawi, suggest that they could be very ancient, however.

Uapaca, and in swamp forests there are also trees which develop pneumatophores, notably *Symphonia globulifera* and species of genus *Hallea (Mitragyna)* (PAGE 148).

As to riverine forests, they line water courses and penetrate the savanna zone in some places as much as several hundred kilometres from the main forest massif. They may be inundated during short periods of high-water, but this is not indispensable to their existence. In fact, they live mainly on ground water coming from the stream, from which, by definition, they are never very far.

Finally, between the land and the ocean there exists a type of forest that is at once very special and very specialised: the mangrove or halophile forest (PAGES 138-145). It develops in shallow, sheltered areas of coasts exposed to the tides in tropical regions, essentially between 30° N and 30° S. They occur in pockets all along the coasts of Central Africa, but the most extensive formations are found in Gabon (in the Ogooué delta around Port-Gentil, in the Komo estuary and in Corisco Bay), in Equatorial Guinea (in the Rio Muni estuary) and in Cameroon (in the Douala region and in the Bay of Rio del Rey at the foot of Mount Cameroon).

All the swamp and inundatable forests of the continental wetlands, as well as the mangroves, are very dynamic formations whose multiple facies are often no more than phases of a succession. We shall discuss them in detail in the chapter treating the temporal gradients (PAGES 118-145).

3.8. In high-rainfall mountainous regions, dense formations are limited to the top of slopes and summits, while the lower part of the slope is often occupied by a very open forest with tree ferns (Cy) along the thalweg. The only large trees that manage to stay upright in this habitat are Entandrophragma excelsum *(E) and a few* Cassipourea *(Ca) with pivot roots. They are sometimes accompanied by rapid-growth species such as* Albizia gummifera *(A),* Neoboutonia macrocalyx *or* Dombeya goutzenii, *but they fall over when they become too large. The dense forest of the hilltops is characterised by the presence of* Parinari excelsum *(P) and* Newtonia buchananii.

3.9. A good example of such a forest is found in this small valley of Kibira National Park in Burundi. The sides of the thalweg are covered by a dense tangle of the liana Sericostachys scandens (tomentosa) *and Bracken* Pteridium aquilinum. *In the centre of the picture is an African Banana Tree* Ensete ventricosa. *This habitat could recall a degradation of anthropic origin. In reality, it has completely characteristic species such as the Strange Weaver* Ploceus alienus, *an endemic of the Albertine Rift.*

3 ECOLOGICAL GRADIENTS

3.10. Vast expanses of swamp or inundatable forests occupy the entire centre of the Congolese basin (0). All around, the terra firme forests are divided into ten main formations:
1. Biafran coastal forests of Cameroon,
2. Biafran coastal forests of Equatorial Guinea and Gabon,
3. Caesalpiniaceae forests
4. Mixed or non-differentiated forests of the Congolese basin,
5. Cameroono-Congolese semideciduous forests,
6. Gabono-Congolese semideciduous forests,
7. Submontane forests,
8. Montane forests,
m. Marantaceae forests,
G. Gilbertiodendron dewevrei forests,

The mature terra firme forests

Not only do the gradients that we have just evoked influence each other and are not easily dissociated in nature, but within the three-dimensional continuum that they create, each species occupies a place of its own. In studying a vast transect stretching from the coast of Liberia to the western part of the Ivory Coast, Renaat Van Rompaey was able to show that no two species have exactly the same distribution. Recognition of well differentiated plant associations, as is done in temperate habitats is therefore impossible in tropical forests.

Working with gradients is, however, not easy, yet it is necessary to give a name to the forests that are studied. Nevertheless, for purely practical reasons, biologists and foresters try to classify them by their dominant species or characteristics, knowing that this classification is based on more or less arbitrary choices. Some researchers thus attach a primordial importance just to the humidity gradient. Others take into account the physiognomy and structure of the vegetation. Still others try to take into account physiognomic, structural, floristic, dynamic, historical and ecological criteria all at once.

This difficulty is of course not limited to forests. It is just as real for a number of other formations. In 1956, at the Yangambi conference, a first attempt was made to define a practical and generally accepted terminology for all types of sub-Saharan African vegetation. Unfortunately, this terminology uses words such as "steppe" which are not truly appropriate in sub-Saharan Africa, and it only considers forest formations in very general terms. With time, it has therefore undergone numerous amendments. Finally, in 1983, UNESCO published a general classification of the vegetation of Africa. Prepared by Frank White, and largely based on the concepts proposed by Greenway since 1943, this publication has become an inescapable reference. Unfortunately, it does not always go sufficiently into detail for the forest formations and covers the African continent very unequally.

A need was felt, especially by ecologists and conservationists, to have a classification not only of the vegetation, but also of habitats in general, that was at once more detailed and more exhaustive. Thus was created, at the demand of the Council of Europe, a classification of the habitats of the world, proposed by Pierre Devillers and Jean Devillers-Terschuren and anchored in an ensemble of works of world significance. A part of the volume treating Africa, largely based on the works of Knapp, White and Schnell, is in preparation, notably in the framework of the Regional Environmental Information Management Programme (REIMP).

Among the mature terra firme forests of Africa, six main groups can be recognised.

Lowland evergreen forests

Lowland evergreen forests cover vast areas in the most humid regions of Central Africa, especially in the zone bordering the Gulf of Guinea, the centre and eastern parts of the Congo basin (FIGURE 3.10). By their physiognomy and their structure, these forests represent the very type of the dense humid tropical forest with its numerous buttressed trees, its lianas, its epiphytes and its high, complex canopy. Nevertheless, they include very diverse formations. Depending on the soil, they may comprise rich and complex mixed formations or generally simpler and poorer formations dominated by a single species; depending on the climate, they also separate into humid and hyperhumid formations.

Firstly, amongst the mixed formations are the Biafran coastal forests (Figures 3.11-13). They occupy the coastal plains immediately bordering the Gulf of Guinea, from the Nigerian border to Gabon (FIGURE 3.10), and grow on generally poor soils, mainly sands or sandstones. In Cameroon and in the north of Equatorial Guinea, under an annual rainfall of 2,500 to 3,500 millimetres, they are characterised by an abundance of Azobé *Lophira alata* and Ozouga *Sacoglottis gabonensis*. The Azobé occupies elevated, well-drained, ground while the Ozouga thrives in valleys. The latter is the only African representative of the Humiriaceae, a family very well represented in tropical America. In the forests south of Equatorial Guinea and Gabon, under a slightly lesser annual rainfall, the Azobé is replaced by Okoumé *Aucoumea klaineana*, a tree of the Burseraceae family, a near endemic of Gabon. Among the abundant species are also *Dacryodes buettneri, Desbordesia glaucescens, Calpocalyx heitzii, Scytopetalum klaineanum, Testulea gabonensis, Didelotia letouzei, Irvingia gabonensis* and *Santiria (Santiriopsis) trimera*. The beautiful *Costus violaceus* is common in the understorey of these coastal forests (FIGURE 3.13).

Within the coastal forests, mainly of the humid type, there are enclaves of hyperhumid forests which are subject to a rainfall of at least 3,000 millimetres per year on average, often even more than 4,000 millimetres. These enclaves are located in north-western Gabon, notably around Mondah Bay and near Coco-Beach, at a few points in Equatorial Guinea and in Cameroon. They are distinguished by the presence of some particular species of trees, by their richness in epiphytes, notably orchids, and by the presence of hanging mosses (PAGE 157).

Away from the coasts lie the Biafran and Gabonese Caesalpinaceae forests (FIGURE 3.14). They form a more or less wide band spread equally from the Nigerian border to Gabon (FIGURE 3.10), but they are also found on the isolated hills that emerge from the coastal plain, such as the Doudou mountains, for example. the Caesalpinaceae are represented by numerous genera – *Berlinia* (FIGURE 3.15), *Brachystegia, Gilbertiodendron, Gilletiodendron, Microberlinia, Prioria (Oxystigma), Cynometra, Julbernardia, Copaifera, Didelotia, Aphanocalyx, Bikinia, Tetraberlinia, Stachyothyrsus, Hymenostegia* – the species of which are not always easy to tell apart. Mimosaceae are also well represented by genera *Newtonia* and *Albizia*. Among other families can be mentioned, in particular, the Annonaceae, Anacardiaceae, Burseraceae, Flacourtiaceae, Icacinaceae, Myristicaceae – with Ilomba *Pycnanthus angolensis* –, Orchidaceae, Olacaceae, Rubiaceae. Afan *Panda oleosa*, only representative of the Pandanaceae, is characteristic, as is the palm *Raphia regalis,* typified by the complete absence of a trunk, and very common in slope forests of the Crystal Montains.

The composition of these forests varies, of course, from one place to another and numerous variants have been described. As in the coastal forests, there are

3.11. Lowland evergreen forest in the proximity of the Nyonié tourist camp, just north of Wonga-Wongué Presidental Reserve, Gabon, is clearly dominated by Okoumé Aucoumea klaineana.

3.12. Cola flavolutina, of the Sterculilaccac family, is typical and abundant in the understorey of the coastal forests of Gabon.

3.13. Costus violaceus, of the Zingiberaceae family, is also abundant in very humid coastal forests of Gabon where it flowers mainly in the rainy season.

3.14. Many Caesalpiniaceae species, like the one seen in the Crystal Mountains in Gabon, have young folliage that is yellow, orange-yellow or red.

3.15. Inflorescence of Berlinia grandibracteata, a Caesalpiniaceae common along watercourses in forest.

hyperhumid enclaves. One of the most important of these is found in the area of Korup National Park in Cameroon. There, under an annual rainfall of 4,000 to 6,000 millimetres, exists a very rich *Tetraberlinia-Didelotia* forest very similar to the most humid forests of Liberia.

Mixed Equato-Guinean and Gabonese forests form a third group. They develop in the interior of Equatorial Guinea and the centre of Gabon, under an annual rainfall of 2,000 to 3,000 millimetres, and mainly north of the Ogooué. Among dominant species there are Okoumé, Ozigo *Dacryodes buttnerii*, Ebab *Santiria trimera*, Coula *Coula edulis*, as well as members of the Irvingiaceae including Alep *Desbordesia glaucescens* or Eveuss *Klainedoxa gabonensis*, Myristicaceae species, such as *Coelocaryon klainei*, Niové *Staudtia stipitata* or Sorga *Scyphocephalium ochocoa* and Ojikuna *Scytopetalum klaineanum* of the Scytopetalaceae family.

Lastly, in the Congo Basin itself, most of the mixed terra firme forests are composed of a mosaic of evergreen formations and more or less deciduous formations. The first occupy mainly low areas, the second ridges. Altogether, they are fairly rich forests, well represented in the eastern part of the Congo basin where they can have occasionally around 300 woody species per hectare. In many places, they are characterised by the presence of Divida *Scorodophloeus zenkeri* and Tchitola *Prioria (Oxystigma) oxyphylla*, two Caesalpiniaceae species. Elsewhere, notably in the Ituri, they are dominated by *Julbernardia serretii*, another species of Caesalpiniaceae. Other typical species are *Cola griseiflora*, Otounga *Greenwayodendron (Polyalthia) suavolens*, *Pericopsis elata*, Tola *Prioria (Gossweilerodendron) balsamiferum*, Afan, as well as various species of genera *Entandrophragma*, *Guarea* and *Celtis*. In the undergrowth *Scaphopetalum thonneri* and *Alchornea floribunda* are abundant. These are very dense forests with a canopy 30 to 40 metres high. After heavy rains, they exhale a very strong oder of garlic, produced by the leaves and branches of *Scorodophloeus*.

In addition to mixed formations, Central Africa is also occupied by vast expanses of monospecific Caesalpiniaceae forests. The most important are the Limbali, *Gilbertiodendron dewevrei*, forests (Figures 3.17 and 3.18). Limbali forms nearly pure associations that cover thousands of square kilometres all along the northern margin of the Congo basin, from Cameroon and northeastern Gabon north of the Congo River and the entire northern and eastern parts of the Democratic Republic of the Congo. They are at the foot of the Albertine Rift, in the Okapis Reserve, and in the lower part of Kahuzi-Biega National Park.

This formation was studied in detail by Teresa Hart. Limbali – *Mbau* in the local language – constitutes at least 75% of trees more the 20 centimetres in diameter.

3 ECOLOGICAL GRADIENTS

Because it tolerates shade well, it regenerates under its own cover. In the vast forests of the Ituri, this type of forest alternates with more varied formations, dominated, however, by Alumbi *Julbernardia serretii*, with no clear correlation between edaphic factors and the presence of one or the other formation. This alternation is all the more intriguing since the large amount of charcoal that is found in the soil of these forests shows that, before the passage of fires, 2,000 to 4,000 years ago, Limbali was practically absent. According to Teresa Hart, the two types of formations represent mature forests that are in competition and that share the space in function of historical events. Alumbi forests would be more dynamic and more apt to colonise lost space, but they would not be able to resist invasion by Limbali, a tree of slower growth, greater longevity and higher invasive potential.

On the plateaux, *Gilbertiodendron ogoouensis* forests and *Brachystegia laurentii* forests are also found, but these formations generally do not cover more than small areas. In the eastern part of the Democratic Republic of the Congo and in western Uganda, between 600 and 1,200 metres above sea level, there are forests of *Cynometra alexandri*, Uganda Ironwood. As the name suggests, its wood, very heavy and hard, is almost unexploitable. Its forests have been described in detail by Eggeling. This Caesalpiniaceae tree is also found in association with other species, notably in some

3.16. *Most of the lowland forests of the Congo Basin are characterised by* Scorodophoeus zenkeri, *of the Caesalpiniaceae family, but they are also rich in many other species:* Polyalthia (Greenwayodendron) suavolens *of the Annonaceae family*, Entandrophragma utile *of the Meliaceae family, and* Pericopsis (Afrormosia) elata *of the Fabaceae family.* Cola griseiflora *is common in the mid-stratum. In the understorey* Scaphopetalum thonneri, *of the Sterculiaceae, and* Alchornea floribunda, *of the Euphorbiaceae, are abundant.*

3.17. *In a* Gilbertiodendron dewevrei *forest about 80% of the plant biomassa belong to that single species of the Caesalpiniaceae family. The canopy is 40 to 50 metres high and complex. The mid-stratum and understorey can be quite open, but most of the young trees belong to* Gilbertiodendron.

79

3.18. São Tomé's low-altitude forests have been severely degraded since the 16th century to make way for extensive cultivation. Only large trees were conserved to provide shade for crops. Since the fields were abandoned, the forests came back but they are rich in introduced species.

parts of the Budongo Forest, where it is associated with Meliaceae of genera *Khaya* and *Entandrophragma*. The trees can become gigantic and develop enormous very thin buttresses. In Semuliki National Park and in Maramagambo Forest in Queen Elizabeth National Park, it constitutes more than 80% of the canopy on vast stretches, but large specimens are rare.

Lianas and epiphytes are rare in all the monospecific formations and there are few accompanying species. The lower strata of the canopy are little developed and the understorey is generally fairly open. Only immense fallen trees provide some diversity in this type of forest.

As for the forests of the Gulf of Guinea, they have largely been replaced, first by sugar cane, more recently by cultivations, and plantations of cacao and coffee. In extreme southern Bioko, difficult of access, there nevertheless remains a hyperhumid forest under an annual rainfall of 11,000 to 14,000 millimetres. Rich in lianas and epiphytes, it is characterised by the presence of *Trichilia heudelotii*, *Carapa procera*, *Pycnanthus angolensis*, *Staudtia gabonensis*, *Sarcocephalus esculentus* and species of genus *Eugenia*. In the south-eastern part of the island, under a climate a little less humid, survive several patches of a forest with *Ceiba pentandra*, *Staudtia gabonensis*, *Pycnanthus angolensis* and *Milicia excelsa*. In the south of São Tomé and Principe also survive patches of forest between 0 and 800 metres altitude with *Milicia excelsa*, *Pycnanthus angolensis*, *Ceiba pentandra* and various *Ficus* species (FIGURE 3.18).

Lowland semideciduous forests

All around the core of evergreen forests extends a zone of semideciduous forests (FIGURE 3.10).

In the south, the Gabono-Congolese forests extend from southern Gabon, Kouilou and the Mayombe, in the west, to the Congo basin and notably the region of Kindu, in the east. They are forests rich in Meliaceae of genera *Entandrophragma* and *Khaya*. Among characteristic species are found, notably, *Pouteria* (*Aningeria*) *superba*, the Onzabili *Antrocaryon klaineanum*, *Baillonella toxisperma* (in the west only), *Dacryodes pubescens*, *Newtonia leucocarpa*, *Vitex ciliata*, *Vitex pachyphylla* and, among the Caesalpiniaceae, Agba *Gossweilerodendron balsamiferum*, Anzem *Copaifera religiosa* and Tchitola *Priora oxyphylla* (previously *Oxystigma oxyphyllum*). Limba *Terminalia superba* is locally common.

In the north lie the Cameroono-Congolese forests. From the Nigerian border and Cameroon, they form an enormous band that crosses all of northern Congo, the south-western part of the Central African Republic – they are found in the Ngotto region – and the north of the Democratic Republic of the Congo and reach north-western Uganda (FIGURE 3.19). These forests are rich in Meliaceae of genera *Entandrophragma* (Sipo, Sapelli and Kosipo *E. candollei* of the foresters), *Guarea*, *Khaya*, *Lovoa*; and Sterculiaceae of genera *Cola*, *Sterculia*, *Mansonia*, *Nesogordonia* and *Triplochiton* (Ayous) of Caesalpiniaceae genera *Brachystegia*, *Distemonanthus* and *Prioria* (including former *Gossweilerodendron* and *Oxystigma*); of Mimosaceae genera *Cylicodiscus* and *Piptadeniastrum*. Limba *Terminalia superba* of the Combretaceae family is encountered again and the Annonaceae and Ebenaceae are well represented in the middle stratum.

These semideciduous forests are often very tall, with a canopy that is 30 to 40 metres above the ground, with emergents reaching 60 metres. Lianas and epiphytes are much less abundant here than in the evergreen forests, sometimes even essentially absent. The percentage of trees with acuminate leaves diminishes. That of trees with small coriaceous leaves – species called microphyl-

3.19. The further we are away from the equator and its high rainfall areas, the more the number of trees losing all their leaves at some moment of the year increases. The transition between evergreen forests and semi-deciduous forests is very gradual, however, and nobody can decide where is an objective limit between these two forest types.

3.20. The liana Combretum racemosum *is often abundant in the mantle and upper canopy of semi-deciduous forests.*

3.21. Dry forests of Olea europaea *and* Euphorbia dawei *extend over the hills bordering the Akagera Valley in Tanzania and Rwanda. In the forest edge may be found the large* Aloe volkensi *(below on the right) which can reach six metres in height. Farther east, in Tanzania, dry forests still cover entire hills and form a nearly continuous massif with an area of some 25,000 hectares.*

3.22. Aerangis kotschyana, *a beautiful white, delicately perfumed, orchid with a long spur is abundant in these dry forests.*

lous – increases without reaching the proportions observed in sclerophyllous forests. The total number of species diminishes. Well to the north of the main forest block, in the Central African Republic, there are remnants of very dry forest. The dominant tree is *Anogeissus leiocarpus*, of the Combretaceae family, with a dense foliage of small leaves.

Sclerophyllous forests

In Uganda, the Democratic Republic of the Congo and Rwanda, sclerophyllous forests are still found in the region of the Great Lakes and the Albertine Rift. However, the most important massif of these forests, covering an area of the order of 25,000 hectares, is found in north-western Tanzania.

In Queen Elizabeth National Park, in the western part of Maramagambo Forest, is found a forest of *Euclea et Cynometra alexandri*. It is located at about 900 metres altitude, on the edge of the Lake Edward plain. Towards the east, it gradually blends into tall nearly monospecific *Cynometra* forests, which, in their turn, melt into submontane and montane forests of *Parinari excelsum*.

In Burundi, on the edge of the Rusizi plain, there were fairly similar forests where *Cynometra alexandri* was associated with *Strychnos potatorum* and *Euphorbia dawei*. Despite being included in Rusizi National Park, the last patches disappeared during the 1980's.

In Akagera National Park, in the crater area of Queen Elizabeth National Park and in the National Park of the Virungas, under an average annual rainfall of 600 to 800 millimetres and installed on shallow, rocky soils of steep slopes and crests, is a forest of *Olea europaea* and *Euphorbia dawei*. In Akagera National Park (FIGURE 3.21), the driest formations, which take on the appearance of thickets, are rich in *Haplocoelum galaense*, while *Sanseviera cylindrica* and *Aloe macrosyphon*, two succulent plants, are abundant in the understorey. Still in Akagera National Park, but on deeper soils along the lakes, *Strychnos usambarensis* becomes the dominant species and *Sanseviera parva* replaces *Sanseviera cylindrica*. In spite of the lesser rainfall, these dry forests are rich in epiphytes. Not only are many lichens of genus *Usnea* found there, but there are also many species of orchids, including *Cyrtorchis* and the very beautiful *Aerangis kotschyana* whose white, very perfumed flowers are decorated with a long spur (FIGURE 3.22).

These dry forests are at present very fragmented and often reduced to hardly recognisable relicts. Surrounded by savannas or cultivation, they are completely isolated from the great humid forests of the Albertine Rift and no one would think that they are part of the great forest biome. A detailed study of the remaining massifs shows that there existed – only a short time ago – a perfect continuity between the driest formations and the humid forests. Between the humid forests and the sclerophyllous formations, stretched a great forest belt of *Pterigota mildbraedii*, a huge deciduous tree of the Sterculiaceae. Only isolated trees still survive here and there in the centre of Rwanda and in Burundi. In a sense, the only transition that survives intact, in Queen Elizabeth National Park, between the *Euclea* thickets and the montane forests of the Albertine rift, somewhat represents a unique natural monument.

Coastal sclerophyllous forests

A narrow, more or less discontinuous band of very characteristic forests stretches along the Atlantic coast and forms the transition between pioneer formations of the beaches (PAGE 137) and evergreen coastal forests. They are characterised notably by *Manilkara lacera* (FIGURE 3.23), *Chrysobalanus orbicularis* and *Syzygium littorale*. These species all have more or less coriaceous leaves resistant to the salty wind from the ocean. Quite common also are *Hibiscus tiliaceaceus* (FIGURE 3.26), *Barteria nigritana* (FIGURE 3.25) and *Fegimanra africana* (FIGURE 3.24). Most of these species have a wide distribution along the tropical Atlantic coast.

Submontane forests

At medium altitude, on the mountains bordering the Gulf of Guinea, grow submontane forests. They are particularly well developed on Mount Cameroon, some high mountains farther north, and on the islands of Bioko, Principe and São Tomé. We find them again on Monte Alén in Equatorial Guinea and on the Crystal mountains, the Doudou mountains and the Chaillu massif in Gabon. Some isolated patches survive on hill tops near Yaoundé.

Usually submontane forests cover summits and high ridges above the altitude of 600 to 800 metres, but above 1,600 to 1,800 metres they give way to montane forests. Their total area has been strongly reduced by Man, and

3.23. The coastal sclerophyll forest just south of Pointe Ngombé, next to Libreville in Gabon. The twisted trees are Manilkara lacera *of the Sapotaceae family.*

3.24. Fegimanra africana *is often abundant in these coastal forests. This tree of the Anacardiaceae family, is a close relative of the Common Mango Tree* Mangifera indica, *and is easily seen in June or July when it is covered with orange or red new leaves.*

3.25. Barteria nigritana *is a small tree of the Passifloraceae family, also common in the mantle of coastal forests.*

3.26. Hibiscus tiliaceus, *a small shrub of the Malvaceae family, is one of the pioneers that grows on sand dunes. It is also found in the mantle of the coastal sclerophyll forests.*

3 ECOLOGICAL GRADIENTS

3.27. In the submontane forests of the Albertine Rift region trees are often loaded with epiphytes.

many remnants are badly disturbed. Except for the studies of Letouzey, and more recently the work of Gaston Achoundoung on Mount Nta Ali and on the hills around Yaoundé, little is known about these forests.

Their aspect is characterised by the fact that most of the trees are small and heavily loaded with numerous epiphytes: mosses, ferns, Gesneriaceae, Piperaceae, Begoniaceae and Orchidaceae (FIGURE 3.27).

Floristically, they harbour some species which are unknown at lower altitude: many Begoniaceae (Figures 3.28-3.31), and in the Lope National Park in Gabon *Ocotea gabonensis*. Locally they harbour also some species which have definitely a montane distribution, like *Prunus africanus*. Most of their trees, however, exist in lowland forests, often very sparsely, and only become dominant at higher altitude. On Mount Cameroon and Mount Kupe *Cola* and *Garcinia* are most abundant. On Mount Nta Ali also the forest is dominated by *Garcinia*. On the hills around Yaounde, it is *Lasiodiscus fasciculiflorus* and *Garcinia lucida*. This last species is also common on hills in Gabon. According to Letouzey, these submontane forests are mainly *Cola* and Clusiceae forest. Etom *Syzygium standtii* can be very abundant, however. Since they develop on shallow soils, their trees are often small and, as they are frequently enveloped in clouds, they are rich in epiphytes. Some species, such as *Ocotea gabonensis*, do not occur in neighbouring forests at lower levels and Begoniaceae species are abundant (Figures 3.28-31).

In the Albertine Rift region, submontane forests are located between 1,000 or 1,200 and 1,650 metres above sea level. They still cover important areas in the east of the Democratic Republic of the Congo and, in Uganda, they constitute the most common type of forest. They are well conserved in Kibale National Park. Everywhere, however, they suffer strong pressure from farmers who convert the lands at great speed. In Rwanda and Burundi, these forests have, in this way, totally disappeared and their situation is alarming in the eastern part of the Democratic Republic of the Congo.

By their structure, these forests much resemble the low altitude forests, but tree height and species diversity are less. Epiphytes can be abundant and, as in the lowland forests, they are represented by ferns and orchids. Many species are the same as in low altitude forests, but some montane elements appear. Some species, such as the Mountain Umbrella Tree *Musanga leo-errerae*, *Carapa procera*, *Lebrunia bushaie*, *Pentadesma lebrunii* or *Ocotea michelsonii*, occur only in the submontane forests, but they are not really numerous. Among birds, also, some species are linked to this transition habitat, in particular the birds, Chapin's Flycatcher Babbler *Kupeornis chapini* and the Yellow-crested Helmet Shrike *Prionops alberti* are typical.

In Uganda, these forests divide up into evergreen and semideciduous formations, depending on rainfall. The evergreen forests are very similar to low altitude forests. They are rich forests with many lianas and some trees develop spectacular buttresses. In the Sese Islands, they are *Piptadeniastrum* and *Uapaca guineensis* forests. On the shores of Lake Victoria, between Mengo and Masaka, they are *Piptadeniastrum*, *Albizia* and *Celtis* forests. Farther west, in the region of the rift, notably in Kalinzu forest, they are *Parinari excelsa* forests, found again at higher altitude.

Among the semideciduous forests, *Celtis* and *Chrysophyllum* forests are encountered north of Lake Victoria, notably in the Mabira forest, crossed by the Kampala-Jinja road; forests of *Olea capensis* var. *welwitschii*, *Pouteria* (*Aningeria*) *altissima*, *Newtonia*

3.28-31. The Begoniaceae are essentially herbaceous plants, often more or less fleshy. They grow in the undergrowth of humid forests and some species are epiphytes. Their flowers – red, orange, pink, yellow or white – are unisexual. The male flowers have coloured sepals that resemble petals, the female flowers are mounted on a large angular or winged ovary. Most of the species require humidity and coolness. Thus, among the 80 known species of Central Africa, most inhabit submontane or montane forests. Begonia subaccata (3.28) is endemic to São Tomé. The three other species (3.29-3.31) grow in the Crystal Mountains in Gabon.

buchananii, Diospyros abyssinica, Celtis and Chrysophyllum in the central part of Kibale National Park and forests dominated by a Pterygota, a species of the Sterculiaceae family, in the southern part of the same national park. In other words, this national park shows a perfect transition between the humid forests of the north and the much drier forests of the south.

3.32. In the understorey of the montane forest, at around 2,300 metres altitude, in Kibira National Park in Burundi, Mimulopsis of the Acanthaceae family is often the dominant plant.

Montane forests

In the region of the Albertine Rift, above 1,600 to 1,700 metres altitude, montane forests constitute most of the forested areas within the national parks of Bwindi, on the Volcans, on the Virunga, on the Rwenzori, on the Kibira (FIGURE 3.32) and the upper Kahuzi-Biega. They also form the forests of Nyungwe and Bururi. In Cameroon, they are found on Mount Cameroon and the high formations of the north-west where forests survive on, notably, the Rumpi Hills, Idjim-Kilum, Mount Kupe, Nlonako and even Tchabal Mbabo. Lastly, they are found on the Isle of Bioko and São Tomé.

Trends appearing in the transition forest are accentuated in these communities. Species richness continues to diminish, trees dwindle in size, buttresses disappear and the canopy progressively reduces to a single stratum. Up to about 2,000 metres, however, a main stratum is still distinguishable, with a few emergents and a middle stratum. Above 2,400 metres, only one upper stratum remains, with a lower stratum. Nevertheless, these high-altitude forests are more than a simple impoverished version of forests below, for their species belong mostly to a different flora, related to the flora of other mountains on the continent and often with no parental links with the lowland forest flora.

Epiphytes are often very abundant and include many mosses and lichens. They are omnipresent and, in the fog forests, cover everything from the ground to the finest twigs above. Contrary to the lowland forests, mosses even cover the soil. This exuberance comes from the frequent fogs that graze the crowns of the trees, especially at night or after rain. The higher the altitude at which a forest is located, the more apparent the effect of the fog becomes. In some privileged places, such as interior valleys, fogs can even stagnate for hours or days. Fine droplets of water cling to the leaves and bark, allowing mosses and lichens to grow. Soon others follow, each in their turn considerably increasing the surface of entrapment and therefore the forest's power of water absorption.

Montane forests vary considerably according to soil conditions, rainfall and drainage. At high altitude, species diversity diminishes so much that finally formations are dominated by one, two or three species only. This simplifies their structure and facilitates their designation and identification in the field.

In regions of high rainfall of the lower part of montane forests – to around 2,100 or 2,200 metres –, *Parinari excelsa*, already encountered in the submontane forests, is often the dominant species. This large tree of the Chrysobalanaceae family has a dense coppery green foliage and from a distance its crown resembles a gigantic broccoli. It is abundant in the eastern part of the Democratic Republic of the Congo, in the lower parts of the Bwindi forest in Uganda and mostly in the Nyungwe forest in Rwanda. It is often accompanied by *Newtonia buchananii*, *Entandrophragma excelsum* – usually limited to the most open formations of interior valleys – and *Carapa grandiflora*, a smaller tree of the Meliaceae family recognisable by its bouquets of young bright red leaves. The wood of *Parinari*, rich in silica, is difficult to exploit. It is nevertheless used to reinforce the galleries of copper mines of Kilembe.

The *Parinari* also constitutes a "mystery". It is abundant in the entire Great Lakes region, but young specimens are nowhere to be found, not even in the fallen trees. This means that when the mature trees of today disappear – by age or exploitation – they will inevitably be replaced by other species. Therefore, these are not true forests. According to Dale, an English botanist, they were formed 400 to 600 years ago when climatic conditions were different than they are today. Thus the *Parinari* forests are rather relict forests, witnesses to a bygone epoch.

As we have seen previously, dense *Parinari excelsa*

forests, often accompanied by *Newtonia buchanani*, are concentrated on hilltops and upper slopes, while valley bottoms are often occupied by more open formations (PAGES 74-76). Some botanists or foresters look here for the signature of an ancient human presence. It is true than in the dry season these clearings are sometimes burned by honey gatherers, but they are a natural element and are even characteristic of these very humid montane forests. They constitute the habitat of predilection for the wild banana *Ensete ventricosa* and the tree fern *Cyathea manniana* (FIGURE 3.39). They often harbour, especially along the thalweg, thickets of *Mimulopsis arborescens*, a large Acanthaceae that flowers gregariously at six- or seven-year intervals and then dies after flowering. Lastly, they are the exclusive habitat of some bird species, including ones endemic to the region such as the Strange Weaver *Ploceus alienus*, which never penetrates closed canopy forests but is also not encountered away from the forest massif.

Contrary to the low-altitude forests, montane forests are often a mosaic of high, dense canopies and more open areas punctuated by isolated trees – this phenomenon is striking in Figure 4.35 (PAGE 116). Some plants even keep these clearings from closing. Bracken *Pteridium aquilinum*, often very abundant in the clearings, apparently emits toxic substances by its roots that limit the development of other plant species, while the liana *Sericostachys scandens (tomentosa)*, of the Amaranthaceae family, covers everything with its dense foliage and prevents young trees from developing. Even though omnipresent it is little noticed at first glance since its white flowers appear only at long intervals, at times as much as 10 or 20 years. According to local beliefs in the Great Lakes region, its flowering announces war or famine. At the end of the 1980s, the liana flowered for three consecutive years in Rwanda and Burundi.

In less humid regions, up to 2,400 or even 2,500 metres altitude, *Olea capensis* forests are encountered. This olive tree is generally accompanied by *Chrysophyllum gorongosanum*, *Ilex mitis* and *Syzygium guineensis* var. *parvifolia*.

On the steepest ridges and slopes, where the soil is usually very shallow, sometimes even stony, in any case well drained, forests of *Faurea* and *Ficalhoa* form. The trees here are of smaller size. They are no more than 15 metres high and drought-resistant species are abundant. These are *Ficalhoa laurifolia*, *Faurea saligna* and *Dichaetanthera corymbosa*, a beautiful Melastomataceae. This forest is vulnerable to fire and in many places it has already been replaced by fern heaths or sparse meadows of *Eragrostis*.

With the juniper *Juniperus procera*, that grows only in East Africa, the podocarps represent the only native tropical African conifers. They are strictly montane.

Mosses and liverworts

Mosses (FIGURE 3.33) and liverworts (FIGURE 3.34) represent two very ancient groups of non-vascular plants, in other words, without cells specialised for the transport of water and nutrients, and therefore close to the very first forms of plants that colonised dry land. Their systematics are less well known than those of vascular plants and many species remain undescribed, but there are in the world, at present, around 10,000 species of mosses and 6,000 species of liverworts.

They are good colonists, often the first to colonise new substrates. They are found at every latitude and longitude, but the humid tropical forests possess more species than all the other habitats in the world. Once again, however, Africa is poorer and has fewer species of its own than America or tropical Asia. It has only a single endemic family and shares species as much with South America as with tropical Asia. As in tropical Asia, the epiphyllic species (PAGE 157) are much more diversified than in South America.

The immense majority of mosses and liverworts of tropical forests are epiphytes and some species develop forms totally unknown or very rare in temperate regions, such as hanging mosses (PAGE 157), bracket mosses and feather mosses. Montane forests are also much richer than lowland forests and around 60% of the families represented in the tropics are even exclusively montane while only 5% are limited to lowland forests.

Mosses and liverworts therefore require both moisture and cool temperatures. But, while the lowland forest species belong in general to groups that occur in all the tropical regions of the world, those of montane forests are often more restricted. Among them are found even very ancient relict elements of Gondwanaland or Laurasian origin.

Like the Gymnosperms, the Bryophytes represent therefore a group of "archaic" plants that, in tropical regions, "survive" mainly in the montane forest. Most present-day species are nevertheless very modern and, according to Gradstein and Pocs, they have evolved at the same time as the tropical humid forests were born, as were the orchids and other epiphytes.

The Pteridophytes: ferns, horsetails, lycopods and selaginellas

These vascular plants comprise an ensemble of groups that reproduce by spores. They were very diverse in the Carboniferous, 300 million years ago, and some species were 40 metres high. They suffered a sharp decline at the end of the Permian, but still today, around 12,000 species of ferns and 1,000 species of lycopods are known. Some Pteridophytes are very old, like the Lycopodiaceae (FIGURE 3.35), the Selaginellaceae (FIGURE 3.36), the Gleicheniaceae (PAGE 105) or the Marattiaceae (FIGURE 3.40). Others such as *Platycerium* (PAGE 158) represent recent genera.

Many species have a very wide distribution, explained in part by their great age, in part by the fact that their spores, like those of the Bryophytes, can travel over very great distances. The appearance of ferns on very recent volcanic islands is proof of this. Although very old, they are still capable of very rapidly colonising new habitats and keeping a good place in the competition. However, as for many other groups of plants, Africa is poor in ferns. Not only is the number of species less than on other tropical continents, but there are also fewer genera.

Within lowland tropical forests, ferns do not generally constitute a predominant element except in the canopy. On the ground they are often limited to transitional areas along roadsides, in disturbed areas, and on fallen trees (FIGURE 3.37). Especially on forest edges, one encounters climbing forms, such as *Selaginella myosurus* (FIGURE 3.36) and *Lygodium smithianum* (FIGURE 3.38). One species, *Dicranopteris linearis* of the Gleicheniaceae family, is very abundant in Gabon, notably in the Lopé National Park reserve, where it covers vast surfaces on forest edges, on steep slopes and everywhere where the ground has been stripped. (PAGE 105). It is found again in the rift region up to 2,000 metres altitude, but it remains more localized there.

Ferns are abundant and diverse in the cool, humid montane forests. In this environment they occupy all the levels, from the ground to the canopy. Along streams, in the shade of large trees, the large *Marattia fraxinea* is found (FIGURE 3.40). In clearings and well lit places, tree ferns of genus *Cyathea* often form dense masses (FIGURE 3.39). In some places, especially in the mountains, there is also an abundance of Bracken *Pteridium aquilinium*, a cosmopolitan species that plays somewhat the same role as *Dicranopteris*. It grows in clearings, on landslides, and in all the disturbed areas where the dense cover of trees has been perturbed.

3.35. Lycopodium cernuum *of the Gabon coastal forests.*

3.36. Selaginella myosurus, *a climbing species common in the forest edge.*

3.37. A large fern of the genus Nephrolepis, *also in the forest edge.*

3.38. Lygodium smithianum, *a climbing fern, also abundant in forest mantles.*

3.39. Tree ferns are very diversified in Oceania, but Africa harbours only three or four species. Cyathea manniana *is very abundant in all the humid valleys of Nyungwe Forest and of Kibira National Park. The largest specimens reach five and a half metres. In fact, their trunk is only an upright rhizome. This species often settles on the edge of streams, but in more humid areas it also inhabits the base of steep slopes.*

3.40. The large fern Marattia fraxinea *is also found along stream edges in montane forests. Its fronds are nearly as large as those of* Cyathea, *but this species does not have a trunk.*

Genus *Podocarpus* contains several African forms presently arranged among three species: *Podocarpus latifolius* (including *milanjianus*); *Podocarpus gracilior* (including *usambarensis*), limited to eastern and southern Africa, and *Podocarpus mannii*, endemic to São Tomé (PAGE 53). The first, often somewhat twisted and with a globular crown, is moisture loving. It is found in regions of high humidity, in the thalwegs and along the edge of marshes, in general between 1,750 and 3,000 metres altitude. In some regions, it can form nearly pure forests, *Podocarpus* forests. This was the case in the southern part of the Nyungwe forest between 2,500 and 2,700 or 2,800 metres altitude, as well as in the Gishwati forest between 2,600 and 2,900 metres. The second, larger and more elongated, can tolerate drier conditions and usually is a part of mixed forests. It is found at around 2,000-2,500 metres altitude, among the large emergents, and some individuals can attain spectacular dimensions. In the Nyungwe forest, not far from Gisovu, there was one that reached 70 metres, in spite of the fact that its head had been broken off by lightning. Everywhere, these two conifers have suffered much since the beginning of the 20th century from unrestrained exploitation and today essentially no intact *Podocarpus* forests remain.

On the Rwandan slope of Sabyinyo volcano, at around 2600 metres altitude, there is a forest of *Afrocrania volkensi*. It covers no more than a few tens of hectares and is probably unique in Central Africa. This type of forest is, however, much more frequent in eastern Africa, particularly in Tanzania. We mention it here as a matter of curiosity.

Lastly, at high altitudes, *Hagenia abyssinica* forests remain. This tree of the Rosaceae family is encountered

Lichens

In spite of the fact the some forms resemble liverworts, lichens are not plants but mushrooms mostly of the Ascomycetes group – like truffles and morels – which live in symbioses with green algae or cyanobacteria. Some 10,000 species are known to this day but their systematics are still rather poorly known, especially of the forms inhabiting tropical forests. For some families and genera, knowledge dates back practically to the end of the 19th century, an epoch in which present methods of investigation were still unavailable. In other words, there remains much to discover in the in the world of lichens.

In the forest, most of lichens are epiphytes. In tropical lowland forests, they are, moreover, very abundant. They are totally absent from the ground, but fairly frequent on tree trunks (Figures 3.41-43), where they are less abundant than mosses. Most species belong to very discrete types, sometimes hardly differentiable from the bark of trees. It is only on the high branches of the canopy that they are omnipresent and well developed. They are represented by numerous species of which some – the epiphylls – live like some mosses, on the leaves of vascular plants (PAGE 157).

As altitude increases, lichens become more abundant and there are more showy, even exuberant, forms, such as *Usnea* recognisable by their long beards (PAGE 160). Finally, starting at 4,000 metres, there is nothing left but lichens (PAGE 98). Lichens are in competition with mosses. In very humid forests mosses push lichens back to the canopy, in drier forests, lichens push mosses to the ground.

Lichens have a wide distribution. All lichen families and most of the genera are pantropical and it is only at the species level that more restricted forms are found. Since their power of dissemination is very weak, this type of distribution attests to the great age of the group and it is highly probable that most of the lichens that are found today in tropical forests are much older than the plants on which they live.

Lichens of old forests are different from those of young forests. And since their dispersal is extremely slow, recolonisation of a forest degraded by exploitation can take centuries.

at 1,800 to 3,500 metres altitude. In the lower limit of this range, it is nowadays a pioneer species, found on the forest edge, in degraded areas and even in the middle of fields. In the eastern part of Nyungwe forest, between 2,300 and 2,500 metres altitude, true thickets of shrubby *Hagenia* occur that cover vast surfaces along the edge of some bogs and with which *Cliffortia*, another shrubby Rosaceae, may be associated (PAGE 53). On the Rwenzori, *Hagenia* is often accompanied by *Rapanea rhododendroides*, a Myrcinaceae. On the Virunga, it is accompanied instead by *Hypericum revolutum* of the Clusiaceae family, recognisable by its large yellow flowers. The most beautiful *Hagenia* forests occupy the saddle between the Karisimbi and Bisoke volcanoes, on the border between Rwanda and the Democratic Republic of the Congo. They are spectacular formations, apparently very ancient and relict (FIGURE 3.44). They constitute a true living monument to the region of the Albertine Rift. Their understorey, made up of shrubs of *Hypericum*, is very open and in many places transformed into a short meadow, often a little marshy. Today, it is probably essentially buffaloes who maintain these open meadows, but not so long ago, elephants also played an important role. Until the 1960s, these habitats were even visited by cattle raisers and their herds.

Lastly, there remain the high altitude swamp forests. They form in inundated basins or valleys. In several places in Rwanda and Burundi, at around 1300-1500 metres altitude, more or less secondarised swamp forests are found, with *Macaranga schweinfurthii*. At around 1,950 metres, in the basin of the Kamiranzovu in the middle of Nyungwe forest, extends a vast swamp forest with *Syzygium rowlandi*, *Carapa grandiflora* and *Anthocleista grandiflora*. It covers nearly 1300 hectares and constitutes probably one of the most ancient plant "monuments" of the rift. According to palynological work by Hamilton, this forest has not changed much for at least 40,000 years: at some times species have appeared, while others have vanished, but the dominant species have remained the same.

3.44. The Hagenia abyssinica *forest, near 3,000 metres altitude in the saddle between the Karisimbi and Bisoke volcanoes, represents a true living fossil. The large trees are all several hundred years old, but there is practically no regeneration. This forest is therefore a witness to a period during which the climate must have been different from what it is nowadays.*

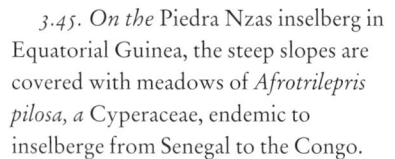

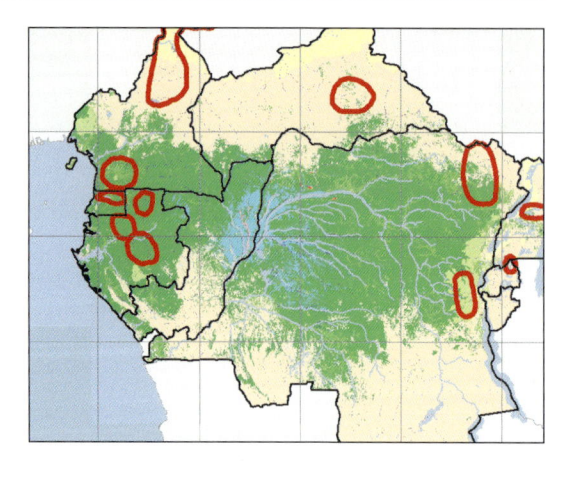

3.45. On the Piedra Nzas *inselberg in Equatorial Guinea, the steep slopes are covered with meadows of* Afrotrilepris pilosa, *a Cyperaceae, endemic to inselberge from Senegal to the Congo.*

3.46. Between open meadow and forest there forms a very dense forest-edge vegetation, in which the cactiform euphorbia Euphorbia letestui *abounds.*

3.47. The distribution of inselberge in Central Africa, according to Barthlott and Porembski (1988), cited by Ingrid Parmentier.

Beyond the limits

Apart from some values, any environmental parameter can become a limiting factor. If rainfall is insufficient, the dry season too long or the soil unfavourable, the forest gives way to savanna. The least change in these conditions can, however, engender a progression or a regression of the forest. If soil is absent, as on the inselberge, or if frost regularly occurs, forest formations are totally excluded.

Inselberge

Rid of the looser materials that shrouded them for hundreds of millions of years, the granitic intrusions born near the end of the Proterozoic today form inselberge, islands of very hard rocks – granites, diorites, gabbros and dolerites – that abruptly rose tens or hundreds of metres above the surrounding countryside. (Figures 1.6 et 3.45). In arid regions, under the effect of great temperature fluctuations, they alter, crack, split up and form piles of chaotic rocks often very picturesque and characteristic of so many landscapes in sub-Saharan Africa. In forested regions in southern Cameroon, Equatorial Guinea, Gabon, eastern and northeastern parts of the Democratic Republic of the Congo or even central-western Uganda, they form immense compact rocky masses that rise above the forest.

From a biological point of view, these inselberge harbour particular faunas and floras, often very different from those that surround them. Although the flora of inselberge in forested regions are poorer than that of savanna formations, it introduces into the forested region very particular species. Some are savanna species or very strongly resemble them. Others occur in no other habitat.

The steep slopes of an inselberg are often bare or may also be covered by a very short closed sward almost exclusively of the Cyperaceae *Afrotrilepsis pilosa*. This plant occurs from Senegal to Gabon, but only on inselberge. During dry periods it yellows and seems to die off, but once the rains begin, it turns green again. By creating its own litter between its roots, it manages to cover the rock.

On less steep slopes, where a thin layer of true soil manages to stay in place, this meadow is enriched by other herbaceous plants such as the white orchid *Habenaria procera* (FIGURE 3.48) or a red gladiolus *Gladiolus sp.*. (FIGURE 3.49). Where the soil is thicker, notably on top of the inselberg, dense brushes develop that often form the edge or the transition with forest strictly speaking. This vegetation contrasts sharply with that of the surrounding forest. It is most conditioned by the shallowness of the soil and the occurrence of more or less long dry periods, as well as by the frequency of fogs. This vegetation, with twisted, stunted trees and shrubs, is rich in lichens, ferns and epiphytic orchids. It is much richer than meadow vegetation, and the species that constitute it often have a very local distribution. Some are known only from a particular group of inselberge or, sometimes, even from a single inselberg. The cactiform euphorb *Euphorbia letestui* for example (FIGURE 3.46), only inhabits inselberge in Cameroon, Equatorial Guinea and Gabon, while *Aloe mubendensis* is strictly limited to a few inselberge in the vicinity of Mubende in Uganda. *Polyscias aequatoguineensis*, an Araliaceae described in 1999 by Lejoly and Lisowski, is known from only two inselberge in eastern Equatorial Guinea, from Assoc, and an inselberg in Gabon.

The vegetation of inselberge, particularly those of forested regions, is a very interesting topic of study for biogeographers. Each inselberg is an island lost in an ocean of forest and each group of inselberge behaves like an archipelago. It is indeed an ideal subject to test hypotheses on species distribution. With the successive retractions and transgressions of the forest massifs, inselberge are alternately surrounded by savanna or by forest. As a consequence, their vegetation retains some relict species that become witnesses to a more or less distant past. The study of inselberge can teach us much about the history of a region. For that, however, it is necessary to start by protecting them. Their vegetation is quite fragile. It resists grazing by the Buffalo *Syncerus caffer* or the Bongo *Tragelaphus euryceros* fairly well, but Man can very easily degrade them. Happily, the main inselberge in Equatorial Guinea are now included in the protected areas of Monte Alen National Park, Piedra Bere Reserve, Piedra Nzas Reserve (FIGURE 3.44) and Monte Temelon Reserve.

3.48. The orchid Habenaria procera, *recognisable by its extremely long spur.*

3.49. The red gladiolus Gladiolus sp. *has still not been identified with certainty.*

3.50. The large white orchid Angraecum eichlerianum *is typical for some inselberge.*

3.51. Above the last Hagenia *extend thickets of tree heather. Philippia trimera constitutes the dominant species on the Rwenzori. It occupies humid bottoms along streams, as well as steep crests, and forms dense thickets more than 12 metres high, draped in long fringes of the lichen Usnea barbatus. Their twisted trunks emerge from a thick layer of mosses punctuated with scapes of a pink orchid, Disa stairsii, and other small, more unobtrusive orchids of genus* Polystachya.

3.52. Erica kingaensis, *known by several local races, is often abundant in marshy areas were it forms low dense heaths.*

Afro-alpine habitats

On the highest mountains of the Albertine Rift, above the last *Hagenia* at around 3,300 or 3,400 metres above sea level, lies the Afro-alpine vegetation belt, a world mainly conditioned by temperature. Since temperature lowers by 0.6°C every 100 metres, inevitably it ends up falling below zero, at least during the night. From about 3,000 metres, its influence is greater than that of rainfall. At the same time, its daily fluctuations amplify, to the point where the Swedish botanist, Olov Hedberg, was able to write that on the high summits of equatorial Africa "summer reigns every day and winter every night". Seasonal fluctuations being very weak, climatological conditions of the Afro-alpine zone differ very noticeably from those of temperate regions and nocturnal frosts can occur at any time of year. It is much easier to protect against a winter of several months than against near-daily nocturnal frosts.

As for rainfall, there is rather little. After a maximum at around 2,500 metres altitude, it lessens progressively and the high summits receive no more than 800 to 1,000 millimetres of rain per year. In addition, nocturnal frosts render the water inaccessible and create a relative aridity. To which must be added an intense radiation, ultraviolet as well as infrared. Above 3,000 metres, many plants show adaptations that cause them to sometimes resemble those of hot desert regions, in others, plants of temperate or even subarctic regions. Hence the word "Afro-alpine". In reality, the flora of the high mountains of Africa resemble most the flora of the high plateaux of the Andes and could be called *paramo*.

By the number of species, this high altitude flora is not rich and White considers the Afro-alpine region as a zone of extreme impoverishment. Nevertheless, for a total of 278 species recorded by Hedberg, 81% are endemic. Nearly all are characterised by an enhanced adaptation to harsh climate conditions. Some have simply eliminated the problem of water transport by reducing their stem, thereby bringing leaves and roots closer together. Lobelias and senecios have developed rosette forms that efficiently protect the central bud. (FIGURE 3.53). The rosettes of lobelias can reach more than a metre and a half in diameter. During the day, they open widely to the heat of the sun, but during the night they close again. They protect the central bud not only by surrounding it with many layers of leaves but also by submerging it in water that stands in the leaves. Often a fine film of ice develops on the surface, but deep inside the water remains fluid and its temperature remains above 0°C. The giant groundsels, arborescent senecios of genus *Senecio*, wrap up in a mantle of persistent dead leaves or have a thick cork bark (FIGURE 3.53). Still others cover themselves with woolly hairs or have extremely reduced leaves, like the heathers of genera *Erica* and *Philippia* (Figures 3.51 and 3.52). In giant lobelias, such as *Lobelia wollastonii* (FIGURE 3.53) and *Lobelia bequaerti*, these mechanical differences are even reinforced by chemical protection: these plants contain sugars whose molecules transform with the temperature and act as antifreeze.

The summital vegetation of the Albertine Rift is encountered again on Mount Kahuzi and even on the summits of the Itombwe, but it is best developed on the Rwenzori and the Virunga. There it is differentiated into two great altitudinal zones. The first zone extends from 3,000 to 4,000 metres and is dominated by arborescent heathers, the second zone, at 4,000 to 4,500 metres is characterised by giant lobelias and dendrosenecios. Above 4,500 or 4,600 metres nothing grows any more except a few lichens (FIGURE 3.55), and the high peaks of the Rwenzori, above 5,000 metres, are nothing but deserts of stone and ice. In reality, all these limits are not necessarily neat horizontal lines. They are strongly influenced by the nature of the soil, the steepness and exposure of the slope, and rainfall. In practice, it is possible to distinguish five essential types of vegetation that are found in varying proportions on the different mountains.

Just above the forests lie the tree-heather thickets, particularly luxuriant on the Rwenzori mountains and the Virunga. On the Rwenzori, the large heather *Philippia trimera* is the dominant species (FIGURE 3.51). The only trees accompanying it are *Rapanea rhododendroides*, the giant senecio *Senecio johnstoni* var. *ericirosenii*, yellow-flowered *Hypericum keniense* or red-orange flowered *Hypericum bequartii*. In the lower parts of this heather zone, species of the understorey recalling the last forests, are again encountered. *Mimulopsis elliotii*, a beautiful Acanthaceae, mingles with diverse brambles *Rubus sp.* to form impenetrable masses. Also found there are balsams *Impatiens sp.*, a pale blue violet *Viola eminii* and a pinkish blue bittercress *Cardamine obliqua*. In the highest places, the giant senecio *Senecio johnstoni* var. *adnivalis* and the giant lobelia *Lobelia lanurensis* are also found, while the understorey is formed by a dense silver-grey carpet of lady's mantle *Alchemilla*. Mosses, lichens, liverworts and lycopods abound. They cover the soil as well as exposed roots, dead wood, trunks and the main branches of trees.

In the volcanoes, *Philippia johnstoni* is the dominant heather, but on Muhabura, drier than the other volcanoes, *Erica arborea* is also found. In areas where drainage is more difficult and where the soil remains soaked, the heather *Erica kingaensis* forms bushy heaths (FIGURE 3.52).

Above 3,500 metres, the large glacial valleys of the Rwenzori are occupied by vast peat bogs. They are a characteristic element of the chain and they form unforgettable landscapes. The dominant vegetation of the bogs is constituted of tussocks of *Carex monostachya* or *Carex runssoroensis* one metre high and, in places, just as wide. In between the tussocks, a silver carpet of lady's mantle *Alchemilla johnstoni* forms and the liquid-peat filled depressions are hidden by thick layers of golden green *Sphagnum*. All the peat bogs occupy ancient lakes formed by moraines and later sealed off. The Rwenzori mountains and some areas of the Virunga are so humid that, in places, bogs manage to form on steep slopes. In is on the edge of these marshes, whether in the bottoms or on the slopes, that *Lobelia bequaerti*, probably the most beautiful of the giant lobelias, is found. Contrary to the giant senecios, giant lobelias grow relatively fast, reach maturity, flower and die within two or three years.

Between 3,500 and 4,500 metres, extend heaths or thickets of giant lobelias and giant senecios (Figures 3.53 et 3.56). They, too, prefer humid zones and need a minimum of deep soil. *Senecio johnstoni* var. *adnivalis* and *Lobelia wollastonii* are the two dominant species. The understorey is carpeted with *Alchemilla johnstonii*, and includes many grasses, as well as mosses and a giant umbellifer *Peucedanum kersteni*, that resembles an enormous celery. On abrupt rocky slopes with little soil, low, silver-grey brushes are formed almost exclusively of the interlacing long stems of *Helichrysum* (FIGURE 3.54). This heath is also found on other moun-

3.53. Above 3,500 metres altitude, the two most typical components of Afro-alpine vegetation are the giant senecios *Senecio sp.* with rosettes of wide leaves and the giant lobelias *Lobelia sp.* (here a *Lobelia wollastonii*) with rosettes of much narrower leaves.

3 ECOLOGICAL GRADIENTS

tains of East Africa, but nowhere is it as abundant as on the Rwenzori. It occupies steep rocky slopes and well drained soils in the upper part of the heather zone and in the giant lobelia and giant senecio zone. It depends on both high rainfall and good drainage.

Where rainfall is insufficient, an herbaceous vegetation forms. This type of vegetation is rare on the Rwenzori range. In the Virunga, it is limited to the eastern slope of Mount Muhabura. On Mount Cameroon, it is the dominant formation. Most of the grasses encountered in this zone belong to other genera than the grasses of the savannas. Many are endemics of African mountains, but some also occur in temperate latitudes. Amongst the dense tufts of grasses orchids can be found, notably of genera *Disa*, *Habenaria* and *Satyrium*.

Independently of their floristic interest, the Afro-alpine vegetation zones are inhabited also by animals endemic or near-endemic to these habitats. Among birds, there is the superb Red-tufted Malachite Sunbird *Nectarina johnstoni*, a specialist of giant lobelias, and the more unobtrusive Brown Woodland Warbler *Phylloscopus umbrovirens*. Both are common in the rift, while the Black-capped White-Eye *Speirops melanocephalus* is limited to Mount Cameroon.

3.54. Helichrysum sp., *an everlasting composite, at nearly 3,500 metres altitude on Bisoke volcano in Rwanda.*

3.55. A tuft of lichens at 3,700 metres on Bisoke's summit. On the Rwenzori, it is the only plant to survive above the altitude of 4,600 metres.

3.56. High slopes, at between 3,500 and 4,500 metres altitude, on the western flank of the Rwenzori mountains in the Democratic Republic of the Congo, are occupied by dense thickets of Senecio johnstoni *var.* adnivalis *and* Lobelia wollastoni. *In ravines, they can be very dense and nearly impenetrable. On steep slopes and ridges, they are much more open.*

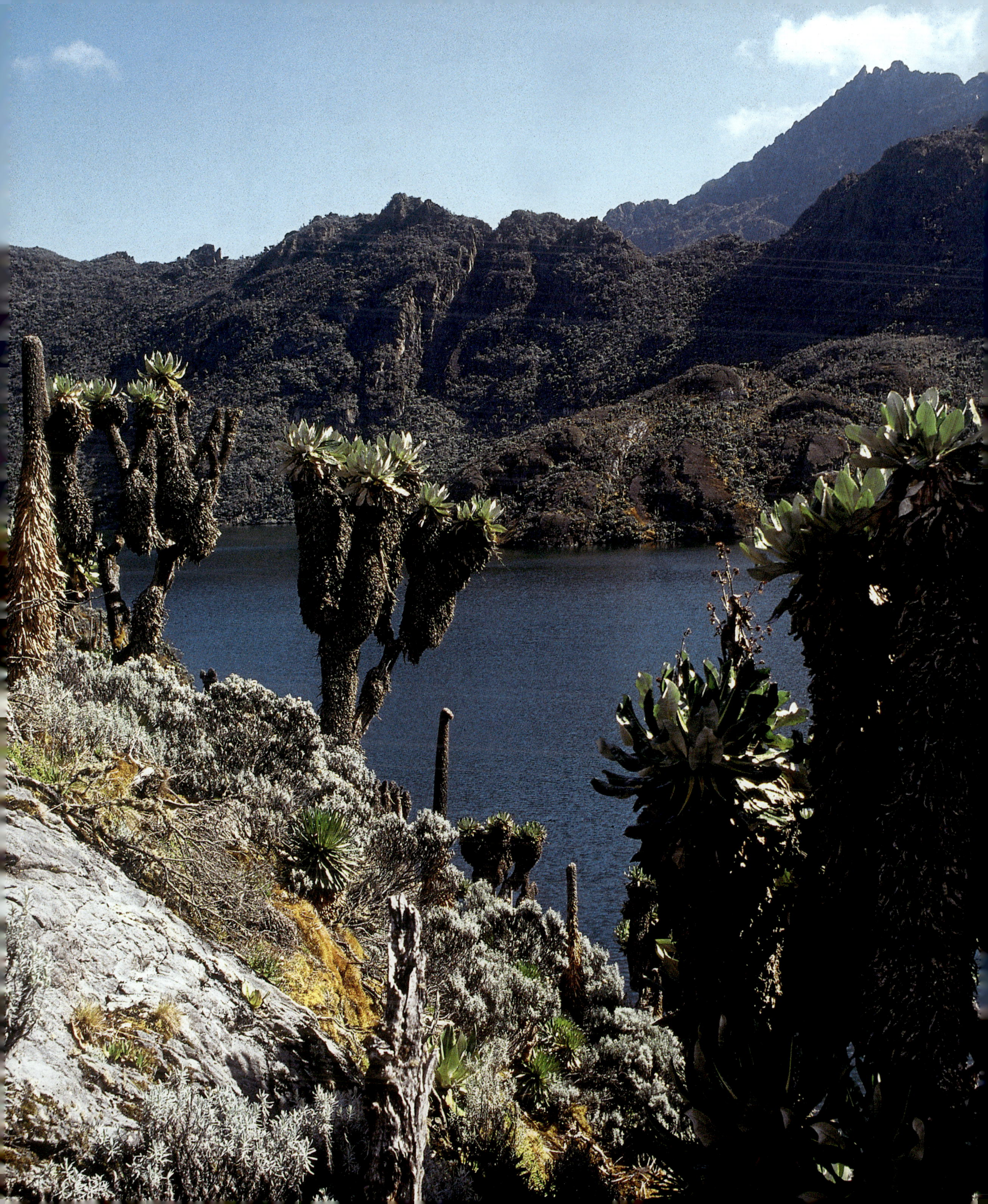

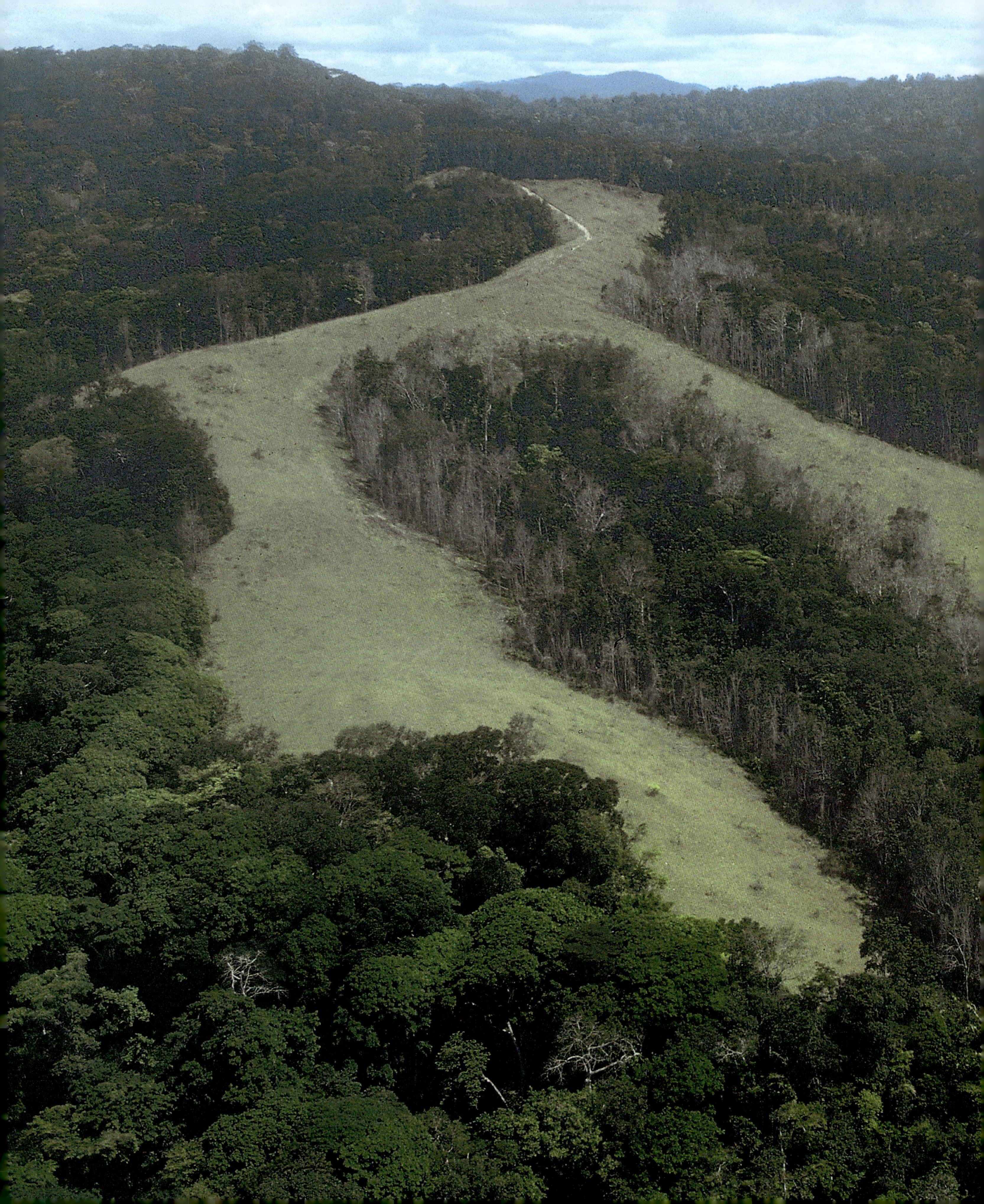

4 Temporal gradients
Tree-fall gaps and successions

The range of tropical forests has fluctuated throughout the course of history and many of the present-day formations are no more than 2,000 years old. Fossil pollen analyses in Cameroon show that their composition evolved constantly and that they were subjected to cycles of 2,300 to 2,500 years (PAGE 57). Tropical forests are thus dynamic habitats and each formation is necessarily the result of a more or less long process the beginning and the end of which are rarely known. A forest undergoes modification as long as it lives, even when it manages to reach "maturity" – the term certainly comes from the foresters – it continues to change. The impression of immobility that it gives is only an illusion linked to the transience of human life. The study of temporal phenomena that affect forests comes up against the fact that human life is measured in tens of years, while that of a tree is counted in centuries.

The transformations that a tropical forest undergoes are of two orders, depending on the spatial scale considered. On the one hand, every forest block undergoes constant internal perturbations: the falling of a tree killed by infirmity or felled by lightning, the impact of large mammals or some herbivorous insects, a landslide or a tornado. As hurricanes do not occur in the Congolese region, all these perturbations cause no more than gaps in the forest cover, drowned in an ocean of more or less mature forests. On the other hand, it happens that the forest is destroyed over vast surfaces, either by Man, by the climate, or by a conjunction of the two. If given a chance, the forest is then capable of reconstituting itself through the succession of a series of increasingly complex plant communities. This succession leads to a secondary forest. It goes on for several tens of years, even several centuries, and takes much longer than the restoration we can observe at the scale of a tree fall gap. Lastly, the forest may also move onto a new substrate or to areas where no forest has been in the recent past. If climatic conditions, soil, fire, animals and Man allow, every forest massif can colonise savannas, humid habitats, rocky surfaces or lava fields. It can even move out onto maritime beaches, a process somewhat comparable to a primary succession. It leads to the establishment of a primary forest and covers centuries or millennia.

All of the temporal processes that affect the forest cause it to decompose, independently of ecological influences, into a mosaic of formations of different ages. Unfortunately, the appetites of Man are such that, day by day, the old forests are being reduced, giving way to younger and younger formations.

4.1. Ridges and hill tops in the northern part of Lopé National Park are often occupied by grasslands. This is essentially the result of soil factors, but the limit between forest and grassland is not stable. It fluctuates according to the importance of fires, and the fringe of dead trees is a definite sign of forest regression.

4.2. In the coastal forest of Gabon, a recent gap resulting from the fall of a large tree. In between the pieces of wood lying on the ground appears a mass of seedlings and shoots needing only light to shoot to the sky. They will have to contend with lianas and young trees on the edge of the clearing which will also throw themselves into the race, but with a good head start. Even though such clearings cover very small surfaces, their frequency is such that after 1,000, 1,500 or 2,000 years – depending on composition and the winds subjected to – the entire surface of the forest will be reshaped.

Tree-fall gaps

Great trees of the canopy can reach an age of several centuries, maybe even a millennium, but inexorably they age and end up dying. Mushrooms and insects attack their wood and in little time they become more and more fragile and finally fall. When this happens they bring with them many smaller trees, often bound together by lianas, and create an open space, the tree-fall gap (FIGURE 4.2). Other tree falls can be provoked by a landslide, a tornado or lightning, even by elephants. Paradoxically, this phenomenon, that has its origin in death, engenders renewal and life.

Whatever the cause, the tree-fall causes a sudden hole in the canopy where light rushes in. But light is a rare resource in the understorey of dense tropical forests (PAGE 149). Herbaceous species are adapted to this condition, but, for many of the large trees, it is a limiting factor. Their seeds germinate, but the seedlings do not manage to develop and many perish. But some form a few leaves, then go into waiting, sometimes for years. In this way, the understorey is an immense tree nursery, ready to come into action whenever the canopy is perforated. The formation of an opening after a tree falls immediately sets off a race for the light. In addition, the disappearance of the dead tree's thick roots reduces competition for nutrients in the soil. In very small openings, the forest can close up again by the simple lateral expansion of neighbouring crowns. This is what happens notably after the fall of one of the main branches from a large canopy tree. In larger openings, even the most insignificant plantlets develop a stem very rapidly that grows as high as possible and often does not divide until it reaches a height of at least ten metres. Dormant seeds sprout and lianas, fallen with the trees, send up shoots.

In two or three years, the opening is closed by a "cork" of inextricable vegetation that slowly grows to repair the canopy and plunges the understorey into the shade again. It is only in tree-fall gaps caused by lightning that regeneration can be greatly retarded by the destruction of seedlings, young plants and stumps.

In small openings, the amount of light that reaches the ground can be fairly limited and not inhibit the young sciaphilous, or shade-loving, trees. In the larger openings, the first trees to shoot up – the pioneers – are invariably the heliophilous, or light-loving, species. They are species that do not grow in old dense forests and whose seeds, generally carried by wind, birds or bats, can survive for a long time before germinating under the effect of light. Their growth is rapid but their life is short. The most typical are the Umbrella Tree *Musanga cecropioides* (FIGURE 4.3) and *M. leo-errerae*, of the Moraceae family. The first inhabits low altitude forests; the second submontane forests. Both grow with disconcerting speed and rapidly form a very broad crown. Other species, such as those of genus *Anthocleista*, of the Loganiaceae family, rapidly form a long stem surrounded by enormous leaves (FIGURE 4.4). These pioneer trees soon monopolise much of the light. However, they also offer protection to the more delicate species which prefer a softer light. After 15 or 20 years, they are replaced by slower growing species and, after 50 years, the opening is obliterated.

For simple reasons of probability – the number of tree species in tropical forests is so high – this patch of new forest is, however, rarely similar to the one it replaces. While regeneration inevitably depends on the seeds and plantlets present in the opening, many scientists think that renewal of the canopy by the process of openings created by tree falls is essentially a random process. However, as we have seen, climatic conditions oscillate continually, and it is highly probable that factors that favour or disfavour a particular species vary from one year to another according to variations of the climate. Moreover, the growth of large canopy trees is so slow that the composition of an old forest does not necessarily reflect present climatic conditions but rather those that prevailed a few centuries ago. In other words, when a very old tree disappears, climatic conditions are probably no longer the same as when the tree germinated and it is natural that the young trees called to replace it are of other species. Tree falls thus rejuvenate the forest, augmenting or maintaining its diversity by creating an intricate mosaic of species of different ages. Imperceptibly, they also change its composition.

This phenomenon was already observed by Aubréville in 1938 and most recent studies confirm it. The composition of the canopy of a forest is not constant, neither in space nor in time. The entire association of dominant species is destined to be replaced by a different association. A vast surface of forest will in this way be constituted of a mosaic of islets that each represent an "occasional" variant of a greater theoretical ensemble. This vision is in opposition to the concept of forest climax advocated in the past. Indeed, in many places one finds associations dominated by one or another species but with few or no young individuals of the species present in the lower strata. Although the associations are generally considered as being climax, it is obvious that, at the death of the present dominants, other species will replace them. The *Parinari excelsa* forests of the rift are a good example. Born 500 to 600 years ago, they show no sign of regeneration at the level of the dominant species that characterises them.

Only forests dominated by one or another species of Caesalpiniaceae apparently escape this reshaping. Mixed forests seem condemned to evolve. However, this phenomenon should be seen in a historical perspective. It may perhaps be based on the fact that most of the Congolese forests were subjected to very profound perturbations of a climatic origin 2,000 to 2,500 years ago, to which today is added human activity.

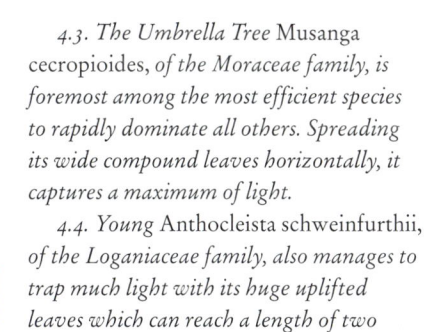

4.3. *The Umbrella Tree* Musanga cecropioides, *of the Moraceae family, is foremost among the most efficient species to rapidly dominate all others. Spreading its wide compound leaves horizontally, it captures a maximum of light.*

4.4. *Young* Anthocleista schweinfurthii, *of the Loganiaceae family, also manages to trap much light with its huge uplifted leaves which can reach a length of two metres. It grows on dry substrates, while a very close relative,* Anthocleista vogelii, *occupies marshes and* Anthocleista grandiflora *inhabits high altitude areas in the east.*

4.5. A ten-year old forest of Umbrella trees Musanga cecropioides *with an understorey of Zingiberaceae and Marantaceae species in the centre of Gabon.*

4.6. Like some trees of damp places, the Umbrella Tree is supported by more or less developed stilt-roots.

Are tropical forests fragile?

To answer this question, we need to know how long they take to recover after destruction. If they regrow quickly, they are resilient and not fragile. But to rebuild an old forest it takes many centuries, even more than 1,000 years, providing reliable seed sources are available. Globally we can say that some forest type are resilient but also that old grown forests are fragile.

Secondary successions

When extensive areas of forest are destroyed simultaneously by fire, by unrelenting clear-felling, by agricultural deforestation or by a violent tornado, a secondary succession may be triggered – if Man and nature permit. As in the case of the tree-fall gap, and contrary to what happens in cases of primary succession, which we will consider next, the soil contains seeds, plantlets and stumps capable of sprouting new shoots of species of the destroyed forest. It also contains a fairly high quantity of organic material. Regeneration could thus be fairly rapid. But the destruction involves a much greater area than the opening made by a tree fall. Exposure of the soil to sun and rain is generally more prolonged and foreign or exotic species brought by Man often dominate and hinder the initial stages of this succession.

Restoration of the forest cover, clearly longer than in a gap made by a tree fall, usually begins with a pioneer shrub stage. Alongside the stumps and shoots of the cleared forest and remains of cultivated plants – especially cassava – shrubs become established and will, for two or three years, cover the available space. These are usually widespread species, such as *Trema orientalis*, *Vernonia conferta* or *Harungana madagascariensis*, which seem less influenced by the soil and climatic circumstances than species of old forests. *Harungana madagascariensis*, for example, a shrub of the Clusiaceae family, is common among the pioneering brushes of Gabon, but is found, often in a more elongated form, at 2,400 metres altitude along roads crossing forests of the Albertine Rift.

After a year or two, a forest dominated by a single species emerges from these brushes. However, this does not constitute more than an avant-garde since, being incapable of regenerating in the shade, it is destined to give way to other species. Depending on ecological conditions – available humidity, soil or altitude – young secondary forests will take on various appearances. They will rarely be dominated by a single species. Their composition will continue to evolve with time and they will slowly evolve into old secondary forests. They will also appear to stabilise, but this will be no more than an impression related to the extreme slowness of their evolution. They will then become forests called climax, mixed and thus very diversified on rich soils but sometimes dominated by a single species – generally a Caesalpiniacea – on poorer soils.

4 TEMPORAL GRADIENTS

Lowland secondary forests

After two or three years, from dense thickets of *Harungana*, *Trema*, *Vernonia conferta*, *Caloncoba welwitschi*, *Chaetocarpus africanus* or *Rauvolfia vomitoria*, trees emerge, including *Anthocleista schweinfurthii* (FIGURE 4.4) and, mostly, the Umbrella Tree *Musanga cecropioides*. This tree reaches its optimal development after eight to ten years (FIGURE 4.5) and constitutes very homogeneous formations 15 to 20 metres tall, its parasol-shaped crowns accounting for 80 to 90% of the cover. The accompanying complement is composed of species such as *Buchnerodendron speciosum*, *Oncoba glauca*, *Croton mubango*, *Lindackeria dentata*, *Macaranga monandra* and *Macaranga spinosa*, *Maesopsis eminii* and *Myrianthus arboreus*. Trees of genera *Musanga*, *Myrianthus* and *Macaranga* are characterised by stilt roots (FIGURE 4.6). Epiphytes are rare or absent. The herb and shrub layer, three to four metres high and very dense, nearly impenetrable and mainly composed of Zingiberaceae of genera *Costus* (FIGURE 4.7) and *Aframommum* (FIGURE 4.8), Marantaceae (FIGURE 4.9), usually of genera *Haumania* and *Marantochloa*, and some Commelinaceae of genus *Palisota*. These forests are very widespread around villages and along roads. After about ten years, they begin to give way to more diverse secondary formations or, more and more often, to new cultivation, but the Umbrella Tree remains dominant for some 20 years.

In the lowland evergreen forest zone, the oldest secondary forests are characterised by an abundance of Ilomba or False Nutmeg *Pycnanthus angolensis*, species of genus *Zanthoxylum* (or *Fagara*), the Kapok Tree *Ceiba pentandra* and, notably, *Funtumia africana*, *Holoptelea grandis*, *Pterygota macrocarpa* and *Pentaclethra macrophylla* – to mention a few species. In the driest areas, old secondary forests are characterised instead by an abundance of Limba *Terminalia superba*, Ayous or Whitewood *Triplochiton scleroxylon*, Iroko *Milicia excelsa*, Aiélé or Incense Tree *Canarium schweinfurthii* and Erimado or Essesang *Ricinodendron heudelotii*. But these species are also abundant in semi-deciduous forests and it is not easy to decide if their presence is strictly climatic or the consequence of an ancient disturbance.

Young secondary forests, regardless of the altitude or climatic zone in which they occur, are easily recognised. But the older a forest is, the harder it is to determine whether it is of secondary or primary origin. Considering the enormous perturbations the Congolese forests underwent 2,000 to 2,500 years ago, this distinction does not, perhaps, make sense, but it continues to bother conservationists.

Primary forests

Strictly speaking, a primary forest is a forest resulting from a primary succession, what ever its age. In common language, they are forests which have not been disturbed by Man for a very long period of time. They can result from a primary succession or a secondary succession.

4.7. *Costus lucanusianus*, of the Zingiberaceae family, is abundant in most of the very disturbed forests of the Congo region, Gabon and Burundi.

4.8. An *Aframommum sangaris* flower in the understorey of a small forest in Burundi. The fruit of some of the species in this genus are very appreciated by both animals and humans.

4.9. A clump of Marantaceae on the edge of a secondary forest in Gabon. The stems of various species of this family are used to make bindings. The leaves serve to wrap and cook some foods or to cover roofs.

4.10. The extensive recent lava fields of the Virunga volcanoes and Western Cameroon are mineral deserts. Even if some trees remain standing, everything is dead, and all life has to start from nothing.

Submontane and montane secondary forests

In the submontane forest zone, *Musanga cecropioides* is replaced by *Musanga leo-errerae*. These formations constitute forests very similar to the lowland forests but they rarely cover large surfaces. They apparently require rather abundant rainfall and are fairly frequent in the eastern part of the Democratic Republic of the Congo. They are also found in the western part of Nyungwe Forest, in Rwanda, and in some areas bordering Kibira National Park, in Burundi. Being situated at higher altitudes, they evolve more slowly and can persist for as long as 30 to 40 years.

In the most arid parts of Uganda, Umbrella Trees are absent and young secondary forests are dominated by *Maesopsis eminii*, a member of the Rhamnaceae family, which can reach a height of nearly 30 metres. Its crown is fairly light and its trunk straight and slender. Its fruits resemble olives and are much appreciated by monkeys, touracos, hornbills, green pigeons and many other birds. These animals widely contribute to the dispersal of such a species. Because it produces a fairly good wood for carpentry, it has also been introduced in many places by Man, which explains why it is found in some parts of Rwanda and Burundi.

Above 1,750 or 1,800 metres altitude, Umbrella Trees disappear and secondary formations are characterised by other species. In montane forests of the Albertine Rift, *Neoboutonia macrocalyx* is thus very common. It belongs to the Euphorbiaceae family and is easily recognisable by its large, very rounded leaves that are always covered with holes made by insects. Elsewhere is found instead *Polyscias fulva*, of the Araliaceae family. These formations usually cover only small surfaces, but vast forests of *Neoboutonia* do exist on Mount Gahinga in Volcanoes National Park and used to exist in the Gishwati forest, in Rwanda.

Primary successions

A true primary succession is a process that leads to the establishment of a forest on new terrain. True examples are not frequent. They are encountered only on islands recently arisen from the sea or on islands where the vegetation has been completely destroyed by volcanic eruption. The classic example is the recolonisation of Krakatoa, in Indonesia, described in detail by Richards.

The colonisation of lava fields, in the Virunga or on Mount Cameroon, most resembles a veritable primary succession in Central Africa (FIGURE 4.10), even though it happens that pockets of vegetation escape from the lavas and hasten colonisation. In spite of that, it remains very slow.

The very first pioneers are usually lichens, mosses and ferns. They belong to "archaic" groups of plants, close to those that colonised emergent lands during the Ordovician, 380 to 400 million years ago. The evolution of this vegetation runs up against the strictly mineral nature of the rock it occupies, and must first allow bacteria and the first pioneers, acting in symbiosis with mushrooms, to do their work before a soil is created and an angiosperm flora can move in. Many years are necessary – a little less in high-rainfall regions than in arid regions – before shrub formations or a forest appear.

A succession somewhat comparable to that of lava fields can be seen on large landslides that strip away the humus layers of the soil. These are sometimes fairly frequent in mountainous regions with high rainfall, loose soils and steep slopes, especially when they are submitted to earthquakes. This is the case in the Rwenzori. As a rule, the organic earth is completely torn away and the bed rock, although often already strongly altered, is laid bare. Here, also, it is mosses and ferns that move in first. The first stages of this succession can evolve fairly rapidly given that the first colonists fix seeds and organic soil carried from above by rainwater runoff. Later, the establishment of trees may be problematic because the steep slope and loose soils compromise their stability. In many places this succession remains more or less blocked at the stage of brushes mixed with Bracken *Pteridium aquilinum*. The phenomenon contributes in this way to the fragmentation of forest cover in mountainous regions with a very humid climate.

In regions with fragile soil, as in the coastal plain of Gabon or on the Batéké Plateau, immense cirque-like erosion scars develop. There, also, surfaces stripped bare and cleaned of any organic earth are invaded by ferns, mainly *Dicranopteris linearis* (Figures 4.11-13). This widespread species is abundant in Gabon and is found on recent terraces and roadside embankments. Considering the poverty and unfavourable texture of the soil, this succession evolves very slowly.

Vast clearings with *Dicranopteris* are found in several places in Lopé National Park, particularly on the steep slopes of some mountains of the central part. Their origin is uncertain, but by analogy with what is observed elsewhere, it may have to do with ancient landslides where forest recolonisation has remained blocked at the fern-heath stage. As in the case of Bracken in the mountains of the Albertine Rift, it appears that the *Dicranopteris* prevents the growth of other plants by the emission of substances from its roots. Fires that periodically sweep these clearings must also prevent the establishment of trees.

4.11. Beginning of colonisation of a land slide by the fern Dicranopteris linearis, *very abundant in Gabon.*

4.12. This fern, of the very old Gleicheniaceae family, is resistant to sun and temporary water shortage. It is characterised by the dichotomous branching of its fronds.

4.13. It is very abundant also in the large erosions cirques on the east flank of Mount Mikongo in the Lopé National Park, comparable to those of the coastal region and the Batéké plateau.

4.14. The coastal savannas in front of Libreville lie on recent white sands. Appearing very flat a first sight, they in fact comprise a succession of ancient flattened dune cordons separated by vague depressions. During the rainy season, these hollows are frequently flooded and attract migrating water birds, such as Purple Heron Ardea purpurea *or Great Snipe* Gallinago media – *especially in March and April. Rosy Bee-eaters* Merops malimbicus *establish colonies on the dune cordons in September-October and mingle with African River Martins* Pseudochelidon eurystomima, *an endemic swallow of the Congolese region, recognisable by its red bill.*

Enclave savannas and their colonisation

When the dry season becomes too long, rainfall too weak, soil too shallow or too permeable, the forest gives way to savannas. Not only do these factors determine the exterior limits of the forest massif, they also preserve, at the very interior of this massif, islets of open habitats, at first glance unexpected.

Closely intertwined with forest formations, these savannas not only increase biological diversity on the scale of a landscape, but they also play a substantial role in the biological processes of the forest. Many studies, notably those of Brugière, Sakom and Gautier-Hion, show to what extent these marginal habitats are important, not only for elephants and buffaloes, but also for primates. At other epochs, when the forest was much more fragmented than nowadays, these savannas must have played an even more important role: zones of contact or margins were then much more widespread and interactions between dense forest and open habitats must have been much more important.

Although the existence of savannas inside the forest belt depends mainly on climate or soil, the relative importance of diverse factors varies from one place to another and fire plays a primordial role almost everywhere. On the periphery of the forest zone, the forest-savanna limit is conditioned principally by fire and by climate, as much by the length of the dry season as by the amount of rainfall. Inside the forest massif, with rainfall no less than 1,500 millimetres per year on average and dry seasons no more than three months long, the savannas have their origin in fire and the soil. Climate acts only as a facilitator. These savannas are qualified as edaphic. In the western part of Central Africa, they are divided into three main zones – savannas of the coast, those of the Nyari Valley and those of the Batéké Plateau – with an island in the very centre of Gabon: the savannas of the Lopé National Park.

The poorest are the "steppic" savannas of the coastal region (FIGURE 4.14). They are developed on very permeable sands that are often flooded during the rainy season. The grass layer is formed of fairly short species, coriaceous and growing in distinct tufts. Most of the grasses belong to the generas *Ctenium*, *Rhynchelytrum*, *Loudetia*, *Elyonurus*, *Anadelphia* (*Pobeguinea*) and *Hyparrhenia*. This savanna is punctuated with small Fabaceae and Cyperaceae species. Orchids of genera *Eulophia*, *Habenaria* (FIGURE 4.15) and *Platycorine* are abundant, as well as Hypoxidaceae. In places there are tussocks of the tall grass *Loudetia arundinacea* or *Clappertonia ficifolia* (FIGURE 4.16). Shrubs are absent.

Regarding their fauna these savannas are just as poor and their bird life is rather disappointing. The only omnipresent species are the Long-legged Pipit *Anthus pallidiventris*, the Grey-rumped Swallow, the Pectoral-patch Cisticola *Cisticola brunnescens* and the Red-billed Quailfinch *Ortygospiza gabonensis*. The African Red-tailed Buzzard *Buteo auguralis* is seen throughout the year. In September and October, the savannas harbour numerous colonies of the African River Martin *Pseudochelidon eurystomina*. This river martin, endemic to the Congolese region, known from Gabon and the centre of the Congo basin, hunts most of the year in large congregations above dense forests, most often in the proximity of an important water course. It is common in the Makokou region, on the Ivindo River, from January to August. During the breeding season, it associates with Rosy Bee-eaters *Merops malimbicus* and its nest holes are dug into ancient dune cordons.

The savannas of the Lopé National Park, completely isolated from any other savanna, are also very poor. Their exact origin is not known, but everything leads one to believe that they are very old and that they existed already well before the retreat of the forests during the last glaciation. In July 2,000, Richard Oslisly discovered an archaeological site occupied nearly without interruption from 30,000 BP to the recent Iron Age, a thousand years ago at most. The populations involved were obviously savanna people. Of course, the extent of savannas has considerably varied in function of the climate. If, despite the forest transgression underway today, they have not completely disappeared, it is probably thanks to a combination of factors at once edaphic and climatic.

The vegetation of the Lopé savannas is dominated by the large grass *Anadelphia (Pobeguinea) arrecta* in association with *Andropogon pseudapricus*, *Schizachyrium platyphyllum* or *Hyparrhenia diplandra*. On the poorest soils, this savanna remains essentially grassy (FIGURE 4.17). Elsewhere, it can have a shrubby stratum comprising *Nauclea latifolia* (FIGURE 4.18), *Crossopteryx febrifuga* (FIGURE 4.19), two Rubiaceae species, as well as *Bridelia ferruginea*, a Euphorbiaceae. Non gramineous species are uncommon. In humid areas, the flora is richer and orchids are fairly abundant, notably the large pale pink *Eulophia oedoplectron*.

As for the fauna, these savannas are nearly as poor as the coastal savannas. To the Long-legged Pipit, the Pectoral-patch Cisticola and the Red-billed Quailfinch are added the Whistling Cisticola *Cisticola lateralis* and the Croaking Cisticola *Cisticola natalensis*, but the large birds of open grasslands, such as the francolins or the guineafowls, are curiously absent. Savanna butterflies, also, are nearly absent or represented by a few widespread, often migratory species. It indeed seems that these savannas are highly impoverished. Perhaps they have been even more constricted than they are today?

The other savannas, in contact with those that extend to the south of the Congolese forest massif, are clearly richer. At the level of the cataracts, the Batéké plateau is occupied by essentially open grasslands of *Aristida dewildemanii*, *Anadelphia arrecta* or *Loudetia arundinacea*. Towards the centre of the plateau, areas of polymorphic sandstones are occupied by *Loudetia demeusii* grasslands forming a low, loose, sparse grassy carpet. The shrubby vegetation is usually very scanty. The grass *Elyonurus brazzae* becomes dominant on very leached soils, *Loudetia simplex* on stripped soils

4.15. This orchid of genus Habenaria *is very abundant in the coastal savannas, on white sands where it forms vast populations. Because they live in symbiosis with fungi, orchids are well adapted to growing in habitats extremely poor in nutrients.*

4.16. Clappertonia ficifolia, *of the Tiliaceae family, grows in somewhat more favourable places, often on the edge of forests. Its flowers are visited by shining, metallic green cetonid beetles.*

4.17. *On the shallow, often stony soils of the northern part of Lopé National Park grows a very open short grass savanna.*

4.18. *The white flowers of* Nauclea latifolia, *of the Rubiaceae family, strongly attract bees, but its fruit attracts elephants.*

4.19. *The white inflorescences of* Crossopteryx febrifugum, *also a Rubiaceae, attract flies, instead.*

and *Hyparrhenia diplandra* on organically enriched soils. *Hymenocardia acida* is practically the only woody plant.

In the central region of the plateau, characterised by its tabular relief, one finds *Trachypogon thollonii* and *Hyparrhenia diplandra* grasslands with the shrubs or trees *Annona senegalensis, Albizia adianthifolia, Psorospermum febrifugum* and *Vitex madiensis*. To the north, in the area of low hills, these savannas become poorer. The *Hyparrhenia* species disappear and the shrub layer becomes scantier. Still further north, *Trachypogon* is replaced by *Andropogon schirensis*. These are the savannas found at Odzala National Park (FIGURE 4.21). They constitute the ultimate northern extension of the Batéké plateau. *Andropogon schirensis* forms a fairly dense, homogeneous grass carpet 1 to 1.80 metre high. *Digitaria uniglumis, Hyparrhenia diplandra* and *Panicum phragmitoides* are also frequent. The shrub layer is represented essentially by *Hymenocardia acida*, often accompanied by *Annona senegalensis*. *Nauclea latifolia* is rarer. *Bridelia ferruginea* and *Maprounea africana* are occasional. In places there are tufts of *Clappertonia ficifolia*.

Nearer to the Atlantic coast, other savannas advance in the Niari valley and reach the Moukalaba-Dougoua Faunal Reserve in Gabon, now included in Moukalaba-Doudou National Park. They are essentially composed of high grasses of genus *Hyparrhenia* in association with *Digitaria, Andropogon* and *Pennisetum*. The grass cover reaches a height of two to three metres. In places a shrub layer develops with *Annona senegalensis, Crossopteryx febrifuga, Bridelia ferruginea, Nauclea latifolia* and *Vitex madiensis*. In the Congo, there is also *Hymenocardia acida, Ficus capensis, Syzygium macrocarpum* and *Psorospermum febrifugum*. In the savannas of the Nyanga in Gabon a small population survives of the Defassa Waterbuck *Kobus ellipsyprimnus*, of the Angolan form.

The persistence of savannas in the Odzala region is fairly unexpected and has caused a lot of ink to spill. The poverty and permeability of the sandy soils certainly play a role. According to some, these savannas are essentially man-made. But, curiously, even though Man's impact on the forests is greater than ever, there is no sign anywhere of forests being transformed into savannas, neither in the Mayombe (South-West Congo) nor in Cameroon. Everywhere where the phenomenon has been studied in detail, forest is encroaching on the savanna, in the Congo as well as in Cameroon, Gabon and the Central African Republic. According to Maley and Foresta, these savannas have an ancient climatic origin, and the openings that punctuate the forests of Odzala today are just vestiges of the enormous deforestation that occurred 2,000 to 2,500 years ago.

Whatever their origin, these savannas harbour a rich fauna. On the Batéké plateau, alongside the common birds of the coastal areas and the Lopé region, are also found bustards, Finsch's Francolin *Francolinus finschi*, the Congo Moorchat *Myrmecocichla tholloni* and the Black-chinned Weaver *Ploceus nigrimentum*. Not so long ago, these habitats also harboured large savanna mammals. In addition to the Bushbuck *Tragelaphus scriptus*, there were the Common Duiker *Sylvicapra grimmia* and the Common Reedbuck *Redunca arundinum*. This population of ungulates maintained the presence of the Lion *Panthera leo*, of which a few individuals survive in the Odzala National Park.

4.20. Ouratea, *a shrub of the Ochnaceae family with bright yellow flowers, is abundant on the abrupt ridges dominating the savannas of Lopé National Park.*

4.21. *The tall grass savannas of Odzala-Kokoua National Park are scattered with large termite mounds. Some evenings lions can still be heard roaring and the savannas were still travelled by Black Rhinoceros* Diceros bicornis *more than a century ago. Since it is unlikely that these animals came from the south by crossing the Congo River, they could only have come from the north during the period when a vast north-south corridor of savannas isolated the forests of the Congo basin from those of the Ogooué basin and the Atlantic coastal region. The existence of a small population of Congo Moorchats* Myrmecocichla tholloni *north of the forest massif in the south-western part of the Central African Republic – which remains to be confirmed – would be another proof of this old savanna corridor.*

4.22. Duparquetia orchidacea is a liana of the Caesalpiniaceae family encountered essentially in forest mantles. It can grow to a length of more than 10 metres.

4.23. A yellow Nephila *spider in its web in the mantle of a forest.*

4.24. Newbouldia laevis *is a small tree of the Bignoniaceae family.*

4.25. Ipomea *species, of the Convolvulaceae family, resemble very much* Convolvulus *species of temperate regions. They are twining plants, often abundant in mantles.*

Mantles

Between the forest and the savanna a mantle forms. This belongs neither entirely to the forest nor to the savanna, constitutes a very heterogeneous habitat and is difficult to describe in a classical way. Mantles have often been ignored and few studies have been devoted to them. They represent no more than a narrow margin of vegetation, but they extend over thousands, even tens of thousands, of kilometres and their importance is immense on an ecological scale. In large part, their width depends on fire. In its absence, they can be fairly wide, but when there are frequent fires, they are generally very abrupt.

Despite their appearance, mantles are well structured. On a very short horizontal distance, but along the entire height of the vegetation, a mantle constitutes a gradient of micro-habitats that form a screen preventing or greatly reducing the lateral penetration of light.

Mantles also have their own array of species. Climbing plants and lianas are very abundant and they form the most effective screen. Some climbing plants of the Convolvulaceae and Cucurbitaceae form vast draperies that cover all the vegetation and cling to any remains of dead trees and shrubs. Among the Caesalpiniaceae there is, for example, *Duparquetia orchidacea* (FIGURE 4.22). Among the trees and upright shrubs, some are rare within the forest and instead come from the savanna, such as *Vitex doniana*, *Newbouldia laevis* (FIGURE 4.24) or the Sausage Tree *Kigelia africana*. Others are truly forest species, such as the African Tulip Tree or Flame Tree *Spathodea campanulata* with large bright red flowers, some *Xylopia*, a few Apocynaceae, including *Rauvolfia vomitoria*, a very sought-after species in traditional medicine, or *Barteria nigritiana* of the coastal forests (PAGE 83).

Mantles are very often the centre of an active dynamism. In a progressive mantle, where the forest is encroaching on the savanna (FIGURE 4.26a), pioneering species become established in more or less shade-loving herb fringes that spread under the large trees. In a first stage, it is mostly Zingiberaceae species or ferns, perhaps one or another of the Marantaceae, that replace the grasses. Next come ligneous plants, such as *Harungana madagascariensis* or, in Gabon at least, by Okoumé or Gaboon Mahogany. These ligneous plants accentuate the shade and prevent the Zingiberaceae and Marantaceae species from covering everything. Towards the interior of the forest, the same Zingiberaceae and Marantaceae are encountered again in the first few tens of metres of the undergrowth, but they die out as the mantle advances towards the savanna and give way to an open understorey.

In a regressive mantle (FIGURE 4.26b), fire penetrates only a few metres into the interior of the forest. In destroying the fringe of trees and shrubs that screen out the light, it allows ferns, Zingiberaceae and Marantaceae to proliferate and to spread towards the interior where they can supplant young ligneous plants of the understorey and retard or prevent forest regeneration.

The ecological importance of mantles goes beyond the simple framework of forest communities. For many animal species, they represent a preferred habitat. The abundance of flowers attracts numerous insects – bees, wasps, butterflies – and therefore numerous predators, spiders (FIGURE 4.23) as well as birds. Among the latter, sunbirds are particularly abundant. They feed both on the nectar offered by the flowers and the insects and spiders that they attract. For other birds, the mantle represents a good place to hide. This is where to search for the Black Bee-eater *Merops gularis* and the spectacular Black-headed Bee-eater *Merops breweri*. It is also where there is the best chance of observing forest goshawks and to surprise a Congo Serpent Eagle *Dryotriorchis spectabilis*. It is also along these same mantles that one can occasionally encounter birds of the upper canopy at lower levels, such as the Yellow-fronted Tit *Anthoscopus flavifrons*.

Colonisation of savannas

Independently from the expansion they can undergo within the framework of the mantles, forests can also spread by colonising savannas over vaster areas. This progressive transformation is related in some way to primary succession because all the elements that feed the process come from outside and are carried out on non-forest soils.

In Uganda, for example, the open grasslands that surrounded the Budongo forest in 1970 were at first transformed into wooded *Terminalia* grasslands. This phenomenon was probably set off by the drastic reduction of the elephant populations. Today these wooded grasslands are in their turn being colonised by forest species, notably trees of genus *Entandrophragma*. In this region there is a progressive succession that leads from the most open grass communities to dense semi-deciduous forest.

In the much more humid regions of Gabon and the Congo, a very different and at first somewhat disconcerting succession is observed. The savannas are first of all colonised by a pioneer dense forest, which transforms into a more open Marantaceae forest before becoming a mature forest, again much denser.

Pioneer forests

In Gabon, the development of pioneer forests was followed in detail by Lee White in the Lopé National Park where they cover enormous surfaces. Generally, they are born in a bushed grassland with *Crossopteryx febrifuga* and *Nauclea latifolia*, whose origin, as we have seen in the preceding pages, is at once linked to the poverty of the soil, the lesser rainfall, and the occurrence of annual fires.

If, for one or another reason, fires do not occur, this savanna evolves, however, invariably towards a forest habitat. Okoumé, Gaboon Mahogany, is the dominant element, but it is often accompanied by Azobé *Lophira alata*. These two species are dispersed by the wind and can thus be transported over several hundred metres. In places, the Liberian Cherry or Ozouga *Sacoglottis gabonensis* joins in, dispersed by elephants. These three species, also typical of evergreen forests of the coastal region of the Gulf of Guinea, end up forming dense, almost pure stands. Gabon Mahogany is usually domi-

4.26. *In a progressive mantle (a), as at the Lopé National Park, a nearly continuous column of foliage "closes" the forest laterally and, in the shade of overhanging branches, a more or less sciaphile herb layer develops from which ligneous pioneers emerge. Not very far inside the forest this herb layer gives way to a very open understorey.*

In a regressive mantle (b), as can be seen in the Ngotto region, fire destroys the trees of the mantle and the forest is not very "closed" laterally. Under the dead trees, a dense thicket of Zingiberaceae and Marantaceae species develops which is able to extend far from the interior of the forest and retard the regeneration of ligneous plants.

nant, but sometimes it is replaced by Liberian Cherry. At maturity, this species-poor forest reaches 40 metres in height. From the ridges of Mount Brazza it can be recognised from afar by the very uniform aspect and the fairly dull greyish green colour of its canopy. It is only when the Gabon Mahogany leafs out that the forest takes on a little colour thanks to the red-orange colour of the young leaves. In December and April-May, also, they can take on remarkable colours when the *Lophira*, whose young leaves are bright red, comes into leaf.

In the region of Odzala National Park, mainly around the Mboko camp, a similar succession can be observed. Gabon Mahogany and Liberian Cherry are absent, but Azobé reappears. These forests have not been studied in as much detail as at Lopé, but they seem to be richer, with species such as Wengé *Millettia laurentii*, False Nutmeg *Pycnanthus angolensis*, Dahoma *Piptadeniastrum africanum*, *Staudtia stipitata*, Pagoda Tree *Alstonia boonei* and Engona *Pentaclethra eetveldeana*.

Marantaceae forests

These very open formations are known from much of Central Africa but they are particularly abundant in Gabon and the Congo. Consisting of an almost continuous entanglement of Marantaceae and Zingiberaceae species where more or less isolated bushes and trees emerge, they intrigue botanists and foresters alike and have provoked much speculation on their origin.

According to studies by White in the Lopé National Park, these sparse forests clearly represent an intermediate stage between the pioneer and the mature forest. According to this model, the pioneer forest, having reached a certain stage of maturity, is no longer able to regenerate itself because the main species are strictly heliophile and are thus unable to grow in the shade. As these large trees are also all nearly the same age, they age and tend to die at the same time. Little by little, openings appear in this way in the canopy. These are immediately invaded by large Zingiberaceae and Marantaceae plants that, in a very short time, form dense, hardly penetrable thickets three to five metres high. Some liana-forming species of Marantaceae, such as *Haumania liebrechtsiana*, even climb over small trees which they suffocate with a column of dense opaque foliage (FIGURE 4.29). The remaining large trees, however, continue to develop and, little by little, emerge from the understorey of other species, better adapted to the shade.

At first, the Marantaceae forest is composed of the same species as the pioneer forest. Much later, it becomes richer and slowly alters its composition. This evolution can be very slow, perhaps even stopped at times, by elephants. In fact, these animals adore this habitat, especially in the dry season, and knock down

4.27. *Crossopteryx and* Nauclea *shrub savannas are floristically, as well as faunistically, very poor. They harbour species typical of open habitats, including the Long-legged Pipit* Anthus pallidiventris *and several species of cisticola, notably the Siffling Cisticola* Cisticola brachypterus, *the Whistling Cisticola* Cisticola lateralis *and the Croaking Cisticola* Cisticola natalensis, *as well as the Black-headed Puff-back* Batis minor *and the Little Button Quail* Turnix sylvatica, *although some groups, such as the francolins* Francolinus *and the savanna guinea-fowl* Numida, *are absent.*

4.28. *Stages of forest colonisation according to Lee White:*
(a) *early stage with* Crossopteryx, Nauclea *and* Lophira,
(c) *pioneer forest with* Aucoumea, *15 years old,*
(d) *pioneer forest with* Aucoumea, *50 years old.*

many small trees. In this way they keep the canopy from closing up. Sooner or later, it ends up being dense enough to extinguish the Marantaceae, which are unable to survive very dense shade. The forest then becomes a dense mature forest. In the Lopé National Park, the savannas are surrounded by a band of pioneer forest, themselves surrounded by a band of Marantaceae forests. The mature forests do not follow until some twenty kilometres from the savanna, except in the case of the mature forests rich in Caesalpiniaceae species that follow water courses and are in fact old forest galleries submerged in the younger formations by forest recolonisation.

In its typical form, the understorey of the Marantaceae forest is made up of a tangle of several species of Marantaceae and Zingiberaceae. At the Lopé, the dominant species is *Haumania liebrechtsiana*, but alongside this are found many other species, some of which are established in humid habitats. In these thickets, visibility is null and passage is impossible except along the paths of the buffaloes and elephants. The more or less widely separated large trees, permit an abundant light, but many small trees are covered over with columns of climbing Marantaceae that can reach ten metres high.

In Odzala-Kokoua National Park, considering their enormous distribution, these Marantaceae forests (FIGURES 4.31, 4.34) vary from one place to another and Rollet had already distinguished three principal types. In the south-west of the park, at altitudes of 500 to 700 metres, they are characterised by the presence, notably, of *Piptadeniastrum africanum*, *Pentaclethra eetveldeana*, *Canarium schweinfurtii* and *Dacryodes buettneri*. In the low central zone between the Mambili and the road going from Sembé to Ouesso as well as in the north-eastern corner of the park, *Markhamia tomentosa* and a *Morinda* species are very common, also with, notably *Entandrophragma candollei*, *Alstonia congensis*, *Millettia laurentii*, *Funtumia elastica*. In the regions situated between the Mambili and the south of Ouesso, *Entandrophragma utile*, *Entandrophragma cylindricum*, a *Guarea*, *Terminalia superba*, *Milicia excelsa* and *Pycnanthus angolensis* are encountered instead.

In the light of observations at Lopé, it would thus seem that the Marantaceae forests could give rise to very different mature forests. At Odzala, however, it remains not at all certain that this happens and that these forests would have the same significance as those of the Lopé. Of course, in progressing from the open savanna towards the interior of the dense forests (FIGURE 4.34), the same sequence as at Lopé is found. In the very first place, a dense pioneer forest with a single Marantaceae, *Sarcophrynium schweinfurtianum*, in low density, next, a more open Marantaceae forest, dominated by *Haumania liebrechtsiana* and, finally, a mature dense forest.

4.29. *A Marantaceae forest in Lopé National Park, with columns of the climbing* Haumania liebrechtsiana *which suffocate small trees.*

4.30. *Transect view of a Marantaceae forest in the Lopé National Park.*

4.31. A path in the Marantaceae forest in Odzala-Kokoua National Park.

4.32. The Marantaceae Ataenida conferta grows in damp places, but not in water. It is recognisable by the red bracts that surround its inflorescences. It is very common in the forests Ipassa reserve near Makokou in Gabon.

4.33. The bright red fruits of the Marantaceae Megaphrynum, found in the coastal forest of Gabon, are eaten by elephants and apes which contribute to the dispersal of the species. Other Marantaceae, like Haumania liebrechtsiana, produce dehiscent capsules from which seed dispersal is achieved "all by itself", but is not effective beyond a few metres. In fact, most Marantaceae and Zingiberaceae species disperse mostly by vegetative reproduction, either by rhizomes or by the formation of plantlets at the end of stems or inflorescences.

According to David Brugière and his colleagues, the Marantaceae forests apparently also colonise mature dense forests which they finally reduce to islets. This expansion seems to begin in tree-fall gaps situated not far from Marantaceae formations. These gaps are then "infected" by the *Haumania* which would suppress the regeneration of ligneous plants, and would open them up. By the confluence of "infected" tree falls, the dense forest would become fragmented and then reduced to smaller and smaller patches.

The role of large and medium-sized mammals in the progression of Marantaceae species is in fact not obvious and probably completely secondary. At most one could imagine that elephants, as at Lopé, favour the Marantaceae by breaking small trees. The role of Marantaceae forests in faunal impoverishment, on the contrary, leaves no doubt. Except for gorillas and elephants, which may be locally more abundant in Marantaceae forests than in dense forests, and forest duikers, which do not display significant population trends, most animal groups decrease, both in species diversity and in densities reached by individual species. In the case of small primates, this decrease appears to be of the order of 50% and to result both from a structural transformation of the habitat and from a reduction in nutritional resources. For birds, the decrease is apparently even more severe, since the population is reduced to the point where only the communities of the dense undergrowth and those of the upper canopy remain. Birds of the intermediate strata, on the other hand, disappear completely, a situation referred to by Chris Wilks as an "empty sandwich".

If this hypothesis proves to be correct, then there is reason to be concerned about the biodiversity of the national park. The last word has not been written about these forests, and it is obvious that more in-depth studies on the entirety of Odzala National Park are necessary and it might perhaps be interesting to see what carbon 13 could tell us (PAGE 55).

4.34. (Page 115) The satellite photograph of Odzala-Kokoua National Park shows very well the distribution of the dense mature forests, the enclave savannas around Mboko, and the Marantaceae forests. The dark formations along the rivers are swamp forests. In the north-west of the national park, the dark line is a north-south abrupt escarpment more than 70 kilometres long. The white spots are clouds and the black spots are their shade on the ground.

4 TEMPORAL GRADIENTS

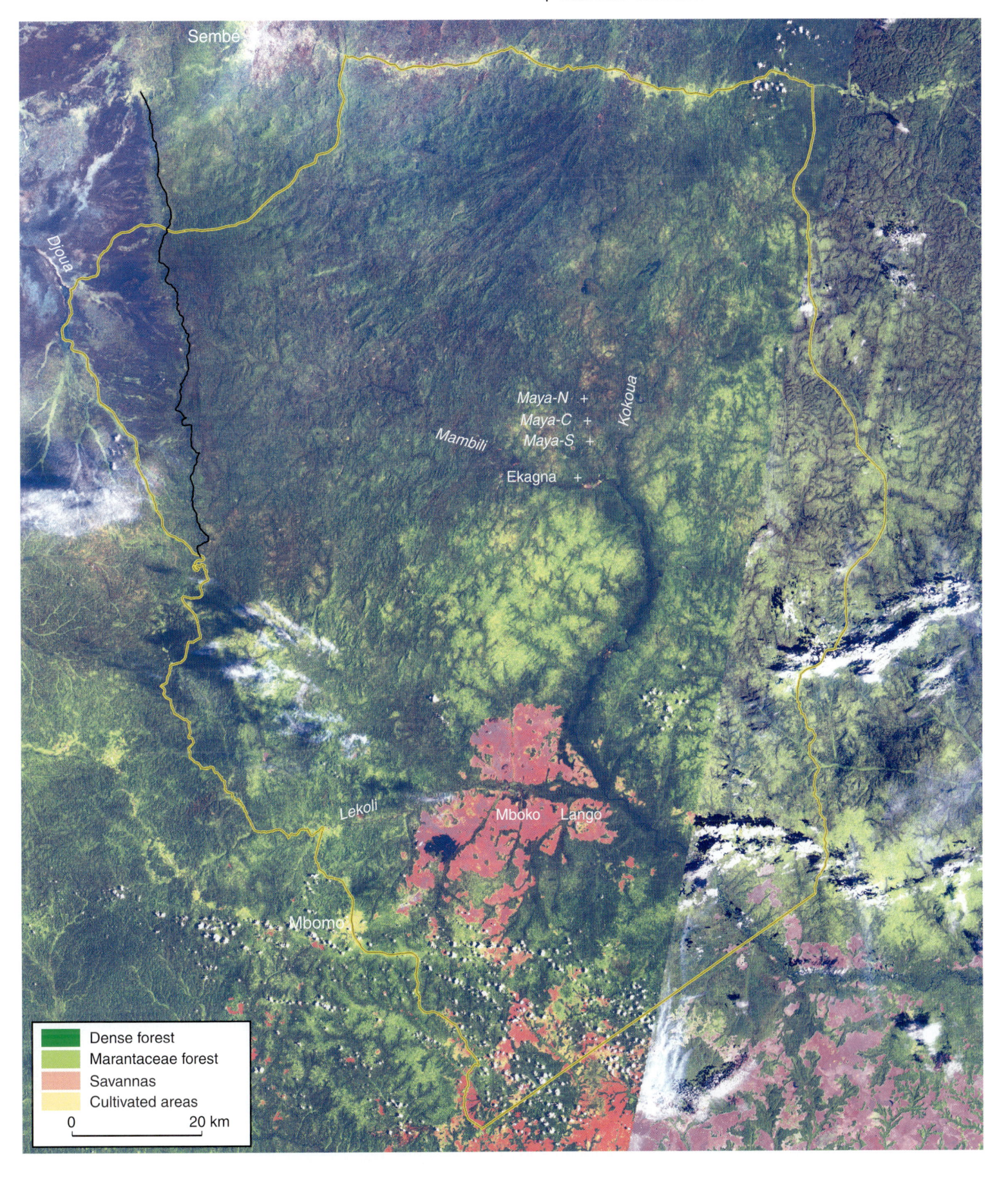

Dense forest
Marantaceae forest
Savannas
Cultivated areas

4.35. On the hill tops and upper slopes of the eastern part of Nyungwe Forest, where most of the ridges reach 2,500 to 2,700 metres altitude, forests of Macaranga capensis are generally fairly dense and are composed of trees having about the same age. At the bottom of glades, where the soil is deeper and often water-soaked, formations are usually very open, with a lower stratum of Bracken Pteridium aquilinum and Sericostachys scandens, a very invasive Amaranthaceae creeper.

The montane pioneer forests

On the mountains of the Albertine Rift, between 2,000 and 2,600 metres altitude, there are vast areas of forests of *Macaranga capensis*, a tree of the Euphorbiaceae family. They constitute the dominant landscape in the eastern part of the Nyungwe forest (FIGURE 4.35), in Kibira National Park, in most of Bwindi Impenetrable National Park and Kahuzi-Biega National Park. They are constituted of dense, monospecific stands of even-aged trees. Under their quite simple canopy, there is a dense understorey of Acanthaceae of the genus *Mimulopsis*.

In a few places these forests are clearly secondary, but then they cover only small areas, often along roads. Everywhere else these forests are pioneer formations – not unlike Okoumé forests in Gabon – and they represent a transitional stage that normally evolves towards more varied formations, notably by the establishment of *Syzygium guineense* or of *Olea capensis*. This evolution seems to be quite slow, but in many places of the Nyungwe forest it is clearly visible. It is not at all clear, however, how they became established and the origin of vast homogeneous surfaces remains a mystery. Some stands are definitely older than other stands, but in most cases they are probably between 100 and 200 years old. Evidently these forests have replaced some other vegetation in quite a short period of time.

Some observations from Nyungwe forest suggest that these *Macaranga* forests can replace thickets of Mountain Bamboo *Sinarundinaria (Arundinaria) alpina*. This bamboo, a giant grass like all the bamboos, is encountered at altitudes between 1,800 and 3,300 metres. In many places, it grows only in isolated clumps, but it can also form a band of dense, nearly monospecific vegetation. In this way, in the Rwenzori, it grows in almost uninterrupted thickets between 2,200 and 3,200 metres altitude; in the Virunga, between 2,300 and 3,000 metres (FIGURE 4.36); in the centre of Nyungwe Forest and at Kahuzi-Biega between 2,300 and 2,700 metres. This distribution could lead one to believe that bamboo thickets represent a form of climax vegetation, appearing between 2,200 and 3,200 metres altitude. In reality, they merely constitute a successional stage in the still poorly known montane forest system.

Bamboo thickets are a very poor habitat, but it is possible to find there, mostly in the openings, some herbaceous plants of the adjacent forests, notably *Mimulopsis solmsii* and *Mimulopsis elliotti*. These Acanthaceae species flower only once every three to seven years, gregariously, as do plants of *Mimulopsis arborescens* of very humid montane forests. After flowering, these plants die massively and a new cycle starts again. Curiously, bamboos also flower and die gregariously, but after much longer periods, from 15 to 30 years or even much more. In fact, in some areas no one has ever seen them bloom.

In Nyungwe Forest, in Rwanda, several flowering periods have been observed. Each time several hundreds of hectares have been affected that were transformed into vast clearings rapidly colonised by inextricable coppice. In one case, the bamboo gave way to pure populations of *Macaranga capensis*. In another, it made room for mixed stands of *Hagenia* and *Macaranga capensis*.

These observations testify to the disappearance of bamboo and show how this plant can suddenly be replaced over large areas by pure stands of *Macaranga* or something else. More precisely, they suggest that the vast *Macaranga* forest we know today were bamboo thickets one or two centuries ago. However, these observations leave not a clue as to the establishment of bamboo.

In many places, large isolated trees are found mingled with the bamboo. Below 2,500 metres altitude, these are montane mixed forest species: *Olea capensis*, *Chrysophyllum gorungosanum*, *Syzygium guineense* or *Faurea saligna*. Above, it is generally the conifer *Podocarpus latifolius*. These trees are all old. It might be that they represent vestiges of a forest that could have preceded the bamboo and that disappeared as a result of human activity, or by climatic events followed by extensive fires. In the vast bamboo thickets of Rwanda and Burundi, however, there is no trace of human presence during the last 2,000 years. Indeed, the current populations of these high-altitude regions are not peoples of the forest but savanna peoples: they do not inhabit forests but convert them into permanent cultivation. It is also possible that these trees germinated at the onset of development of the bamboo thicket. Their spread and sometimes twisted carriage – Podocarps, especially, often have a very short, gnarled crown – suggest that they developed in an open habitat.

Inevitably we are tempted to establish some parallel between the succession from bamboo thicket to pioneer *Macaranga* forest and the successions involving Marantaceae forest. The more so, that in the very centre of the Nyungwe forest there are extensive stands of open forest with an understorey of Bracken *Pteridium aquilinum*. Just as in the lowlands, there could be a succession of pioneer or colonising forests in the uplands involving open types of forest. Unfortunately we too often qualified these open stands as man-made, "disturbed" habitats, and thus ignored their real origin. All these "unexpected" types of forest probably are part of complex vegetation cycles, and should be much better investigated. Modern means should be able to allow this.

4.36. *At 2,700 metres altitude, thickets of Mountain Bamboo* Sinarundinaria (Arundinaria) alpina *of the Rwandan slope of Gahinga volcano are low and fairly dense, with an understorey of* Mimulopsis. *Bamboo stems with a diameter of more than five centimetres are rare. In the Kahuzi-Biega National Park region and in Burundi, there are much taller formations with much thicker canes. Bamboo thickets, tall or small, are poor habitats but nevertheless they harbour some endemic species, like Rockefeller's Sunbird* Nectarina rockefelleri *and the butterfly* Charaxes bohemanni.

4.37. Not far from Mboko camp in Odzala-Kokoua National Park, the shores of the Lékoli River show all the stages of colonisation of humid habitats. Here a bay, shallow in the dry season, is bordered successively by muds, thickets of Cleome afrospina *and stands of Dwarf Date Palm* Phoenix reclinata.

4.38. Nymphaea lotus, *a Nymphaeacea, is recognisable by its white flowers and leaves with crenate edges.*

4.39. Water Lettuce Pistia stratiotes *is a small floating plant of the Araceae family, with roots hanging freely in the water. It has a very wide distribution in tropical Africa, but is plentiful mostly where the water is laden with nutrients.*

Colonisation of humid habitats

Nearly 20% of the surface of Congolese forests is occupied by various humid habitats, essentially represented by water courses, the Ogooué delta with its associated lakes and mostly the enormous marshy expanses of the centre of the basin, whose ramifications reach all the way to Odzala National Park. To all this can be added the lakes and innumerable anonymous marshes hidden away in the forest and dispersed throughout the entire region, from sea level to over 2,500 metres altitude. All these habitats are extremely dynamic and most represent no more than stages in a succession, that could be qualified as primary and of which the ultimate result is a terra firme forest.

Pioneer formations

At the onset of this type of succession, there is only open water and the very first stages consist of submerged vegetation: *Ceratophyllum* or *Utricularia* beds. These formations need light, however, and therefore cannot develop except in sufficiently limpid waters. In water loaded with alluvium, as in the rivers in Odzala

118

National Park, and in shallow areas, only rooted vegetation with leaves and flowers floating at the surface can develop. These are mainly water lily beds of *Nymphaea lotus*, recognisable by their white flowers and leaves with crenate edges (FIGURE 4.38). Elsewhere there are *Nymphaea nouchallii*, with pink or blue flowers and leaves with entire edges, or rosettes of Water Lettuce *Pistia stratiotes*, a small floating Aracea (FIGURE 4.39).

In Odzala National Park, beds of *Cleome afrospina* develop on unencumbered but regularly flooded banks of the Lékoli and the Mambili rivers (Figures 4.40), together with *Mimosa pigra*, associated with floating meadows of Water Lettuce or tall grasses – mainly *Vossia cuspidata* or *Echinochloa* species. These floating meadows can break up, detach from the banks, drift off and go on to colonise other reaches. The huge draperies of roots hanging in the muddy water from under this floating carpet slow the current, trap alluvia in suspension and eventually attach.

The first stages of this succession to announce the forest are dominated by *Pandanus* (FIGURE 4.41) or the palms *Phoenix reclinata* (FIGURE 4.37), *Elaeis guinensis* and especially genus *Raphia*. Very rapidly they are accompanied by creeping palms and rattans of genera *Laccosperma* (PAGE 154), *Calamus* and *Eremospatha*. Dicotyledonous trees will not become established until

4.40. In Odzala-Kokoua National Park, dense masses of Cleome afrospina *occupy the exposed silt of the Mambili River's alluvium banks.*

4.41. Not far from there, along the Lékoli River, essentially impenetrable thickets of Pandanus candelabrum *stand up on their stilt-roots. This small, little-branched tree is about ten metres high. Its trunk is covered with fine spines, as are the edge of its long leaves. In spite of this, it is eaten by elephants. The Pandanaceae have few species in Africa, yet they inhabit all the tropical regions of the world.*

much later. *Uapaca*, with its spectacular stilt roots, is usually installed along the immediate edge of the water course (PAGE 125). Inundatable forests (PAGES 124-126) or sometimes swamp forests (PAGE 126-127) form further inside.

A very different succession takes place in the Ogooué delta in Gabon. As we have seen (PAGES 32-33), this delta covers 500,000 hectares and therefore represents one of the main fresh water humid areas of sub-Saharan Africa (FIGURE 4.44). Nowadays, maritime influences are, in fact, limited to the area immediately adjacent to the ocean. Originally this vast complex probably constituted an estuary, however, somewhat like the Komo and the Bokoué estuaries farther north (PAGE 142). Unlike these two water courses, the Ogooué carries along much more substantial volumes of sediments. Having progressively filled in the entire central part of this estuary and its lower course between

4.42. *North of lakes Evaro, Ezanga and Onangué, immense floating meadows of* Vossia cuspidata *extend over the Ogooué delta. They attract many birds, notably the Little Bittern* Ixobrychus minutus, *Allen's Gallinule* Porphyrula alleni *and the Winding Cisticola* Cisticola galactotes, *as well as the Hippopotamus* Hippopotamus amphibius *and the West African Manatee* Trichechus senegalensis. *During the boreal winter, more or less open areas, or places degraded by large mammals, are sought out by White-winged Black Terns* Chlidonias leucopterus, *some Whiskered Terns* Chlidonias hybridus *and the local form of the Common Tern* Sterna hirundo.

4.43. *The channels that feed the lakes of the Ogooué delta are bordered by dense thickets of* Alchornea cordifolia, *a shrub of the Euphorbiaceae family with which mingle reedbeds of* Echinochloa, *a tall grass. This inpenetrable place is the favourite habitat of the Reed Warbler* Acrocephalus arundinaceus, *a Eurasian migrant. Kapok Trees* Ceiba pentandra *inhabit banks of alluvium. They offer an excellent support for colonies of African Darter* Anhinga rufa *and Pink-backed Pelican* Pelecanus rufescens.

4 TEMPORAL GRADIENTS

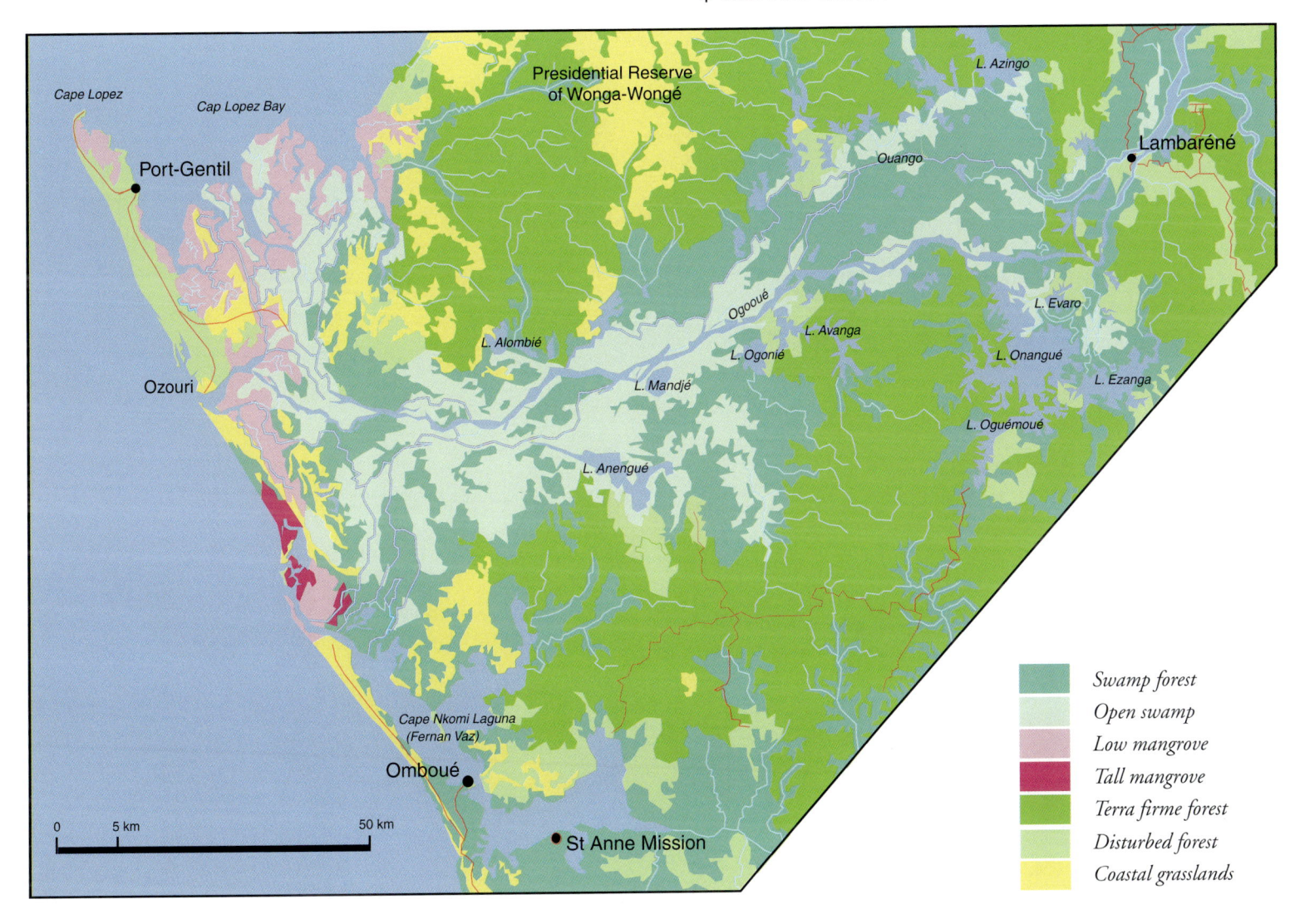

Lambéréné and Port-Gentil, it is now clearing a complex and tortuous passage through its own alluvial deposits.

Only the "bays" farthest away from the central axis of the river have escaped filling in and to this day form a series of lakes with capricious contours separated from one another by hills of high ground – the shores of the ancient estuary – or by alluvia from the Ogooué. There are some to the north and the south of the river, but the most important ensemble is made up of lakes Ezanga, Evaro, Onangué and Oguémoué, two or three hours by dugout south-west of Lambaréné. Their southern reaches, with deep bays, innumerable islets and very indented shores, usually covered with dense forests, resemble landscapes of drowned valleys. Their northern shores, on the contrary, make up part of a vast interior delta, or a multitude of small deltas, fed by secondary branches of the Ogooué that come to deposit their sediments.

The vegetation of the pioneer front is formed by immense floating meadows of *Vossia cuspidata* (FIGURE 4.42). Upstream from these are stands of Papyrus *Cyperus papyrus*, reedbeds of *Echinochloa*, semi-floating thickets of *Alchornea cordifolia*. Still further upstream, Kapok Trees *Ceiba pentandra* grow, mostly on banks of alluvia.

When the water level is low, as in July-August, or in January, water reaches the lakes by narrow, deeply cut channels (FIGURE 4.43). Water from lakes Ezanga and Evaro then filters towards Lake Onangué, where another channel drains it north-westwards and carries it back to the Ogooué. When the water level is high, as in November or April, the water spills over the deposit banks and reaches the lakes directly by filtering across the inundated forests.

4.44. The Ogooué delta with its main vegetation types. During the dry season in July and August the main course of the river is largely blocked by sand banks. During other periods the flow is strong enough to push back the salt water of the sea. Consequently, 85% of the complex is made of fresh water habitats and the marine influence does not enter more than 15 to 20 kilometres inland.

4 TEMPORAL GRADIENTS

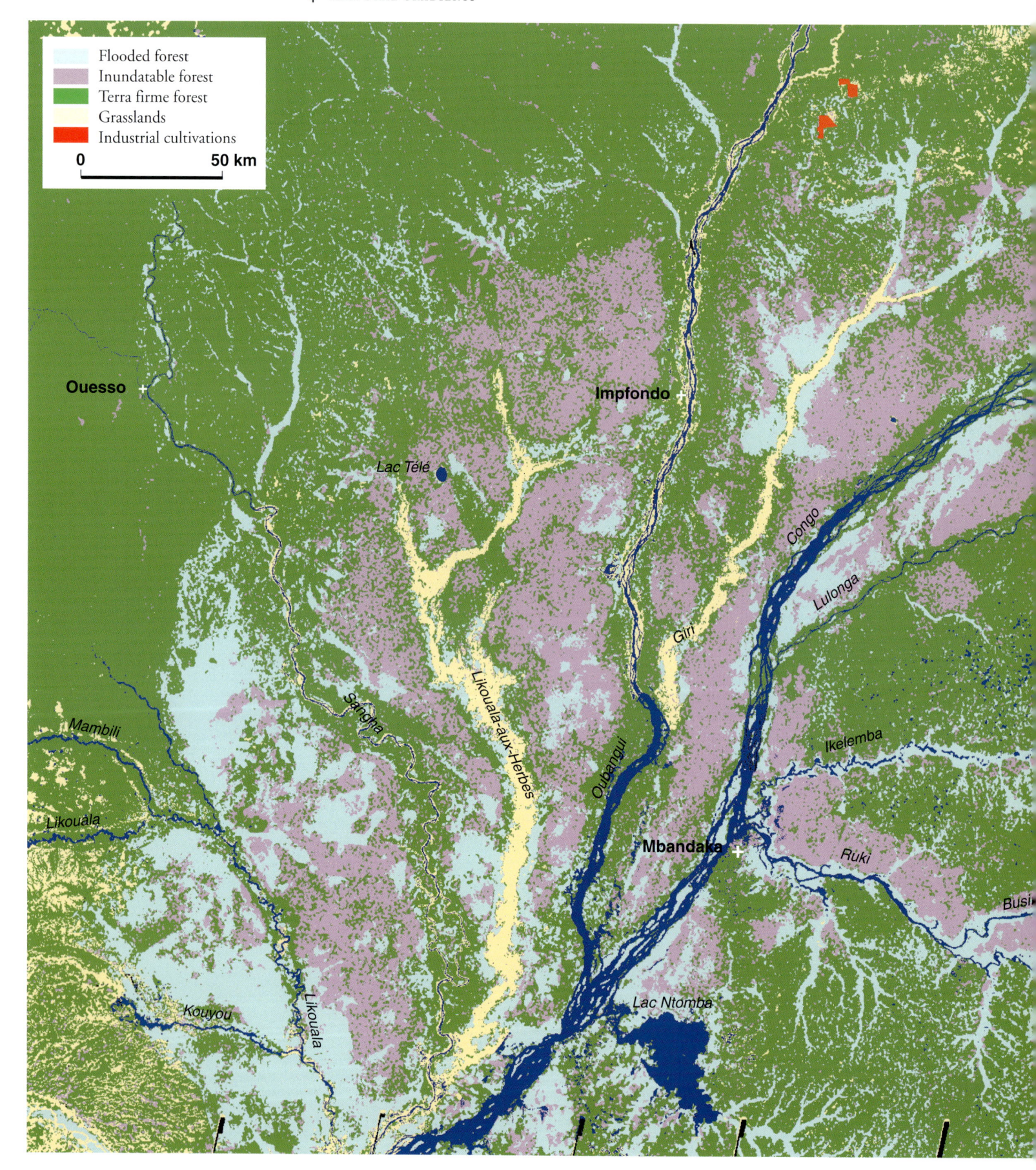

Legend:
- Flooded forest
- Inundatable forest
- Terra firme forest
- Grasslands
- Industrial cultivations

0 — 50 km

Ouesso, *Impfondo*, *Mbandaka*

Lac Télé, *Lac Ntomba*

Rivers: *Mambili*, *Likouala*, *Kouyou*, *Sangha*, *Likouala*, *Likouala-aux-Herbes*, *Oubangui*, *Giri*, *Congo*, *Lulonga*, *Ikelemba*, *Ruki*, *Busi*

122

4 TEMPORAL GRADIENTS

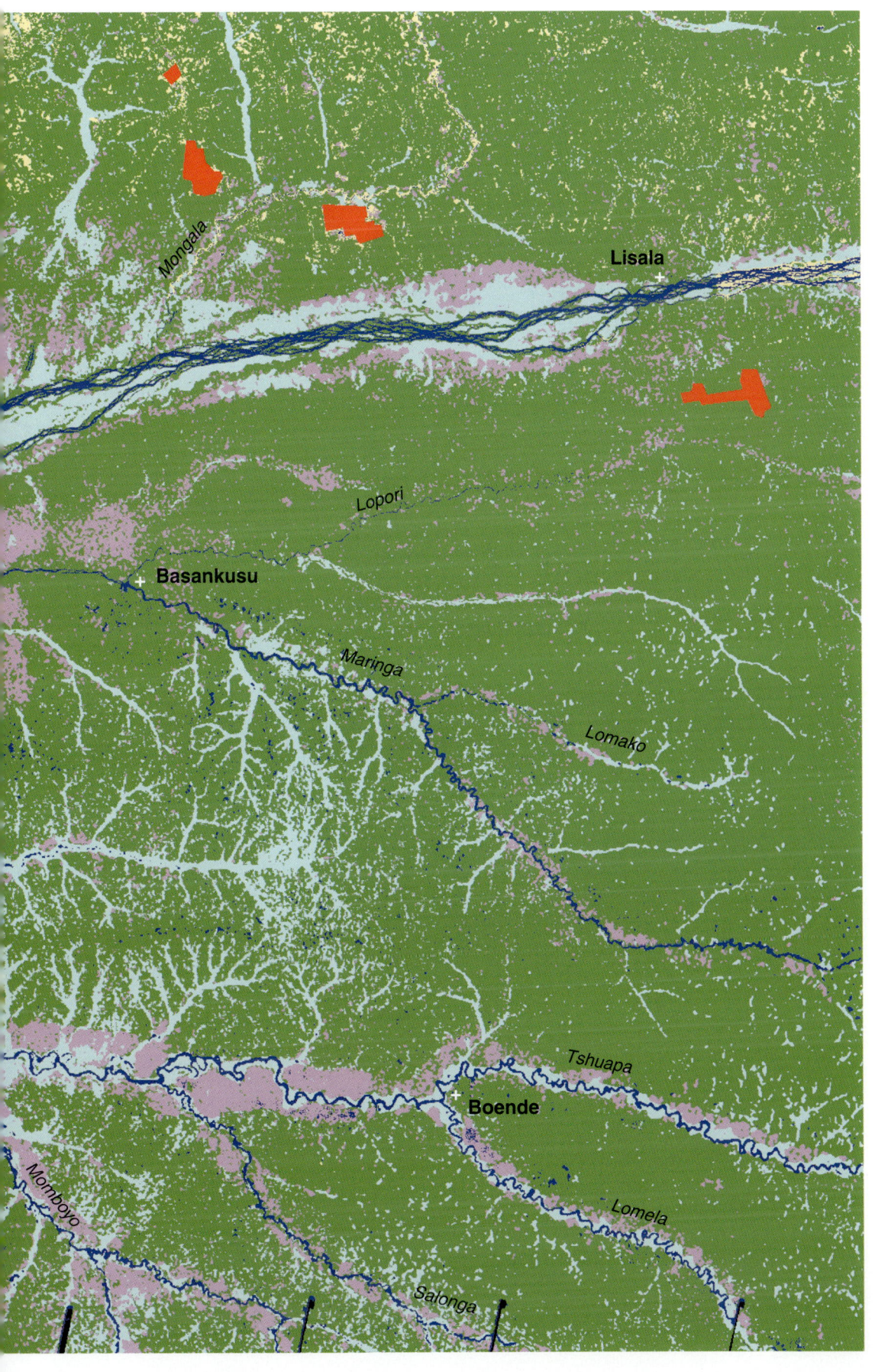

4.45. In the centre of the Congo basin, the confluent of the Congo, the Ubangi, the Likouala-aux-herbes, the Sangha and the Likouala rivers engender one of the largest marshes in the world. Covering nearly 200,000 square kilometres it is about the same size as the Pantanal in South America. Thanks to the combination of classic satellite photos and radar photos, the Joint Research Centre of the European Commission in Ispra (Italy) was able to produce a detailed picture showing the real extent of the main vegetation types of that nearly inaccessible area, especially the swamp forests and inundatable forests. The grassy habitats along the Likouala-aux-Herbes and the Giri rivers are in fact extensive floating meadows, subjected to seasonal fires.

4.46. An inundated forest bordering the Ogooué River, more precisely, between the river and Lake Evaro.

Inundatable forests

When fluctuations of the water level are great, inundatable forests form. During periods of strong flooding, travelling is made by dugout (FIGURE 4.46), but during low-water periods it is easy to walk on foot (4.48). Depending on the composition of the water that floods them, these forests are divided into two very distinct types.

Forests of the *varzea* type, a name of Amazonian origin, develop along watercourses heavily loaded with silt (FIGURE 4.47). These are the most widespread type in Central Africa. They cover enormous surfaces in the centre of the Congo basin (FIGURE 4.45) as well as along some distant tributaries, such as the Mambili in Odzala-Kokoua National Park. Among the most characteristic species are *Uapaca heudelotii*, recognisable by its spectacular stilt roots, *Cathormion altissimum*, *Pterocarpus santalinoides*, *Haplormosia monophylla*, *Anthonotha vignei*, various species of *Berlinia*, *Chrysophyllum africanum*, *Cynometra megalophylla*, *Drypetes afzelii*, several *Gilbertiodendron* species, *Gluema ivorensis*, *Manilkara argentea*, *Millettia chrysophylla*, *Myrianthus serratus*, *Nauclea pobeguini*, *Parinari robusta*, *Pierrodendron kerstingii*, *Plagiosiphon emarginatus*, *Strephonema pseudocola* and *Syzygium rowlandii*.

Very similar forests occur in the delta of the Ogooué, between Lambaréné and Port-Gentil, as well as further upstream, notably along the Ngounié. Several *Uapaca* species (FIGURE 4.48) are to be found here, as well.

4.47. *During the rainy season, the high, reddish brown waters of the Mambili River flood varzea type forests.*

4.48. *During periods of low water, these inundatable forests dry out and the immense stilt-roots of the* Uapaca *can be admired in all their splendour.*

Forests of the *igapo* type, another name of Amazonian origin, are subjected to floods by "black waters", without silt and oligotrophic. In Atlantic regions, these waters are limited to the middle course of some rivers of Cameroon, such as the Nyong, and of north-eastern Gabon, such as the Ivindo and the Zadié (FIGURE 4.49). Small stretches are also found in the interior delta of the Ogooué. In the Congo basin, they exist mainly in region of Lake Mai Ndombe and along the Fimi River. They engender forests rich in endemic species, particularly of the Caesalpiniaceae family. Among characteristic species are *Sindora klaineana*, *Berlinia bracteosa*, *Gilbertiodendron brachystegioides*, *Gilbertiodendron klainei*, *Monopetalanthus microphyllus*, *Marquesia excelsa*, *Guibourtia demeusei*, *Oxystigma buchholzii*, *Haplormosia monophylla* and *Irvingia grandifolia*.

Swamp forests

When conditions are such that water level fluctuations are weak and the water level never very far from the surface of the soil, swamp forests develop. Often these forests are dominated by palms of genera *Phoenix*, *Raphia* and even *Elaeis*. The palms of genus *Raphia*, of which several species are known from Central Africa, are found up to nearly 1,200 metres altitude, notably on the banks of Lake Victoria. In many places, they constitute nearly pure formations: raphia forests (FIGURE 4.50). Also found are vast forests of Dwarf Date Palm *Phoenix reclinata*, notably at Odzala and bordering the mangroves of the Gulf of Guinea. On the eastern shore of Mondah Bay, for example, this palm forms a dense band 20 to 100 metres wide between the mangrove and the raphia forests. They are found along the edge of the immense papyrus beds of Lake Victoria, up to an altitude of 1,300 metres. Elsewhere, this species may mix with the Oil Palm *Elaeis guineensis*, notably along the shores of the Semliki River, on the border between the Democratic Republic of the Congo and Uganda, or on the Gabonese coast, for instance near Cap Estérias (FIGURE 4.51).

4.49. The Zadié River, affluent of the Ivindo, drains waters from north-eastern Gabon. Since the forests of its catchment basin are still intact and substrates are mostly of crystalline origin, it is a "black water" river. These waters are transparent, but very acid and with high tannin levels, giving them a blackish or even reddish colour.

4.50. Palms of genus Raphia, *known locally under the name «bambou», are encountered in all the humid forest bottomlands. In places, however, it forms dense palm stands, notably along the Ivindo River, downstream from Makokou. These of the Liboumba, a small tributary of the Ivindo, serve as a roost for Grey Parrots* Psittacus erithacus *which assemble there in the thousands. The palm is found in dense formations along Mondah Bay or in the Gabon estuary, especially between stands of* Phoenix reclinata *and the terra firme forest.*

4.51. *The* Phoenix reclinata *swamp forest in the coastal region of Gabon. In the Lake Victoria region, in Uganda, its trunk, very rot-resistant, is used as construction material.*

River banks

Just as in between forest and savanna, a margin forms between the forest and the free water of lakes and water courses. Generally, margins form a veritable wall of greenery that closes off the forest laterally. In many places, this wall is "consolidated" by draperies of Cucurbitaceae or Convolvulaceae climbers (FIGURE 4.52).

Like the mantles of savannas, mantles overhanging lake shores and river banks are not unchanging. Their dynamism is not, however, determined by fire, but by the force of the water: where the current nibbles at the bank it recedes, laying bare the tangle of tree roots (FIGURE 4.53). Where the river deposits its alluvia, the bank advances. In fact, it is precisely in these zones that the different successive stages are found: *Vossia* rafts, *Alchornea* thickets, and thickets of *Pandanus*, *Phoenix reclinata* or *Calamus*.

As in savanna mantles, shores have their own species. not only the ones that characterise the successions we just mentioned, but many others, as well. Among lianas, can be mentioned the Cucurbitaceae *Cogniauxia podolaema*, recognisable by its large yellow flowers and its five-lobed leaves. Many species of shrubs can be named, but we will note only *Millettia griffoniana*, with its superb lilac inflorescences (FIGURE 4.54), or the *Garcinia*, of the Clusiaceae family, whose vivid yellow fruits are eaten by fish when they fall into the water.

Finally, among trees, *Uapaca heudelotii*, recognisable by its spectacular stilt-roots, is obviously very characteristic (PAGE 2), but otherwise the arborescent vegetation of shores is often dominated by numerous Caesalpiniaceae species. Among them can be noted sev-

4.52. A small river in the Ogooué delta system not far from Lambaréné. Banks are often invaded by curtains of Cucurbitaceae or Convolvulaceae climbers which cover the vegetation and close up the forest laterally.

4.53. Along the Ogooué, roots exposed by erosion of the banks.

eral *Berlinia* species or *Baikiaea robynsii*, recognisable from afar by its large white flowers. There is also the Moraceae *Treculia africana*, or the Rubiacea *Nauclea diderrichii*, identifiable at a distance by its tiered silhouette.

As is the case of savanna mantles, lake and river shores have their own fauna.

Among mammals, we can mention otters, the Water Chevrotain *Hyemochus aquaticus* and the Sitatunga *Tragelaphus spekei*.

Among birds, there are Peters' Finfoot *Podica senegalensis*, that threads its way under vegetation overhanging the water and among the jumble of bare roots along the eroded bank; Cassin's Grey Flycatcher *Muscicapa cassini*, a small, unobtrusive bird that hunts by lying in wait on dead branches or stones emerging from the water; the White-throated Blue Swallow *Hirundo nigrita*, that hunts at the surface of the water but spends long periods resting on dead branches; the Mountain Wagtail *Motacilla clara*, that mostly frequents rocky shores; the kingfishers *Alcedo quadribrachys* and *Corythornis cristata*, the observation of which is often reduced to a flash of blue streaking over the water. Among the Nectariniidae, the Violet-tailed Sunbird *Anthreptes aurantium* and Reichenbach's Sunbird *Nectarinia reichenbachii* are especially attached to shores. Finally, among the Ploceidae, there is the Black-throated Malimbe *Malimbus cassini* that is strictly linked to the presence of the rattan *Laccosperma secundiflora* where it hangs its nest fitted with a long entrance tube.

4.54. Millettia griffoniana, *a more or less creeping Fabacea with pale lilac flowers, overhangs usually somewhat stony banks.*

4.55. *Stony banks along the Ogooué River, at the level of the Lopé National Park, are covered with dense thickets capable of withstanding the violent currents of the highest floods. Among the typical plant species of that habitat type, are* Mostuea brunonis, *from the Loganiaceae family, and* Donella ogowensis, *from the Sapotaceae family.*

4.56. Biophytum zenkeri, *of the Oxalidaceae family, forms small, miniature thickets between rocks on the banks of the Ogooué.*

4.57. Lango clearing, near Odzala National Park, is often under water, when one gets around by dugout. Formerly, this clearing was the site of traditional salt exploitation, from which the name saline.

4.58. The Green-backed Heron Butorides striatus *is a regular visitor of flooded salines, as is the Little Egret* Egretta garzetta, *the Purple Heron* Ardea purpurea, *the Cattle Egret* Bubulcus ibis *and Hartlaub's Duck* Pteronetta hartlaubi.

Swampy glades or bais

In south-eastern Cameroon, eastern Gabon, northern Congo, the south-western Central African Republic and the western Democratic Republic of the Congo, more or less marshy clearings occur in the forest. They appear under various names in the literature. In the central Congo basin, they are spoken of as *ésobés*. At Odzala, they are *salines* or *bais* – a term used by the Pygmies. Sometimes they are also simply called swampy meadows or elephant baths.

Remarkable from a botanical point of view, these unexpected open habitats are true poles of attraction for the fauna of the surrounding forests. At Odzala, for example, the Maya-Nord clearing was visited at the end of the 1980s by a community of large mammals, including essentially Elephant, Buffalo, Sitatunga, Gorilla, Red River Hog and Giant Hog; also, occasionally, Peters' Duiker *Cephalophus callipygus* and Blue Duiker *Cephalophus monticola*, as well as Bongo *Tragelaphus euryceros*. At night, added to the list are Swamp Otter *Aonyx congicus* and Spotted Hyena *Crocuta crocuta*. In addition, the great number of individuals of some species that visit this clearing – more than 600 elephants and 400 gorillas – gives us an insight on the impact of these clearings on vast surfaces.

Apart from that, the clearings represent a unique opportunity to observe and study the behaviour of some of the great forest mammals and they are a major asset from the tourism point of view, all the more so because of the fact that forest national parks are very difficult to valorise in this field. Consequently, it is understandable that the ECOFAC programme would consecrate time and means to a more in-depth study of these clearings.

Professor Jean Lejoly's team, of the Université libre de Bruxelles, with the help notably of Professor Lisowski of Krakow, studied botanical aspects. Florence Magliocca, of the Université de Rennes, studied the fauna under the supervision of Annie Gautier-Hion.

At first glance, these clearings comprise two types of formations, different by their relationship with the hydrographic system (PAGE 118). Some of them are located along the Lékoli and Mambili rivers, either near their confluence with small tributaries or in depressions located behind banks of alluvia. The others are completely inside the forest massif, far from any important water course.

Whatever their position, their size varies from a few hectares to nearly 100 hectares. Some, such as the Maya-Nord clearing, comprise a small swampy zone in the centre and are not flooded except for short periods. Others, such as the Lango clearing, are under water more often (FIGURE 4.57).

At a strictly botanical level, all these clearings are fairly different, but in the end they represent variations on the same theme. The Lejoly team recognises seven main plant associations, described in detail at the bai de Mouébé by Lisowski.

The aquatic vegetation of small pools of shallow stagnant water, strongly tinted black by humus and situated in the middle of peaty areas, is characterised by the presence of waterlilies *Nymphaea maculata* and *Nymphaea lotus*. While this vegetation covers no more than very small surfaces, it brings to the entirely forested zone plants that are rather rare in Central Africa, such as *Nymphoides forbesiana* of the Menyanthaceae family, *Laurembergia tetrandra* of the Haloragaceae, *Hydrocharis chevalieri* of the Hydrocharitaceae, *Heteranthera callifolia* of the Pontederiaceae – like Waterhyacinth *Eichhornia crassipes* – and *Hydrolea glabra* of the Hydrophyllaceae. Some of these pools, notably near the Ekagna camp, are almost entirely covered over by Water Lettuce *Pistia stratiotes*, indicating a strong eutrophisation after the presence of large mammals.

In the peripheral area of these pools, notably at Mouébé, the damp peaty soils support a meadow of sphagnum *Sphagnum planifolium*, *Mesanthemum radicans*, of the Eriocaulaceae family, and *Selaginella congoensis*, a small pteridophyte. To these few dominant plants are often associated numerous other species, among including *Drosera pilosa* of the Droseraceae family and several species of genus *Xyris* of the Xyridaceae. This vegetation is typical for oligotrophic humid habitats of Central Africa.

Very different is the *Bacopa crenata* meadow that occupies areas strongly marked by the passage of animals. The soil, regularly laid bare by the treading of elephants and buffaloes, is composed of a blackish mixture of clay and humus material to a depth of at least 20 centimetres. Everywhere, the vegetation is opened up by the impact of the animals, which creates spaces of bare soil colonised by semi-aquatic annual plants such as the Scrophulariaceae species *Bacopa crenata*, *Bacopa egensis*, *Lindernia nummulariifolia* and *Torenia thouarsi*;

the Cyperaceae *Rhynchospora breviorostris*, *Elaeocharis geniculata* and *Cyperus difformis*; the Rubiaceae *Pentodon pentandrus* and the Lythraceae *Ammania baccifera*. This meadow is typical of semi-aquatic vegetation that develops on boggy soil, damp throughout the year and submerged during short periods only during the rainy season.

In slightly elevated, less often flooded areas, a meadon of *Kyllinga erecta* and *Diodia scandens* forms. Very often this association comprises nitrophilous ruderals, introduced by the dung of animals.

Apart from these three types of low vegetation, there are also taller semi-aquatic formations of large perennial Cyperaceae, such as *Rhynchospora corymbosa*, with which is associated the fern *Cyclosorus tottae*; thorny thickets of *Mimosa pigra* and *Cleome*

4.59. The Langoué bai in eastern Gabon is included in the Ivindo National Park. It is a large clearing, sharply separated from the surrounding forest and crossed by a small river. Like the bais of Odzala-Kokoua National Park it is regularly visited by large mammals, including the Elephant, the Gorilla, the Sitatunga and the Red River Hog.

4.60. The Maya-Nord saline with its "elephant bath".

afrospina (PAGE 111), as well as pre-forest brushes of *Ixora brachypoda*, a Rubiacea, and *Jasminum dichotomum* of the Oleaceae family.

In the Maya-Nord clearing, Florence Magliocca found nearly the same associations. On the edge of the pools and rivulets, in the areas most frequented by animals, there is a meadow of *Bacopa crenata* and *Ludwigia stolonifera,* an Onagracea. In areas of soft soil, where large animals do not venture very much for fear of sinking in, meadows of *Enydra fluctuans,* an Asteracea growing on nearly permanently flooded soils, sometimes in association with some Cyperaceae, notably *Pycreus mundtii*. Non-peaty meadows bordering streams and flooded in the rainy season are occupied by the grass *Acroceras zizanioides* and various Cyperaceae. In somewhat higher zones, but nevertheless regularly flooded, meadows form that are dominated by the two Cyperaceae *Kyllinga erecta* and *Rhynchospora corymbosa*. The first is dominant in the dampest places, the second in rather dryer areas. As for higher ground, out of reach of flooding, this is mainly covered by the grass *Paspalum conjugatum*.

All these associations are part of the successions that lead in the end to terra firme forests. It should be pointed out that there are two types of succession. The first takes place in stagnant water where the accumulation of organic matter ends up forming peat. It is characterised by the *Mesanthemum* and *Selaginella* association. The second, much more frequent, takes place in waters with a weak current. It is characterised by Cyperaceae instead.

Because of the concentration of large mammals, these successions are, however, blocked. The stripping of the soil favours the development of *Bacopa* meadows. These attract gorillas, antelopes, elephants and buffaloes which maintain them by grazing and trampling. *Kyllinga* meadows, slightly dryer, are also very frequented by buffaloes, which slows their evolution to *Ixora* thickets. They can nevertheless be colonised by thorny plants such as *Cleome afrospina*. Not pastured by the buffaloes and antelopes, these end up forming veritable uninterrupted thickets about one metre high.

Thus, if the clearings or bais of Odzala are of obviously hydrological origin, then their maintenance in the middle of the forest massif is the result of the activities of large mammals. In other words, to understand the functioning of this distinctive ecosystem and situate it with respect to the formations that surround it, it is necessary to know how and why it attracts large mammals.

A first factor that comes to mind is the presence of minerals. The Lango clearing has, for that matter, been the object of traditional exploitation of salt – from which its name *saline* – and the work of Florence Magliocca confirmed that as the rate of periodic flooding increases, the mineral and salt content of the soil as well as of the vegetation increases. These high levels of chlorides justify the term saline.

The origin of these minerals not being marine, two mechanisms could be imagined. Either the depressions, occupied today by clearings, were able to concentrate salts during dry climatic phases, notably 2,000 to 2,500 years ago, by collecting runoff water from the surrounding forests and evaporating on the spot, or the increase in mineral salts in the soil and vegetation originates from large quantities of droppings and urine brought by animals. The latter factor is not at all negligible.

The attraction of minerals does not explain everything, particularly not why some animals spend entire days in these clearings. In fact, only elephants visit them just for their minerals. All the other species find a more or less substantial part of their food there. Gorillas, that feed essentially in the Marantaceae forests, periodically visit the clearings to consume the four plants highest in minerals: *Enydra fluctuans*, *Rhynchospora corymbosa*, *Pycreus mundtii* and another small Cyperacea. Sitatungas, more eclectic in their choice, consume much *Enydra* and also Cyperaceae plants. Buffaloes prefer grasses and *Kyllinga*. Giant Forest Hogs visit the most humid areas and Red River Hogs seek out the formations of *Rhynchospora*.

Along with these herbaceous resources, Sitatungas and Red River Hogs also search on the ground or at the bottom of pools for *Strychnos camptoneura* and

4.61. The Chattering Cisticola Cisticola anonymus *is common in bais, as it is in some enclave savannas. Without actually inhabiting forests, it is, however, widely linked to the forested regions of Central Africa. In some bais, like the bai of Langoué (Page 131), the Dja River Warbler* Bradypterus grandis *is found also. It is a small brownish grey warbler, more often heard than seen. It depends on the presence of* Rhynchospora corymbosa, *from the Cyperaceae family, and is endemic to Gabon, South Cameroon and the south-western part of Central African republic.*

4.62. On the edge of a pool, encounter between a Hartlaub's Duck Pteronetta hartlaubi *and a Long-snouted Crocodile* Crocodylus cataphractus.

Strombosia pustulata seeds, often abundant in elephant dung.

By means of herbaceous vegetation or seeds in elephant droppings, the clearings thus very clearly constitute an important and stable food source that complements the resources of the surrounding forest. In fact, the forest offers fruit in abundance only twice a year and, in regions near the equator, as is the case at Odzala, there is even a regular inversion of seasons that renders fruit production unpredictable. Observations confirm that frequentation of clearings, by gorillas well as by elephants, diminishes during fructification in the forest.

Finally, these clearings also seem to have a profound influence on the social behaviour of the species that visit them. In the case of Sitatungas, for example, the clearings favour the appearance of large temporary gatherings, while in the forest these animals are solitary, instead. A very similar phenomenon was observed in the Bongo, in the Central African Republic. These observations might be explained by the abundance of food, perhaps by a greater security linked to the open nature of the habitat and perhaps also by social behaviour. Adult females reject young and "foreign" females, which reduces sexual competition.

One could ask how all these animals manage to coexist. Of course, each species uses the clearings in its own way and there is a temporal and spatial separation of resources, but there is, after all, a substantial overlapping, as much in the use of space as in the use of resources. There should, then, be some competition for them. In reality, few signs of direct competition for food have been detected so far. This is perhaps explained by the fact that the animals of the clearings also use resources in the forest.

Whatever it might be, a question still remains: to what degree are the clearings of Odzala responsible for the high densities of large mammals observed throughout the national park? According to Florence Magliocca, the answer could be positive. In most of the species that frequent the open habitats, their population structure indicates high reproduction and survival rates. This may be linked to the presence of minerals, whose physiological importance is well known but remains to be demonstrated in the case of the clearings. It could also be related to social phenomena engendered by the opening up of habitat and the intensification of intraspecific contacts.

For the conservation of Odzala National Park, the clearings, «salines» or bais are therefore, until proof to the contrary, of capital importance and any reduction of their frequentation, as a result of poaching or excessive tourism, must be prevented at all cost. Such a lowered use of these clearings could lead to a reduction of mineral levels, destabilise the present vegetation – perhaps mostly by a reduction of grazing and trampling – and provoke rapid reforestation. As early as in 1951, Léonard had noticed this process in the Belgian Congo.

4.63. A buffalo Syncerus caffer *lounges in the pool of a bai while a Yellow-billed Oxpecker* Buphagus africanus *rids it of its ticks and Cattle Egrets* Bubulcus ibis *hunt the innumerable insects attracted by the water, vegetation and odours of the animals.*

4.64. A lonely elephant in a flooded bai at Odzala.

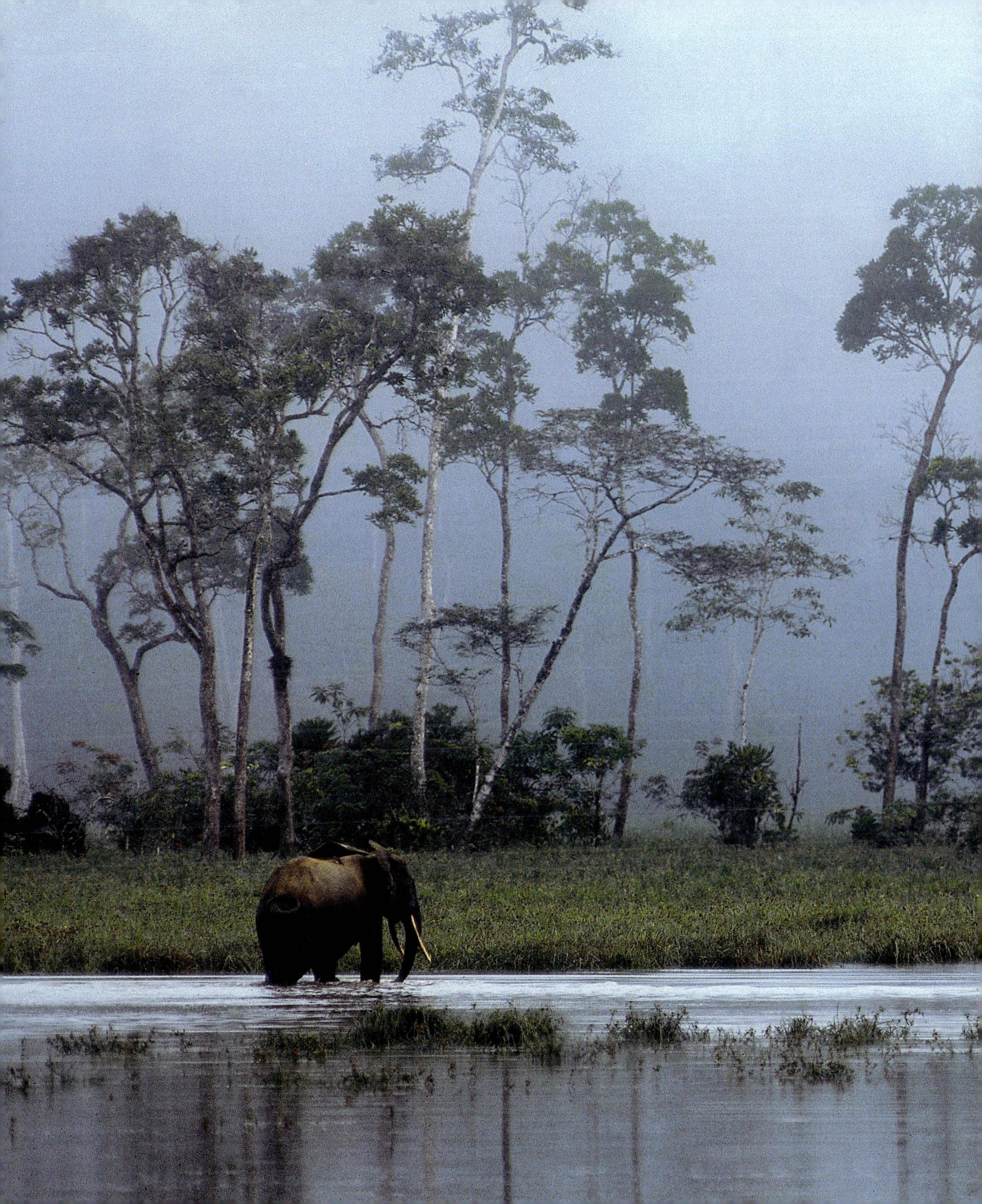

4 TEMPORAL GRADIENTS

Coastal successions

4.65. Like many others that empty directly into the ocean, this small river of the Gabon coast not far from Nyonié does not have a flow strong enough to oppose the deposition of a sand bar that obstructs its mouth. It is only after heavy rains that this bar can be temporarily carried away. Mangroves are readily visible near the ocean.

In the absence of rock cliffs worthy of the name, the roughly 1200 kilometres of the Atlantic coasts of Central Africa have two very different aspects: sand beaches back to back to terra firme forests and mud beaches up against mangroves.

Sand beaches

For nearly 1000 kilometres, especially in Equatorial Guinea, southern Gabon and the Congo, the coast is formed of a narrow beach of white sand, generally fairly abrupt and rectilinear, cut here and there by the mouth of a small river (FIGURE 4.65). In southern Gabon, it is also cut by the much wider outlets of the large Ozouri, Fernan Vaz and Iguéla lagoons. Backed up directly against the forest, these beaches are subjected to constant modification. Sometimes a very big tide carries off sand, laying bare the roots of trees and causing the forest mantle to retreat. Sometimes great quantities of sand are deposited on shore and form a low dune cordon. Most of the time, the mouths of the small rivers are, for that matter, cut off from the sea by a beach that only heavy rains or high tides manage to break through. This type of coast is in this way often constituted by a succession of old dune cordons between which swampy areas form.

In spite of its apparent stability, the coast is therefore the seat of a very active dynamism at the vegetation level. Above the driftline, white sand beaches are in effect rapidly invaded by a pioneer vegetation dominated by ipomeas, notably *Ipomea pes-caprae* (FIGURE 4.67), and a *Fabacea* with large purple flowers *Canavalia rosea* (FIGURE 4.68). These plants, all characterised by long creeping stems, very rapidly form a short but more and more dense carpet, in which other herbaceous plants become established. Soon they are followed by a few shrubs, notably *Dalbergia ecastaphyllum* (Figures 4.66 and 4.70) and *Hibiscus tiliaceus* (PAGE 83). Often this vegetation is invaded by a parasitic plant: *Cassytha*

filiformis of the Lauraceae family (FIGURE 4.69). These different zones of vegetation, that can be found a few metres from each other, constitute, in fact, phases of a succession that leads from the sand beach to the Atlantic sclerophyllous forest (PAGE 83).

As long as the coast remains stable, this succession is checked by the rigidity of the gradients and the different stages represent nothing more than adapted edaphic climaxes. If, on the other hand, the coast collects substantial deposits of sand, succession can start up. Pioneer plants fix the sand and activate the formation of dune cordons which will be colonised by shrubby and arborescent formations.

This tropical coast pioneer vegetation is found on the beaches of the Indian Ocean, the Pacific and the Antilles. Everywhere, it has the same appearance, differing essentially in function of the configuration of the beaches. On wide, gently sloping beaches, a very progressive gradation, from open herbaceous to ligneous formations, is established. On steep narrow beaches, the transition can be extremely abrupt.

4.66. *A beach near Cape Estérias, near Libreville, colonised by pioneer vegetation whose long creeping stems form a dense resistant carpet in which shrubs become established.*

4.67. *Ipomea pes-caprae has dark-centred pale pink flowers and leaves that resemble a goat's hoof. It is a cosmopolitan species, encountered as much on the beaches of the Atlantic as those of Lake Tanganyika.*

4.68. *Canavalia rosea is a creeping Fabacea with bright pink flowers.*

4.69. *Cassytha filiformis, of the Lauraceae family, is a very abundant parasitic creeper in the littoral vegetation. It much resembles Dodder Cuscuta sp., of the Convolvulaceae family.*

4.70. *Dalbergia ecastaphyllum, of the Fabaceae family, is one the first woody pioneers of the sandy beaches. Its white flowers are much visited by bees.*

Mangroves

Contrasting with the white sand beaches, the immense flat coasts with shallow turbid waters of rias are strewn with mudflats up against immense mangrove forests. In some places, rocky formations emerge. They are vestiges of sandstone sediments from the Cretaceous, exposed by the uplifting of the western edge of the continent. These formations are very extensive in front of Libreville and in the Cape Estérias region, further north. Most of the bays and rias remain, however, the domain of immense mudflats "of unstructured soil", often difficult of access, that constitute the unchallenged habitat of great concentrations of migrating or wintering Palaearctic waders.

Behind these mudflats and muddy beaches halophilous tree formations develop: the mangrove forests, a very special, and very specialised, habitat, eminently picturesque, poor in species but highly productive, located at the limit between the terrestrial world and the marine world. Most animal life takes place in the water and at the surface of the mudflats. The mangroves serve, in fact, as spawning grounds for many marine fishes. Crabs of many, very different, species proliferate (FIGURE 4.72), and mud skippers (FIGURE 4.73), fish capable of breathing out of water that are seen skipping across mudflats at low tide and at the surface of shallow pools. They even manage to climb on the thick roots of mangrove trees.

Mangrove swamps of the Gulf of Guinea are very similar to those of the coasts of the Americas and the Antilles. They are composed of three Rhizophoraceae species: Red Mangrove *Rhizophora racemosa*, *Rhizophora mangle* and *Rhizophora harrisonii*. According to work by Frans Breteler, the latter is apparently no more than a hybrid, for that matter fairly variable, between the two first species. Mangrove swamps also include the Great White Mangrove *Avicennia germinans*, of the Avicenniaceae family and two Combretaceae, the Grey Mangrove *Conocarpus erectus* and the White Mangrove *Laguncularia racemosa*. These mangrove forests are

4.71. The immense mudflats of the Mondah bay, and the leading edge of mangrove forest colonisation, a marine habitat where most animal life goes on in the water and at the surface of the mudflats.

4.72. Crabs are omnipresent and are represented by many different species.

4.73. More spectacular are the mud skippers *Periophtalmus barbarus*. These fish are capable of breathing out of water and skip about on the mud during low tide.

4 TEMPORAL GRADIENTS

therefore clearly poorer than those of the Indian or Pacific oceans, where more than 30 species have been censused in some places.

The *Rhizophora* are the commonest species. They are immediately distinguished by their long, strongly arched stilt roots. They can reach 10 to 30 metres in height, or even more, but often remain much smaller, especially when they develop on shallow soils for example covering a submerged rock slab. Facing the sea, in the strictly marine mangrove formations, *Rhizophora mangle* is abundant. In Cameroon, it is the first pioneer facing the sea, but farther south in the Gulf of Guinea it is rarer. *Rhizophora racemosa*, on the contrary, prefers brackish waters. *Rhizophora harrisoni* occupies an intermediate place.

The Great White Mangrove can also reach 30 metres in height and even a little more. It generally grows behind the *Rhizophora*. Being less dependent on salt water, it also reaches much farther upstream in the rivers.

Mangrove trees are very specialised. The *Rhizophora* develop spectacular stilt-roots and hanging roots that assure their stability in still mobile soils (FIGURE 4.76). As for the *Avicennia*, they have, just like the *Hallea* (*Mitragyna*), pneumatophores (PAGE 148). *Rhizophora* also have long fruits which form "darts", 30 to 40 centimetres long, capable of planting themselves ver-

4.74-75. Rhizophora *fruits measure 30 to 40 centimetres in length and resemble a dart. Their seeds begin to develop already on the tree and send out a stem which can form roots as soon as it drop into the mud.*

4.76. *Along the edge of an internal channel of the mangrove forest – a "rivière", as they say in Gabon – rise large mangrove trees of genus* Rhizophora. *Upright on their stilt-roots and draped in their long aerial roots, they often form no more than a narrow screen masking the immense, inaccessible expanses of lower mangrove forest, such as the one easily seen in figure 4.77.*

139

tically in the mud (Figures 4.74-4.75). The seed starts to germinate while it is still on the tree and sends out a stem that will be able to plunge its roots immediately into the substrate. The young mangrove is then able to resist the tides.

Mangroves are transitory formations capable of engendering substantial land formation and, sometimes, regression, depending on the action of the waves and current. This constant dynamism is translated by the fact that, in spite of the few species involved, they form very different association.

First, there are the *Rhizophora* mangrove stands. They are pioneer formations *par excellence*. Their very dense root system locally slows the marine currents and allows the deposit of sediments. They participate in this way in the slow raising of the coasts. They are present in two forms. A dense mangrove formation, 15 to 20 metres high but generally fairly narrow, develops along the sea front and borders the interior channels. In fact, it only forms a narrow curtain behind which stretch vast expanses of low and much more open mangroves not more than five to six metres tall (FIGURE 4.77). All the *Rhizophora* can form tall or low mangrove stands: the tall ones trap most of the sediments; the short formations must make do with what remains. Where there is a sufficient supply of sediment, as in the upper part of

4.77. *The mangrove forests of the Mondah bay receive very little fresh water. They are thus strictly marine and composed essentially of low formations, in places even very open. High stands only form narrow galleries along the "rivières".*

4.78. *Very similar mangrove forests are found again in the estuary of the Komo, but those that form more towards the interior are very different. Formations 20, 30 and even 40 metres high extend over a surface several hundred metres, or as much as a few kilometres, wide.*

the Komo estuary, tall mangroves can expand considerably (FIGURE 4.78).

Behind the *Rhizophora* stands, *Avicennia* mangroves develop, in which *Conocarpus* and *Laguncularia* formations mix with, or succeed, each other. Finally, at the limit of mangroves and terra firme forests, *Phoenix* or *Pandanus* palm formations are established. These are often no more than a narrow fringe, but in some places, such as along the east shore of Mondah Bay, between the mangrove formation strictly speaking and the terra firme forest, there are two well-marked belts: the first, 20 to 100 metres wide, is made up of Dwarf Date Palms *Phoenix reclinata*; the second, 20 to 100, even 150 metres wide, of *Raphia* palms. In places *Phoenix* palm stands occupy flooded areas where water lily beds sometimes develop (FIGURE 4.79).

Towards the inland side, it sometimes also happens that mangroves give way to extensive bare spaces: the "tannes" (FIGURES 4.77 and 4.80). These are limited to the marine zone of the mangroves and arise where there

4.79. Stands of Dwarf Date Palms Phoenix reclinata *extend between the mangroves and the terra firme forests. Often they are no more than a few rows of trees grouped into clumps, but in many places they can be more than 100 metres wide.*

4.80. Tannes are typical of marine mangrove forests. They are abundant in the bay of the Mondah and in the western part of the estuary of the Gabon. A colonisation by small Grey Mangroves Conocarpus erectus *can often be seen bordering the completely bare space.*

4.81. Libreville is built on a peninsula narrowly attached to the mainland, and separating the bay of the Mondah, in the north, from the estuary of the Gabon, in the south. These two rias harbour vast expanses of mangrove forest, but differ fundamentally. The 35,000 hectares of the bay of the Mondah are essentially maritime, because this bay receives very little fresh water from inland. The 80,000 hectares of the Gabon estuary divide into a marine zone downstream and a fluvial zone further upstream. The marine zone in the west is quite similar to the mangroves of the Mondah bay with its numerous tannes. More to the east, is a transition zone with wide, tall mangroves (Figure 4.85) and no tannes. Still further upstream are freshwater swamp forests, extending over an area of more than 20,000 hectares. From west to east, the Gabon or Komo estuary is lined with a complete gradient of vegetation types going from strict marine mangroves to continental swamp and riverine forests.

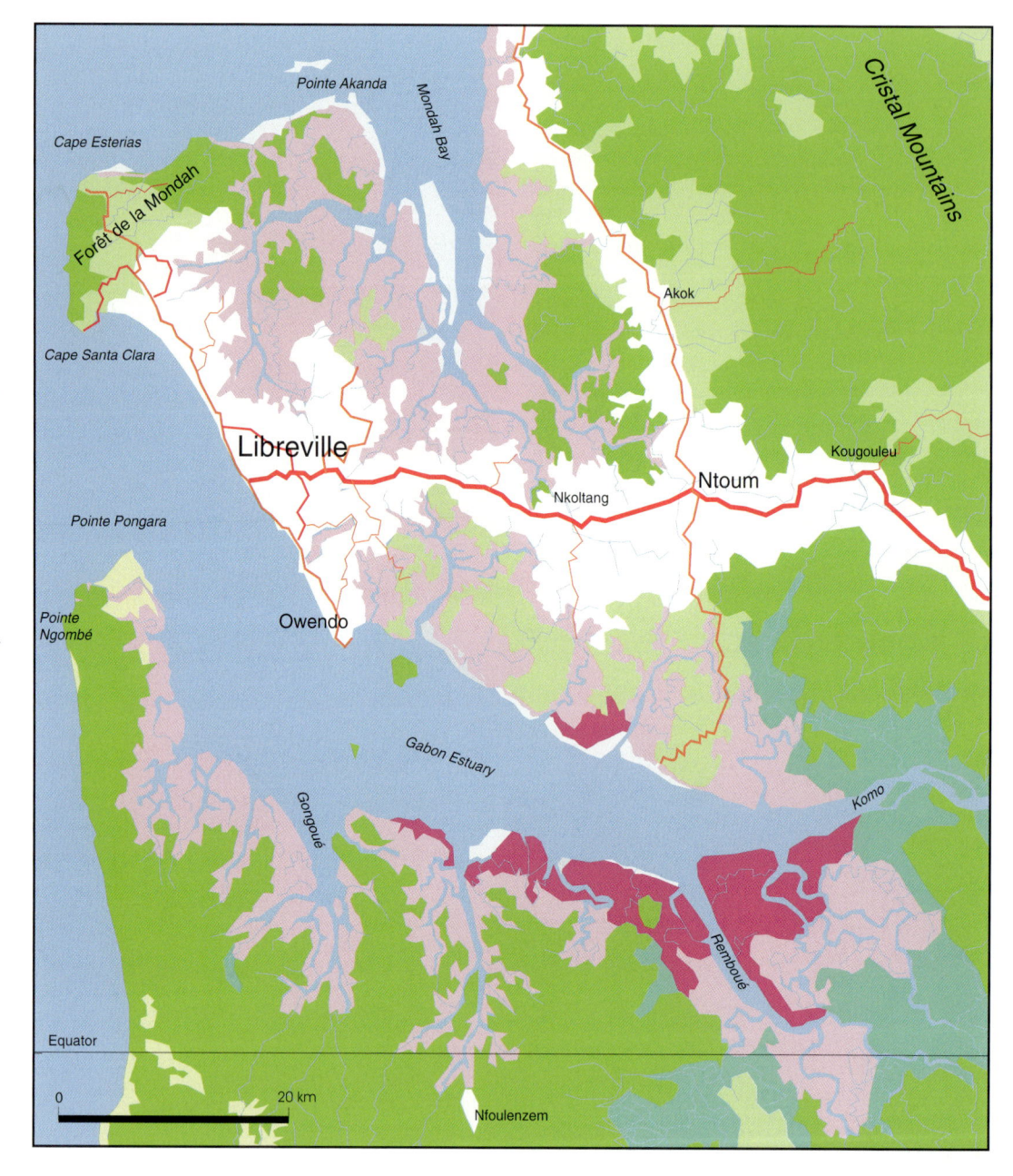

- Terre firme forest
- Disturbed terra firme forest
- Mud banks
- Low mangrove
- Tall mangrove
- Swamp forest
- Man-made habitats
- Coastal grasslands

is a lack of water circulation and evaporation concentrates the mineral salts. Around a completely bare central area, flooded at most during the highest tides, a carpet of small *Conocarpus* often forms (FIGURE 4.80).

On the scale of a bay or an entire estuary, these seven main types of associations which form the mangrove stands are somewhat like the pieces of a vast game. By their arrangement they characterise much wider ensembles and allow the retracing of their history.

Mondah Bay, which forms only part of the large Corisco Bay situated between Equatorial Guinea and Gabon, constitutes a typically marine ensemble. Its waters are frankly salty, the arrival of fresh water being limited to a few minor rivers. Low mangroves with small *Rhizophora* are the dominant element, while high mangroves form only narrow screens, and tannes are very widespread. The transition formations of palms are not well developed except on the eastern shore of the bay where numerous small streams empty their waters from the Crystal Mountains.

The estuary forms an ensemble at once greater and more complex. Its western part, south of Libreville, is strictly marine and the vegetation there absolutely resembles that of Mondah Bay. The tannes are also well developed. Its eastern part is, in contrast, strongly influenced by the fresh waters and sediments carried by the Komo and Bokoué rivers. The low mangroves are entrenched behind the vast expanses of old, very tall mangroves. Tannes are absent. Between the mangroves and the terra firme forests stretch enormous swamp forests of palms and on land the transition is often difficult to follow.

In addition to the typical tree species, mangrove forests also harbour other particular plants. In old *Rhizophora* formations, the epiphytic fern *Microsorium punctatum* can be found clinging to the large stilt-roots (FIGURE 4.82). In the shade of the canopy, encrusted on these same thick stilt-roots, are even a few orchids, mainly of genera *Bulbophyllum*, *Polystachya* and *Genyorchis*. The large fern *Acrostichum aureum* is found on the edge of the tannes or even mingled with *Phoenix* palms or *Avicennia*, *Laguncularia* and *Conocarpus* (FIGURE 4.83).

As picturesque and sometimes spectacular as mangrove forests might be, their terrestrial fauna is very poor. In fact, life in the mangroves takes place in the water, at the surface of the mud and between the roots of the trees. Most visible are the various species of crabs, the clusters of oysters attached to the roots of the mangroves, and mudskippers, with their bulging eyes, capable of breathing out of the water and seen skipping on the mud or at the surface of the water. From time to time a Forest Monitor *Varanus ornatus* takes off in the tangle of roots, but in the forest canopy there is not much. Of course aquatic species are encountered along the channels, such as the Shining-blue Kingfisher *Alcedo quadribrachys* and the Blue-breasted Kingfisher *Halcyon malimbica*, herons – notably the African Tiger Bittern *Tigriornis leucolophus* – and storks, such as the Wooly-necked Stork *Ciconia episcopus*, sometimes a Pink-necked Pelican *Pelecanus rufescens* and, outside of the breeding season, the Rosy Bee-eater *Merops malimbicus,* that hunts above the mangroves. At high tide, many Palaearctic waders – especially the Whimbrel

4.82. *An old mangrove forest near Cape Estérias, north of Libreville. Large epiphytic ferns* Microsorium punctatum *are established on its thick stilt-roots.*

4.83. *In the zone of transition between mangrove forest and terra firme forests, the large fern* Acrostichum aureum *is common, often mingled with Dwarf Date Palms* Phoenix reclinata*. It is probably a very old species, occurring in all the mangrove forests of the world.*

4.84. In the area of Cape Estérias, north of Libreville, the mangrove forest comes and goes. Young formations develop alongside the remains of old mangrove trees drowned by the tides.

4.85. In the south-eastern part of the Komo estuary extend immense expanses of very old Rhizophora *mangrove forests, with trees reaching 30 to 40 metres in height.*

Numenius phaeopus – that feed on the immense mud-flats, take refuge on the dead branches of the great mangrove trees or on their thick stilt-roots.

Apart from these widely distributed species, in no way restricted to mangrove forests, there are a few particular species. There is the Gabon Boubou *Laniarius bicolor* that is not easily seen but is often heard. There is also the Yellow-chested Apalis *Apalis flavida*. This species is widely distributed in the savanna regions of sub-Saharan Africa, but along the Gulf of Guinea, it is strictly linked to mangrove forests. The Brown Sunbird *Anthreptes gabonicus* has its range from the Congo estuary to Gambia and is linked everywhere to the mangroves. The Loanga Slender-billed Weaver *Ploceus subpersonatus,* known from Mondah Bay, in Gabon, to the estuary of the Congo River, is strictly limited to the fronds of Dwarf Date Palms *Phoenix reclinata*.

Being pioneer formations, mangrove forests move mainly with fluctuations of the sea level. This fact is confirmed by the work of Bengo and Maley who show that the mangrove forests underwent major expansion between 130,000 and 120,000 BP and between 40,000 and 30,000 BP. Between 30,000 and 12,000 BP, during the last glaciation, on the other hand, they experienced severe regressions. During the glacial maximum, 18,000 years ago, the level of the ocean was more than 100 metres lower than it is today and, at Libreville, the Atlantic beach was located some 60 kilometres farther west. The island of Corisco was then part of the continent and the Gabon estuary was no more than a wide valley. Its flooding began around 7,500 BP. This entire phase of the rising of the level of the ocean was accompanied by the extension of the mangrove forests.

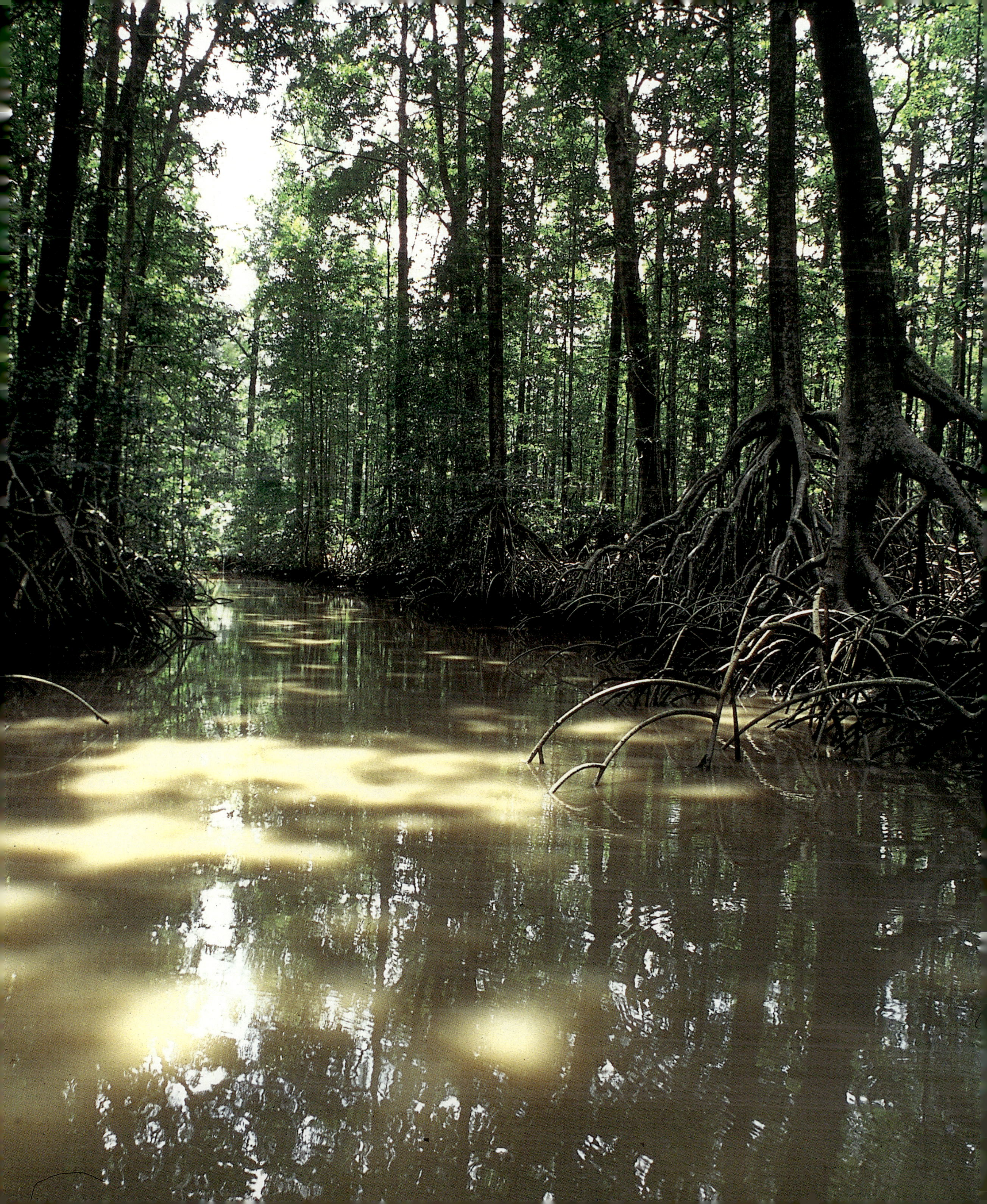

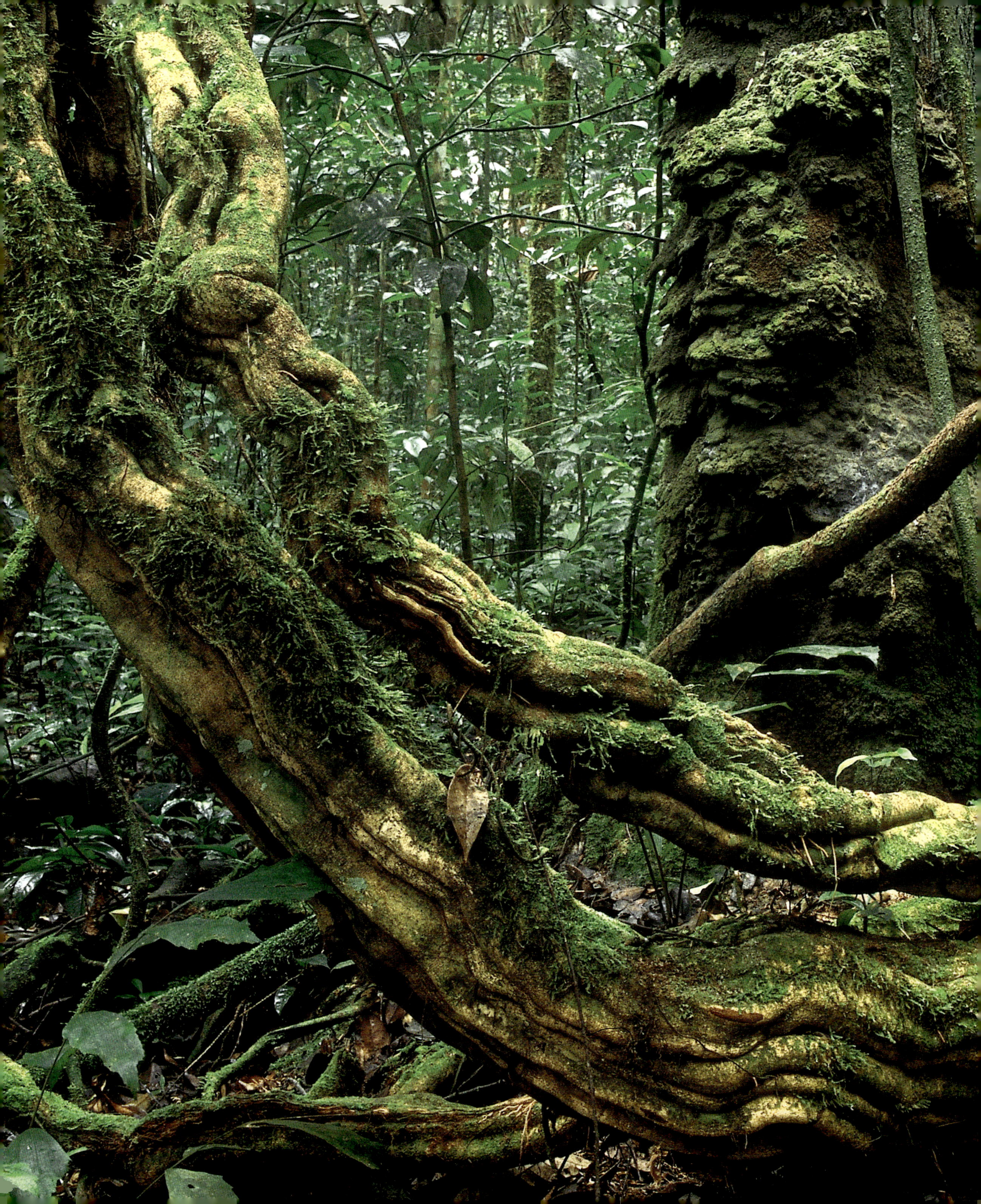

5 The life in the forest
The struggle for resources and reproduction

The immense richness of humid tropical forests has no equal except the complexity of their ecological mechanisms. Their luxuriance and their fabulous diversity, however, go hand in hand with the rarity of many species and makes one forget that some resources indispensable to life are at times very rare. An intense competition arises from this that engenders masses of survival strategies, each more ingenious than the others. Some aim for the direct appropriation of resources. Aeration of the soil poses problems in inundated forests and respiration can be difficult in the understorey of the most humid forests. The hyperabundance of water can be troublesome in the lower strata while the canopy suffers during moments of drought. For light, the problem is inverted: while the large emergent trees bathe in intense light, the understorey plants are plunged in a constant obscurity, brightened up by fleeting spots of sunlight, at most. Between these two extremes, a whole gradient of light ambiences develops which fashion the physiognomy and structure of the vegetation. Plants of the understorey are, in this way, organised to capture the little light available. Lianas rush towards the light by taking support on other plants and epiphytes take root directly in the heights. As for the stranglers, they start their life as epiphytes and finish as true independent trees. The paucity in nutrients stimulates development of very economical nutrient cycles based at once on maximum reduction of losses by rapid recycling of organic material through the action of insects and fungi, and on efficient storage. Because of this fact, tropical forests are capable of living essentially on their own, with a minimum contribution from the exterior, and manage, in this way, to develop and maintain themselves on soils at first glance totally inappropriate. Other strategies look for resources through predation or parasitism. Still others consist of no more than developing defences to protect themselves from predators of all sizes.

But surviving is one thing, it is still necessary to reproduce. In fact, in such a rich habitat most of the species are more or less rare and unable to fecundate themselves. Therefore, reproducing poses serious problems. Plants have solved these problems by means of an intense collaboration with insects, birds and some mammals that are often inescapable assistants in flower pollination and seed dispersal.

5.1. In a tropical forest animal and plant species have established multiple, complex and often unpredictable relations, which indicate an intense competition. While trees and lianas compete for light, organic matter is constantly recycled, especially by termites.

5 THE LIFE IN THE FOREST

Air

In general, air does not pose a problem. In swamp and mangrove forests, the compact and poorly drained soils can, however, be totally without aeration, at least during some periods. Some trees remedy this inconvenience by developing very specialised roots: pneumatophores. There are two types. Those of *Avicennia germinans*, a large tree of mangrove swamps, are made up of small lateral roots that emerge vertically from the substrate without forming a knee or turning back down. They form dense carpets of more or less downy and supple tips 12 to 15 centimetres high (FIGURE 5.2). The pneumatophores of the palm *Phoenix reclinata* are of the same type (FIGURE 5.3). Those of *Hallea* (*Mitragyna*) are very different. They emerge vertically from the surface, form a gnarled knee and plunge back into the depths of the soil (FIGURE 5.4). Several such knees can develop on the same root. *Symphonia globulifera*, when it grows in swamp forests, can produce very similar pneumatophores.

The primary role of pneumatophores seems to be respiration, and studies by Karsten, as early as 1891, show that they produce large quantities of carbon dioxide. It is probable also – at least in mangroves – that they allow the trees to compensate the rapid accumulation of sediments which would otherwise threaten to suffocate them.

Water

All plants need water to live, but they must also be able to transpire. It is actually evaporation at the leaf level that creates the negative pressure necessary to make sap and nutrients rise from the roots to the foliage. In dense tropical forest – as in any forest – atmospheric humidity follows a strong gradient with the highest values in the understorey and the lowest values in the upper canopy. Plants of tropical forests are therefore confronted with very different situations depending at which stratum they develop their foliage.

Species of the upper canopy have no problem. Their transpiration is usually least during the night and greatest during the morning and afternoon. During the hottest hours, they reduce transpiration by closing their pores. Otherwise they could exceed their water transport capacities and the foliage could wilt. During dry periods, this problem could become worse. It happens, then, that water transport cannot at all compensate evaporation. That is probably in part the reason why some trees briefly lose all their leaves and that some epiphytes develop adaptations similar to those of species found in arid habitats. We will come back to this further on (PAGES 156-160).

5.2. The ground at the foot of the Great White Mangrove Avicennia germinans, *is covered with a dense, supple carpet of pneumatophores that may extend for several metres beyond the base of the tree that produces them.*

5.3. The Dwarf Date Palm Phoenix reclinata *sometimes forms a profusion of pneumatophores all around the base of its stalk or false trunk. At other times, it produces a more spread-out carpet of pneumatophores, somewhat like that of the Avicennia.*

5.4. In swamp forests, one or another of the Hallea (Mitragyna) *often flourishes. These are large trees of the Rubiaceae family with coiled masses of pneumatophores that emerge from the stagnant water.*

5.5. In very humid forests, like those of the Atlantic coastal region near Libreville, many shrubs of the understorey and lower strata of the canopy have leaves with long, acuminate tips to facilitate the draining-off of water.

In the understorey, transpiration is hampered by atmospheric humidity which is always very high and near saturation. Plants of this stratum have a very reduced transpiration, in any case a good deal less than those of the canopy. Respiration briefly augments very substantially when plants are exposed to the rays of the sun which pierce the canopy and cast patches of strong light on the forest floor. Most plants of the understorey, which live almost permanently in the shade, are perfectly adapted to this situation, but they are completely incapable of compensating an increased transpiration in the case of prolonged exposure to the sun. Even though their leaves are protected by a cuticle as are those of other plants, the vessels of their petioles are too narrow and do not allow an increase in the delivery of water to compensate the loss by transpiration. Typical understorey plants cannot survive the opening of the canopy. They wilt immediately and die.

In the understorey it also happens that excess water stagnates on leaf surfaces. Not only does that hinder respiration and transpiration, but it also favours the development of epiphyllous lichens, mosses, liverworts and algae (PAGE 157). Because of this, many plants of the understorey and lower strata of the canopy have leaves with long, acuminate "drip tips" that more easily drain off excess water (FIGURE 5.5).

Light

Light intensity in the understorey is rarely more than 1 % of the value above the canopy. Intensity decreases 99% from the tops of the highest trees down to the understorey. This gradient divides the forest into two very different worlds: that of the high canopy and emergent crowns, flooded by light or nearly so; that of the middle and lower strata, perpetually bathed in half-light.

This light gradient is at the origin of many structures and phenomena that characterise tropical forests. It regulates the form of crowns and the structural organisation of the canopy. It profoundly influences vegetation of the understorey and engenders the development and diversification of very particular plant forms, such as lianas, stranglers and epiphytes.

This gradient is variable from one place of the canopy to another according to the types of crown and foliage. It is not always smooth and very often light intensity drops abruptly just below the crown of the main canopy trees, emphasizing in this way the subdivision of the canopy.

5 THE LIFE IN THE FOREST

Canopy structure

The complexity of the tropical forest canopy fascinated scientists very early and its often chaotic appearance was felt to be a challenge: it must certainly hide a structure responding to strict laws. Man has always liked to understand and how is one to understand the organisation of crowns in a canopy if they do not obey some rule.

P.W. Richards postulated, in 1952, in his capital work on tropical forests of the world, that the canopy of mixed tropical forests is composed of five independent strata: three arborescent strata, a partially shrubby stratum and a stratum formed of herbaceous plants and seedlings (FIGURE 5.6). The height of the strata may vary from one forest to another, but remains within some limits. Upper stratum A is formed by the emergent crowns which may reach heights of 40 to 60 metres or more. It is generally discontinuous and this discontinuity is stronger in semi-deciduous forests than in evergreen forests. Stratum B, located between 20 and 30 metres, on average, may be continuous or discontinuous. Stratum C, situated between 8 and 15 metres, is nearly always continuous. A vertical discontinuity between the different strata may, or may not, be present, well marked or barely perceptible. Often, it is best marked between strata B and C. The limit between strata D and E is, on the contrary, rarely well defined. Each

5.6. According to Richards, mixed forests have five more or less independent strata:

A. the emergent stratum, discontinuous and composed of large trees 30 to more than 50 metres high, with spread crowns (tabular or umbrella-shaped); B. the upper canopy, continuous or discontinuous, 20 to 30 metres high, also composed of more or less spread crowns; C. the lower canopy, 10 to 15 metres high, composed of more spindle-shaped crowns; D. the shrub layer, three to eight metres high; E. the understorey, herbaceous and suffrutescent, two to three metres high at most.

5.7. Many large trees of the two upper strata (A and B) develop impressive buttresses in order to increase their stability.

stratum has its own species cortège, specially adapted to its particular ecological conditions. However, except in strata A and B, a high proportion of individuals comprises immature specimens from higher strata.

Among the various adaptations there is crown shape. Crowns of stratum A are often tabular or umbrella shaped (Figures 5.6-5.8), therefore much wider than high. Very often they have a candelabra-like structure with all the thick branches taking off at the same level. Those of stratum B are generally as high as wide or even a little taller. Those of stratum C are frankly spindle shaped, much taller than wide. Those of the lower strata are again rather globular. Besides this, trees of the upper strata also develop buttresses, more often than those of the lower strata (FIGURE 5.7).

The structures described by Richards are based on observations from Borneo, Trinidad and Nigeria, but other scientists later recognised the same structures in other forests or described very analogous structures. Eggeling, notably, who worked in Budongo Forest in Uganda, can be mentioned. Very soon, however, voices were raised to cast doubt on Richards' observations and conclusions. It was felt that the profiles used for this type of study were inadequate and, because of this, not representative enough of all the forest formations that they were supposed to describe. Many scientists doubt the existence of any formal structure of the canopy in dense tropical forests and opt for its chaotic development. Even Richards has finally started to doubt the objectivity of his propositions. The subject fell out of fashion and there has been no interest for a long time – least of all in Africa where the little forest research carried out over the last few years focused almost entirely on practical applications, such as population dynamics and regeneration. The controversy therefore remains unresolved.

In a 1992 work, John Terborgh, who has worked extensively on South American forests, relaunched the subject. He believes he has found the only thing missing in Richards' model: a mechanism that explains it. Without being able to confirm the objective reality of the five precise strata described by Richards and without being able to prove their independence, Terborgh estimates that the structure of the canopy of tropical forests is essentially linked to the light gradient and to the nature of the luminous atmosphere at each level of the forest. Not only does light reduce greatly in the lower strata, but the incidence of light changes and, below a certain height, the light field becomes more or less uniform.

The large emergent trees, exposed for most of the day and the year to a nearly vertical light, have a great interest in having widely spread crowns. For that matter, there is no obstacle to oppose this growth habit. Trees located in the shade of these emergents are exposed to a more lateral light. Therefore, they have an interest in having a more vertically elongated crown. Finally, the trees of lower strata are exposed to indirect light, produced by a multitude of small openings in the canopy and varying greatly from one moment of the day to the next. A globular crown suits them perfectly.

The observations of Terborgh and his team confirm that the average form of the crowns is clearly correlated with the height of the trees and their place in the canopy. These observations suggest that each species, by its genetic aptitudes for developing one kind of crown rather than another, is therefore adapted to occupy a well defined place in the canopy. As for knowing if well defined and objectively separable strata in the canopy truly exist, as Richards had proposed, it would require carrying out a great number of measurements. But making precise measurements in the canopy of a tropical forest is not an easy matter, and it would still require the deployment of great effort to resolve this controversy definitively.

5.8. In this forest in Kivu, located not far from Kahuzi-Biega National Park, the wide flattened crowns of trees of the upper stratum are very visible after the clearing of the lower strata. In addition to the difference in the form of their crowns, the trees of different strata also differ by the architecture of their crowns. In the upper-stratum species, the main axis, formed by the continuation of the trunk, disappears very rapidly or becomes uncertain. In species of lower strata – lower right in the photo – this axis continues nearly to the top of the tree.

5.9. *Most of the plants of the understorey spread their leaves horizontally in order to trap a maximum of the little light available, like the small Marantaceae.*

5.10. *The Poaceae or grass* Centotheca lappacea *is common in the forests bordering the Gulf of Guinea. It looks more like a species of the Commelinaceae family than a true grass.*

5.11. *A small Melastomatacea with broad leaves, found in the Crystal Mountains in Gabon.*

Plants of the understorey

Herbaceous plants can be abundant in mantles and in open formations, notably in some montane forests and in Marantaceae forests, but they are not numerous under a dense canopy, neither in number of individuals nor in number of species. Most are ferns and selaginellas or belong to the Acanthaceae, Araceae and Commelinaceae families, and, in submontane forests, the Begoniaceae. Grasses, or Poaceae, are not represented except by genera *Leptaspis, Olyra, Guaduella, Centotheca* (FIGURE 5.10), *Commelidinium, Microcalamus, Oplismenus, Puelia* and *Streptogyne*. All have wide leaves and resemble some Commelinaceae more than the classical savanna grasses.

Understorey plants have developed forms that are well adapted to the half light (Figures 5.9 and 5.11). Dense rosettes are rare. Leaves are usually large or average, often very thin, transparent and wide. By their disposition and their development, they orient themselves in a way to best catch the available light (FIGURE 5.10). Many spread themselves out in a perfectly horizontal plane. Some, such as the leaves of the grass *Leptaspis cochleata*, are even somewhat articulated so as to be able to orient themselves according to the time of day.

Lianas

True lianas are ligneous plants whose roots are anchored in the soil but whose foliage develops in the upper parts of the canopy. Their relatively thin, flexible stems – they can nevertheless be 20 to 30 centimetres in diameter – glide almost imperceptibly across the lower strata where they merge with branches of young trees, with only their sinuous form betraying them. In the canopy, their foliage is intimately mixed with that of the trees. Only the furrowed, twisted and coiled loops of old lianas hanging near the ground are noticeable (FIGURE 5.12). The difference between a liana and a lianescent ligneous species is not always very clear-cut and some species can show a variable carriage. Between lianas and epiphytes, also, the distinction is not always clear, notably in the case of the Araceae.

Lianas are not always easily distinguished, but they may constitute up to 50% of the global mass of all upper canopy foliage. This abundance shows up only when lianas flower, which can be spectacular. Among the most striking species are *Combretum racemosum* (PAGE 81), *Combretum paniculatum* and *Combretum platypterum* (FIGURE 5.15), whose mauve, red or orange inflorescences, respectively, yield nothing to the most

5.12. Old lianas in the understorey betray their presence by their coiled stems.

aggressive *Bougainvillea*. As for number of species, lianas can represent up to 20% of the total number of plant species and, in some secondary forests, there can even be as many species of lianas as of trees. It is in the most humid forests that they are the most abundant. Lianas therefore represent a very important element within tropical humid forests, even if they do displease the foresters.

As one could imagine, lianas appeared well after trees during the course of evolution. They were born in the race for light and appeared in a large number of plant families. Among dicotyledons, they are mostly found – in Africa, at least – among the Annonaceae, Apocynaceae, Combretaceae, Connaraceae, Convolvulaceae, Fabaceae and Caesalpiniaceae. Among the monocotyledons, they are found mostly in the Arecaceae.

Thus, they each evolved extremely diverse forms and structures independently from one another. Some are capable of "climbing" by themselves into the canopy, but many cling instead to young trees and let themselves be carried passively towards the top of the canopy. In any case, they have no interest in growing faster than the trees that support them. The way in which they attach themselves varies from one species to another. Some completely entwine themselves around their support. Others have tendrils or hooks, sensitive leaves, petioles, lateral branches or even inflorescences, which wrap around a support as soon as they touch it.

5.13. *One of the big problems lianas must face is anchorage to the vegetation. The rattan palms, such as* Laccosperma secundiflora, *are provided with pointed hooks which arm the extension of their leaves.*

5.14. *Many of the Araceae – the arum family – develop adventitious roots which they use to firmly attach themselves to the bark of trees.*

Still others, such as the rattans or lianescent palms of genera *Laccosperma* (FIGURE 5.13), *Calamus*, *Oncocalamus* and *Eremospatha* – to cite only those occurring in Central Africa – attach themselves with spines or stiff thorns. Some, like the Araceae *Rhaphidophora africana*, *Remusatia vivipara*, *Rhektophyllum mirabile* and various species of genera *Cercestis* and *Culcasia*, attach themselves with adventitious roots, short roots that appear all along the stems and develop adherent hairs when they come in contact with the bark of trees (FIGURE 5.14). Some lianas possess the ability to produce roots which plunge from the top of the canopy to the ground. They manage in this way to survive the cutting or rupture of their main stem. Since they are exposed to full sunlight, they consume much water and are sensitive to a lack of rainfall. To remedy this inconvenience, some species lose their leaves during crisis periods and others develop bulbs. Their most remarkable adaptation, however, is in their stems. In order to be able to transport the necessary quantities of water, liana stems have extremely wide vessels, in spite of being relatively thin compared to trees. The stems of some lianas are thus gorged with water, which people of the forest have learned to exploit: they cut a segment of liana and drink the water that comes from it. Lianas are also very resistant and, when strongly bent or twisted, their vessels do not strangle. This allows them to survive the collapse of their support and explains the often complex pathway they accomplish in the vault.

Just like trees, lianas form several strata, although they are less apparent. In the forests of Uganda, Eggeling thus distinguished three main layers. The upper layer develops above 20 metres and comprises the true ligneous lianas. The intermediate layer forms between 10 and 15 metres. It is made up of vigorous, but non-ligneous, species. The lower layer does not exceed nine metres and is of exclusively herbaceous species. These three strata evolve very differently over the course of the life of the forest. The upper stratum does not exist in young secondary forests. It becomes luxuriant in older secondary forests, but regresses afterwards. The lower stratum disappears during the course of maturation of the forest and only the middle stratum remains more or less stable.

Lianas play an important role in the overall structure of tropical forests. On the one hand, they increase the overall resistance of the forest to violent winds by binding the trees to one another. But, on the other, when a great emergent falls, the lianas inevitably drag other, smaller, trees along with it. Lianas also play an important role in forest regeneration. By forming opaque draperies or enveloping young trees in dense columns they can considerably slow, or even prevent, the growth of trees. The Amaranthaceae *Sericostachys tomentosa*, abundant in montane forests of the Albertine Rift, can in this way actually freeze the development of the forest.

Finally, for the peoples of the forest, lianas represent a far from negligible resource. Some, such as *Smilax kraussiana*, of the Smilacaceae family, can be transformed into ropes of any calibre that are very resistant to termite attacks. In the lowland forests, *Manniophyton fulvum*, of the Euphorbiaceae family, is most often used to make ropes and nets, while the rattans are used in making furniture (FIGURE 5.13). Finally, in many places in Central Africa, lianas are used in the construction of bridges over water courses (PAGES 8-9).

5.15. Combretum platypterum is a very common liana, notably in Gabon. Its bright red flowers are supported by bracts of the same colour and its winged seeds are red tinged with orange (Page 176).

5.16. Branches overhanging a water course form a privileged site for the establishment of many epiphytes.

Epiphytes

In order to reach the light, epiphytes have developed a very different strategy from that of lianas. Rather than taking root in the ground, their seeds germinate in the canopy and they use other plants for support. Unlike parasitic plants, such as the Loranthaceae (PAGE 167), they do not draw off any nutrients. They are therefore obliged to use, as much as possible, the nutrients brought by rain.

All the forests of the world have epiphytes, but the ones of temperate forests are, in general, very unassuming, being limited to mosses, algae or lichens. Those of tropical forests are much more exuberant, much more varied and belong to numerous large plant families, including the most evolved. There are even ligneous epiphytes.

As in so many other faunal and floral domains, African forests are much less rich in epiphytes than forests on other continents. The Orchidaceae, for example, number no more than about 450 species for the entire Democratic Republic of the Congo, while in Colombia they number 3,000 and in Malaysia there are 5,000. Of abundant families in South America, the Bromeliaceae are totally absent and the Cactaceae are represented by only a single species: *Rhipsalis baccifera*. The Gesneriaceae are rare. Epiphytic Asclepiadaceae, Ericaceae and Rubiaceae, abundant in the forests of Asia, are equally absent. As we have seen for many other aspects, this paucity is apparently related to the strong contractions African forests underwent during the glacial phases of the Pleistocene. In spite of this, Eggeling was able to count up to 45 species of various vascular epiphytes on a single tree in Uganda, and the Budongo Forest– also in Uganda – possesses over a hundred species of epiphytes altogether.

Since we can no longer move about in the trees as did our very distant ancestors, we experience great difficulty in studying the epiphyte world, yet we have been able to observe that it is a very structured world that responds to subtle differences of light and humidity. All the species have their preferences and their limitations, while their associations vary in time and space.

The principal studies on epiphytes in Africa, mainly the vascular epiphytes, have been summarised by Dick Johanson. Most of the information assembled here comes from his synthesis.

At the bottom of tree trunks, where light is very weak, one generally finds only mosses, liverworts, a few lichens and small ferns that are very shade-tolerant. In the shadiest places, on the base of the trunk, mosses are the most abundant and extremely varied. Up to a certain height – the moss line – they tend to cover everything. This line may be situated several metres above the ground in very humid forests. In the understorey of the most humid forests, hanging mosses also develop (FIGURE 5.17) and, in the shadiest places, bracket mosses are often found.

In these same very humid forests, there is, finally, a very special epiphyte community, better represented in Africa than in tropical America: the epiphylls (FIGURE 5.18). These are mosses, liverworts or lichens growing on the leaves of vascular plants, often in association with algae. These small communities are, obviously, unable to develop except in the most humid forests and on leaves that live for a sufficiently long time. They are composed of species that are generally not found elsewhere.

5.17. In very humid forests, like some parts of the Mondah Forest in Gabon, there are hanging mosses – an exclusive element of hyperhumid tropical forests. Attached to branches in the understorey, they hang freely, bathed in atmospheric humidity.

5.18. In these same forests, mosses and liverworts grow on leaves in the understorey. They are epiphyllic plants particularly adapted to this way of life and cannot live in other conditions.

5.19. In humid or very humid forests, the base of trunks, dead branches lying on the ground and roots coming out on the surface are generally covered with mosses, sometimes with selaginellas, up to a given height. The more humid it is the higher the mosses grow. They may even reach the tops of the trees, as is the case in some high-altitude forests (Page 160).

Above the moss line, there is a zone poor in epiphytes but, at the base of large branches, a great many are again found. These are mostly orchids and ferns. The ferns accumulate masses of humus between their leaves and roots where Begoniaceae and Araceae plants, as well as Melastomataceae and ligneous Urticaceae, can become established.

5.20. The palm Elaeis guineensis *serves as a very satisfactory support for many epiphytes, particularly the ferns* Micrograma owariensis *(small rounded to lanceolate fronds),* Phymatodes scolopendria *(fronds with five points) and* Nephrolepis biserrata *(pinnate fronds).*

5.21. This fern of genus Drynaria, *observed in Kibira National Park in Burundi, has two types of leaves. Only the upright leaves participate in photosynthesis. They disappear during the dry season not to reappear again until the beginning of the rains. The enveloping leaves serve only to accumulate organic material from which the fern takes its resources.*

5.22. Platycerum angolense, *although belonging to a very different family of ferns and probably much more recent, has also developed two different types of leaves: flat sterile leaves and upright fertile ones.*

Away from the trunk, but always on large branches, possibilities to accumulate humus in a natural way diminish although light increases. It is there that large ferns armed with special leaves for defending themselves against drought are found. The *Platycerum* with their large protective leaves tightly enveloping their support, create a well protected pocket where humus accumulates from which they can draw nutrients (FIGURE 5.22). The *Drynaria* also produce two types of leaves: enveloping leaves, that dry up very rapidly but protect a small mass of humus, and erect leaves, responsible for photosynthesis (FIGURE 5.21). *Rhipsalis baccifera* – the only indigenous cactus in Africa – is found there also, and many orchids. Exposed to full sunlight and drying winds, they are generally well adapted to temporary drought and have developed, like plants of arid regions, structures or organs allowing them to accumulate water and nutrient reserves. The *Bulbophyllum, Genyorchis, Polystachya* and *Ansellia* have developed bulges in their stems (Figures 5.23 and 5.25). The *Chamaeangis, Cyrtorchis, Diaphananthe* and *Tridactyle* have developed very thick roots (FIGURE 5.24). Many have also developed more or less succulent leaves (Figures 5.24 and 5.26), others, still, very narrow, nearly filiform, leaves.

Farther along the trunk, on the finest branches, pioneers become established. These are usually small species, sometimes very specialised, such as orchids of genus *Microcoelia*, reduced to a mass of more or less thick roots from which emerges a tiny floral stem without leaves. Finally, all the way out at the tips of the branches of large emergent trees, exposed to the sun and to desiccation, only lichens are found.

In high altitude forests, less complex and cooler, epi-

5.23. The small orchid of the genus Genyorchis *is frequent in old mangrove forests and swamp forests that border the bay of the Mondah. In spite of the high ambient humidity, it develops pseudobulbs, like* Bulbophyllum *species, that protect it against any temporary loss of water. Its roots are very fine, however, and hang freely in the damp atmosphere.*

5.24. *Epiphytic orchids, such as* Calyptrochilum emarginatum *of the Mondah forest in Gabon, possess roots that cling to their host and also participate in photosynthesis, especially in heavy shade: their tips are then often tinged green.*

5.25. *Orchids of genus* Bulbophyllum *protect themselves during periods of drought by developing a swelling in their stem that stores reserves of both water and nutrients.*

5.26. Angraecum podochiloides *is able to survive occasional drought thanks to its succulent leaves.*

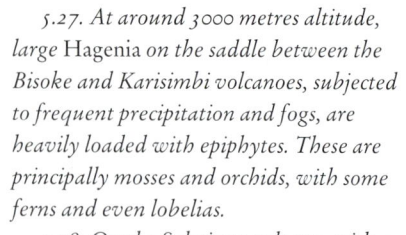

5.27. At around 3000 metres altitude, large Hagenia *on the saddle between the Bisoke and Karisimbi volcanoes, subjected to frequent precipitation and fogs, are heavily loaded with epiphytes. These are principally mosses and orchids, with some ferns and even lobelias.*

5.28. On the Sabyinyo volcano, with a much drier climate, the large arborescent heathers of Philippia johnstoni *are draped in lichens of genus* Usnea, *which withstands long dry periods much better than mosses and orchids.*

phytes of the canopy are mainly mosses and lichens. Mosses are abundant mainly in formations living in conditions of constant high humidity. Like many other epiphytes, they have developed ways to conserve water. In montane forests, they form enormous cushions that retain water and accumulate humus, allowing terrestrial orchids, notably of genera *Satyrium* and *Disa*, to become established (FIGURE 5.27). On the thin branches of the upper canopy, exposed to the sun and to temporary drying, mosses are unable to grow except in the most humid cloud forests. In general, this space is occupied instead by lichens, especially *Usnea*, more tolerant of an alternation of dry and humid periods (FIGURE 5.28).

By their very efficient fixing of mists and fogs, the role of epiphytes is enormous, especially in montane forests. The cryptic precipitation they trap can, in fact, exceed measurable rainfall by 50%. After having been trapped, some of the water evaporates, but most is retained. The surplus falls drop by drop to lower levels and gives life to an entire flora of terrestrial mosses. Only montane forests have mosses on the ground.

Although epiphytes cannot, generally speaking, draw upon nutrients in the plants that support them, they nevertheless compete with them, not only for light, but for some nutrients, as well. They intercept much of the nutrient material that streams over the canopy. Sometimes they favour attack by fungi. Epiphytes can hamper their support and threaten its life by their weight and volume. At times epiphytes can be beneficial by wrapping a sheath of decomposing vegetal matter around the host plant into which it can send fine roots and absorb nutrients. Some epiphytes – mostly algae – have a harmful effect on their host and behave as real parasites. Others seem to facilitate some nutrient exchanges. Finally, epiphytes also create a multitude of micro-habitats harbouring throngs of worms, arthropods, amphibians and even reptiles. Indirectly, they also condition the presence of some birds: the babblers *Kupeornis ruficinctus*, *Kupeornis gilberti* and *Kupeornis chapini*; some weavers, including the Brown-capped Weaver *Ploceus insignis*, and, to some extent, the wood hoopoes *Phoeniculus castaneiceps* and *Phoeniculus bollei*. These species are well adapted to hunting arthropods or worms among the masses of epiphytes.

Stranglers

In the competition for light, a particular group of plants has found an original solution: the stranglers. They begin life as epiphytes, perched in the canopy, and end up as independent trees. They are found among the Araliaceae, notably in genus *Schefflera*, although most are figs *Ficus* of the Moraceae family. They occur in all tropical forests and are even more abundant in Africa than in South America. They are also more abundant in the drier forests.

The seeds of stranglers, deposited with the droppings of a monkey or a bird, germinate preferably in a fork formed by a large branch and the trunk of a tree at the level of the middle stratum or lower canopy. As the seedling develops, it sends long, fine roots downward (FIGURE 5.30). These roots, closely applied to the host trunk, reach the ground and ramify. As the fig grows, new roots appear, unite and reticulate in anastomotic fashion, and eventually form a veritable yoke, fibrous but solid, around the trunk of the host tree (FIGURE 5.29). The tree slowly dies and one day disappears, giving way to the fig, hollow but independent. Some figs are more aggressive than others, although the end of the process remains the same. The *Schefflera*, on the other hand, are much less aggressive species that allow their host to survive.

In natural forests, strangler figs represent an important source of food for the fauna. In industrial plantations, they can become a real nuisance, notably in plantations of oil palm *Elaeis guineensis*.

5.29-30. Two strangler figs: the first in a forest near Brazzaville (left), the second in the Mondah forest near Libreville (right).

Nutrients

Because of intense sun shine, high temperature and abundant water, the primary production of a tropical forest is nearly twice that of temperate forests. Yet its soils are often very poor because their essential nutrients have long been used up or washed away by frequent, heavy rains. Humid tropical forests – especially lowland forests – have therefore necessarily had to develop a high-performance nutrient cycle, limiting losses to the maximum and mostly independent of the soil (FIGURE 5.31). This cycle comprises two essential centres of activity: the vegetal mass – mostly foliage – the centre of primary production, and the litter, the centre of decomposition and recycling.

The litter is supplied mainly by the fall of dead leaves, fruits, branches and fallen tree trunks, as well as by droppings and corpses of herbivores and carnivores. To which is added rain. Not only does rain drain free organic matter from the canopy by passing through the foliage and running off the masses of epiphytes, along branches and trunks, it also brings dissolved mineral material external to the system. The organic material that makes up litter covers the ground. This litter is inhabited by throngs of more or less specialised organisms which rapidly decompose it thanks to the constant humidity and high temperature. Much of the nutrient material thus liberated is immediately reabsorbed by tree roots. Some is carried off by water, however, either into the soil or to the nearest stream. Inversely, some minerals migrate up from the depths of the soil to the surface, where they enter the circuit. The crucial part of the nutrient cycle is at the soil surface, which explains why tropical forest plants usually develop shallow, extremely far-reaching root systems. Some recycling has also taken place in the canopy itself, at the level of the bark of trees and among the masses of epiphytes, whose roots directly absorb substances of interest to them before they ever reach the ground. Competition for light is therefore not the only reason to be an epiphyte.

Soils

Altogether, it can be affirmed that the tropical forest is a habitat that in some ways nourishes itself from its own substance. Given that its nutrient cycle is almost closed, in any case hardly dependent on the soil, the correlation between its characteristics and those of the soil that support it are often not very marked, sometimes even non-existent. In other words, the humid tropical forest is a poor soil indicator and has misled more than one colonist. This is not to say that these soils have no importance. Like all soils, those of tropical forests result from complex interactions between vegetation, climate and underlying rocks. Their formation is nevertheless influenced by the slope, which conditions drainage and erosion, as well as by animals and man, which act on the vegetation. In humid regions, alteration of the rocks is mainly a chemical process. The sun's action does not enter in to play except at the very beginning when the bare rock is still exposed to an alternation of daily warming and nocturnal cooling, as in the case of icebergs.

With temperatures of 25-27°C at the soil surface and high humidity, rocks, whatever their nature, are rapidly degraded and the altered materials may accumulate to a depth of 10 to 15 metres. Biological soil – where vital processes develop – does not exceed a depth of one or two, exceptionally three, metres. On steep slopes, this soil is even much less deep as a result of erosion. Intense rains can thus bring about fairly serious erosion, even on relatively gentle slopes.

The minerals liberated by the degradation of the bedrock move about in function of the movements of

5.31. A humid tropical forest ecosystem gets its nutrients from the soil and from the rain. The excess water running to the rivers takes nutrients away. Most life activities are located in the canopy, where the leaves, through photosynthesis, take up the carbon from the air and provide food to the herbivores and, indirectly, also the carnivores. At the level of the litter, leaves, fruits, wood, excrements and dead animals are recycled. The mineralised organic substances are rapidly absorbed in the superficial soil by roots and their mycorrhizae. They are sent back to the canopy. The wood is not only a support for the foliage; it is also the place where nutrients are accumulated (after Golley, 1983).

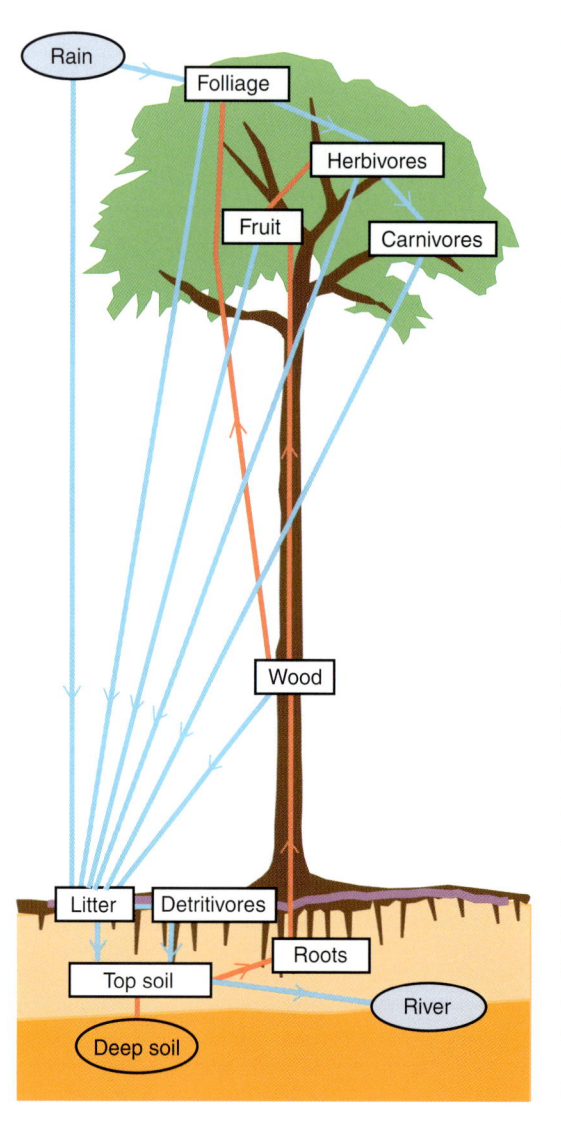

water in the soil. In regions where rainfall exceeds evaporation and evapotranspiration, these movements are always downward. Where precipitation is less, movement towards the surface can develop. In general, downward movements are dominant. Nearly all the soils of humid tropical forests are thus strongly leached. In their upper horizon, rain water carries only a little carbon dioxide and organic material, bases being very soluble and usually rapidly absorbed by the vegetation, these soils are mainly acid.

Pedologists distinguish two main types of soil. The most frequent are the oxysols. Profoundly altered, they have almost no distinct stratification. At depths of around two or three metres, they melt into the saprolithic mass. The degradation of primary rocks, by extraction of potassium, magnesium and other soluble minerals, ends up with a concentration of ferrous and aluminium oxides– more exactly, sesquioxydes – which give to soils their typical reddish coloration. Nutrients coming from litter decomposition are rapidly absorbed by the vegetation and are thus rare. Kaolinites – inert clays arising from the combination of aluminium and silicates – are, on the contrary, abundant and give to these soils a more or less clayey texture. In spite of this, they remain fairly permeable. They also remain strongly acid, which can bring out the toxicity of aluminium and interfere with the absorption of phosphorus.

Ultisols form in regions where humidity varies according to the seasons and where, at times, soils have a tendency to dry out, as is the case in semi-deciduous forests. They comprise a brownish-yellow or reddish clay horizon because of high concentrations of ferrous sesquioxydes. As in the oxysols, these ultisols are poor in organic material and in nutrients. Because of their structure, they are, however, less well drained, more susceptible to erosion and more apt to form layers of laterite, which can harden and form compact slabs.

In addition to the oxysols and ultisols, there is still a whole array of very different, but often very localised, soils. At high altitude, because of the fact that the organic matter does not decompose completely, one finds, for example, peaty soils. At the upper altitudinal limit of forests, there are soils with a very thick humus-bearing horizon up to a metre or more deep. Elsewhere, notably in the plains bordering the Gulf of Guinea, there are completely leached white sands, very acid and of an extreme poverty.

5.32. In many places in Central Africa the soils have at some depth a more or less continuous stone line. This phenomenon is the result of ancients erosions which have taken away the superficial layers and left in place only the stones. Later these stones have been covered again by soil, notably through the activity of termites. The presence of a stone line shows that at some moment in the past, forest was absent.

5.33. Part of the dead wood of a forest is consumed "in place" by saprophytic fungi and disappears before even falling to the ground.

5.34. On the ground, a dead trunk is rapidly consumed by termites, fungi and other decomposers, including some beetles, but its decomposition can still go on for a year or two or even more.

5.35. Millipedes are a part of the fauna that actively participates in decomposition in the litter of humid tropical forests. The largest are about 20 centimetres long.

Decomposition

Although much dead wood is decomposed in the canopy itself (FIGURE 5.33), most of the nutrients to be found in tropical forest soils come from the decomposition of organic material at the level of the litter. Supplied throughout the year mainly by the constant fall of dead leaves, fruits, twigs, branches, even entire trunks (FIGURE 5.34), secondarily by the droppings and carcasses of animals, speed of decomposition is a key factor in the regulation and productivity in tropical forests. It is so rapid because of very warm, humid conditions, and dead leaves disappear within three to six months. In semi-deciduous forests, a temporary accumulation of the litter appears in the dry season as a result of the acceleration of leaf fall and reduction of activity of the decomposers. In montane forests also, the accumulation of incompletely decomposed organic matter may be substantial because of lower average temperatures.

As in temperate forests, decomposition is the work of a great variety of organisms: bacteria, fungi, protozoa, nematodes and oligochaetes, arthropods of all sizes (FIGURE 5.35) and many other animals, including throngs of predators. Since the litter remains very thin, the total biomass of this very diverse and variable fauna remains low, but its activity is very high.

The individual role of the different types of organisms remains just about unknown and except for some floristic or faunal studies, nothing has been done on this subject in Central Africa. The most active "visible"

decomposers are termites, social insects mostly limited to tropical and subtropical regions and highly specialised in the decomposition of wood (PAGES 190-191). Their digestive system contains protozoa and bacteria that digest cellulose. Some species feed directly on dead or decomposing wood. Some attack living wood. Others consume fungi that they cultivate by feeding them from the litter. Still others feed on dead leaves. Some, less numerous, except in Asia, directly consume organic matter found in the soil. Their impact is thus enormous, which is linked both to the strength of their overall biomass and to their diversity. On the basis of his studies in the Democratic Republic of the Congo, Michel Maldague has estimated that around half of the leaves fallen from trees were consumed by termites.

Decomposing fungi are equally extremely important, but they have been little studied. Other groups, such as saprophytic bacteria, actinomycetes, myxomycetes and protozoa, have not been studied at all. Yet, in the end it is these micro-organisms that complete the job of decomposition. Concerning tropical forest bacteria, we know they are very diversified and that some have very specific activities while others are much more opportunistic. The number of species, apparently enormous, is on a scale of the diversity of the flora. Each species of plant develops its own chemical substances which require, in their turn, bacteria adapted to destroying them. This microscopic world escapes our perception and it is ignored in most studies on the functioning of forest ecosystems. And yet, it is omnipresent. Even the air of tropical forests is full of bacteria and it is probable that much of the decomposition of wood in place in the canopy could be the work of bacteria.

As for animal communities, they vary greatly from one type of forest to another. Except for termites, their role is poorly known. It could well be, as Frank Golley suggests, that decomposition in tropical forests is essentially the business of bacteria and fungi, contrary to the temperate forests, where the role of the macro- and microfauna, notably earthworms, seems much more important.

5.36. Orange saprophytic mushrooms on an old stump in the Nyungwe forest of Rwanda.

5.37-38. Some saprophytic mushrooms in the Lopé National Park in Gabon.

5.39. *In order not to lose nutrients freed by decomposition, trees develop a dense network of very superficial roots. The soil of a tropical forest is very often covered with a thick interlacing of root systems, developed mostly within the first 20 or 40 centimetres of the surface of the ground. The more humid a forest is, the more visible this phenomenon becomes.*

5.40. *Basidiomycetes mushrooms are very diverse and abundant, especially in forests with many Caesalpiniaceae.*

Absorption: roots and mycorrhizae

Given the great speed of decomposition and the enormous risk of leaching by the rains, nutrients must be absorbed as soon as they are liberated. This objective, essential to the survival of a tropical forest ecosystem, is attained through two strategies: the development of a very dense system of superficial roots (FIGURE 5.39) and the facilitation of absorption by mycorrhizae, fungi that live in and around roots.

Since the mid-19th century, it has been known that the roots of many higher plants live in association with such fungi. But it was not until the second half of the 20th century, that it was discovered that trees whose roots are infected by mycorrhizae grow faster than those that do not have any.

Endomycorrhizae develop partly in the interior of the roots, but they have extensions that infiltrate the soil with which they greatly increase the absorption of nutrients. In return, their host furnishes carbon. These endomycorrhizae belong to more than 100 species of zygomycete fungi which are not selective, pass easily from one host to another and are often dispersed by rodents.

Ectomycorrhizae develop around the roots. They are fungi of the basidiomycete group (FIGURE 5.40) whose reproductive organs are well known to us in the form of boletuses, russulas, chanterelles or agarics. They surround small roots with fungal masses and also send filaments into the intercellular spaces of their host. Very efficient for the absorption of phosphorus, they are often strictly linked to a single genus of host tree. Their spores are dispersed by the wind.

Some trees can live with several species of mycelia. Others are very specific. A tree can even establish a symbiosis with several species of mycorrhizae at once.

Given the poverty of soils in tropical forests, mycorrhizae are particularly frequent there and trees that live obligatorily in association with a mycorrhiza are clearly favoured. In mixed forests, usually found on the richest soils, the majority of trees have endomycorrhizae. In monospecific forests (PAGES 79-80), on the poorest soils, ectomycorrhizae are dominant. Trees that can develop without mycorrhizae are not numerous. They are often transient pioneers.

The relations between trees and their mycorrhizae are certainly very complex and still poorly known, but it seems that it is mycorrhizae that influence or even determine the composition of a forest rather than the inverse. The existence of mycorrhizae is very ancient. It goes back to the epoch where the first plants started to colonise terrestrial habitats and it is probable that without them, no plant would have ever been able to establish itself on absolutely virgin ground.

5 THE LIFE IN THE FOREST

5.41. Phragmantera capitata, *common in the Lopé National Park, belongs to the cosmopolitan Loranthaceae family. Like the European mistletoe* Viscum album *of the Viscaceae family, all these plants are semi-parasites that draw resources from the tissues of their hosts, but are able to perform photosynthesis as well. Their flowers are often very brightly coloured and attract sunbirds or Nectarinidae. Their fruit – small berries – are an important food resource for barbets or Lybiidae, which are important propagators of these species. As for the foliage, it feeds the caterpillars of some Pieridae butterflies.*

5.42. Thonningia sanguinea, *of the Balanophoraceae family, is a true parasite. This plant without chlorophyll lives on the roots of trees deeply buried in the ground and shows only its scaly bright red inflorescences.*

Parasites

When resources are rare, one can understand that any means are justified for procuring them. It is therefore completely predictable that some species would parasitise others.

Among these parasites, some are no more than semi-parasitic, however. In other words, they take some of their resources from a host plant and produce the rest themselves. This is the case of the Loranthaceae (FIGURE 5.41) and the Viscaceae – mistletoes – that pump sap from the tree that supports them, but also possess some chlorophyll and are thus perfectly capable of insuring photosynthesis. Among the Scrophulariaceae, also, there are species that parasitise herbaceous plants while being able to accomplish their own photosynthesis.

Some plants, however, are entirely parasitic. They have no chlorophyll, therefore usually not even leaves, and live buried in the soil on the roots of some trees. This is the case of the Hydnoraceae, found more in humid savannas, and the Balanophoraceae, very widespread in dense humid forests (FIGURE 5.42). Curiously, the species of these two families have struck the imagination of human populations who systematically attribute to them curative properties for cardiac diseases.

The conservation of nutrients

When the nutrient cycle of a tropical forest operates in isolation, with a minimum contribution from the exterior, it obviously is important to limit losses at every level. At the litter and soil level, this constraint is limited by the speed of absorption. Mycorrhizae are an indirect means of conserving nutrients. Once absorbed, most of the nutrient material is stored in trunks and main branches, which represent nearly 80% of the vegetal biomass above the ground. At this level, the conservation of nutrients rests on the great longevity of the main species, of the order of several hundred years, at least.

But not everything can be immobilised. It is necessary to live! Nutrients going to the foliage – the active part – therefore must be managed very parsimoniously. Of course they will later be recycled when the leaves fall and come back again to nourish the productivity of the forest, but they also run the risk of being directly washed out by the rains and sent to the litter before even having been utilised by the tree that absorbed them. Some elements dissolve fairly easily in rain water on the surface of leaves – notably potassium – especially when they are damaged by herbivores. Many leaves possess mechanisms that limit losses: long acuminate points to drain off water as rapidly as possible, thereby reducing the chance of leaching (FIGURE 5.5), hairs, waxes or thick cuticles to avoid the dissolution of elements. Losses are also reduced by the return of most of the nutrient material present in leaves to the branches just before leaf fall. The best system seems to be to not lose leaves at all, or at least to keep them as long as possible. Seen from this angle, being evergreen should contribute to the conservation of nutrients.

Nutrient conservation should not only be seen at the level of a single tree. It must also be conceived at the scale of an entire ecosystem. It is there that all the importance of the swamp and riverine forests appears. By filtering the waters draining from terra firme forests, they recuperate much "escaped" nutrient material (FIGURE 5.43). Unfortunately, this is a totally unknown domain. In South America, studies have compared terra firme forests with igapo and varzea type forests, but nothing comparable has been done in Africa. In fact, this should be a fascinating world where at times terra firme elements, at times aquatic elements, interact. All that we can suppose is that swamp and riverine forests constitute major systems of nutrient recuperation on the scale of the entire Congolese forest ecosystem.

In general, the opening of forest cover by Man can very easily bring with it irreversible damage. Not only are part of the reserves stored in branches and trunks burned or removed from the system during deforestation, but also litter disappears and the upper soil layer is eroded, the very instruments of nutrient recycling are destroyed and scarce residual material is carried off by the rains. Restoration of the original nutrient cycle may thus take tens of years. Sometimes the damage is even irreversible and "secondary" ecosystems never again manage to rival the system that they replace. Without being biologists, most of the traditional farmers of forested regions – at least those who truly live in the forest, not the mountain farmers who are, in fact, no more than forest "converters" – have learned this hard truth. To deal with this problem, they have developed itinerant agriculture and, until 10 or 20 years ago, they avoided returning too soon to surfaces already cultivated. Nowadays, with the increase of all sorts of pressures, the rest time allowed to forest soils unfortunately has become shorter and shorter.

5.43. In inundatable forests, many plant species – not only Uapaca, *recognisable by its spectacular stilt roots – develop abundant free-hanging roots in order to catch nutrients carried by the water.*

Herbivory

In a habitat rich in species, and where most of the nutritive resources are in the litter or in the vegetation – in the form of roots, wood, foliage and fruits – it is not surprising that nearly 60% of the animal biomass feeds on the litter or the fungi that decomposes it and about 20% on wood. The latter are mainly termites, beetles and the larva of some other insects. Only 5% of the biomass seems to be feeding on leaves. As for flowers and fruits, they are of interest to only an insignificant fraction of the consumers.

Even though foliage represents no more than about 1% of the organic biomass of a tropical forest, it is the centre of photosynthesis. Foliage constitutes an absolutely vital part of the vegetation and merits special protection. Some methods are purely mechanical and very general. Trees lose their old leaves and some eliminate their bark in plates to rid themselves of parasites and other undesirables. Even if these do not necessarily bother the trees, they can weigh them down and sometimes facilitate the installation of more dangerous parasites. Some species cover themselves with sharp spines to discourage large herbivores – a strategy rarer in the forest than in the savanna – or incorporate crystals of silica in their tissues to discourage smaller herbivores.

The majority of defences are chemical. Unlike savanna vegetation, many foliages in the forest contain repulsive or toxic substances. Some alkaloids affect the enzyme systems of herbivores. Phenols, tannins and other organic acids combine with proteins and form indigestible compositions. Terpenes, comprising some gums, oils and resins, block digestion physically or release toxic substances during their degradation. Numerous plants – manioc *Manihot esculenta* for example – contain glucosides which produce cyanides during the degradation process.

These toxic alkaloids are, however, very unevenly distributed. They are much more abundant in lowland forests than in montane forests, often more concentrated in plants growing on poor soil than those growing on rich soil and less frequent in species with rapid growth. These toxic materials are not limited to the leaves. They can be found in some barks, such as in *Warburgia ugandensis* whose bark is hot like pimento. Other species, such as many Euphorbiaceae and Sapotaceae, contain a toxic latex.

In spite of all this, predators find ways to adapt either by metabolising the chemical substances or by incorporating them into it (FIGURE 5.50). In this way they become specific to particular plants and many insects are, in fact, strictly linked to a single plant species. Some have a more varied diet, but none is capable of consuming more than 10% of all the plant species present. Relationships between the composition of a

5.44. Young leaves of the Okoumé or Gaboon Mahogany *Aucoumea klaineana* are red and loaded with resins. In spite of this, there is a small butterfly whose caterpillar attacks this species.

5.45. The young leaves of a Gilbertiodendron *on the banks of the Mambili in Odzala National Park are orange-red.*

5.46. The young leaves of a lianescent shrub on the edge of a forest in Gabon are transparent yellowish green.

5.47. *Independently of insects, a mycosis can also attack leaves, like this one of* Alchornea cordifolia.

5.48. *The three caterpillars of* Pyrrhochalcia iphis, *the largest Hesperid butterfly of Africa and probably in the world, do not leave much of this leaf.*

5.49. *The production of leaves, vital organs of plants, consumes energy and nutrients. To let them be eaten by insects thus has a "cost", but this depends on which part of the leaf is destroyed and at which moment of its development damage is done. In defending itself, a plant must therefore avoid threatening its reproduction. In the end it is the only thing that matters.*

tropical forest and its entomological fauna are thus very complex. They bring on a sort of "chemical arms race" that transforms the tropical forest into an infernal laboratory.

This laboratory has a cost and the energy that a plant employs to defend itself is lost for reproduction or defence against climatic or edaphic adversities. Thus, tropical forest vegetation can never protect itself completely against defoliation (FIGURE 5.49), which strikes secondary species more often than those of old forests.

The most common defoliators are caterpillars (FIGURE 5.48). Often damage done by them is not very noticeable because they attack a single species of plant. Defoliation can be total, but usually the leaves disappear by means of innumerable holes and notches (Figures 5.48-5.49). In tropical American forests, it is estimated that more than 12% of the foliage is destroyed annually by insects, without taking into account sap sucked and buds destroyed before even having had the time to develop. Besides, without being completely destroyed, a leaf can be rendered inoperative following a reduction of its surface by curling or by being coated with various materials. Herbivores thus constitute a crucial problem in tropical forests. However, they remain poorly known in Central African forests. As for insects that attack flowers and seeds, they are an unexplored domain.

In addition to insects, there are many vertebrates that eat leaves. Except for colobus monkeys, they are, in general, less selective. Unlike insects, vertebrate herbivores search more for a varied diet offering them all the necessary elements, while avoiding being poisoned by toxic leaves or being surprised by a carnivore. Most of the defences developed by leaves do not target vertebrates but rather invertebrates. All the same, the presence of large herbivorous mammals, such as the Elephant and the large Bovids, have certainly affected the vegetation of tropical African forests.

However, within a population of a particular plant species, not all individuals have the same thresholds of response to environmental factors, and not all individuals will produce flowers, and later fruits, at the same time. In this way, individuals can be separated genetically without being separated geographically. This could be a mechanism allowing sympatric speciation.

Reproduction

In a forest of some 300 to 500 species of trees, and where 50 to 90, or even 150 species can be found on a single hectare, it is inevitable that many of these species are rare or very rare. This rarity, the unavoidable corollary of diversity, offers some advantage in predator protection, but it is also a serious inconvenience in terms of reproduction.

Not only is vegetation inevitably immobile, but many species are incapable of pollinating themselves or else have developed mechanisms to prevent it. In order to reproduce, plants of the tropical forest are thus not only confronted with problems of pollination, as are all the plants in the world, but also with problems of synchronisation of their flowering, not to mention problems linked to seed dispersal.

Some plants get around these problems by practising vegetative reproduction. An extreme case is that of the parasitic plant *Thonningia sanguinea* (FIGURE 5.42) whose male and female flowers have never been observed at the same time and whose reproduction cannot take place except vegetatively. In the case of Zingiberaceae and Marantaceae species of the understorey, reproduction by seeds occurs, of course, but these plants nevertheless propagate mostly by means of stolons and rhizomes. In the case of some species of genus *Costus*, a Zingiberaceae, young plantlets develop on the inflorescences after flowering.

Flowering

Some species, mostly in secondary forests, flower very precociously, from the age of two or three years. Others, notably among the large emergent trees of the canopy, do not experience a first flowering until the age of 40 or 50 years, and the Moabi *Baillonella toxisperma* does not even flower before about 70 years.

Some species flower only once and then die. This is the case with the bamboos, such as the Mountain Bamboo *Sinarundinaria alpina* (PAGE 117), some Acanthaceae of the understorey, such as *Mimulopsis arborescens* of montane forests, and the palms of genus *Raphia*. Others, mostly understorey species, flower fairly continuously. Most tropical forest plants flower once or several times a year at more or less regular intervals, however.

Some species do not flower except every three to 13 years. Their flowering is then often massive and well synchronised. This gregarious flowering seems to be set off by climatic "anomalies": a prolonged drought or an exceptional lowering of the nocturnal temperature. Very often, it accompanies phenomena of the *El Niño* or *La Niña* type. By influencing flowering, these phenomena could in this way have an impact on the regeneration of forests and, in the longer term, on their composition. For the peoples of the forest and neighbouring areas, these massive and unpredictable flowerings have a magic signification. Thus, in the montane forests of the Albertine Rift, notably in Rwanda, the flowering of the liana *Sericostachys scandens*, an Amaranthacea, announces famine and war. Curiously, soon before the genocide of 1994, the liana bloomed three years in a row. Whatever the explanations, it has to be asked, what is the advantage of such a strategy? Is it perhaps a protection against seed predators? A gregarious flowering brings with it a massive fruiting that that would take predators off their guard.

A great majority of plant species of humid tropical forests flower each year, more or less seasonally: some at the beginning or at the end of the rainy season, others rather in the dry season. Many flower in this way around the same period each year which seems to indicate that their flowering may be linked to one or another climatic phenomenon. The importance of climate is, in addition, confirmed by the fact that seasonality is more marked in the canopy than in the understorey. But which climatic factor could be responsible? Temperature? Rainfall? In some cases an answer was able to be given to this question: Caroline Tutin and Michel Fernandez discovered that some species of the Lopé National Park, notably *Cola lizae*, flower following a lowering of the nocturnal temperature to below 19° C that is observed during the dry season. In other species and in other places, flowering is set off by the first rains.

In still other cases, links between flowering and climatic factors are not at all obvious and, according to Richards, some plants seem to respond to an internal rhythm. As an example, he indicates the case of the Flamboyant or Flame Tree *Delonix regia* which flowers every seven to ten months and whose flowering shifts from year to year. There is still much work that remains to be done in order to understand when and why one or another species flowers.

In Central Africa, the phenology of flowering has not, unfortunately, been much studied. That is why a monitoring of the main tree species has been set up by the ECOFAC programme under the supervision of Professor Jean Lejoly. The most important observations have been obtained in Ngotto Forest in the Central African Republic. They involve 200 trees of 20 species over three consecutive years (see table). Ngotto is located in the semi-deciduous forest zone, submitted to a well marked seasonality. Observations from this site are therefore not representative of most of the Congolese forests. It is nevertheless interesting to compare with them other ECOFAC sites, such as Monte Alén in Equatorial Guinea and Odzala in the Congo.

Since there are no common species among the most important species of Ngotto and Monte Alén, only a

5.50. *Butterflies of the Acreinae subfamily, like this* Acaea, *are protected by toxins originating from plants that their caterpillars have accumulated.*

5 THE LIFE IN THE FOREST

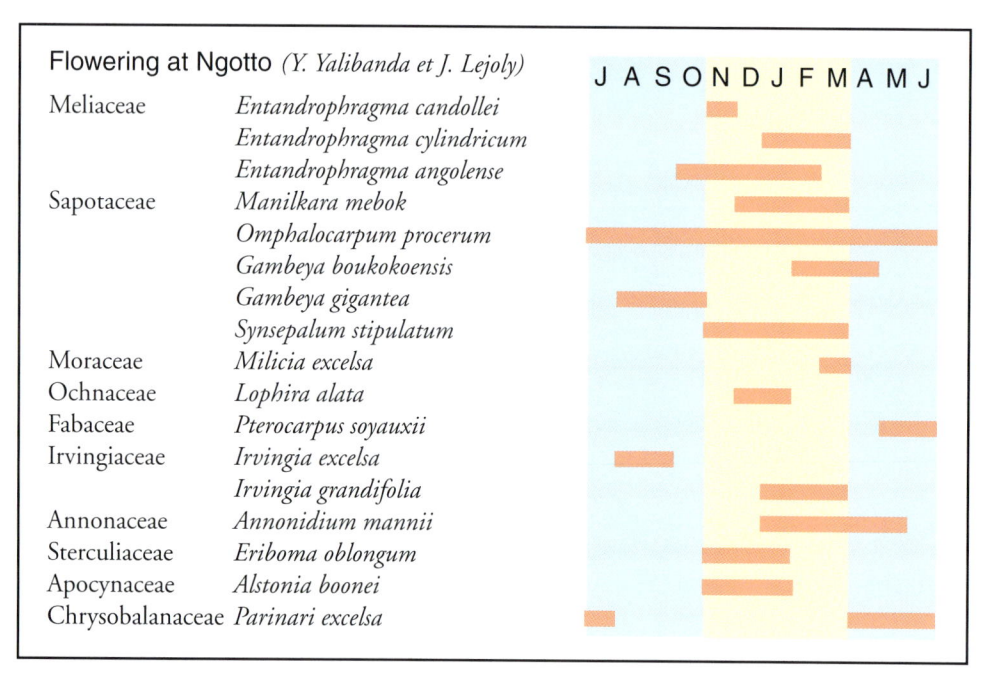

Flowering at Ngotto (Y. Yalibanda et J. Lejoly)

Family	Species	J A S O N D J F M A M J
Meliaceae	*Entandrophragma candollei*	
	Entandrophragma cylindricum	
	Entandrophragma angolense	
Sapotaceae	*Manilkara mebok*	
	Omphalocarpum procerum	
	Gambeya boukokoensis	
	Gambeya gigantea	
	Synsepalum stipulatum	
Moraceae	*Milicia excelsa*	
Ochnaceae	*Lophira alata*	
Fabaceae	*Pterocarpus soyauxii*	
Irvingiaceae	*Irvingia excelsa*	
	Irvingia grandifolia	
Annonaceae	*Annonidium mannii*	
Sterculiaceae	*Eriboma oblongum*	
Apocynaceae	*Alstonia boonei*	
Chrysobalanaceae	*Parinari excelsa*	

5.51. Gloriosa superba *is a climbing Liliacea that grows in forest mantles and savanna woods. Its pendulous flowers change colour throughout their existence and their stamens are held far from the pistil.*

very general comparison is possible. At Monte Alén, a very humid evergreen forest with a dominance of Burseraceae and Anacardiaceae, the principle species flower in March-April, during the main rainy season, and fruit therefore in July-August, during the dry season. At Ngotto, the forest is rich in Meliaceae and Sapotaceae. The first group flowers mainly from November to May, in the dry season. The second flowers nearly year-round but with a marked preference for the dry season and the beginning of the rainy season.

The forests of Monte Alèn and Odzala, on the other hand, share two species. Ebap S*antiria trimera*, a Burseracea, flowers and fruits at the two sites between January and April, but the False Nutmeg or Ilomba *Pycnanthus angolensis*, of the Myristicaceae family, shows no seasonality in Equatorial Guinea, while in the Congo it flowers gregariously in January-February in the main dry season

As for Odzala and Ngotto, these two sites have four important species in common. For the African Oak or Azobé *Lophira alata*, an Ochnaceae, there is no difference between the two sites and flowering takes place between December and January. In the case of the Kosipo *Entandrophragma candollei*, of the Meliaceae family, the very short flowering period takes place in the dry season: November at Ngotto, January at Odzala. At Odzala, trees flower one year out of two, but at Ngotto only 10 to 30% of the trees flower each year. Barwood or Padouk *Pterocarpus soyauxii*, of the Fabaceae family, loses all its leaves at Ngotto for about a month during the season of heavy rains, that fall between September and November, but loses only part of them at Odzala. Despite this difference, flowering takes place in the two sites in May or June, thus during the dry season at Odzala. As for *Annonidium mannii*, of the Annonaceae, flowering goes on in the two sites between January and May, that is, in the dry season and at the very beginning of the rainy season.

These few observations show that the same species can adapt its seasonality to local conditions, that many possible schemes exist, that the problem is thus very complex and that it is much too soon to draw general conclusions.

Pollination

In spite of their scarcity, many plants of tropical forests cannot pollinate themselves. Many species are, in fact, dioecious – some of the plants carry male flowers, others female flowers – and, among monoecious species, many have mechanisms that prevents pollen from pollinating the flower from which it comes or even another flower of the same plant. Sometimes, self pollination is prevented by the very structure of the flowers, as can be seen in *Gloriosa superba*, a climbing Liliaca of the mantle (FIGURE 5.51). In other species, the stamens reach maturity before the pistils are receptive. This mechanism is frequent in the Fabaceae and related families. Various *Parkia*, of the Mimosaceae family, are a good example. In general, self pollination is rendered impossible by a self incompatibility. In other words, the pollen of an individual cannot pollinate the ovule of the same individual. This phenomenon relies on a great many very different mechanisms. In the *Mussaenda*, for example (Figures 5.52-5.53), all flowers have stamens and pistils, but some individuals have flowers with long pistils, others with short pistils. In the former, only the female organs are pollinated, in the latter, only the male organs.

This «aversion» to self pollination can be explained by the fact that only cross pollination allows the maintenance of genetic diversity. But self incompatibility is not always permanent and it happens in some species that it disappears in the course of flowering, that it is present in some populations and not in others or that it exists only in some individuals. The fact that self incompatibility is frequently overridden enables us to make out that the plant world is torn between two tendencies: to increase genetic diversity by cross pollination or to fix some "successful" genotypes by self pollination. Only natural selection allows a decision on which strategy is the best, but the response is not necessarily the same in all cases. If cross pollination allows the maintenance of great genetic diversity, self pollination should facilitate the implantation of a species in marginal areas of its range, where individuals are still very far apart and where chances of cross pollination are therefore rare.

Whatever the reasons and advantages, cross pollination implies the intervention of a pollinator. Given that the density of individuals of a species are often very low and that the wind is insignificant inside the vegetation, species of tropical forests that make use of the wind are rare. This type of pollination occurs, nevertheless, in the Rhizophoraceae mangroves, in some palms and in some Moraceae.

The large majority of plants necessarily depend on animals to carry their pollen: insects, birds or bats. In order to attract pollinators and afterwards ensure a successful pollination, flowering plants have had to develop a whole series of strategies involving the shape of the flowers as well as their position and the production of nectar. Pollinators have also developed all kinds of strategies to optimise their search for food. Plants have adapted to these. Examples of co-evolution are not lacking in the world of angiosperm reproduction and are often particularly sophisticated and diversified in tropical forests.

Butterflies are attracted by white, yellow, pink and red flowers, never by blue flowers. Not all of them seek the nectar of flowers and are not, therefore, pollinators. Many Nymphalidae, notably in genera *Euphaedra* (PAGE 186) and *Bebearia*, feed instead on fruit rotting on the ground. As for moths, they search for white or pale flowers. Many species pose as day-flying butterflies, but the Sphingidae, very abundant in tropical forests, hover in front of flowers and suck the nectar with their long flexible proboscis. They especially seek out fragrant flowers in the evening and at night and are capable of sucking nectar from the bottom of the deepest corollas and spurs (FIGURE 5.57). Generally speaking, flowers that attract moths and butterflies produce a very liquid, lightly sugared nectar, easy to take up with a fine proboscis.

Bees are attracted by flowers of all sorts of colours and are as a rule very sensitive to those reflecting ultraviolet. Social bees, of genera *Apis*, *Melipona* and *Trigona*, certainly play an important role in pollination, but large solitary bees, although less abundant, have perhaps a proportionally greater role. Their effectiveness as pollinators arises from the fact that these insects have the habit of visiting a series of plants placed along a more or less fixed itinerary. They may fly twenty or so kilometres a day in this way. In addition, each species exploits a particular stratum of the forest. Wasps are equally good pollinators. Often, they are attracted by purple flowers, but they also visit white flowers, notably those of several species of *Cordia*. Flowers that attract Hymenoptera usually produce a thick, concentrated nectar. Among the Hymenoptera, the most curious example of co-evolution is that of the small species of the Agaonidae family that pollinate figs *Ficus*. For each species of *Ficus* there is a corresponding species of wasp and the two species – *Ficus* and wasp – cannot reproduce without each other (Figures 5.58 and 5.59).

5.52. Mussaenda tenuifolia, *of the Rubiaceae family, has yellow flowers and large, white sepals looking like leaves.*

5.53. *The flowers of* Mussaenda erythrophylla *are white but surrounded by wide red sepals. Both* Mussaenda *species are lianas able to reach 10 metres in length and are fairly frequent in mantles and disturbed forests. Their seeds are eaten by gorillas.*

5.54. *The flowers of Araceae, like those of* Cyrtosperma senegalense, *a tall species, common in inundatable areas, form a perfect trap for beetles attracted by their bad smell.*

5.55. Fruits of Maesobotrya, *a small tree of the Euphorbiaceae family, are encountered in the coastal forest of Gabon.*

5.56. In the understorey of the same very humid coastal forest, the fruit of a Brazzeia, *a small tree of the Scytopetalaceae family – one of the few families endemic to African forests.*

Both species, in very different families, have flowers appearing directly on the trunk or on the main branches, as is the case in Omphalocarpum *(Page 178) or Cacao* Theobroma cacao *(Page 262). This phenomenon, called cauliflory, is frequent in tropical forests and it affects mostly shrubs and small trees. The reason for it is not, however, well established. Perhaps there is a link with pollination by ants, which continuously search trunks and larger branches, or with the later dispersal of the fruit. Perhaps it rests on a physiological phenomenon.*

Flies are usually attracted by small, inodorous white flowers, but meat flies preferentially visit brownish-purple flowers that give off a nauseating odour. It is the same for beetles, that are often attracted and captured by Araceae flowers, malodorous and funnel-shaped.

The bats that intervene in pollination in Central African forests are exclusively frugivorous bats. They preferentially visit dark flowers with an insipid or frankly nauseous fragrance, that do not open except at night and present themselves at the end of long stems hanging in such a way as to make them more accessible. This is the case with the *Parkia*. Bats also visit flowers of the Kapok Tree *Ceiba pentandra*, grouped at the ends of branches. Pollination by bats is more frequent, however, in trees that lose their leaves. It is therefore more usual in semi-deciduous forests on the edge of the forest massif.

As for birds, it is mostly sunbirds or Nectariniidae that play the roll of pollinators. They are usually attracted by red or orange flowers, often more or less tubular in shape, like those of the Loranthaceae. Plants that depend on birds for pollination sometimes have large, cup-shaped flowers, like those of the Flame Tree or African Tulip Tree *Spathodea campanulata* of the Bignoniaceae family. Its flowers are so large that they can serve as a drinking trough.

Some small, even insignificant, flowers, are surrounded by brightly coloured bracts or leaves to attract insects, but the most remarkable adaptations are found in the orchids where many species have very complex flowers which not only attract insects very selectively but also have structures destined to facilitate the transport of pollen.

5.57. In some Orchids, like this Cyrtorchis, *pollination is performed by moths. In others it is carried out by diurnal Hymenoptera, often of a single species. For this their plants have developed very complex flowers.*

5.58-59. Many figs are cauliflorous, bearing flowers direct from the main stem or older branches. However, the most peculiar trait of these Moraceae is their system of pollination.

A fig is no more than a receptacle, closed or linked to the exterior world by a narrow channel. Inside this receptacle, numerous small flowers develop. Male flowers produce pollen, female flowers produce seeds and sterile flowers receive the eggs and larvae of wasps. For each fig species there is a corresponding wasp species whose wingless males never quit the interior

Fruit dispersal

In order for a seed to become a mature plant, it is necessary for it to be dispersed. To accomplish this, 5 to 10 % of the species of the canopy rely only on themselves. Thus, the pods of the Caesalpiniaceae and Mimosaceae burst and scatter their seeds far and wide.

Other species use the wind. For this, orchids produce enormous quantities of extremely fine seeds. Some lianas, notably some of the *Combretum* (FIGURE 5.61), and some large trees such as the *Entandrophragma*, produce winged seeds. Other species, such as *Ceiba pentandra* and *Bombax buonopozense*, of the Bombacaceae family, or *Funtumia africana* – the African rubber tree – produce feathery seeds. In old lowland forests, species whose seeds are transported by the wind are nearly absent from the understorey and do not represent more than 15 % of the canopy species. They are more numerous in semi-deciduous forests, such as at Ngotto, than in evergreen forests, such as those of the Dja. They are most abundant in secondary forests and montane forests, where they may represent up to 37 % of the species.

A small number of plants have their seeds dispersed by water. They are generally found along streams. Among these, there are probably even a few whose dispersal is assured by fish.

The great majority of tropical forest plants – 70 to 80% in lowland forests, 50 % in montane forests – are dispersed by animals, especially birds and bats, as well as by primates and some ungulates, such as forest duikers and elephants.

The birds, diurnal and equipped with excellent colour vision, are especially attracted by red or orange fruits. Bats, being nocturnal, are attracted more by

while the females, once they have been fertilised, leave the fig of their hatching and go out in search of another fig tree, enter its fruit and lay their eggs in the sterile flowers. At the same time, they transport some pollen and fertilise the newly invested female flowers.

5.60. Oncoba (Lindakeria) dentata, *a shrub of the Flacourtiaceae family, protects its seeds by spiny envelopes and relies only on itself for their dispersal.*

5.61. *The winged fruits of the liana* Combretum platypterum *are dispersed by the wind.*

5.62. *The white and pink berries of* Anchomanes difformis, *a common Araceae of the understorey, have a very powerful astringent quality. Nevertheless gorillas like them.*

5.63. *The fruits of* Palisota, *a genus of the Commelinaceae family, common in the understorey, are probably mainly dispersed by birds.*

strong odours. Each species, however, has its preferences. Fruit-eating birds of the canopy, such as hornbills of the family Bucerotidae, often seek large fruits, rich in fats and proteins. More opportunistic birds, such as bulbuls of the Pycnonotidae, are satisfied with small seeds, often not very nutritious but very abundant in the understorey (FIGURE 5.64). These birds complete their diet by consuming many insects.

In order for dispersal to be effective, the disperser must also move about after having fed. Depending on their habits, different species have a variable impact on dispersal, and it is birds and bats that make the best dispersers because of their mobility. This quality can even be enhanced as a result of prudence vis-à-vis predators. In fact, predators, notably diurnal raptors, may have an indirect but important influence on the process of dispersal. Annie Gautier-Hion observed, in the high canopy, that birds rapidly gather a few fruits and then fly off, probably out of fear of raptors, while in the lower dense strata they have a tendency to eat on the spot.

In spite of everything, and whatever the mode of dispersal, a great majority of the seeds fall to the ground not far away from the tree that produced them. Usually birds and bats carry them the farthest, but there are exceptions to this. Lee White was able to observe that the best, if not the only, disperser of the seeds of *Cola lizae* – a tree of the Sterculiaceae family, near endemic to the Lopé National Park – is the Gorilla. It is the only animal that feeds on fruits of this tree, swallowing them whole and eliminating the seeds intact at a good distance from the tree from which they originated. All other species destroy the seeds or spit them out practically under the producing tree.

Fruit-eating is a very complex phenomenon that sets plants before a dilemma. On the one hand, they must attract the frugivorous animal, therefore offering it

5 THE LIFE IN THE FOREST

appetising fruits. Some fruits give off an odour that attracts the consumers. On the other hand, they must also protect their seeds against mastication and very aggressive gastric juices. Some plants protect their fruits by chemicals that discourage at least some of the predators. In this complex situation, birds and bats seem to be most efficient at dispersal. Many species of birds and bats have become exclusive frugivores and even specialised. Nevertheless, some plants depend on mammals. Very heavy fruits, notably those of Oboto *Mammea africana*, Alen *Detarium macrocarpum*, Moabi *Baillonella toxisperma*, Afane *Panda oleosa* or Afo *Poga oleosa*, are dispersed exclusively by elephants.

Germination

The seeds of most tropical trees lose their power to germinate after a few weeks or within two to three months, but there are some that last much longer. These seeds, said to be dormant, do not germinate except when they are put in the light, for example in a clearing made by a fallen tree. The immense majority of seeds that have escaped predation germinate in the days or weeks following their production. Sometimes mass germination can be observed in the understorey of dense forests, but, in the end, very few of the plantlets will conquer a place in the canopy and reach maturity. For that to happen, local conditions must be favourable. Many species of the upper canopy cannot develop except if they have enough light. Therefore, they have to depend on the occurrence of tree falls.

5.64. Many shrubs of the understorey produce abundant, small, blackish or reddish fruits, which are eaten by birds.

5.65. The Connaraceae family essentially comprises lianas with brightly coloured fruits whose seeds are much appreciated by animals. Most of the species are large lianas of the canopy, but some, like this *Cnestis* species, grow in mantles and also isolated savanna thickets. Because these plants produce fruit for most of the year, they have an important role in the forest ecosystem.

5.66. When the bright red fruits of this *Cola* species, seen in the understorey of the Crystal Mountains in Gabon, open, shiny black seeds appear surrounded by a sweet substance which attracts birds.

5 THE LIFE IN THE FOREST

In small tree-fall gaps, initial regeneration is performed essentially by dormant seeds or by plantlets already present before the clearing formed. It is only in large tree-fall clearings, and man-made clearings, that regeneration is accomplished by seeds arriving after the opening is made. A good part of these seeds are wind-carried. Others are brought by animals. The role of animals is not limited just to the dispersal of seeds. It is also implicated in the germination process and many seeds do not sprout except after they have transited the digestive system of one or another animal (Figures 5.67-5.68). This phenomenon has been known for a long time but remains poorly understood. Subsequently, we have very few objective indications of what is happening to a forest when its animals disappear.

5.67. Omphalocarpum is a small tree of the Sapotaceae family, quite common in the Marantaceae forests of the Lopé National Park in Gabon. Its heavy caulinary fruits are compared by the local people to the breasts of a young girl, but they can be as large as a human head, and contain about 30 to 40 large seeds. They are very hard, however, and only elephants can open them. Like around 70% of the tree species of the Lopé National Park – according to Lee White and Kate Abernethy – Omphalocarpum can only be dispersed by animals, more specifically by elephants. If these animals disappear, the fruits become rotten at the foot of the trees, but the seeds never germinate.

5.68. In the Lopé National Park, some young plantlets emerge from a pile of elephant dung.

5.69. A stump in decomposition: many trees decompose before they fall down.

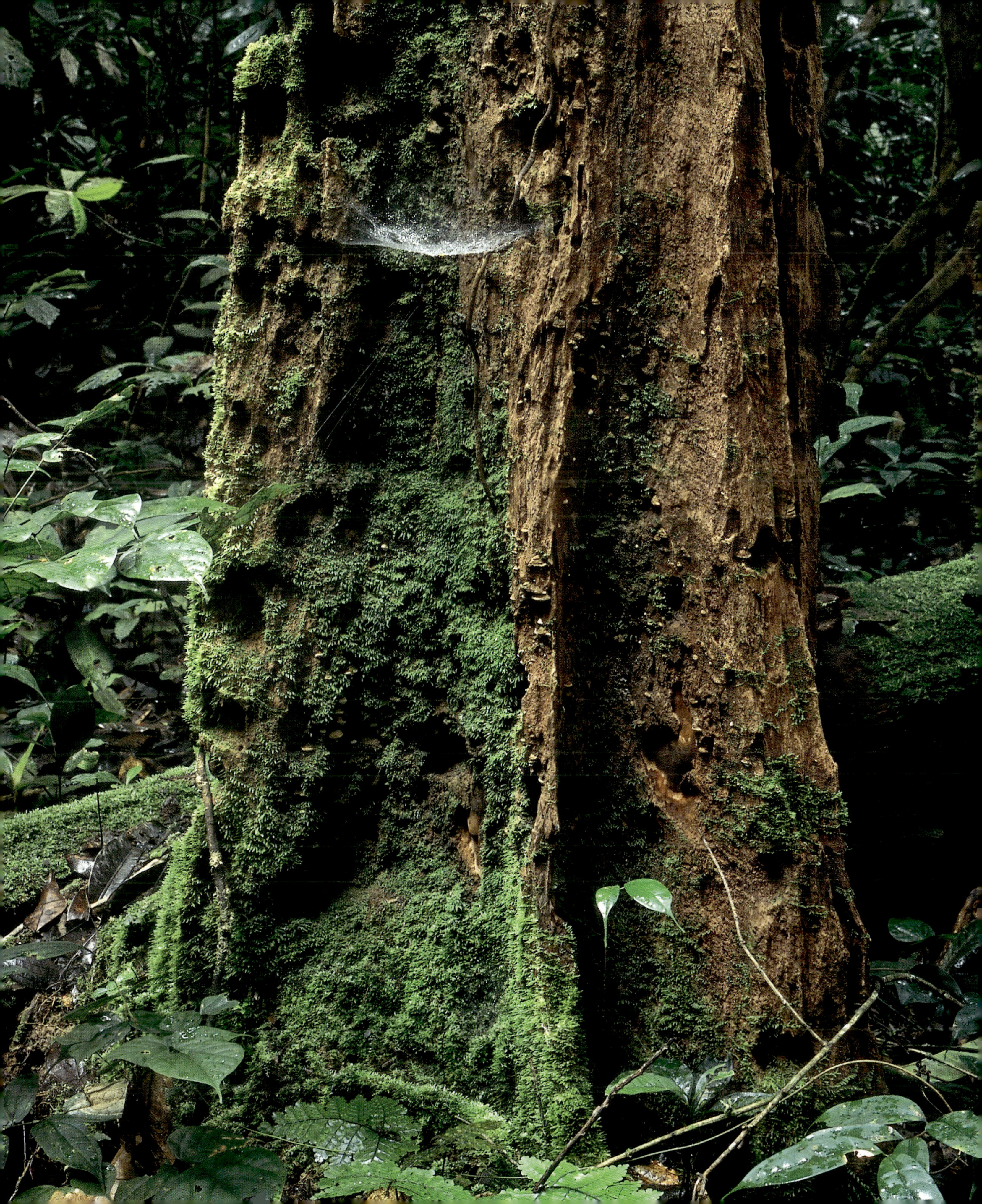

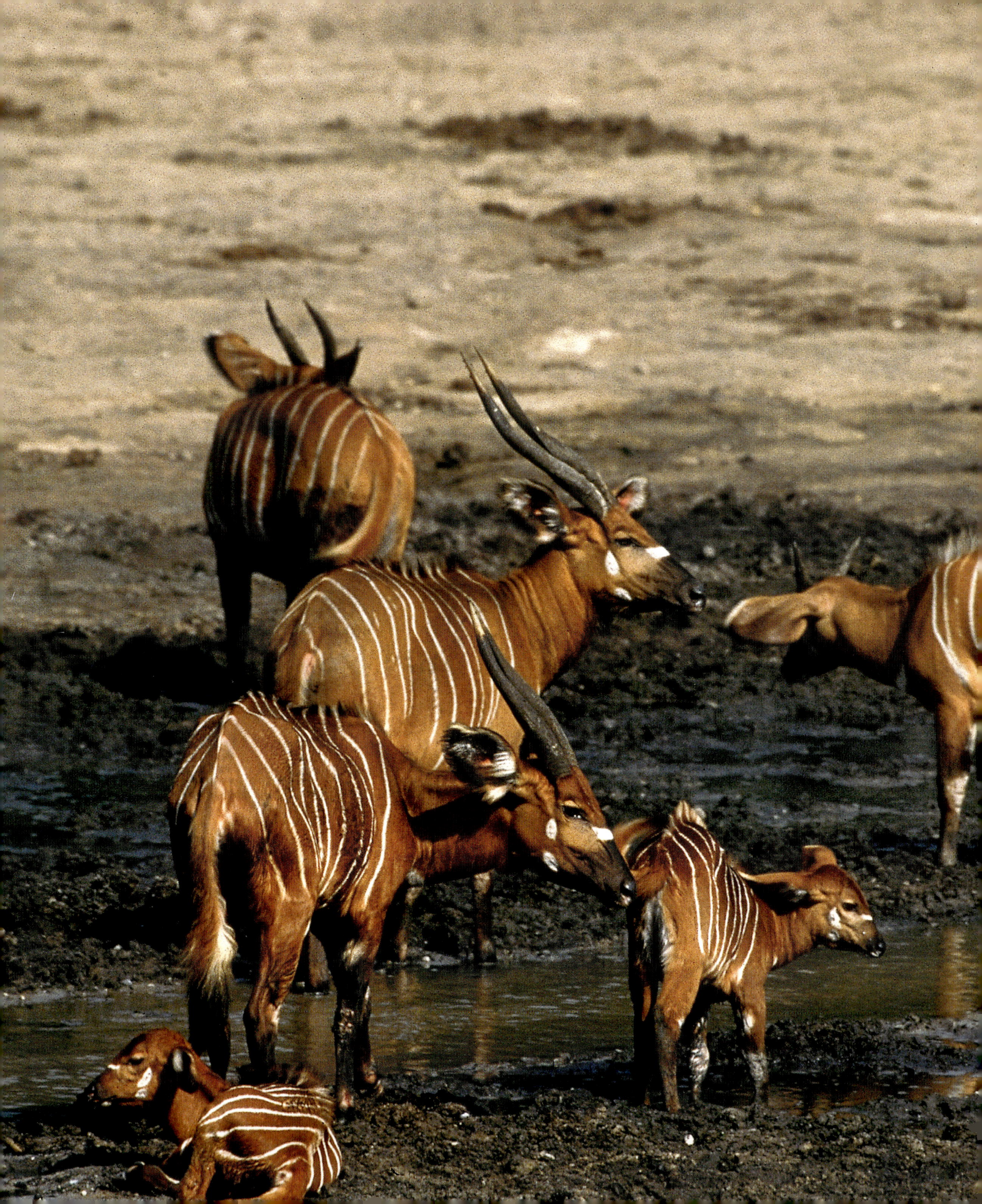

6 | The Fauna

The tropical forest is above all a world of plants and fungi. Its animal biomass is very small, but the diversity of animal life is just as rich as that of the plant world and any tropical forest harbours more animal species than the grasslands that surround it or than any other forest. Perception and measurement of this diversity confront problems of methodology and scale, but, apart from these obstacles, there arise more fundamental questions concerning the coexistence of all these species. Elements of the answers are to be found in their spatial distribution, in the mechanisms that govern competition between species and in the relationships of dependence between species – in a word, the complexity of the forest environment. Generally speaking, tropical forest animal species are, however, represented by a low, or even very low, density of individuals. The great majority of tropical forest animal species therefore find themselves confronted with a contradictory situation: on the one hand it is necessary to be as discreet as possible in order to escape predators, on the other it is necessary to advertise one's presence to congeners in order to meet and reproduce. That is why some species have opted for a nocturnal life and others for mimicry, while still others have developed brilliant costumes, vocalisations and powerful clamouring or penetrating scents. Many have developed a subtle blend of several of these strategies, sometimes using one, sometimes another.

In most classes of animals, the forests of Central Africa are poorer in species than those of South America or of Asia, but some families or groups are better represented than elsewhere. Altogether, the fauna of tropical African forests is typical of the Old World and shares some very distinctive groups with Asia such as, for example, the elephantids, pteropodids or fruit bats, chevrotains, bovids, cercopithecids and hominids. It also has endemic families and sub-families, notably the guinea fowl, mousebirds, wattle-eyes, bushshrikes, helmetshrikes and touracos among the birds or the galagos, giraffes, anomalures, elephant shrews, otter shrews, hippopotamuses and duikers among the mammals. Some families show clear convergences with non-related families of South America: among amphibians, the African arthroleptid frogs resemble the South American dendrobatids so much that you cannot tell them apart easily; among birds, the African libiid barbets resemble the American capitonid barbets, and the African muscicapid flycatchers the American tyrannid flycatchers.

6-1. The Bongo Tragelaphus euryceros *is the largest and the most prestigious of the forest antelopes. It inhabits intermittently the forests from Sierra Leone to Kenya. It is a solitary animal but, in the clearings of Dzanga-Sangha National Park, it is not rare to encounter small temporary gatherings of as many as a dozen individuals.*

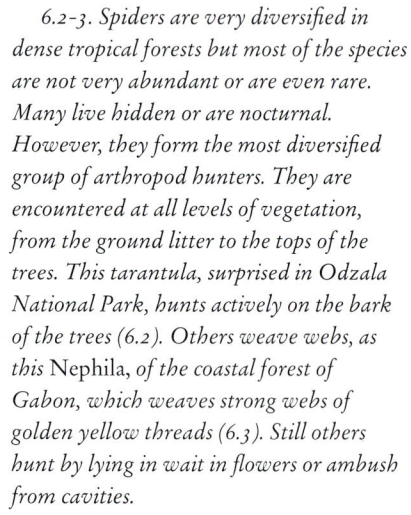

6.2-3. Spiders are very diversified in dense tropical forests but most of the species are not very abundant or are even rare. Many live hidden or are nocturnal. However, they form the most diversified group of arthropod hunters. They are encountered at all levels of vegetation, from the ground litter to the tops of the trees. This tarantula, surprised in Odzala National Park, hunts actively on the bark of the trees (6.2). Others weave webs, as this Nephila, *of the coastal forest of Gabon, which weaves strong webs of golden yellow threads (6.3). Still others hunt by lying in wait in flowers or ambush from cavities.*

6.4. Centipedes, like this one observed in Mondah Forest near Libreville, are even more hidden than spiders. Generally nocturnal, they are attached to damp places. Some are more venomous than others and their sting can be very painful.

Diversity, biomass and coexistence

The tropical forest is essentially a world of plants and fungi. The animal biomass is very small and represents no more than 2 to 3% of the total biomass. The diversity of animal life is just as great as that of the plant world and any tropical forest harbours more animal species than the surrounding grasslands or any other terrestrial habitat in the world. Of course just as is the case with plant species, African forests are overall less rich than those of Asia or South America and some trophic groups are differently represented. The number of fruit-eating birds is, for example, relatively small, while fruit-eating mammals – monkeys, squirrels and bats – are more abundant than on other continents.

Tallying this diversity objectively is, however, impossible: one would have to know both all the species and the number of individuals of each species. Besides, diversity varies according to the scale considered: a habitat, or a landscape including several habitats. It can also be expressed by the rate of replacement of species between two habitats. Finally, there would have to be agreement on the notion of habitat, for this is obviously not the same for an ant in the litter, an earthworm, a butterfly of the canopy, a bird or an elephant. Therefore, all that we can do is estimate the number of species in the most well known groups, which is called species richness, all the while knowing that this constitutes no more than an "indicator" of diversity.

Even this is not simple: very few sites have been inventoried in a thorough way and some groups escape observation too easily. How, for example, does one inventory amphibians and lizards of the canopy or the innumerable insects living inside wood? Finally, very few sites in Central Africa have been followed for long periods, and the founal composition, especially at the arthropod level, may vary considerably from one year to the next.

All the same, no one doubts the faunal richness of the forests of Central Africa and the actual number of species is of little importance. As for the flora, scientists are rather intrigued by fundamental questions that we have already evoked before (PAGE 58): where does this richness come from and how do the numerous species manage to coexist?

The answers to the first question are not very different from those that we gave for the flora. The fundamental mechanisms of speciation are about the same for animals and plants. However, different groups respond in a different way to different constraints of habitat, and develop sometimes very different patterns of distribution. Unfortunately, too few groups are sufficiently well known to be able to make valid comparisons.

Let us look rather at problems raised by their coexistence. First of all, this relies on the fact that many species of animals encounter each other only rarely. Many birds are diurnal, while most mammals come out only at night. Even within a limited group all the species are not active at the same time. In the case of bats, for example (PAGE 222), some species are active at the beginning of the night, others do not fly except in the middle of the night.

Many species are also separated in space. In birds, for example, swifts fly above the forest; hornbills and parrots live in the upper canopy; trogons, flycatchers and paradise flycatchers live in the middle strata; guinea fowl and francolins in the understorey. In addition, among the birds of the understorey, some prefer the dense brush of tree-fall gaps or mantles, others inhabit the open understorey of old forests and still others the proximity of watercourses. Finally, not all species exploit the same resources and those that do use similar resources differ very often in size. Altogether, each species has a unique ecological niche.

The complexity of the forest habitat, from the point of view of the fauna, does not stop there, however. Foliage and fruits, which constitute the main food for animals, are not all edible or nutritious. Some even contain repulsive or toxic substances (PAGE 169). Our human perception being essentially visual, we see the macroscopic structure of the forest, but for most of the insects this constitutes only a gigantic chemical mosaic in three dimensions. Relationships between the presence and abundance of resources on the one hand and consumer species on the other, therefore certainly do not follow simple rules.

Lastly, whether direct or more diffuse, competition between different groups of animals probably plays a far from negligible role, although perhaps less important than one thought 20 years ago. Between monkeys, fruit-eating birds and squirrels, for example, there should exist a multitude of relationships that influence the composition of their communities. Some of the differences that we observe in tropical forests from one continent to the next could thus be linked to the presence or absence of some groups of animals and the African forests would probably not be what they are if they had not been inhabited by the elephant. Inversely, one could imagine that they would inevitably be transformed if this animal were to disappear. In the Kivu, in Uganda and Rwanda, this hypothesis can even be easily verified. Only the time scale keeps us from appraising its extent.

These close ties between the plant world and the animal world are the expression of millions of years of co-evolution. Most of the angiosperms thus depend on some insects, birds or mammals for the pollination of flowers or the dispersal of seeds (PAGES 171-180) and many animal species have become dependant on some plants, such as, for example, the small wasps that live in figs (PAGE 175). Obligatory links have also developed between animal species. Numerous butterflies of the Lycenidae family live in symbiosis with ants.

All these mechanisms allow a great number of species to coexist, probably while narrowing the ecological niche of each of them. As for predation, its role is difficult to evaluate and we do not know to what extent it could really affect the density of populations. There are, however, phenomena that can seem bizarre: spiders (FIGURES 6.2, 6.3 and 6.5) are astonishingly diverse and sometimes abundant while their prey seem rare; many leaves show heavy damage yet one sees no or few caterpillars. It is therefore evident that our perception is incomplete or biased. According to some scientists, predation reduces competition between species and there seems to be a subtle connection between the great diversity of the animal world and its small

6.5. In tropical regions, many spiders live in more or less large colonies. This phenomenon has been observed in several different families. At the research station of Makokou in Gabon, R. Darchen, B. Krafft and S.E. Richert have studied the systematic of the species, their ecology and behaviour, the organisation and evolution of their communities, and finally also the reasons of this phenomenon. The species in the picture has established a colony on the branches of a fallen tree in Mambili River in Odzala National Park.

Insects

By the number of species, tropical forests are the uncontested kingdom of the insects. The most diversified group are the beetles (FIGURES 6.6 and 6.7). They are even so diverse that they are impossible to list. At most, can we say that many species have larvae that live inside wood and therefore play an important role in the recycling of organic matter. We will only devote a few pages to the ants, the termites and the butterflies, groups that attract the attention of all visitors.

Knowledge of forest insects has made enormous progress thanks to the use of powerful insecticides that are active for only a short time, such as pyrethrin. Canopy fogging, a method used since the beginning of the 1980s in America, is obviously not free from bias, however. Some groups of insects are more easily collected than others and those that live inside the vegetation can remain completely unnoticed. Butterflies, dragonflies and cicadas with a strong flight can escape and some insects, although killed by the insecticide, remain attached to the vegetation. In spite of everything, this technique has given a much more realistic idea of the composition of arthropod communities of the canopy. Most of all, it has led to the discovery that the real number of species is much greater than was possible to imagine. Unfortunately, the method has not often been used in Africa.

Butterflies and moths

Of the insects that strike the visitor to tropical forests – at least those that know how to use their eyes – foremost are the butterflies. The interest of these insects often with such shimmering colours, that make the dreams of collectors, goes beyond purely aesthetic aspects, however. The adults of some species are important pollinators and the caterpillars represent the main herbivores of the forest. The impact of butterflies on forest ecosystems is therefore not negligible.

Having appeared at the end of the Cretaceous, the butterflies and moths, or lepidoptera, underwent an astonishing diversification during the Tertiary, at the same time as the angiosperms diversified. Today, some 170,000 species have been described in the world and many remain to be discovered. Among the 18,000 diurnal species, 4,000 inhabit sub-Saharan Africa, mainly the dense humid forests. By its composition, this fauna is very different from that of tropical America: in the family Lycaenidae the Riodininae are nearly absent, and in the family Nymhalidae the Heliconiinae, Brassolinae, Itominae and Morphinae are totally absent. Furthermore there are two endemic subfamilies: the Liptenine in the family Lycaenidae, with over 600 species, and the Pseudopontiinae in the family Pieridae with only one species. Overall, nearly 98% of the

6.6-7. *The beetles constitute the most diversified order of insects with more than 360,000 species described and probably more than a million existing species, of which a great many inhabit tropical forests. However, many species are small, rare, nocturnal or very hidden. Also, beetles are not what visitors to the forest notice first. The Goliath Beetle* Goliatus goliatus *(6.6), as big as a fist, is one of the largest beetles in the world. It belongs to the Scarabaeidae family and more exactly to the sub-family Cetoniinae. This group is very diversified in the humid forests of Central Africa and is composed mainly of diurnal, often brightly coloured insects, but with many species inhabiting the high canopy. The Cerambycidae or longhorn beetles (6.7) are equally very diversified in dense humid forests where their larvae live inside wood. Most of the species are nocturnal.*

biomass. For some scientists, however, the very low density of most species could not be a consequence of predation but a survival strategy. This hypothesis rests on the fact that, especially among insects, the abundant species are often nocturnal and that the rate of mimicry and camouflage is high.

species and 75% of the genera are endemic to the continent.

The Papilionidae number some 70 species and are thus less diverse than in Asia. They are, however, more diversified in forests than in grasslands and are represented by the genera *Papilio*, whose caterpillars feed on Rutaceae plants, and *Graphium* (FIGURE 6.10), whose caterpillars live on some species of Annonaceae. Moreover, this family has two spectacular species: *Papilio antimachus*, the largest species on the continent, whose males have a wingspan of up to 28 centimetres, and *Papilio zalmoxis*, a large sky-blue butterfly with reddish undersides (PAGE 1) with a wingspan reaching up to 18 centimetres in the males. Both species have a vast but discontinuous distribution in Central Africa. Among the ECOFAC sites, they are common only at Ngotto in the Central African Republic. They are species of the canopy with only the males descending to the ground to drink and take up minerals from the wet soil along roads and watercourses.

Nearly half of the species belong to the family Lycaenidae – the "little blues", that can, however, be blue, white, yellow, brown or many-coloured. Many live in association with ants, but in a very variable way, even inside a single sub-family. Some species in the genus *Iolaus*, from the sub-family Theclinae, and some Lipteninae are totally independent. Some Polyommatinae and a few Theclinae are associated with ants, but not always. Most Lipteninae, however, are associated with the canopy ants of the genus *Crematogaster*. Some species feed on leaves, but cannot survive without ants. Finally, the Lyphyrinae, a sub-family known only from Africa and Asia, are entirely predatory on ant larvae. Among the most characteristic genera, we mention *Hypolycaena* whose males have long tails (FIGURE 6.8), *Epitola*, of which many species are metallic blue and live high in the canopy, *Mimacraea*, of which most species look remarkably like Acreinae, and *Lepidochrysops*, with 122 species, entirely endemic to Africa.

The Pieridae, generally with the dominant colour white, comprise about 150 species in Africa. The majority live in grasslands, but some are characteristic of lowland forests or montane habitats (FIGURE 6.9). Among the forest dwellers, several feed on plants of the Loranthaceae.

Nymphalidae number more than 1,000 species, of which a large majority live in dense forest. Sub-family Charaxinae (FIGURE 6.12) includes strong flyers of the canopy and mantle, with often brilliant colours, but which readily descend to the ground: the males to feed on excrement, the females in search of fermented fruits. Typical of the understorey of dense forests are the genera *Bebearia* and *Euphaedra* (FIGURES 6.11 and 6.17), whose numerous species fly just above the ground in search of patches of light and fermented fruits. Species

6.8. Hypolycaena lebona *belongs to the sub-family Theclinae in the Lycaenidae family. Its caterpillar is a leaf eater, but lives in association with ants of the genus* Pheidola. *The long tails, with an eye on the hind wing, divert predators.*

6.9. *Forest-dwelling Pieridae, like this species of the genus* Appias *in the Lopé National Park, often haunt the middle and upper levels of the canopy, but then readily come to the ground to drink or look for minerals they need.*

6.10. *Two Papilionidae,* Graphium latreilianus *(on the left) and* Graphium polycenes *(on the right, with its long tails) on the ground with some lycaenid butterflies of the genus* Leptotes.

6 THE FAUNA

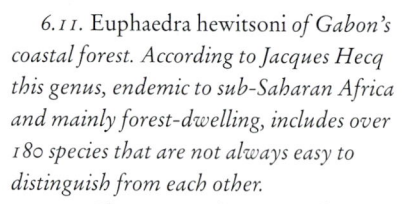

6.11. *Euphaedra hewitsoni of Gabon's coastal forest. According to Jacques Hecq this genus, endemic to sub-Saharan Africa and mainly forest-dwelling, includes over 180 species that are not always easy to distinguish from each other.*

6.12. *Charaxes ameliae is mainly a species of the high canopy in moist evergreen or semi-deciduous forests. Males are large black and blue butterflies, but females are even larger and blackish brown and white.*

6.13. *A female of* Harma theobene, *a species known from throughout the guineo-congolian forests.*

6.14. *One of the many* Bicyclus *species in the sub-family Satyrinae.*

6.15. *The Central African forests are the habitat of the genus* Catuna – *here* Catuna oberthuri – *nymphalids of the understorey. They are very mimetic and go unnoticed when resting on the dead leaves of the litter.*

of genus *Cymothoe* on the other hand prefer the intermediate and upper strata of the canopy (FIGURE 6.12). Sub-family Satyrinae includes essentially unobtrusive species, generally brown with or without spots, that haunt the forest mantle and dark understorey (FIGURE 6.14).

As for sub-families Acraeinae (FIGURES 5.50 and 6.17), near-endemic to Africa, and Danainae (FIGURES 6.16 and 6.17), they include some much more colourful, but more or less toxic, species. In the Danainae, the toxins are derived from plants on which their caterpillars live: Asclepiadaceae species. In the Acraeinae, toxins can come from food plants – often Passifloraceae – or be synthesised by the insect. This toxicity causes these butterflies to be rejected by predators, especially birds, and their type of coloration is imitated by a number of non-related species among the Nymphalidae, in particular genus *Pseudacraea*, the Lycaenidae, notably genus *Mimacraea*, and the Papilionidae. In order for this manner of mimicry – called Batesian mimicry – to remain an effective protection, the imitators must not be more numerous than their models.

This is why in the Papilionidae it is often only the females who manifest it. Curiously, in *Papilio dardanus*, which inhabits forests as well as grasslands, there are even several kinds of females who imitate different species of Danainae (FIGURE 6.16). Sometimes the imitator is itself more or less toxic (FIGURES 6.17a and 6.17b). It can then be just as abundant or even more so than its model. The advantage of this type of mimicry, called Müllerian, lies in the fact that the sacrifice of a smaller total number of individuals may be sufficient to disgust the predators.

Altogether, and as is the case for many other groups of organisms, the guineo-congolian lowland forests are richer in butterflies than the savannas, and many species have a wide distribution from Liberia to Uganda. Several faunal regions can be identified. They seem to be separated by areas of high rainfall. In Central Africa the richest areas are Lower Guinea and eastern Congolia. The fauna of Uganda is less rich, but has nevertheless several endemic species. Lower Guinea has over 1,000 species, and seems to be the center of dispersal of the genera *Charaxes*, *Euphaedra* and *Pseudacraea*.

Montane forests are definitely less rich and their butterfly fauna is very similar throughout. The Cameroon highlands are very poor and most of their species also exist in the Albertine Rift forests. These are the richest montane forests of Africa and harbour many endemic species: *Papilio leucotaenia*, *Graphium gudenusi*, *Charaxes opinatus*, several species in the genus *Acraea*, many lycenids and some hesperids. As with other groups, the number of species decreases with altitude, and no species inhabit permanently the afro-alpine belt.

Some day-flying butterflies are active nearly all year long, others do not appear except during short periods, some come out during the rainy season, others rather during the dry season. The abundance of butterflies varies thus from one month to the next, but the periods of maximum abundance are not always predictable. Often butterflies are observed at the very end of the rainy season and the beginning of the dry season. In bimodal climate regions, this can be in December-January or June-July. In the coastal lowland forests of Gabon, spectacular emergences are rare, however. They are much more frequent in semi-deciduous forests, such as those of the Central African Republic.

Moreover, the rarity of some species constitutes an enigma in itself. Some butterflies seem rare, perhaps because they are very difficult to capture – this could be the case of species living in the high canopy – or because no one has prospected the regions where it could be abundant. Some species are truly rare because of the disappearance of their habitat, but some are rare without any apparent reason. The subspecies of *Charaxes fournierae*, for example, described from Nyungwe Forest, in Rwanda, around 1970, has never been found again, while the habitat remains practically unchanged. Yet it involves a large orange-brown butterfly that is not easily overlooked and which in addition is easy to

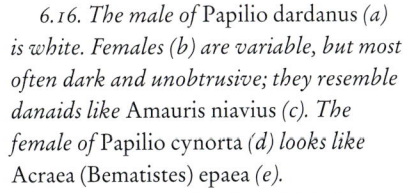

6.16. *The male of* Papilio dardanus *(a) is white. Females (b) are variable, but most often dark and unobtrusive; they resemble danaids like* Amauris niavius *(c). The female of* Papilio cynorta *(d) looks like* Acraea (Bematistes) epaea *(e).*

6.17. Mimacraea landbecki *(a) of the subfamily Liptenininae in the family Lycaenidae looks like* Acraea epaea *(b);* Mimacraea kraussi *(c) like* Acraea bonasia *(d). Since species of the genera* Amauris *and* Acraea *are more or less toxic, these are examples of Batesian mimicry.*

6.18. The Cossidea are heavy-bodied moths with narrow wings. Their caterpillars live inside wood, and some species, like the one depicted, can easily be mistaken for sphingids.

6.19. The Saturniid and Brahmaïd (like the one depicted) moths can be very large, but they live for only a very brief period. The males, moreover, do not even have a digestive system and their only function is reproduction. The caterpillars of some species, notably one that lives on the sapelli Entandrophragma cylindricum, *can be very abundant and are sought after by human populations, for whom they represent a far from negligible addition of proteins.*

6.20. Most moths have mimetic colour patterns in order to remain unseen during daytime. Some, like this large noctuid moth, have bright colours on the under wing, perhaps allowing them to frighten potential predators.

capture in traps. Is it perhaps a relict species on the edge of extinction? Is it perhaps a phenomenon linked to a particular population dynamic? Many butterflies can in fact remain very rare for years, suddenly experience a veritable demographic explosion and disappear again for an indeterminate time.

As for moths they are less visible than diurnal butterflies, but much more diversified. Even though many species remain to be described, there should be at least 10 times more than butterflies. The adults often do not appear except during the rainy season and many species do not even fly except during rainy nights. They protect themselves in this way from bats, which are less active then and mostly, less skilful, because the rain scrambles their echolocation system. Some moths, however, have sensitive organs on the thorax or abdomen which allow them to detect bats. Many moths also have developed colour patterns which can be very bright and contrasted, but at the same time very mimetic in order to protect them during the day (FIGURES 6.18 and 6.19).

Among the spectacular species there are the Sphingidae. These rapid-winged moths do not alight, but gather nectar from flowers with their long proboscis while hovering in place. The largest moths, however, are encountered among the Saturniidae of which many species are recognised by the large "eyes" that decorate their wings (FIGURE 6.19). Contrary to the Sphingidae, they have a rather ephemeral life and are not even capable of eating. Their range is sometimes very local.

The great majority of moths are small or very small in size, but it is those which are often the main forest defoliators. Some species live inside the leaves which they mine with their galleries (FIGURE 6.20). Others live in stems, flowers, fruits or seeds. Some are predators of aphids or hemiptera. Still others dig into the wood of trunks, feed on detritus, notably animal wool or horn, fungi, lichens or mosses. Moths thus play an important role in the nutrient cycles of the forest, all the more so that their caterpillars constitute a very large part of the food of birds and that some species, notably among the Saturniidae, are also very sought after by the human populations of some regions.

Diptera

Of the innumerable species of insects that inhabit tropical forests, there are some that are not very spectacular, sometimes even almost invisible, but which draw attention to themselves by conveying grave human diseases. The champions are diptera: flies, mosquitoes and related species.

Some species have vast ranges in tropical regions, sometimes even in temperate regions, but they can be more frequent in dense humid forests because the climate is favourable to them.

First of all, there are the numerous species of mosquitoes of genus *Culex*. They are the most common stinging insects whose females feed on human blood, mostly at night and at twilight. Most of the time they are more annoying than dangerous. Occasionally, they can transmit viruses and provoke encephalitis in this way, but cases are rare in Central Africa. They can also transmit filariasis *Wuchereria bancrofti*, a nematode threadworm that lives in the lymphatic vessels and provokes elephantiasis, a strictly human, and mostly urban, disease, very frequent in tropical and subtropical Africa.

Next there are the mosquitoes of genus *Anopheles*, vectors of malaria, a disease that touches hundreds of millions of humans in the world and is responsible for several million deaths each year. They are represented by numerous species that inhabit tropical, subtropical and temperate regions. Some inhabit village or urban localities, others marshes, humid or swamp forests. Some species reproduce in water held in the axils of large leaves, like those of banana plants, or in abandoned recipients. Adults are rather shy and are not active except at night. They are therefore not often seen.

As for the mosquitoes of genus *Aedes*, they transmit yellow fever. This grave viral disease, whose symptoms are fairly variable, owes its name to the fact that it generally provokes a jaundice or icterus. In South America, it is very frequent, but more urban. In Africa, it is rarer and hits mostly persons working in the forest. *Aedes* mosquitoes live, in effect, in the canopy where they feed on the blood of primates and pass the virus in this way from one animal to another. As this virus penetrates the eggs of the mosquito, it can survive from one rainy season to the next. The reservoir of the disease is therefore rather the mosquito than the primate. In Africa, tree cutters are particularly vulnerable because of the fact that in felling trees they bring mosquitoes from the high canopy to the ground. In spite of the existence of an effective vaccine, the disease is progressing.

The small black fly *Simulium damnosum* is around two millimetres long. It is attracted to the vicinity of torrential watercourses, in the forest as well as in grasslands, and its larvae live in rapids. Females are active throughout the day but, especially at nightfall, they feed on human blood and can transmit onchocerciasis, a disease provoked by a parasitic worm *Onchocerca volvulus* that causes river blindness in some cases.

Sand flies *Phlebotomus* are also small diptera about two millimetres long that suck human blood. They inhabit forest edges and the outskirts of villages where they are especially active at twilight. In Eastern Africa, they can transmit leishmaniasis, a disease provoked by the protozoan *Leishmania*. It is essentially a zoonosis or disease of animals that is transmitted only occasionally to Man. In Central Africa, sand flies, locally called fourous, can be very frequent and annoying but not dangerous.

There are also myases. The most frequent is provoked by the fly *Cordylobia anthropophaga*, whose larvae constitute Cayor's worm. The eggs are laid on the ground or on clothing and the larvae infiltrate under the skin of their victims where they develop, forming a sort of boil.

Other diptera are more forest-living. In the humid forests of Central Africa, the fly of genus *Chrysops* transmits the worm *Loa loa*. In the region around the Gulf of Guinea, 3 to 30% of the human population is affected. The Tsetse fly *Glossina palpalis* transmits trypanosomiasis, or sleeping sickness, provoked by the protozoan *Trypanosoma gambiense*. This very resistant brownish fly inhabits forest edges and gallery forests throughout Central Africa. It constitutes the principal obstacle to the progression of cattle raising into the forested regions and is responsible for the depopulation of some vast areas of the Congolese basin.

6.21. *In order to move from their nest to feeding places, very safely and out of the light, termites arrange galleries made of earth, saliva and excrement.*

6.22. *A nest of* Procubitermes *attached to a tree trunk in the forest in the Crystal Mountains in Gabon. Its shelved structure and the chevrons protect it from water running down the trunk.*

Termites

Hidden in the ground, in dead wood, in leaf litter or in the shelter of their protective galleries (FIGURE 6.21), termites, or Isoptera, are omnipresent but not very visible. They live in colonies of a few hundred to a few million individuals, including one or several reproducing pairs, together with nymphs, larvae, workers and usually soldiers. Most species construct nests made of soil, saliva and droppings. The forms that feed on wood also build "paper" nests.

The winged males and females emerge from the nest only briefly, to fly away, mate and found new colonies. These are the insects that beat against lights on rainy nights, lose their wings after having formed a pair and go off in tandem in search of a good place to establish a new colony.

Tropical forests harbour more species of termites than all other habitats in the world. They are divided into seven families, of which five occur in Africa. The first six constitute the primitive termites, whose intestine is inhabited by flagellate protozoans able to digest cellulose and some optional bacteria that fix nitrogen from the air. These species are not very visible and live mainly in dead wood.

The great majority of termites belong to the family Termitidae or superior termites, that do not live in symbiosis with protozoans but with bacteria or fungi. They are divided into six sub-families, five of which occur in Africa: Macrotermitinae, Apicotermitinae, Amitermitinae, Nasutitermitinae and Termitinae. They appeared in the Cretaceous or even the Jurassic. The Apicotermitinae, of Afro-tropical origin and absent in Asia, live on organic material in the soil, but they are occasionally found in rotting wood. Among the Termitinae there are species that feed on wood. They are not very numerous, but nevertheless comprise some genera endemic to Africa. Most of the species feed on organic material in the ground. Some, more numerous in Africa than in South America, have soldiers that bite with their wide mandibles. Others, common everywhere, have soldiers that clack their mandibles without biting. The Macrotermitinae are limited to the Old World and probably even originate from Africa where they are more diversified than in Asia. They live thanks to mushrooms of genus *Termitomyces* that grow on stacks of their droppings. These termites feed on wood and dead leaves. The Nasutitermitinae, probably of South American origin, can have soldiers that bite, but these do not exist in Africa. However, the most evolved forms possess a sort of "nose" secreting an irritating substance that is somewhat sticky and locally toxic. In this case the mandibles are atrophied. Clearly less abundant in Africa than in South America and Asia, where the other termites are rarer, they feed on organic debris, humus, dead wood, rotting leaves and dry leaves.

Termites are most abundant in dense and humid equatorial forests where it is not rare to find 30 to 40 species in the same locality. In a forest reserve near Mbalmayo in Cameroon, 136 species were even censused. They are much less abundant in montane forests, where they hardly go beyond 2,000 metres altitude, and in swamp or inundatable forests.

In fact, termites do not like water, which dissolves and washes away their earth galleries and can infiltrate their nests. To keep this from happening, some species build their nest in the trees or protect it with structures specially made to carry off water. Species of genus *Cubitermes* thus build their nests in the form of shelved mushrooms, stuck onto the trunks of large trees in the middle of the forest. Termites of genus *Procubitermes* add chevrons of earth above their nests to deviate water that runs down from the canopy (FIGURE 6.22). Contrary to grassland termites, those of dense humid forests only rarely build large nests. This phenomenon remains unexplained to this day.

All termites live on cellulose from wood, bark, leaves, litter or soil, but most species are capable of using at least two types of food. As one can imagine, species feeding on grasses are absent in the forest. Those that feed on the soil are particularly diversified and abundant in African forests, from which they probably originated. Studies conducted in Cameroon and in the Democratic Republic of the Congo show in fact that the *Cubitermes* species living on soil can constitute 30% of the total biomass of termites which itself can reach 11 grams to the square metre or 110 kilos per hectare.

Even if their ecological impact is little known in detail, we do know that it is enormous on the forest. Not only do termites decompose dead vegetable matter, but they activate decomposition by hastening the fall of dead trees, aerate and transform the soil and fix atmospheric nitrogen. Their nests constitute sources of nutrients or shelters for a good many other organisms (PAGE 203). In addition, the termites themselves, the workers as well as the winged forms, constitute an important source of food for a number of specialised or opportunistic predators, from the size of an ant to that of the pangolins and the Aardvark *Orycteropus afer*. Their productivity can attain 30 grams per metre per year, which is the equivalent of the production of one cow to the hectare.

6.23. *A termite nest built in the fork of a mangrove tree.*

6.24. *A column of termites working under a pile of dead wood. Soldiers are recognisable by their large heads.*

Ants

Ants represent a very ancient group of hymenoptera that already existed in the Cretaceous, 80 million years ago. From the Eocene, 45 million years ago, they developed a great diversity of ties with the angiosperms, of which most were of interest as much for the plants as for the ants implicated.

They can be found today at all levels in the tropical forest. However, they are divided into three distinct ecological groups. The first comprises ants of the litter. It is the most diversified group, but also the least well known. The second is made up of species that hunt mainly on the ground, although some times also in the vegetation. It is well known through the driver ants of genus *Dorylus* that hunt in columns (FIGURE 6.26). They are redoubtable predators, but stories that circulate about them are often exaggerated. The third group is made up of ants of the canopy. They are less rich in species than those of the litter, but in lowland forests they probably constitute the most important group of arthropods, as much by their biomass as by the impact that they have on the other arthropods. In fact, they touch all the trophic levels through multiple associations with other species of animals, plants or mushrooms. One could consider a fourth group: that of the ants inhabiting the litter of the canopy, which collects under the mass of epiphytes or in cavities.

A synthetic survey of present-day knowledge about ants was published by J. E. Tobin. In spite all the difficulties of observation inherent to their way of life, they are presently the best known of arthropods of the canopy. Most of the species are small and unobtrusive. The dominant species, aggressive and living in vast colonies, do not, in fact, represent more than a small part of the total diversity of ants. Nevertheless, the family Formicidae alone contributes more to the total animal biomass of the canopy than any other animal family.

If the ants of the forest floor and the litter are content to hunt and to glean organic waste, those of the canopy exploit more varied resources. In addition to hunting and the collection of dead organisms, they consume sugary substances from the honeydew secreted by aphids which they raise, nectar from flowers and extra-floral organs, juice from fruits and sap from plants.

The raising of Homoptera or even of caterpillars is frequent among the dominant species and perhaps it is in order to defend this resource that they have developed territorial behaviour. Aphids or caterpillars pump sap from plants by filtering amino acids and other nutrients. They share a part of these resources with ants which in counterpart offer them protection against predators and parasitic organisms such as ichneumon wasps that lay their eggs in caterpillars. This resource is also very much used by ants of the litter.

6.25. *The ant* Tetraponera aethiops *inhabits hollow twigs of* Barteria fistulosa, *a shrub or small tree of the Passifloraceae family. Normally, they are not visible, but as soon as a branch is touched, the workers come out of their cavities through small holes and attack everything that moves. Armed with a sting, like bees, they inflict painful stings and protect their tree against predators of all sorts. In some regions of Central Africa, women having committed adultery were attached to* Barteria fistulosa *for punishment.*

6.26. *Ants of genus* Dorylus *live essentially on the ground. Active by day and night, they come out from their nests, buried deeply underground, in dense columns protected by soldiers with fierce mandibles. They resemble Driver Ants of South America.*

Many ants also defend the trees and bushes that they colonise. A striking example is that of *Tetraponera aethiops* that inhabits *Barteria fistulosa*, a shrub or small tree of the Passifloraceae family, common in most of the forests of the interior of Gabon (FIGURE 6.25). Not only does this ant protect the shrub's foliage, but it regularly cuts back lianas that attempt to invade it or grass that grows around it. There are many other less spectacular examples.

Generally speaking, the fact that a tree or a shrub attracts ants by offering them food constitutes an excellent defence against herbivores that could attack the leaves and buds. Ants, common in South America, that cut leaves from trees to feed fungi in vast subterranean nests are unknown in Africa. Gardens of epiphytes maintained by ants – ant gardens – frequently encountered in South America, are also unknown in Africa.

Ants of the canopy make large nests in carton, occupy natural cavities or hollow parts of stems, as in *Barteria*, invade abandoned termite nests or sometimes living but weakened termite nests, inhabit piles of dead leaves and accumulated humus caught in the hollows of large branches or held by epiphytes or associate with domatia – organs specially developed to attract insects – of some epiphytes. Ants of genus *Oecophylla*, endemic to the Old World, sew leaves to form closed nests (FIGURES 6.27-6.28).

Thanks to the technique of canopy fogging, we know that ants represent on average 2 to 4% of the arthropod species of the canopy. Few studies have been made in natural forests of Africa, but those made in cacao plantations in Ghana seem to confirm the observations in America and Asia. In tropical America, ants represent 25 to 50% of the total biomass of invertebrates of the canopy, in Asia, 20 to 40%. As for termites, their abundance diminishes when rising in altitude or moving away from the equator, but they seem less sensitive to temperature than termites.

Typical for the canopy of tropical forests is the development of a three-dimensional mosaic of large dominant species that mutually exclude each other and in large part determine the distribution of dominated species that they tolerate. These mosaics, fairly stable in time, considerably influence other arthropods. One can observe in fact that many other species of insects are positively or negatively associated with ants.

This observation is the basis of attempts to use ants in pest control, but results have not always been very encouraging. The introduction of the South American ant *Wasmannia auropunctata* into Africa – as in many other tropical regions of the world – has even turned out to be an ecological catastrophe whose amplitude has only just begun to appear (PAGE 308).

6.27-28. Ants of the genus Oecophylla *are known only from the tropical regions of the Old World. Africa has only a single species that lives in the shrub layer or the canopy where it feeds partly on honeydew of scale insects (picture left). Sometimes it is the only dominant ant. Thanks to a very close collaboration between the workers, they manage to bring together leaves and to sew them together. Each nest is in this way made of a mass of leaves joined by silk threads secreted by the larvae. The nest contains hundreds or thousands of workers, eggs and larvae as well as organisms that live in symbiosis with the ants. A colony is made up of a great number of similar nests, often very spread out.*

Amphibians

African forests are distinctly poorer in amphibians than those of South America and salamanders are totally absent, as is the family Dendrobatidae, characteristic of Amazonian forests. In spite of this, Thierry Frétey and Charles Blanc have catalogued 336 species for the islands of the Gulf of Guinea and the six countries of Central Africa located between the Atlantic and the Albertine Rift: nine species of Gymnophiona – the legless amphibians (FIGURE 6.29) – and 327 species of Anoura – the toads, frogs and tree frogs. Among them six genera and 114 species are endemic to the region, but this endemism involves members of the Gymnophiona much more than those of the Anoura. Moreover, how the seven endemic species of the islands of São Tomé and Principe, in particular the two Gymnophiona, crossed the ocean remains a mystery. These animals have in fact a very limited power of colonisation and do not tolerate salt water.

Obviously, not all of these species live in the forest, but a very large proportion of them do. Also, the forests of south-western Cameroon are extremely rich in amphibians and this country is altogether the richest in the sub-region with 194 species, of which 26% are endemic. The region of Korup National Park and Banyang Mbo alone has 90 species and is, in fact, one of the richest in the world. This zone of great diversity extends to Equatorial Guinea, which counts 74 species in spite of its small area. It extends very probably even to Gabon, but this country has been insufficiently investigated. Globally, this great diversity of amphibians on the edge of the Gulf of Guinea indicates that this region must have always had pockets of very humid forest, even during the driest phases of its history. There are other centres of endemism, however, in the centre of the Congo basin and in the Albertine Rift region.

As for their ecological distribution, forest amphibians are found everywhere: in the water, underground and from the litter of the forest floor to the upper canopy.

Among aquatic species, there are mainly the Pipidae of genera *Xenopus* and *Hymenochirus*, and the Ranidae of genus *Rana* (FIGURE 6.32).

Among those that live on the surface of the ground, the Bufonidae – the true toads – are very common and diversified (FIGURE 6.30). The Arthroleptidae replace the South American Dendrobatidae and the two families show very clear convergences. Males of the Arthroleptidae differ, however, by a very elongated third digit which they use during mating. Otherwise, in the most humid regions of Central Africa this family comprises some primitive species such as the *Cardioglossa* – the only representatives of the family having conserved a larval stage – and the curious Hairy Frog *Trichobatrachus robustus*. On the other hand even among the Ranidae terrestrial species are found. In the very humid forests of Cameroon and Equatorial Guinea, there occurs the Goliath Frog *Conraua goliath*, the largest frog in the world, reaching up to 30 centimetres in height. There are also numerous terrestrial species in genera *Phrynobatrachus* and *Ptychadena* throughout the entire sub-region. However, most of these terrestrial amphibians are only partially so and even the Bufonidae depend on water for their reproduction.

6.29. *Order Gymnophiona is made up of the legless amphibians.* Geotrypetes seraphini, *of the Caeciliaidae family, is found in Cameroon, Equatorial Guinea, Gabon and the Democratic Republic of the Congo. It lives in the mud along creeks or streams in the forest and feeds on small crustaceans, shrimp larvae, mud worms and other small aquatic invertebrates.*

6.30. *The giant toad* Bufo superciliaris *is found in Central Africa, as much in old forests as in recent secondary formations.*

The burrowing amphibians are represented essentially by the Caeciliaidae, much less numerous than in South America, with only four species in continental Central Africa and three species on the islands of the Gulf of Guinea. Family Scolecomorphidae is represented by only two species of genus *Crotaphatrema*, endemic to Cameroon. Some *Arthroleptis* and *Hemisus* of the Hemisotidae family also dig.

In the vegetation, arboreal tree frogs of the Hyperoliidae family (FIGURES 6.31 and 6.34) are found, as well as *Chiromantis rufescens*, sole representatives of the Rhacophoridae in Africa (FIGURE 6.33). Some species inhabit the dark understorey, others the canopy. Many species are bright green, more or less brilliant, brownish or greenish grey. Some are a nearly transparent pale green. Curiously, in the *Hyperolius*, the males are often dull while the females are brightly coloured. Arboreal life has also induced the development of specialised "techniques" of reproduction. In this way, *Chiromantis* construct foam nests attached to large leaves (FIGURE 6.33).

6.31. The small, arboreal leaf-folding frog *Afrixalus dorsalis*, of the Hyperoliidae, is brown at night, but is covered with cream-coloured spots during the day, under the effect of the light. They are found along small creeks in the forest in most of Central Africa and are even known from Sao Tomé. The species does not occur in the Democratic Republic of Congo, however.

6.32. The frog *Rana lepus* belongs to the Ranidae family.

6.33. The Old World tree frog *Chiromantis rufescens*, unique representative of the Rhacophoridae, does not lay its eggs in the water. Instead, it deposits them in a foamy mass attached to the leaves above a small stream or a puddle – in this case a Marantaceae leaf. Several males can participate in their fertilisation. The tadpoles fall into the water where they remain until their first metamorphosis.

6.34. An arboreal, green bush frog of genus *Leptopelis*, family Hyperoliidae.

6 THE FAUNA

Reptiles

This group comprises animals as diverse and distant as turtles, crocodiles, snakes and lizards. In all, Thierry Frétey and Charles Blanc were able to census 438 species for the seven countries of Central Africa situated between the Gulf of Guinea and the Albertine Rift, the two richest groups being the saurians (lizards, geckos, agamas, monitors..), with 164 species, and the ophidians (snakes) with 223 species. Four genera and 33 species are endemic to the sub-region, a level of endemism of only 7.5%, therefore clearly less than in the amphibians. These animals occupy nearly all habitats and the forest species are divided accordingly into aquatic, burrowing, terrestrial and arboreal species.

Among aquatic species, obviously, figure three crocodiles, two are associated with watercourses in the forest: the African Slender-snouted Crocodile *Crocodylus cataphractus*, very common notably along the Mambili River at Odzala National Park (FIGURE 6.35), and the African Dwarf Crocodile *Osteolaemus tetraspis*. The latter, wrongly called "Caiman" in Gabon, is very widespread, mostly in swampy areas and small streams. Because of its nocturnal habits, it is only rarely seen. The Nile Crocodile *Crocodylus niloticus*, at the origin of an entire repertory of horrible stories, also occurs or used to occur on practically all the large watercourses, but is not dependent on the forest milieu. In Gabon, it seems to have disappeared from most of the country, but it can still be found in the delta of the Ogooué River and at the level of the Gamba complex of protected areas where it even frequents the sea.

Without speaking of the five large marine species that all frequent the beaches of the Gulf of Guinea during the reproduction period, most of the species of turtles are also linked to water. They are represented by the genera *Pelomedusa* and *Pelusios* of the Pelomedusidae family, as well as genera *Cyclanorbis*, *Cycloderma* and *Trionyx* of the Trionychidae – the soft-shell turtles, endemic to Africa.

Among the snakes, many species can be occasionally encountered in the water when they cross a stream or a creek. It is the case with the Rock Python *Python sebae*, the largest of the African snakes, which can reach a length of seven metres. However, some inoffensive colubrids in genera *Grayia*, *Helopsis*, *Hydraethiops* and *Natriciteres*, as well as two extremely venomous elapids – the aquatic cobras *Boulengerina annulata* and *Boulengerina christyi* – are entirely aquatic.

Finally, among the lizards, there is the Forest Monitor *Varanus ornatus*, that while not strictly aquatic is nevertheless often seen near water.

Among the burrowing species are found mainly snakes of the Typhlopidae and Leptotyphlopidae families, as well as the beautiful *Charina reinhardtii* of the

6.35. A Slender-snouted Crocodile *Crocodylus cataphractus* rests on an emergent trunk in the Mambili River in Odzala-Kokoua National Park.

6.36. The turtle *Kinixys erosa* lives in the forest, old dense forests as well as habitats modified by human activity. Its carapace is articulated and its plastron is equipped with a projection that allows it to attack other turtles that it can then overturn. Being a vegetarian, it participates in the dispersal of seeds of some plants.

6.37. The Calabar Python *Charina* (Calabaria) *reinhardtii*, apparently "with neither head nor tail", feeds on arthropods and small rodents. At the least danger, it rolls up and plays dead. The black extremity of its body constitutes its tail.

Boidae (FIGURE 6.37). Having very narrow mouths, these species feed mainly on very small ground arthropods – ants and termites – but are themselves often the prey of venomous burrowing snakes such as the *Elapsoidea*, *Polemon* and *Atractaspis*. Until very recently, this latter genus was considered as part of the viperids, but it is presently included in a separate family, the Atractaspididae, with, notably, genera *Amblyodipsas*, *Aparallactus*, *Poecilopholis* – an endemic of the sub-region, known only from Cameroon and Equatorial Guinea – and *Polemon*.

Most reptiles are terrestrial. First of all there are two species of tortoises, *Kinixys homeana* and *Kinixys erosa* of the Testudinidae family (FIGURE 6.36). Then there is a whole cortège of snakes. Among African forest colubrids (FIGURES 6.40, 6.42 and 6.43), there are some very specialised forms: the snake-eaters of genus *Mehelya*, the amphibian-eaters of genus *Geodipsas* and the egg-eaters of genus *Dasypeltis*. The former ecological group is an African speciality, without an equivalent in tropical America. Among the terrestrial elapids are found mainly genus *Elapsoidea* and genus *Naja*. Both are venomous. As for the viperids, equally very venomous, they are represented mainly by the genera *Bitis* and *Causus*, the most spectacular being the Gabon Viper *Bitis gabonica* (FIGURE 6.39). As for terrestrial lizards, they are represented mostly by the Amphisbenidae of

6.38. *This arboreal viper* Atheris squamigera *is found in forests from West Africa to Uganda. It is not rare in Odzala-Kokoua National Park.*

6.39. *The Gabon Viper* Bitis gabonica *is incontestably the most spectacular of the viperids. It may reach nearly two metres in length. By its mimicry and its sluggish behaviour it often passes unnoticed and is at the origin of many accidents*

6.40. The arboreal colubrid Toxicodryas (Boiga) pulverulenta – *here ready to strike – inhabits the guineo-congolian forests from Upper Guinea to Uganda. It is particularly fond of low branches above watercourses in the forest. It is about one metre long, and is a good swimmer, feeding on other snakes and birds.*

6.41. The casked chameleon Chamaeleo cristatus *inhabits the great forest, but is more easily found in the mantle where insects are the most abundant.*

As for arboreal reptiles, they are first of all represented by a great many snakes, which are the most numerous of the arboreal reptiles, as well. The colubrids comprise three very characteristic groups of species: the fine green snakes with big eyes and round pupils, such as, among others, those of genera *Gastropyxis*, *Hapsidophrys*, *Philothamnus* (FIGURE 6.42) and *Thrasops*; the opistoglyphic snakes with a large head and vertical pupil, in genera *Boiga* (FIGURE 6.40) and *Dipsadoboa*; the liana-snakes, opistoglyphs, very dangerous, with a nearly filiform body and horizontal pupils, represented by genus *Thelotornis*. They also include the boomslang *Dispholidus typus*, with extremely toxic poison. The viperids are represented by *Atheris*, very venomous and usually green snakes (FIGURE 6.38). The elaphids are represented by the Arboreal Cobra *Pseudohaje goldii* and the Green Mamba *Dendroaspis jamesoni*, one of the deadliest snakes in the world. All these arboreal species feed mainly on insects, amphibians, birds and other reptiles.

Among the arboreal lizards we mention some agamids – essentially terrestrial but which do not hesitate to climb trees – and especially the gekkonids and chamaeleonids (FIGURE 6.41), well adapted to life in trees. The geckos have developed suction cups on the tips of their toes. The chameleons have developed hands and feet in the form of pinchers, a prehensile tail, inde-

genera *Amphisbaena*, *Cynisca* and *Zygaspis* and *Monopeltis*, and some legless specialised Lacertidae, such as *Melanoseps* and *Feylina*. The latter are even practically blind. To these can be added the stump lizard *Mochlus fernandi* that likes dead wood lying on the ground (FIGURES 6.44 and 6.45).

6.42. Colubrids of genus Phylothamnus *like secondary forests and plantations, places with plenty of sunshine. By their green colour and vivacity they are often mistaken for young green mambas* Dendroaspis jamesoni.

6.43. The colubrid Rhamnophis aethiopissa *is frequent around villages where it mainly seeks out plantations and banana plantings. About 1.10 to 1.25 metre long, it is very aggressive but not venomous.*

pendent eyes and a protractile tongue. Even some lacertids have been able to adapt to a life in the trees, including rare species such as *Gastropholis echinata, Gastropholis tropidopholis* – an endemic of the Democratic Republic of the Congo – and *Holaspis guentheri*, able to elongate their bodies to form a sort of membrane which allows them to glide very short distances.

Except for some lizards, reptiles are difficult to see in the forest and are only very occasionally encountered.

6.44. The lizard Mochlus fernandi, *of the family Scincidae, has a liking for old decomposing trunks and stumps, but it is not rare in old plantations.*

6.45. Gastropholis echinata *is a lizard of the Lacertidae family. In Central Africa it is known from Cameroon, Gabon, Congo and the Democratic Republic of the Congo. As a canopy species it is rarely encountered. Its colour is quite variable, but it is easily recognised by the brushy scales at the base of its tail.*

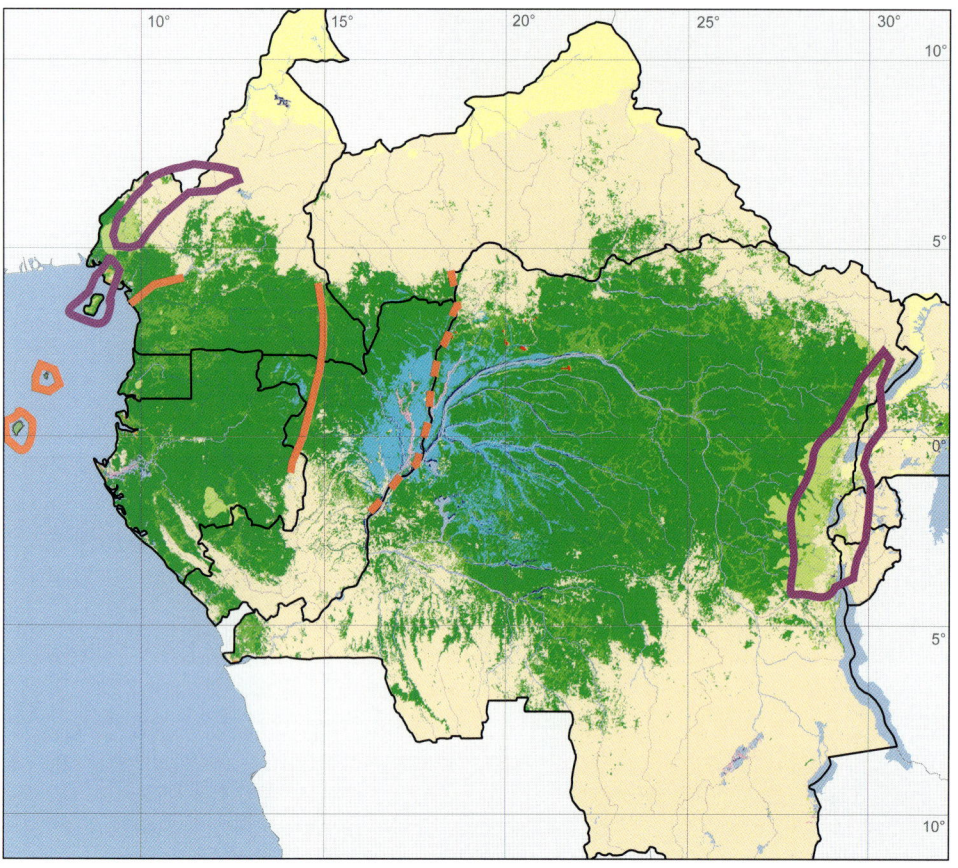

6.46. *The range of distribution of birds is much more well known than that of reptiles and amphibians but, probably because of their mobility, their biogeography is clearly less complex. The montane species are concentrated on the Albertine Rift, in the east, and in Cameroon, in the west, but some, however, have relict populations between these two regions of high relief. The montane region of Cameroon is divided into two distinct entities: the first encompasses Mount Cameroon and the island of Bioko, the second comprises the high reliefs north of Mount Cameroon. As for the low altitude forest species, they show, within the Congolese forests, a first clear-cut limit of distribution at the level of the Sanaga River and a second along a line between the water shed of the Congo basin and that of the coastal rivers. Some species of Lower Guinea (Cameroon-Gabon), however, extend their eastern limit to the swamp forests of the Sangha-Oubangui system.*

Birds

Birds are less difficult to find in the tropical forest than amphibians and reptiles, but it is easier to hear them than to see them. As in all tropical forests, the observation of birds can in fact be fairly discouraging. Hidden by the foliage, they are hard to spot and many remain invisible in the upper canopy. It is therefore necessary, more than in any other habitat, to know their vocalisations well. Unfortunately, many species do not sing except very briefly at dawn. Others sing all day, but sporadically. Still others do not sing except during a short period of the year.

This being said, the forest avifauna is very different from that of the surrounding grasslands. Of course, some opportunistic species, such as the Garden Bulbul *Pycnonotus barbatus* or the Speckled Mousebird *Colius striatus*, penetrate forest massifs along openings caused by streams, paths or cultivated fields. Some forest species, such as the Didrick Cuckoo *Chrysococcyx cupreus* and the Black Cuckoo *Cuculus clamosus*, also inhabit densely wooded grasslands and forest galleries far from large forests. Overall, very few species inhabit both the forest and grasslands, however. The profound dichotomy between these two biomes in Africa is very marked in the avifauna also.

Altogether, 221 species are typical of Central African lowland forests. According to Françoise Dowsett-Lemaire and Robert Dowsett, their distribution suggests a division of the Central African forests into three blocks, separated mainly at the level of the Sanaga River and by the line between the watershed of the coastal rivers and of the Congo basin drainage (FIGURE 6.46).

Montane species of birds number around 110. Some have kept small relict populations between these two extreme regions (PAGE 66), but 22 are endemic to the Cameroon-Nigeria-Bioko region and 30 are of the Albertine Rift. Unlike lowland-forest birds, most of the montane species have a reduced range. Not only do high-altitude forests cover only very small, more or less isolated areas, but some species are limited to particular mountains, especially in Cameroon where the high summits are isolated from each other by vast expanses of lowland or sub-montane forest (PAGE 67). Serle's Bush-shrike *Malaconotus kupeensis* is known only from Mount Kupe. In the Albertine Rift region, the high mountains form nearly continuous chains and birds have a more extended range. The forests located around Lake Kivu are the richest and those at the extremities of the chains are the most impoverished (PAGE 67).

Altogether, this forest avifauna is obviously typical for Old World tropical regions with, as principal passerine families, the sylviids, muscicapids, turdids, timaliids, pycnonotids and nectariniids. It is distinguished, however, by the presence of six families that do not exist in Asia – the numidids, coliids, musophagids, platysteirids, malaconotids and prionopids – as well as 32 genera that do not occur outside of the Guineo-Congolese forests. In comparison with temperate forests, and all other tropical forests, fruit-eating birds are abundant and are represented by the large hornbills, the turacos, some pigeons, the parrots, the barbets, some bulbuls, the starlings and many other species. For the rest, the bird fauna shares the forest habitat mainly by its different strata and according to the age of formations.

Birds are obviously the best known vertebrates, but about the ecology and behaviour of tropical forest species there are still many unanswered questions. In the case of some species we have even to accept that we know nearly nothing. In fact the only detailed ecological and behavioural studies are those of Christian Erard and André Brosset made during the 1960s and 1970s at Makokou in Gabon.

Terrestrial non-passerines

Among terrestrial non-passerines that inhabit the forest understorey there are two forest ibises, the Olive Ibis *Bostrychia olivacea*, whose race *bocagei* is endemic to São Tomé, and the Spot-breasted Ibis *Bostrychia rara* (FIGURE 6.48). More common, however, are the phasianids, represented by Latham's Francolin *Francolinus lathami* (FIGURE 6.49), Nahan's Forest Francolin *Francolinus nahani* in dense lowland forests, the Scaly Francolin *Francolinus squamatus* in the forest mantle, the Handsome Francolin *Francolinus nobilis* in montane forests of the Albertine Rift and the Cameroon Mountain Francolin *Francolinus camerounensis* on Mount Cameroon. In the north-east of the Democratic Republic of the Congo, there is also the Congo Peafowl *Afropavo congensis*, an endemic of the Congo basin, whose status is unfortunately subject to anxiety. It seems to have disappeared from most of its range, perhaps because the males easily remain caught by their spurs in hunting nets. The numidids are represented by the Black Guineafowl *Agelastes niger*, the Plumed Guineafowl *Guttera plumifera* and the Kenya Crested Guineafowl *Guttera pucherani*. These birds are not very visible and are located mostly by the calls that they make at nightfall. Some columbids also feed almost exclusively on the ground on seeds and fallen fruit: the

6.47. The Blue-headed Wood Dove *Turtur brehmeri* *inhabits old forests as well as younger formations. Alone or in pairs, it feeds on the ground without ever associating with other columbids.*

6.48. The Spot-breasted Ibis *Bostrychia rara* *at its nest in the Ipassa Forest in Gabon. This bird frequents muddy areas covered with dense low vegetation at the mouth of small forest rivers. Outside the nesting period, it lives in small groups. It is often heard early in the morning and in the evening when it moves between its feeding grounds and its roosts.*

6.49. Latham' Francolin *Francolinus lathami* *surprised at its nest in Ipassa Forest, now included in Ivindo National Park, in Gabon. It is a bird of old forests where it feeds notably on termites, beetles and berries that it finds in the litter. It can nest at any time of the year, but contrary to many other phasianids, it lays only one or two eggs at a time.*

Non-passerines of the canopy

Among the fruit-eaters, there are the pigeons of genus *Columba* and the green pigeons of genus *Treron*, the parrots or psittacids and the turacos or musophagids. In Central Africa the parrots are represented by the Grey Parrot *Psittacus erithacus* (FIGURE 6.50), Jardine's Parrot *Poicephalus gulielmi*, much rarer, and the Cape Parrot *Poicephalus robustus*, a Miombo species of which a form inhabits some montane forests of the Albertine Rift.

As for the small lovebirds of genus *Agapornis*, they are represented by one species in the forest and one in the mantle. The turacos constitute an endemic family of Africa, of which some ten species are known from Central Africa. The Rwenzori Turaco *Ruwenzorornis johnstoni* is, however, limited to the montane forests of the Albertine Rift. In low-altitude forests this family is represented by a green species of the group of *Tauraco persa*, by the Crested Turaco *Tauraco macrorhynchus*, green also and limited to Lower Guinea, and the Great Blue Turaco *Corythaeola cristata*, much larger and blue.

Other non-passerines of the canopy are both frugivorous and insectivorous. These are the hornbills or bucerotids (FIGURE 6.51) – not to be confused with the tucans or ramphastids of South America – of which nine species inhabit Central African forests. The African White-crested Hornbill *Tropicranus albocristatus* is remarkable in that it lives largely in association with monkeys of the canopy. The barbets or lybiids, another family typical of the Old World, is divided into some 15 species, hardly visible but often locatable by their repetitive, monotonous and tiresome calls that can be heard at any time of day. As for the trogons or trogo-

6.50. *Although it is present in the canopy of the primary forest, the Grey Parrot* Psittacus erithacus *prefers secondary formations. It likes stands of* Elaeis *palms whose fruit probably constitutes its main food source. At night, it seeks out* Raphia *palms to roost. Several thousand individuals may gather to spend the night together. In 1986, André Brosset and Christian Erard recorded more than 10,000 individuals near Bitam, in north-eastern Gabon.*

6.51. *The Black Dwarf Hornbill* Tockus hartlaubi *mostly inhabits the middle and upper strata of the forest.*

6.52. *The Bare-cheeked Trogon* Apaloderma aequatoriale *is a bird of the middle levels of the canopy where it hunts lying in wait for large insects that it captures either in flight or in the foliage.*

Lemon Dove *Aplopelia larvata*, the Blue-headed Wood Dove *Turtur brehmeri* (FIGURE 6.47) and the Tambourine Dove *Turtur tympanistria*. There remain, finally, the rallids which are only exceptionally observable. They are represented by the White-spotted Crake *Sarothrura pulchra*, heard during day time, while the Nkulengu Rail *Himantornis haematopus* and the Grey-throated Rail *Canirallus oculeus* are heard at night or at nightfall only.

nids, only three species are known from Africa, of which two are forest birds: the Bare-cheeked Trogon *Apaloderma aequatoriale* of low-altitude forests (FIGURE 6.52) and the Bar-tailed Trogon *Apaloderma vittata* of montane forests.

A particular group of non-passerines feeds on insects that live in bark. These are the woodpeckers or picids and the wood hoopoes or phoeniculids. The first comprise some ten species that avoid competition by the fact that they are of different sizes and occupy different micro-habitats. Most of them nest in cavities that they dig into tree trunks themselves, but the Buff-spotted Woodpecker *Campethera nivosa* nests exclusively in Nasutitermitinid termite mounds (FIGURE 6.53).

The cuckoos or cuculids are insectivores of the canopy, essentially caterpillar eaters. The three long-tailed cuckoos of genus *Cercococcyx* are typical of the dense forests, of which one species is strictly montane, the African Emerald Cuckoo *Chrysococcyx cupreus* and the Yellow-throated Green Cuckoo *Chrysococcyx flavigularis*. In the understorey there are coucals, "cuckoos" that brood their own eggs, represented notably by the Gabon Coucal *Centropus anselli*.

The kingfishers or alcedinids comprise three distinct groups. the first includes the small African Dwarf Kingfisher *Ceyx lecontei*, that hunts insects in the understorey of terra firme forests (FIGURE 6.54), and the Shining Blue Kingfisher *Alcedo quadribrachys*, that hunts rather for fish in little streams in the forest or in the shelter of large trees along larger rivers. The second group comprises the Blue-breasted Kingfisher *Halcyon malimbicus* and the Chocolate-backed Kingfisher *Halcyon badia*. Both hunt in the canopy. The third group comprises the Giant Kingfisher *Megaceryle maxima* that hunts lying in wait along the banks of small and large watercourses of the forest.

Finally, there remains a very particular family, the honeyguides and honeybirds or indicatorids, also endemic to Africa. These unspectacular birds in grey, brown or olive colours, are recognisable by their undu-

6.53. The Buff-spotted Woodpecker Campethera nivosa *inhabits the lower levels of forests, old and young, where it feeds mainly on ants and some termites. It nests exclusively in termite mounds of* Nasutitermes arborum.

6.54. The African Dwarf Kingfisher Ceyx lecontei *lives in the understorey. Independently of the presence of water, it hunts alone or in pairs by lying in wait for large insects on the ground. It nests in tunnels arranged mainly along tracks.*

6.55. *The Black and White Flycatcher* Bias musicus, *of the Platysteiridae family, is linked to secondary formations and habitats modified by human activity. At Kampala and Libreville, it even inhabits urban gardens. It flits around hunting for insects on the foliage.*

6.56. *The Golden-bellied Wattle-eye* Dyaphorophyia concreta, *also a platysteirid, is rather a bird of old forests with an open understorey and it can be found at altitudes up to 1,900 metres in the Rift region. It often has a liking for the perimeter of a tree-fell gap, without going so far as to actually venture out into the open.*

6.57. *A Grey-necked Bald Crow* Picathartes oreas *near its nest in the Makokou area, in Gabon.*

6.58. *The Grey Ground Thrush* Zoothera princei *inhabits old terra firme forests of Cameroon, Gabon, Equatorial Guinea and the Democratic Republic of the Congo. It has a particular liking for tree-fall gaps and places where low liananescent formations and Marantaceae plants abound.*

lating flight and their white external rectrices. The small species feed mainly on scale insects, the large ones feed largely on the nests of social hymenoptera, such as bees. All lay their eggs in the nests of barbets. The Lyre-tailed Honeyguide *Meligneutes robustus* is a bird of the canopy, very difficult to see, but locatable by the noise it produces by rubbing its tail during its displays.

Passerines

Passerines occupy all the strata of the forest and also have many more insectivorous species than seed- or fruit-eating ones. The bulbuls or pycnonotids include ten species of genus *Andropadus*, mostly fruit-eaters, and nine species of genus *Phyllastrephus*, rather insect-eaters. All occupy different strata of the canopy. The thrushes or turdids, equally very diversified and including many forms that are not met with outside African tropical forests, are insectivorous or omnivorous and hunt on the ground or near to the ground. Typical of Central African forests are the thrushes of genus *Zoothera* (Figure 6.58). Recognisable by their facial mask and by white spots that mark their wings, they are unobtrusive without being truly timid.

Among the insectivores of the canopy, the paradise flycatchers and monarchs or monarchids, as well as the flycatchers or muscicapids catch insects in flight from a perch. The platysteirids also hunt in flight, but catch the insects rather on the underside of the foliage (FIGURES 6.55 and 6.56). The sylviids comprise numerous "warblers", notably in the genera *Prinia*, *Apalis*, *Camaroptera* and *Eremomela*, which glean insects inside the foliage. The timaliids, finally, represent an assemblage of very different species as much in their appearance as in their ecology. The thrush babblers of genera *Kakamega* and *Illadopsis* inhabit dense understories. The Capuchin Babbler *Phyllanthus atripennis* lives in dense liana columns and babblers of genus *Kupeornis* occupy the upper canopy of wet forests where they feed mostly in the masses of epiphytes. The Grey-necked Bald Crow *Picathartes oreas* hunts on the ground (FIGURE 6.57).

6 THE FAUNA

A very particular group of insectivores follows columns of army ants and captures the insects disturbed by them. Some thrushes, babblers and flycatchers do this occasionally. Others, such as the alethes, *Alethe*, have become true specialists of this technique.

Raptors

The 16 species of raptors known from Central African forests all belong to the Accipitridae family. Even though they are fairly numerous, diurnal raptors are difficult to observe. Most often, they are not seen except during their territorial displays when they rise high above the canopy and hurl characteristic vocalisations. The most common is probably the African Harrier Hawk *Polyboroides typus* that is often encountered around villages when it is pillaging weaver colonies. The most imposing raptor is the Crowned Eagle *Stephanoaetus coronatus*, hunter of monkeys. Cassin's Hawk Eagle *Spizaetus africanus* and the African Long-tailed Hawk *Urotriorchis macrourus*, distinctly smaller, hunt mostly for squirrels, while the African Mountain Buzzard *Buteo oreophilus*, limited to the montane forests of the Albertine Rift, hunts lizards, chameleons and amphibians instead. The Congo Serpent Eagle *Dryotriorchis spectabilis*, with the silhouette of a sparrow hawk and the head of a Short-toed Eagle, hunts mostly snakes, lizards and chameleons. It is fairly unobtrusive, but is most often seen perched at the edge of a clearing or a large tree-fall gap. The large sparrow hawks and goshawks of genus *Accipiter* hunt birds, but the smaller ones also consume reptiles and insects (FIGURES 6.59 and 6.60).

A very particular raptor, typical of Central Africa, is the Palm-nut Vulture *Gypohierax angolensis* (FIGURE 6.61). It is mostly abundant along rivers, but can also be encountered in mangrove forests and on beaches. It is a very opportunistic carrion-eater that consumes not only dead fish but also fruit from the Oil Palm *Elaeis guineensis* and the Dwarf Date Palm *Phoenix reclinata*. This vulture can sometimes be seen with an African Fish Eagle *Haliaetus vocifer* along large watercourses. The latter bird is much less common in forested regions than in more open regions of Eastern Africa, however.

Nocturnal raptors (FIGURES 6.62 and 6.63) have 13 continental species in the family Strigidae, of which two are montane: the Albertine Rift Owlet *Glaucidium albertinum*, known uniquely from the forests of Kahuzi-Biega and Itombwe in the Democratic Republic of the Congo and Nyungwe Forest in Rwanda; the Long-eared Owl *Asio otus*, known only at high altitude on the Ruwenzori and Virunga mountains. There is also an endemic species from São Tomé: the São Tomé Scops

6.59. A Red-thighed Sparrow Hawk Accipiter erythropus, *surprised with its freshly captured prey, an Orange Weaver* Ploceus aurantius. *It is a raptor of the mantle and of open formations.*

6.60. *The African Goshawk,* Accipiter tachiro *is represented in the forests of Central Africa by the form* toussenelli, *a very pale bird that can be found in all the forest formations and even around villages. It feeds notably on amphibians, rodents and large beetles, but will not scorn a chicken from the barnyard.*

6.61. *A Palm-nut Vulture* Gypohierax angolensis *surprised at the edge of a clearing in Odzala-Kokoua National Park*

6.62. *Pel's Fishing Owl* Scotopelia peli *is common along the large and middle watercourses of Central Africa. It feeds exclusively on fish.*

6.63. *The Akun Eagle Owl* Bubo leucostictus *hunts mainly in the understorey of old forests. It probably feeds mostly on large insects.*

Bird pictures taken in the wild in African rainforests are very uncommon. Most of the pictures illustrating this chapter, like those of nocturnal primates and many small or medium sized mammals which are difficult to see, have been taken in the forests of north-eastern Gabon by Alain Devez. This French photographer spent many years at the research station of Makokou.

Owl *Otus hartlaubi*. Some species hunt in the understorey, like the Nduk Eagle Owl *Bubo poensis*; others in the canopy, such as the Red-chested Owlet *Glaucidium tephronotum*. The two fishing owls *Scotopelia bouvieri* and *Scotopelia peli* (FIGURE 6.62) are limited to the vicinity of watercourses. The tytonids are not truly forest birds, but Prigogine's Owl *Phodilus prigoginei* inhabits clearings and more or less open areas within montane forest. This bird is not known except in the Itombwe in the Democratic Republic of the Congo, but several observations suggest that it also occurs in Nyungwe Forest in Rwanda and Kibira Forest in Burundi.

Aerial species

Among the non-passerines, there are finally some birds that feed on insects in flight above the forest. The swifts or apodids, are represented by several species, notably migrants from Europe, such as the Common Swift *Apus apus*. Three species are, however, limited to forested regions: Cassin's Spinetailed Swift *Neafrapus cassini*, Sabine's Swift *Raphidura sabini* and the Ituri Mottled-throated Spinetailed Swift *Telacantura melanopygia*. The first two are fairly common and are recognised at once by their very short tail and their strangely notched wings. The latter, all black, is rarer, but can be observed quite easily above the moist clearings in Odzala National Park.

The Blue-throated Roller *Eurystomus gularis*, a member of the Coraciidae family, hunts rather by lying in wait, perched along the forest edge or in some large emergent. As for the bee-eaters or meropids, some species hunt from a hide in the forest mantle, as does the Blue-breasted Bee-eater *Merops variegatus* or the Black-headed Bee-eater *Merops breweri*. Others hunt high above the forest, like the Rosy Bee-eater *Merops malimbicus*, common in coastal areas.

Nightjars, or caprimulgids, also hunt in flight but only at night. In the forest, they are seen only rarely, but they are common along the margin of grasslands in the Lopé National Park.

Multispecific flocks

During the day the forest is often very silent. From time to time, however, this silence is shattered by the eruption of a mixed flock. Then, for a few minutes, the forest seems to be swarming with birds. The reason is that, in the tropical forests, even more than in grasslands, many birds move about in large multispecific groups, including several tens of individuals, from 10, 20 or even 25 different species. These groups comprise mostly insectivorous birds such as bulbuls, warblers, flycatchers, paradise flycatchers, sunbirds, a few weavers, some woodpeckers, wood hoopoes, cuckoo shrikes and small hornbills. All these birds circulate together in the same direction across the canopy. Those liking shade exploit the middle strata and are at times quite visible. Those that prefer the light– and they are often the most numerous – move in the upper strata and the emergent tree crowns. All that can be seen of them are small silhouettes that fly by against the light, from one mass of foliage to another. In any case, after a few minutes, the group passes, pursuing its route, and the forest plunges again into silence.

Mammals

Mammals are not lacking in tropical forests but their density is even lower than that of birds. In most places in Central Africa they are also extremely evasive, probably in part because of the fact that they are submitted to heavy hunting pressure. In the Great Lakes region, where the eastern Bantus subject hunt mostly only ungulates, monkeys and squirrels can be very trusting, even outside protected areas. In Uganda, it is therefore not unusual to see a few Guerezas *Colobus guereza* perched right on the edge of a village peacefully watching the humans going about their activities. In Bururi Forest or in Ruvubu National Park, in Burundi, where there is not even any protection against poaching, it was easy to observe guenons or colobus monkeys at a short distance. Thus if the clearings of Odzala-Kokoua National Park are without any doubt the best place to see gorillas in complete tranquillity, Kibale National Park, in Uganda, remains the best place to see small primates and the Chimpanzee *Pan troglodytes*.

According to Peter Grubb, the forests of Central Africa contain about 270 species of mammals, of which 170 are endemic. More than in other groups of terrestrial animals, their distribution is strongly influenced by the hydrographic system. Many species are, in fact, incapable of crossing large watercourses, such as the Congo, Sangha, Ubangi, or even the Ogooué or Sanaga rivers. Based on the distribution of mammals, the lowland forests of Central Africa are divided accordingly into several faunal regions, separated by the larger rivers and at present linked – we are in a phase of forest transgression – by vast zones of intergradation (FIGURE 6.64).

The Atlantic coastal biogeographical region, or Lower Guinea (PAGE 63), is the richest with 207 species, of which 59 are endemic. According to Marc Colyn, they comprise two faunal regions separated by the Sanaga River. According to Grubb, the region located south of the Sanaga is subdivided in its turn by the Ogooué River.

The Congolese region strictly speaking comprises three faunal regions, according to studies made by Grubb and by Colyn. The richest one is situated northeast of the Congo River with 192 species, including 52 that are endemic. It is subdivided into several zones by tributaries of the Congo. The region located south of the river is clearly less rich with only 99 species including 13 that are endemic. It is divided into three distinct zones: the Lualaba-Lomami interfleuve, the zone lying in between the Congo, Lomami and Kasai rivers, and the zone lying south of the Kasai River. The third faunal region, recently described by Colyn, lies west of the Ubangi River.

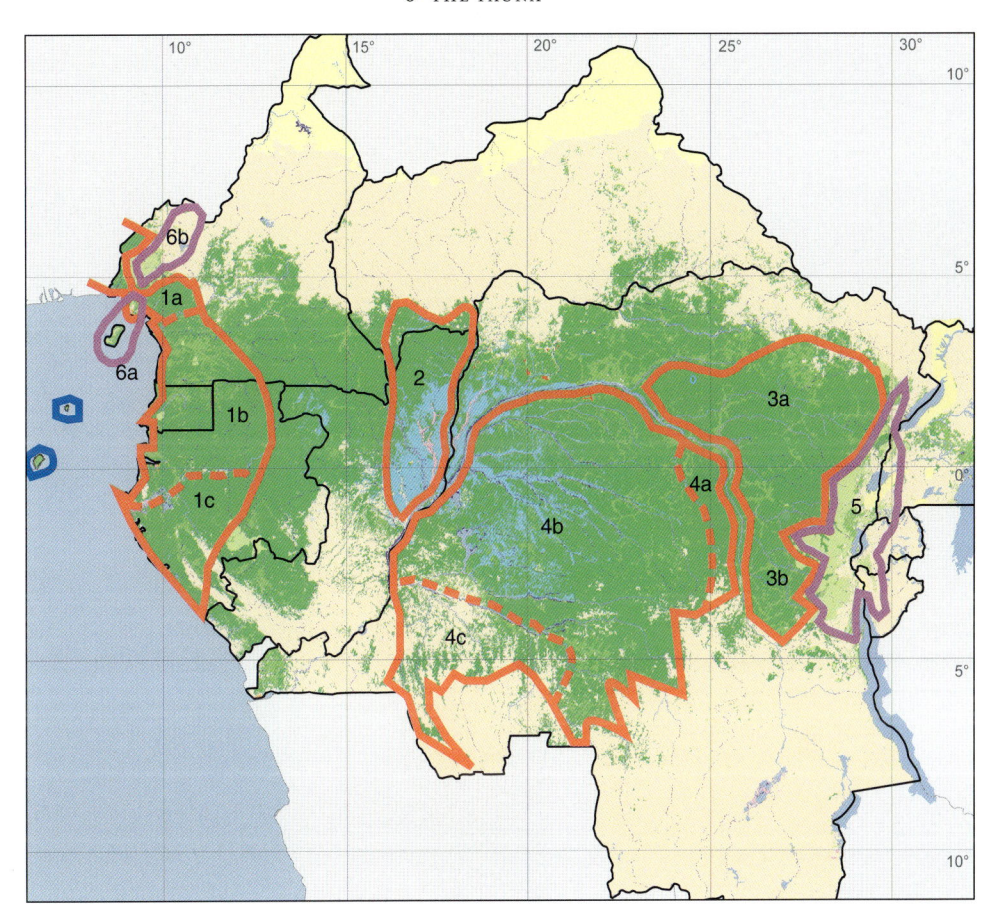

To these faunal regions must be added those related to the montane forests: the Mount Cameroon and Bioko region, western Cameroon and the Albertine-Rift region (PAGE 65).

Primates

The ancestors of the primates, a group of small insectivorous arboreal mammals, lived in Gondwanaland well before the disappearance of the last dinosaurs. Those isolated in the heart of the African continent gave rise to the Old World monkeys. At the very beginning of the Tertiary, 55 to 60 million years ago, they separated into two groups: the prosimians and the monkeys.

In their turn, about 25 million years ago, the prosimians gave rise to the lorisids and the galagonids. The first were able to colonise Asia, the second would remain African. The lorisids and galagonids would keep some of the primitive primate characters, but the species that we know today are modern and successfully occupy the diverse ecological niches available to small nocturnal primates.

Lorisids are solitary, nocturnal, nearly tailless and slow moving. They are represented in the forests of Central Africa by three species. The Potto *Perodicticus potto*, the most common of them, inhabits forests from Sierra Leone to Uganda (FIGURE 6.65). It likes second-

6.64. The lowland forests of Central Africa comprise four faunal regions which are subdivided into a total of nine zones of endemism:
1. Atlantic coastal region with:
a. Cross-Sanaga interfluve,
b. zone north of the Ogooué River,
c. zone south of the Ogooué River.
2. Sangha-Ubangi region.
3. region north-east of the Congo basin with:
a. Ituri-Uele zone,
b. Kivu-Maniema zone.
4. region south of the Congo basin with:
a. Lomami-Lualaba interfluve,
b. zone located in between the Congo, Lomami and Kasai rivers,
c. zone located south of the Kasai River.
Most of these regions are delimited by rivers. As for the montane forests, they comprise three faunal regions:
5. Albertine Rift region,
6a. Mount Cameroun and Bioko region,
6b. western Cameroon region.

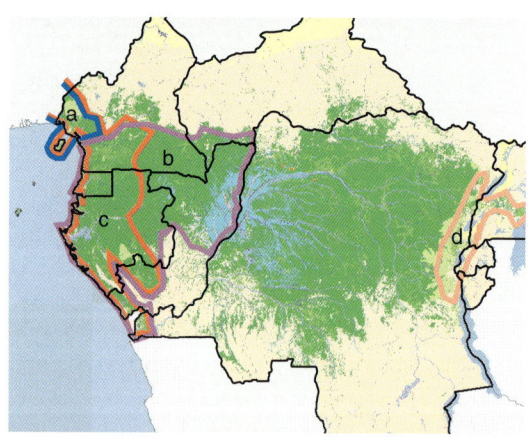

6 THE FAUNA

ary formations, pioneer forests and forest edges. It is encountered even in montane forests up to 2,600 metres altitude. During the dry season its food includes many gums, while during the rainy season it consumes various fruits, snails and insects. Pottos like figs and, with the help of a good flashlight, it is possible to see one at night in a *Ficus* laden with fruit. The Calabar Angwantibo *Arctocebus calabarensis* is limited to forests located between the Niger and the Sanaga rivers, while the Golden Angwantibo *Arctocebus aureus* inhabits forests that lie between the Sanaga, Congo and Ubangi (FIGURE 6.66). These two pottos are mainly insectivorous and mostly eat caterpillars.

The galagonids are an endemic family of Africa and are at least 22 million years old. They are nocturnal, possess a long tail, strong hind limbs, large eyes and well developed ears, allowing them to run and jump from branch to branch with great agility. The 16 to 18 currently recognised species all resemble each other and do not differ at first sight except by subtle details of size, colour and proportions. In nature, however, they are told apart by their ecological specialisation and their very typical vocalisations. Closer examination also shows typical differences in the hands and sexual organs. Genetic studies show that the present-day species are all older than had been imagined. Unfortunately, several species have been confused until not very long ago and, because of this fact, understanding of their biology is somewhat muddled. Since they are nocturnal and their vocalisations are little known, their distribution remains poorly known. However, the seven species that inhabit the forests of Central Africa are divided into five very distinct groups.

The Greater Galago *Otolemur crassicaudatus* inhabits forests of the Albertine Rift and the Great Lakes region, riparian as well as montane and dry forests. It is an animal the size of a domestic cat that walks but does not jump. It feeds on gums, flowers and seeds, fruits, snails and slugs, insects, reptiles and birds.

The needle-clawed galagos, *Euoticus elegantulus* (FIGURE 6.70) and *Euoticus pallidus,* are found only in the forests of Lower Guinea, the first in forests south of the Sanaga River, the second in those of the north and on Bioko. They are small, very agile animals with large orange eyes. Their nails are prolonged by claws and their long lower incisors form a cutting edge. Their food is composed almost exclusively of gums, mainly of the Mimosaceae trees *Newtonia, Albizia* and *Entada.* These galagos are therefore more abundant in young forests than in old ones. They obtain the gum by making small incisions in the bark, as well as from wounds in trees caused by anomalures or flying mice (PAGE 225).

The squirrel-galagos, of genus *Galago,* are small and extremely agile. They are represented by Allen's Galago *Galago alleni* (FIGURE 6.69) which inhabits the forests

6.65. *The Potto* Perodicticus potto *is recognisable by its short tail. It has a vast range extending throughout Central Africa.*

6.66. *The Golden Angwantibo* Arctocebus aureus, *smaller and without a tail, feeds in the lower strata of the forest where, at the slightest alarm, it takes refuge in dense columns of lianas. It regularly comes to the ground to gather fallen fruits or insects and, as it moves through the vegetation, leaves traces of urine on branches to mark its territory. When surprised, it humps its back like this one photographed in north-eastern Gabon.*

6.67. *Among the ten species of nocturnal primates known in Central Africa, three have a wide distribution, from the Nigerian border to Uganda. Six have a much more restricted distribution. The Calabar Angwantibo* Arctocebus calabarensis *and the Northern Needle-clawed Galago* Euoticus pallidus *are found only to the north of the Sanaga River and on Bioko Island (a). The Golden Angwantibo and the Southern Needle-clawed Galago* Euoticus elegantulus *are found between the Sanaga and Congo Rivers (b). Allen's Squirrel Galago* Galago alleni *is found from the Nigerian border to the Congo estuary, but not east of the Ogooué basin. The Spectacled Galago* Galago matschei *(d) is limited to the Albertine Rift highlands and Uganda.*

lying between Cross River and the mouth of the Congo, but its range does not extend beyond the Ogooué basin, in the east. Absent from young secondary forests, it feeds mainly on the ground in old forests with an open understorey. Its food comprises many fallen fruits, as well as amphibians, insects, spiders and snails.

The lesser galagos comprise mostly grassland species, but they are represented in the forests of the Albertine Rift and the Great Lakes region by the Spectacled Galago *Galago matschei* (formerly *G. inustus*). It inhabits lowland forests as well as montane forests, young forests or old forests, but its presence is apparently linked to that of *Parinari excelsa*, one of the dominant trees in these forests. Its food includes insects, fruits and gums, depending on the seasonal availability of these resources.

The group of the dwarf galagos comprises Demidoff's Galago *Galagoides demidoff* (FIGURE 6.68) and Thomas's Galago *Galagoides thomasi*, considered until very recently as conspecific. These two species are in fact practically identical on sight and cannot be separated except by their distinctly different vocalisations.

Demidoff's Galago haunts the understorey of dense secondary formations, forest mantles and tree-fall gaps, while Thomas's Galago inhabits rather the upper canopy of old forests. Their food consists mostly of insects, with some fruits and gums.

Some 25 to 30 million years ago, the diurnal primates gave rise to the Cercopithecidae and the Hominidae. The latter group were the dominant primates in the hot humid forests of the Miocene and diversified into dozens of species, today mostly extinct. The Cercopithecidae would develop later, probably in more

6.68. *A Demidoff's Galago* Galagoides demidoff *in north-eastern Gabon.*

6.69. *Allen's Squirrel Galago* Galago alleni *feeds almost entirely on the ground in old forests with a very open understorey. Males occupy a vast territory, covering the much smaller territories of several females. Individuals communicate by use of croaking calls and mark their territories with scent trails. By day, this species hides inside tree trunks, like savanna monkeys.*

6.70. *The Elegant Needle-clawed Galago* Euoticus elegantulus, *with its brilliantly coloured pelage, can be fairly abundant in secondary forests. A density of 30 individuals per square kilometre was observed in Gabon. Females and their young spend the day in small groups of three to five individuals, well hidden in masses of foliage. From time to time they are joined by the males.*

6 THE FAUNA

6.71. In the case of the Mandrill Mandrillus sphinx, *males are distinguished by their impressive facial mask and by their weight, which can reach 35 kilos, while females weigh no more than 11 kilos on average. Unlike baboons, this species has no more than the vestige of a tail.*

6.72. The number of diurnal primates found in Central African forests can reach 12 species in Lower Guinea and on the eastern edge of the Congo basin between the Lualaba River and the Rift, but is no more than 10 species in the centre of the basin. Many species have a wide distribution, and similar species communities are found all over the Central African forests. Only the Drill Mandrillus leucophaeus *(a) and the Mandrill* Mandrillus sphinx *(b) have no close relatives outside Lower Guinea.*

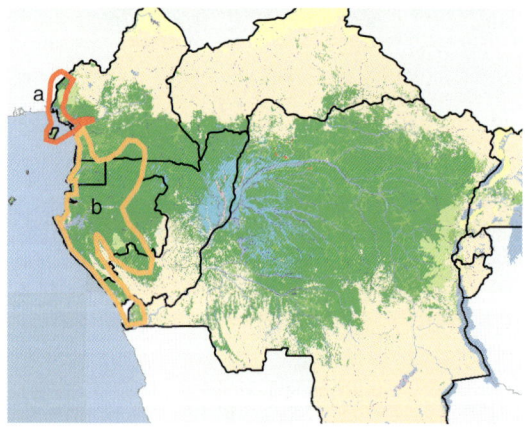

open habitats. Their most ancient forms must have resembled the present-day baboons *Papio* and macaques *Macaca*. About 10 million years ago, they produced a first offshoot that was able to reconquer the forests and the arboreal life style. These were the colobus monkeys, mainly vegetarians. The baboons and macaques diversified and some were able to colonise humid forests. These are representatives of genus *Mandrillus*. Finally, five million years ago, there would appear other more or less arboreal Cercopithecidae: the *Cercopithecus*, *Cercocebus* and *Lophocebus*.

The Macaques being limited to the arid regions of North Africa and Asia, the only ancient terrestrial Cercopithecidae in Central Africa are the baboons. The Sacred Baboon *Papio hamadryas* has a vast distribution and regroups diverse forms that some scientists, among others Jonathan Kingdon, recognise as five distinct species: *hamadryas*, *papio*, *cynocephalus*, *ursinus* and *anubis*. Others recognise only two: *hamadryas* and *papio*; the latter arrangement bringing together all the remaining forms. Given the considerable genetic exchanges between the different forms, many scientists, including Annie Gautier-Hion, Marc Colyn and Jean-Pierre Gautier, now regroup all of them in *hamadryas*. Even though it is mostly a grassland inhabitant, the baboon locally penetrates forests, notably in the Ituri, in the Rift region, and southern Kasai. It is an animal more of open formations and it is fairly frequent in montane forest where it can be encountered up to nearly 2,500 metres altitude. It is essentially terrestrial and mostly frequents the understorey. However, when there are trees bearing fruit, it does not hesitate to go up into the upper canopy. It happens then, notably at Kibale, that baboons can be seen perched at the top of large emergents in the company of typical forest monkeys.

More forest-going are the Drill *Mandrillus leucophaeus* and the Mandrill *Mandrillus sphinx* (FIGURE 6.71). The first inhabits forests lying between the Cross and Sanaga rivers as well as those on the island of Bioko. The domain of the second extends from the Sanaga to Kouilou, but does not go beyond the Ogooué basin in the east, except in the region of the Dja River where it makes a small incursion into the Congo basin. These are terrestrial animals, but they do not hesitate to climb trees. They are observed in old forests, secondary forests, abandoned cultivated areas and galleries in grasslands.

The Mandrill lives in groups of as many as several hundred individuals. In the Lopé National Park in Gabon, a group of 1,350 individuals was even observed, in November 1996. The group studied by Kate Abernethy and researchers at the Gorilla and Chimpanzee Research Station – installed in the reserves 20 kilometres south of the headquarters – numbers around 600 animals. Thanks to the use of radio tracking collars, the group could be followed through the dense forests. From these observations, it appeared that for most of the year, these groups are made up of only females and young. Unlike baboons, male mandrills do not join the females except for the reproduction period, which lasts from June to November. The rest of the year they live alone, and do not even associate with other males. During the period that they live with the females, the males are interested only in those that are in oestrus. In this species, females have only one young every two years. Considering the large size of the groups, they move about a great deal in order to find food and groups may cover daily distances of more than ten kilometres. Solitary males on the contrary remain contained within a more reduced space.

The *Cercocebus* monkeys have an ancestor in common with the drills and mandrills. They are represented by two species in Central Africa. The Red-capped Mangabey *Cercocebus torquatus* (FIGURE 6.73) is endemic to the coastal forests of the Gulf of Guinea, from the Sanaga River to the Kouilou River. The Agile Mangabey *Cercocebus agilis* has a wide distribution in the Congo basin and the north-eastern part of the Ogooué basin. Its nominate form is found from the Dja to the Ituri, passing through Odzala National Park. Its form *chrysogaster*, to which Kingdon attributes the rank of species, replaces it south of the Congo River. These two *Cercocebus* monkeys live in groups of several males and females. Attached to riverine and swamp forests, they move about a lot on the ground. Both forms feed on fruits and nuts and on the stems and roots of plants of the understorey. Sometimes, these animals also eat mushrooms, herbaceous plants and even some animal material.

The Grey-cheeeked Mangabey *Lophocebus albigena* (FIGURE 6.74) seems to have an ancestor in common not with the mandrills but with the baboons. It lives from the Cross River and the Atlantic coast to Rwanda and Uganda. Its form *atterrimus*, which inhabits the central basin south of the Congo River, is considered by some, notably Kingdon, as a species in its own right. Altogether, the *Lophocebus* are mostly monkeys of terra firme forests, haunting the upper and middle strata of the canopy of old forests. In areas where it is not hunted, as in Kibale National Park or Nyungwe Forest, it can descend to very near the ground. As soon as it is disturbed it flees into the canopy. It is rather large and has a long tail that it often holds, like the two *Cercocebus*, curved forward over the body. It lives in groups of 10 to 30 individuals. Small groups, often of short duration, are composed of one adult male, the largest formations of up to four or five males with their females and young. Its movements are calm and measured. It feeds on fruits, often with very hard pits, and seeds, but it also consumes young leaves and on occasion does not shun insects and young birds. From time to time, adult males give sonorous calls to announce their presence and orient the movements of their group. These powerful calls are produced by an enlargement of the respiratory tract at the level of the larynx. Each adult male has its own vocalisations with which it can be identified.

At the centre of the Congo basin lives Allen's Swamp Monkey *Allenopithecus nigroviridis* which has an ancestor in common with the macaques. It occupies a place intermediate between the baboons, the mandrills and the *Cercocebus* species on one side and the *Cercopithecus* species on the other. As in the baboons and macaques, females in oestrus show a swelling of the sexual skin. It inhabits enormous expanses of swamp and riverine forests as well as all the humid, grassy or shrubby habitats associated with them. It does not hesitate to visit open habitats and cultivated areas. During periods of low water it moves about much on the ground. According to Annie Gautier-Hion and her colleagues, the only scientists to have observed this monkey in nature, this occupies at least 60% of its time. Its food is apparently very varied and includes leaves, fruits, invertebrates – notably crabs – and fish. It nor-

6.73. The Collared Mangabey *Cercocebus torquatus is found only in the coastal regions of Lower Guinea. It is subjected to very strong hunting pressure and definitely endangered.*

6.74. The Grey-cheeked Mangabey *Lophocebus albigena is absent from Monte Alén National Park, but occurs in all the other ECOFAC sites.*

6.75. The Northern Talapoin Miopithecus ogoouensis *inhabits riverine forests and swamp forests exclusively. It spends the night in dense vegetation clumps above water. Its food includes fruit, palm nuts, seeds, leaves, flowers and insects.*

mally lives in groups of on average 35 individuals, but at night more than 85 may assemble at a roost. In some regions, such as around Lomako, groups are formed of hardly more than three or four individuals.

The Northern Talapoin *Miopithecus ogoouensis* (FIGURE 6.75) and the Southern Talapoin *Miopithecus talapoin* are two small monkeys whose weight is no more than two kilos in the males. The range of the first species covers the Ogooué River basin and northward to the Sanaga River. The second, slightly larger and with longer limbs, lives south of the Congo River. As in Allen's Swamp Monkey, the females develop a swelling of the sexual skin during oestrus. They inhabit riverine forests and never venture very far from watercourses. In groups of up to 70, or even 100, individuals, they mainly frequent the dense vegetation of the understorey. They swim very well and eat much fruit, but insects constitute around 30% of their diet. At night, more than 100 talapoins can assemble in a roost, invariably located over water.

Most of the monkeys that we will meet in the forest are Cercopithecinae, comprising the guenons and their allies. It is a strictly African group. Some species are grasslands dwellers, such as the Patas Monkey *Erythrocebus patas* and the various forms of the Savanna Monkey *Chlorocebus* (or *Cercopithecus*) *aethiops*, but most inhabit the forest. They all belong to genus *Cercopithecus*, and have much longer hind limbs than fore limbs, move with great agility in trees and show no swelling of the sexual skin of females. The various species all feed on fruits, flowers, leaves and buds, insects, eggs and young birds. They differ in size, in choice of habitat and mostly in their very striking coloration.

The genus *Cercopithecus* is a varied group, comprised of multiple species and multiple forms within the species. The group's systematics are somewhat disconcerting because it is still in full evolution and includes numerous regional forms difficult to place as to species or subspecies. This allocation is not easy because some forms are clearly separated – by a water course, for example – while others show intergradation. We will follow here the classification used by Annie Gautier-Hion, Marc Colyn and Jean-Pierre Gautier in their ECOFAC guide to the diurnal primates of Central Africa, but we will also mention the opinion advanced by Kingdon in his guide to the mammals of Africa.

Among the 12 species that inhabit Central Africa, six are partially terrestrial. Except for de Brazza's Monkey *Cercopithecus neglectus* (FIGURE 6.77), known from the Gulf of Guinea to Kenya and Ethiopia, they all have a fairly limited range (FIGURE 6.78). Also, they have practically no subspecific differentiation. It is therefore a rather less succesful group. Sexual dimorphism is very striking: as in the baboons, mandrills and swamp monkeys, the males are much heavier than the females. The tail, not having to play a balancing role, is proportionally shorter than in arboreal species.

De Brazza's Monkey inhabits the understorey of swamp forests, inundatable forests, semi-aquatic thickets and terra firme forests, as long as water is not far. It swims well and can in this way colonise islands. In spite of its spectacular pelage, it is very difficult to see because it moves about very slowly and, at the slightest alert, it freezes completely and melts into the dense vegetation. However, with a little luck and a lot of patience it is possible to observe it in the small forest gallery that goes along the Mboko camp in Odzala-Kokoua National Park. This species lives in small groups comprised of an adult male and a few females and young.

The Owl-faced Monkey *Cercopithecus hamlyni* (FIGURE 6.76) inhabits forests lying between the Lualaba River, the upper course of the Congo River, and the Albertine Rift. In the east, its domain extends to the eastern part of Nyungwe Forest in Rwanda; in the north, to the Lindi and Nepoko rivers. It thus inhabits low altitude forests as well as montane forests. At high altitude, at Kahuzi-Biega National Park and in the Nyungwe Forest, its form *kahuziensis* seems to have a clear preference for bamboo thickets. Until 1993, this animal could easily be observed along the road that crossed Kahuzi-Biega National Park but, since the events which have stricken the region, its status is unknown. However, there is no reason to believe that it would have disappeared from Nyungwe Forest. This monkey lives in small groups of less than ten individuals, led by an adult male.

The Salongo Monkey *Cercopithecus dryas* remains a mystery. It is known only from two skins collected in 1932 and 1977 and no scientist has ever seen it in the wild. Its domain must be the basin of the upper Maringa (FIGURE 6.78). According to local hunters, it lives in groups of two to 30 individuals with one or several adult males and inhabits the lower stratum of thickets and forests along watercourses. Its relationships within the genus *Cercopithecus* are poorly known, but some researchers, notably Annie Gautier-Hion, Marc Colyn and Jean-Pierre Gautier, think that it could possibly have an ancestor in common with some savanna monkeys "trapped" in blocks of enclaved grasslands.

The three other semi-terrestrial species are closely related. The have long limbs. Males are clearly heavier than females and their black, relatively long muzzle sharply contrasts with their white throat and chin. They live in small groups comprising an adult male and a few females with their young; they are much quieter than the arboreal species and occupy extensive territories. They feed on fruits and seeds, young shoots, leaves, flowers and mushrooms. If the occasion presents itself they do not hesitate to visit fields and, because of this, can become true pests.

L'Hoest's Monkey *Cercopithecus lhoesti* (FIGURES 6.78 and 6.79) has a range that closely coincides with that of the Owl-faced Monkey. It is, however, clearly more common east of the Albertine Rift, in Burundi, Rwanda and Uganda. In Nyungwe Forest, it is much more abundant than *hamlyni*, while in the Kahuzi-Biega region it is the reverse situation. It occupies lowland forests and montane forests up to 2,700 metres altitude. The Sun-tailed Monkey *Cercopithecus solatus*, discovered only in 1984, inhabits the *forêt des Abeilles* and the central part of Lopé National Park in the centre of Gabon (FIGURE 6.80). It can be seen notably along the Offoué River. As for Preuss's Monkey *Cercopithecus preussi*, it essentially inhabits the slopes of Mount Cameroon between 1,000 and 1,800 metres and the island of Bioko, but it is still found on mounts Oban and Obudu. As a result of very heavy hunting pressure, it survives only in isolated pockets and its future is more and more problematical.

These three species seem to have descended from a common ancestor whose range covered most of Central Africa where it inhabited lowland forests as well as sub-montane or montane forests. Today *C. lhoesti* and *C. preussi* can still be seen at high altitude and in lowland forests.

6.76. *Owl-faced Monkey*
Cercopithecus hamlyni.

6.77. *de Brazza's Monkey*
Cercopithecus neglectus.

6.78. *Range of five very localised species of terrestrial guenons:*
(a) Owl-faced Monkey,
(b) Salongo Monkey C. dryas,
(c) Preuss's Monkey C. preussi,
(d) Sun-tailed Monkey C. solatus,
(e) and l'Hoest's Monkey C. lhoesti.

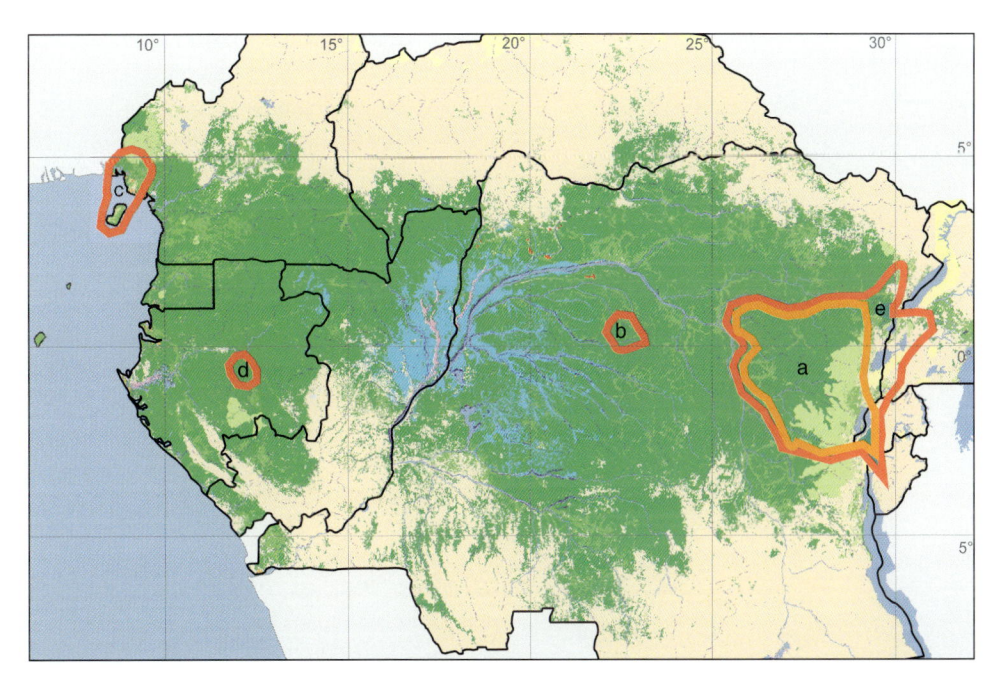

6 THE FAUNA

6.79. *L'Hoest's Monkey* Cercopithecus lhoesti *is the commonest monkey and the easiest to see in the Nyungwe Forest in Rwanda.*

6.80. *The Sun-tailed Monkey* Cercopithecus solatus *was only discovered in 1984, and is endemic to central Gabon. It is known in the Lopé National Park, but seems to be more abundant in the "Forêt des Abeilles" just to the east.*

The seven other *Cercopithecus* species of Central African forests are all arboreal. They have long tails and long hind legs. They often form multispecific groups and readily associate with *Cercocebus* and *Lophocebus* species, or mangabeys, and colobus monkeys. They are represented by a great many regional forms, sometimes very distinctive, which shows that they are on route to speciation. They can be divided into groups, differentiated by size and by their distribution in the forest.

The first group comprises the Red-eared Monkey *Cercopithecus erythrotis*, the Moustached Monkey *Cercopithecus cephus* and the Red-tailed Monkey *Cercopithecus ascanius*. The first lives between the Cross and the Sanaga rivers and on the island of Bioko; the second between the Sanaga and the Congo-Ubangi system; the last occupies the area from south and east of this system to Uganda, Rwanda and Burundi. They are the smallest arboreal guenons and the weight of males is no more than 4.5 kilos. They inhabit essentially the lower and middle strata of the canopy of lowland forests. In the Albertine Rift region, however, the Red-tailed Monkey is encountered up to 2,000 metres, and even somewhat above, notably in the Bururi and Kibira forests in Burundi.

The second group includes the heaviest species, of which males can reach 10 kilos. Absent south of the Congo River, except between the Lualaba and the Lomami rivers, it comprises the Putty-nosed Monkey *Cercopithecus nictitans* (FIGURE 6.82) and the Gentle Monkey *Cercopithecus mitis*. The first is known from the Sanaga River and the Atlantic coast to the Congo and its tributary, the Itimbiri. It inhabits the middle stratum and upper canopy of lowland forests. The second is known from the Itimbiri and Congo rivers to eastern Africa. It, too, inhabits middle and upper levels of the vegetation, but it can also be found in old lowland forests as well as in montane forests up to 2,600 or even 2,800 metres, in riverine forests within grasslands,

Chances to watch monkeys in ECOFAC sites

	Dja	Odzala	Lopé	Monte Alén	Ngotto
Cercocebus albigena	8.33	3.3-14.3	4.5-5.9	–	3.1-25
Cercopithecus cephus	5.9	1.7-5.9	8.3-11?1	3.3-4.2	2-5.5
Cercopithecus nictitans	4.5	1-5	2-3	3-4.2	1.1-1.8
Cercopithecus pogonias	7.1	6.2-10	8.3-50	5.3-8.3	1.7-3.6

These figures give the mean number of kilometres one needs to walk to see one of the species. They show that Odzala-Kokoua National Park and Ngotto Forest offer the best opportunities to watch monkeys; Lopé the least good (according to Annie Gautier-Hion and Marc Colyn).

in semi-floating stands of *Alchornea cordifolia* and in large, extensive papyrus swamps. Its form *kandti*, with its remarkable golden brown back – hence its common name Golden Monkey – is found exclusively in the bamboo thickets of the Virunga, the Gishwati and Nyungwe forests in Rwanda and the Echuya Forest in southern Uganda. The monkeys of this group live in groups of about twenty individuals and feed essentially on fruits, leaves and arthropods.

The third group comprises two species: the Crowned Monkey *Cercopithecus pogonias* (FIGURE 6.81) and Wolf's Monkey *Cercopithecus wolfi*. The first inhabits lowland forests from Cross River and the Atlantic coast to the Congo and Itimbiri rivers. The second inhabits forests located south of the Congo River and extending from the Itimbiri to the Albertine Rift region. This latter species is represented, however, by its nominate form south of the Congo River and its form *denti* to the east. According to Kingdon, these two forms are to be considered as separate species. Whatever the case may be, these monkeys are slightly larger in size than those of the *cephus-ascanius* group, and they also inhabit the middle and upper strata of the canopy. Their food is composed essentially of fruits and invertebrates. They live in groups of five to 20 individuals. In Nyungwe Forest, Wolf's Monkey can be found up to 2.400 metres altitude.

6.81. A Crowned Monkey Cercopithecus pogonias *in the top of a tree in Odzala-Kokoua National Park.*
6.82. A Putty-nosed Monkey Cercopithecus nictitans, *surprised in the same tree.*

6.83. *The Guereza Colobus* Colobus guereza *can be observed along the edge of clearings in Odzala-Kokoua National Park where it often comes down to the ground. At the slightest alarm, it flees into the trees.*

The Colobinae subfamily is older than the guenons and baboons. It includes medium-size monkeys with a long tail and long limbs that are distinguished from the guenons and baboons by three main characters: the absence or strong reduction of the thumb, the absence of cheek pouches and the presence of a stomach made of three to four pouches harbouring a bacterial flora able to digest cellulose. Colobid monkeys are vegetarians and have a slow digestion related to that of ruminants. They feed in part on leaves and buds, but also a lot on seeds. The four species present in Central Africa have nearly the same weight: eight to 13 kilograms for males, six to eight for females. Sexual dimorphism is therefore less pronounced than in the guenons and baboons.

The Guereza Colobus *Colobus guereza* occurs in all of the northern part of the Central African forests, from southern Cameroon – it is known from the Dja – to Uganda and even Kenya and Ethiopia. It inhabits lowland forests and montane forests up to 3,000 metres altitude, but it is absent from the Rwenzori. It is a black monkey with a large white cape, a white face and a long pendulous tail terminating in a thick tuft of long white hairs. (FIGURE 6.83). Because of its very striking costume, the species suffers heavy pressure in many areas.

The Angola Colobus *Colobus angolensis* replaces the Guereza Colobus south of the Congo River, and is also found in the north-eastern part of the Democratic Republic of the Congo, in several forest blocks of the Albertine Rift region, notably the Rwenzori and Nyungwe forests, where it is found up to 3,000 metres altitude. A small population survives in the Sango Bay forests on the western edge of Lake Victoria. It is also black, but with long tufts of white hairs on its cheeks and shoulders. The two species generally live in groups of five to 15 individuals, some of which can be easily observed.

The Black Colobus *Colobus satanas* is known from Atlantic coastal forests from the Sanaga to the Kouilou. They are encountered in the Lopé National Park and at Dja faunal reserve, as well as around Odzala-Kokoua National Park. They also occur on the island of Bioko.

Pennant's Red Colobus *Procolobus pennanti* is distinguished from the black and white colobus monkeys by the colour of its pelage, which contains brownish red, and by the fact that the females have cyclic swellings of the sexual skin. In Central Africa, it is known from the isle of Bioko and from the centre of the Congo to East Africa. It is represented by a dozen forms, some of which are elevated to the rank of species by some scientists. Kingdon, for example, recognises four for Central Africa alone: *tholloni* south of the Congo River, *preussi* in Cameroon, *pennanti* on Bioko and on the right bank of the Congo, *oustaleti* to the north and east of the river. In general, this species inhabits the upper strata of the canopy of lowland forests, especially old forests. Some forms also inhabit riverine forests or submontane forests up to 1,500 metres. Most forms live in groups of a few tens of individuals with several adult males. In the case of *tholloni* and *tephrosceles* – the eastern African form found at Kibale and in Burundi – larger groups, of more than 100 individuals, are found. More mobile and noisier than black and white colobus monkeys, red colobus monkeys feed on seeds and leaves, but the form *tholloni* is apparently more insectivorous.

The hominid family, presently represented in Africa by the Gorilla *Gorilla gorilla*, the Chimpanzee *Pan troglodytes*, the Bonobo or Pygmy Chimpanzee *Pan paniscus* and Man *Homo sapiens*, appeared 25 million years ago, thus at the beginning of the Miocene. Until 10 million years ago, it was the dominant family among Old World primates. Since then, it has given way to the cercopithecids and the number of species has been greatly reduced. The remaining three apes that people the forests of Africa are about the same age and apparently separated from the human branch seven or eight million years ago. Formerly, these apes were placed in a separate family: the Pongidae. But molecular biology has recently shown that their relationship with Man is so close that they are clearly part of one and the same family: the hominids. This relationship is characterised by the absence of a tail, a more or less thickset body, a large round head with a voluminous brain, long arms equipped with hands and a relatively long gestation. The Pongidae family is reduced to the Orang-utan *Pongo pygmaeus*. Even though this species has survived only in Asia, it nevertheless represents a branch of African origin, separated around 17 million years ago.

The heaviest of the apes is the Gorilla (FIGURES 6.84-87). It is distributed discontinuously from eastern Nigeria to Uganda and Rwanda (FIGURE 6.88), but it does not occur south of the Congo River. It comprises three main geographical forms.

The western lowland Gorilla *Gorilla gorilla gorilla* (FIGURES 6.84-85) lives west of the Congo and Ubangi rivers. They are found therefore in southern Cameroon, in the south-west of the Central African Republic, Equatorial Guinea, Gabon and the Congo. Small isolated populations also survive on the border between Cameroon and Nigeria. Recently, it apparently was also seen on the east bank of the Ubangi and, at the very end of the 19th century, it still occurred along the Uele River north of the present Democratic Republic of the Congo. According to Annie Gautier-Hion, Marc Colyn and Jean-Pierre Gautier, the entire population of this form is estimated at 50,000 individuals.

> At Ngotto, in the Central African Republic, research conducted within the framework of the ECOFAC programme shows that the Red Colobus *Procolobus pennanti* is strictly limited to riverine forests. This particularity, which merits further study, could be based on the higher nutritional quality of foliage found at the water's edge. It could also indicate that this population is genetically different from others and constitutes a separate subspecies. It could also be the result of heavy hunting pressure, since colobus monkeys are in fact very sensitive to hunting.

6.84. *Two females and a young western lowland Gorilla* gorilla gorilla *on the edge of a clearing at Odzala-Kokoua National Park.*

6 THE FAUNA

6.85. An adult male Gorilla of the nominate form in Odzala-Kokoua National Park. The study conducted by the ECOFAC programme shows that Maya-Nord clearing was probably visited by about 500 Gorillas over eight months' time. Given that the density of Gorillas at Odzala is of the order of ten individuals per square kilometre, Maya-Nord must drain Gorillas from an area of around 5,000 hectares. Groups can contain as many as 30 individuals, but, on average, they are made up of six. Each of these groups is thought to have a vital domain of 700 to 1,400 hectares. They are composed of 56% immatures, which indicates a very high rate of reproduction, probably related to the abundance of food offered by Marantaceae forests and to low hunting pressure.

The eastern lowland Gorilla *Gorilla gorilla graueri* is more massive and blacker than the preceding form (FIGURE 6.86). It inhabits the eastern part of the Democratic Republic of the Congo, from the Itombwe, in the south, to Maiko National Park, in the north. Its very fragmented population was of the order of 17,000 individuals, of which some 4,000 inhabit Kahuzi-Biega National Park. This population has nevertheless suffered from the war and, in the upper part of the park, a great many individuals have been massacred during recent years, including groups habituated to the presence of Man.

The mountain Gorilla *Gorilla gorilla beringei* is somewhat smaller, but is immediately distinguishable by its long black pelage and the imposing helmet of the males (FIGURE 6.87). This form inhabits the Virunga volcanoes in the Democratic Republic of the Congo, Rwanda and Uganda, as well as Bwindi Impenetrable National Park in Uganda. It thus exists only on the eastern edge of the Albertine Rift, separated from the eastern lowland Gorilla by lava fields and the low plains of the Rutshuru. In all, its population is about 600 to 650 individuals. In spite of the armed conflicts of the Great Lakes region, it seems to have increased by 10% over the past ten years. According to data collected by the International Programme for the Conservation of Gorillas (IPCG) – a joint initiative of the African Wildlife Foundation (AWF), Fauna and Flora International (FFI) and the World Wide Fund for Nature (WWF) – and by the Dian Fossey Gorilla Fund-International, the population of the Virunga has gone from 320 to 355 individuals. Yet during this same period at least 15 Gorillas were killed as a direct result of the war. In addition, the last complete censuses dates from 1989 and it is very probable that there are at present more than 355 Gorillas since some groups probably escaped the eye of recent observers. Without the combined efforts of the guards and trackers, of whom many were killed in the exercise of their function, and the responsible conservation bodies of the Democratic Republic of the Congo, Uganda and Rwanda, the status of this subspecies would certainly be much less brilliant. The small differences that can be observed between the mountain Gorillas of Bwindi Impenetrable National Park and those of the volcanoes, arise from the fact that these two populations have probably been isolated for around 500 years and subjected to different environmental pressures.

Gorillas inhabit mainly lowland forests, preferably mixed forests, where they are especially fond of open formations with a rich herbaceous vegetation. Their distribution is therefore fairly irregular and the density of populations is very variable. They are thus abundant in the Marantaceae forests of Odzala National Park (see box page 220) and in some places also inhabit riverine and swamp forests, notably raffia palm groves of the Dja, where they reach a density of around five weaned individuals per square kilometre. This preference for humid forests is perhaps a result of hunting pressure, but Liz Williamson thinks that this tendency can be linked to an abundance of construction material for nests. Annie Gautier-Hion also estimates that the main factor is the presence of an abundant herbaceous vegetation. The western Gorillas willingly visit clearings, but pass very little time in them. The eastern lowland Gorilla inhabits lowland forests as well as montane forests. The mountain Gorilla is strictly montane. The Virunga population lives at altitudes between 2,300 and 3,000 metres, sometimes even as high as 3,700 metres, in bamboo thickets, *Hagenia* forests, *Hypericum* thickets and arborescent Ericaceae thickets. Those of Bwindi Impenetrable National Park inhabit transition forests and montane forests.

The diet of the Gorilla is vegetarian, but occasionally includes a few ants or termites. In lowland forests, it includes many fruits, and also herbaceous plants, notably Marantaceae and Zingiberaceae species. Mountain Gorillas feed nearly exclusively on herbaceous plants which they find in clearings. During some periods of the year, they also consume bamboo shoots. It is even only when the bamboo sprouts that Gorillas appear in Mgahinga Gorilla National Park in Uganda. The search for minerals seems important: in the clearings at Odzala, Gorillas feed essentially on plants containing high levels of minerals and they often lick the ground.

Gorillas live in family groups, guided by an adult male, recognisable by its silvery back, very rarely by two males. He is accompanied by several females with their young. The size of the groups can attain 15 to 25 individuals, exceptionally thirty, but many groups are smaller and solitary males may be encountered.

Each group occupies a vital space whose extent can reach 2,000 hectares in lowland forest, but is sometimes not more than 300 hectares in montane forest. Even though the silver-back males vigorously defend members of their group, they do not defend their vital space against neighbouring groups. Generally, groups limit conflict by avoiding each other or by using intimidation behaviours. Battles are rare. In the Odzala clearings, it is not rare to see several groups file by in the course of the same day and two or three groups sometimes may be seen feeding simultaneously.

The Gorillas' day comprises a period of activity in the morning and one in the afternoon, broken by a period of rest in the middle of the day. At the end of the day,

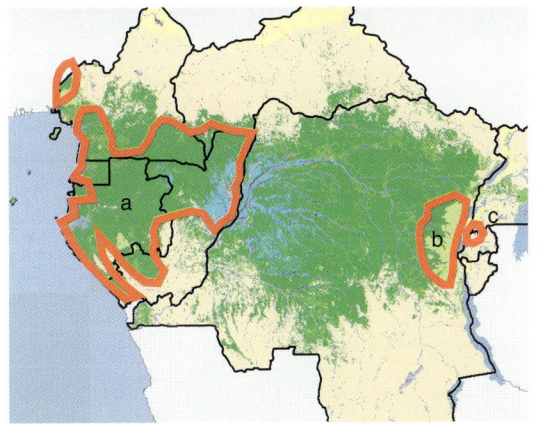

6.86. *A male eastern lowland Gorilla* Gorilla gorilla graueri *in the region of Kahuzi-Biega.National Park.*

6.87. *A group of females of the Mountain Gorilla* Gorilla gorilla beringei *in the Virunga.*

In these two eastern forms, groups do not contain more than 36 to 43% of immatures, which may be the result of their instability following human pressures. In fact, these animals live in regions where the density of human populations can be a hundred times higher than in the Congo or in Gabon.

6.88. *The range of the Gorilla:*
(a) *Western lowland Gorilla* Gorilla gorilla gorilla,
(b) *Eastern lowland Gorilla* G. g. graueri,
(c) *Mountain Gorilla* G. g. beringei.
Recent studies have attempted to describe other forms. Gorillas of Bwindi Impenetrable National Park were separated from those of the Volcanoes and other populations were separated in Lower Guinea.

6.89. A Chimpanzee *Pan troglodytes* in a forest of north-eastern Gabon.

Densities of Chimpanzee and Gorilla populations

In the ECOFAC sites the following densities have been found (in numbers of individuals per square kilometre):

	Gorilla	Chimpanzee
Dja	1.02-2.86	0.6-1.04
Odzala	1.04-10.3	1.6-4.6
Lopé	0.3-1	0.2-1.1
Monte Alén	0.3-4.5	0.04-1.06

These estimates are based on the counting of the nests which these animals build to spend the night. They are subjected to some errors, since it is difficult to know the exact duration of the nests, which can change from one place to another according to the ambient humidity and the materials used. Some animals also build more than one nest in a night, and tree nests of gorillas, frequent in Ngotto, can easily be mistaken for Chimpanzee nests.

the groups stop their activities and construct the rudimentary nests of branches or plants in which they will spend the night.

The Chimpanzee occurs from Senegal to Uganda, Rwanda, Burundi and Tanzania but, just like the Gorilla, it remains confined to the north of the Congo River. From the Cross River to the Congo and Ubangi rivers, it is represented by the nominate form *troglodytes* (FIGURE 6.89); east of the Ubangi by the form *schweinfurthi*. The latter is somewhat smaller, but apart from of that, the differences are minimal.

The first inhabits not only the large forest massifs where it prefers the mixed forests but also forest galleries in grasslands or wooded grasslands, and it does not hesitate to cross open spaces, even fields. The Chimpanzee has a more homogeneous distribution and is more numerous in old forests with open understorey than are Gorillas. In the mountains of the Albertine Rift, it occurs up to nearly 3,000 metres altitude, notably in the Rwenzori. Outside of forests where some groups have become more or less habituated to human pressures, it is quite rare to see Chimpanzees and most of the time their presence is betrayed only by their clamouring which can be heard at a great distance.

The Chimpanzee often moves about on the ground leaning on the articulations of its fingers. In spite of its weight, it is largely arboreal and most of its food comes from the canopy. Half of the diet is composed of fruits, but it is completed by leaves, flowers, buds, some bark, insects – mostly termites – nestlings and eggs, and occasionally mammals. Sometimes chimpanzees also hunt other primates, mostly when fruit is scarce. In some places this hunt is solitary and opportunistic. Elsewhere, it is practised in a concerted and co-operative way. Some males join up into groups in order to attack red colobus monkeys, guenons or even baboons. They mostly catch young animals that they capture by the hind feet and kill by hitting the head against a large branch or against the ground. Sometimes they also catch rodents or young forest duikers and cases of cannibalism have been observed. More rarely, they kill their prey by hitting them with stones or sticks. Termites, ants and bees are extracted from their nests with the aid of twigs. Nuts are cracked by use of clubs or stones. This diet varies, however, from one season, and from one forest, to the next.

In general, chimpanzees live in communities of 15 to more than 100 individuals that defend a vast territory extending to more than 20 square kilometres. Depending on the season, these split up, however, into smaller groups – mostly in the dry season when the diet is mainly fruits – but a good many other factors intervene. Females with their young generally remain at the centre of the territory, while the males often move to its limits to watch over the neighbouring groups. Very noisy, chimpanzees call from morning to night, after the rain, when groups meet or split up, when they hunt and often for reasons that escape us. They feed mostly in the morning and rest around the middle of the day.

As for the Bonobo, also called Pygmy Chimpanzee (FIGURE 6.90), it has long been considered as a geographic race of the common Chimpanzee. Today, we know that it represents a species in its own right, separated from the common Chimpanzee for around 1.5 million years. Less heavy than the Chimpanzee, the Bonobo is not smaller but distinctly more elongated. Its limbs are long and its skull more slender; its brow ridges and its muzzle less prominent. Its coat is black, just like most of its face and, the long hair lying to the sides of its head give it a characteristic appearance.

It inhabits the Congo basin south of the river, but its distribution is presently very fragmented. A small relict population survives in the Marungu Mountains on the edge of Lake Tanganyika (FIGURE 6.91). Its habitat is constituted of terra firme forest, inundatable forests not being visited except seasonally. Very recently, a group was also discovered in grassland.

Like the Chimpanzee, the Bonobo feeds mostly on fruits and seeds, but it also consumes many leaves, buds and stems. It likes Marantaceae plants. Although less carnivorous than Chimpanzees, it also eats insects, myriapods, snails, small rodents, shrews, squirrels and occasionally forest duikers. It does not hunt in a concerted way, however.

Like the Chimpanzee, the Bonobo lives in communities that can have as many as 120 individuals and exploiting a territory of the order of 5,000 hectares. This is visited in its entirety by both males and females. Often, territory is shared with neighbouring communities. Individual groups number generally 5 to 15 animals and are relatively stable. Groups made up only of males or of females are rare. Males maintain rather distant relationships of a hierarchical type. Females, on the contrary, develop very close relationships and because of this assure the cohesion of the group. When neighbouring communities meet, their contacts are generally peaceful and they can even associate with each other temporarily. Unlike the Chimpanzee, Bonobo females can mate throughout their cycle, independently of the state of swelling of their sexual skin. Face-to-face copulation is not rare and homosexual relationships between females are fairly frequent.

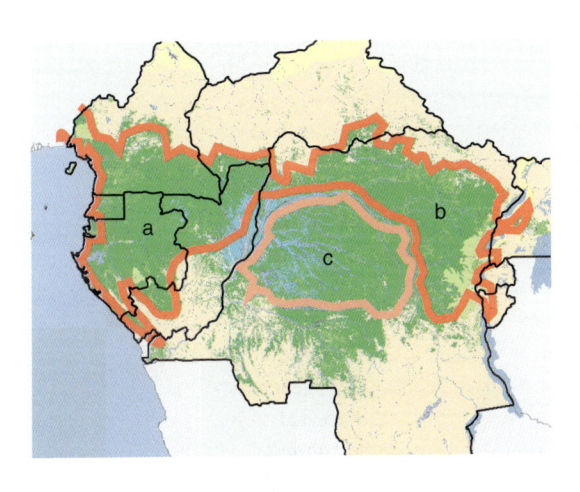

The status of the Bonobo is worrying: its populations are diminishing rapidly and are becoming more and more fragmented. It suffers intolerable hunting pressure, mostly because of the heavy commercialisation of hunting and the decline of all activities which, at other times, furnished alternative resources to the human populations of the central basin of the Congo. Unfortunately, this species is only weakly represented in Salonga National Park. Of course, there are Bonobos in captivity and they reproduce, but in the long term this is not what is going to save the species.

6.90. *The Bonobo* Pan paniscus *is one of the most endangered primates, and only a small fraction of its remaining population is inside Salongo National Park.*

6.91. *The Chimpanzee* Pan troglodytes *inhabits the forests north of the Congo River and the Albertine Rift region. West of the Ubangi it is represented by the nominate form (1a); in the east by the form* schweinfurthi *(1b). In West Africa, its range extends to the south of Senegal. The Bonobo (2) is confined to the central basin of the Congo south of the river, but a small population apparently existed on the edge of Lake Tanganyika in the Marungu Mountains.*

6.92. Males of the Hammer Bat Hypsignatus monstruosus, *the largest African bat, have developed special organs for amplifying their voice: an enormous larynx that invades nearly the entire thoracic cage, inflatable pouches covering the nasal cavities and sides of the neck, enormous lips transformed to control the emitted sound. Jonathan Kingdon compares them to veritable flying loudspeakers.*

6.93. Insectivorous bats hiding during day-time.

Bats

Being very abundant in tropical forests, more so than in any other habitat, bats have an important biological role. They are poorly known, however, because of their nocturnal and air-borne life. In fact, for many tropical species, we know only that they exist, but we have little or no knowledge of their distribution, ecology or behaviour. And yet, bats are important pollinators and dispersers of forest plants. Some are redoubtable insect predators. And some even attack vertebrates. As is the case for many other groups of animals, sub-Saharan Africa is less rich in bats than South America, however, especially in frugivorous species; moreover, the fauna of Africa is more closely related to that of Asia than to that of America.

Living species of bats divide into two very different groups and probably have totally different origins. They are: the large frugivorous bats or Megachiroptera and the small insectivorous bats or Microchiroptera.

All the Megachiroptera belong to the Pteropodidae, a family strictly limited to the Old World. According to Kingdon, they appeared 30 million years ago, perhaps from lemurines capable of gliding in the same way as flying mice. Today, 13 species are known from the forests of Central Africa, including some that frequent large closed forests less than grassland-forest mosaics. One species, *Myonycteris brachycephala*, is endemic to São Tomé and Principe.

Their food, located by well developed sight and smell, consists exclusively of fruits and flowers, from which they consume only the juice, pulp or nectar. Only the Rousette Bat or Egyptian Fruit Bat *Rousettus aegyptiacus*, which lives in caves during the day, has developed a primitive system of echolocation based on tongue clicking.

By visiting flowers on trees these bats often act as effective pollinators. When they tear off seeds in flight, they effectively contribute to the reforestation of cleared spaces. Some species live in small groups. Others form assemblages numbering several tens or even hundreds of thousands of individuals. At nightfall these groups disperse over large territories in search of food. Part of the assemblage of Straw-coloured Fruit Bats *Eidolon helvum*, which resides during the day in

Kapok Trees near the port of Bujumbura, cross Lake Tanganyika to feed in the Democratic Republic of the Congo more than 25 or 30 kilometres away. This species, which reproduces in equatorial regions, migrates far north and south outside of the breeding period.

Species that live in small groups are very vocal and the males announce their presence by distinctive calls that can be confused with the sounds of amphibians (FIGURE 6.92).

The Nectar Bat *Megaloglossus woermanni* is a very particular species that feeds only on nectar. It belongs to the subfamily Macroglossinae, characterised by small size, a long pointed muzzle and a long tongue.

The Microchiroptera (FIGURE 6.93) are probably descended from insectivores. They probably appeared 70 to 100 million years ago and horseshoe bats very much like our present-day forms already occurred 40 millions years ago. They are therefore much older animals than rousette bats. They are divided into several families of which the best represented are the Molossidae and the Rhinolophidae. Many African genera also occur in Asia and the African fauna does not include blood eaters, like the vampires of South America nor hardly any carnivorous species. Only the Heart-nosed Bat *Cardioderma cor*, of the Megadermatidae family, occasionally attacks other bats.

Most bats capture their prey with their feet or caudal membrane. Some capture insects on leaf surfaces. These are mainly the Nycteridae or slit-faced bats, endemic to Africa, Rhinolophidae of genera *Rhinolophus* and *Hipposideros*, as well as some Vespertilionidae of genus *Myotis* and the continent's two megadermatid species. Other bats also hunt flying insects. These include mainly species of Vespertilionidae, Emballonuridae, Molossidae and Rhinolophidae. How these animals divide up the aerial space and available food between them remains very poorly known, however.

Insectivores

In the dense forests of Central Africa, this very heterogeneous order of very ancient small mammals is represented essentially by the shrews or Soricidae. In sub-Saharan Africa, they number at least 140 species of nine genera, with the great majority of species belonging to genus *Crocidura*. Many shrews prefer humid habitats. Some are burrowers, while still others go up into trees. They are thus well represented in dense humid forests, where seven to ten species can in general be found at the same site. They live alone, however, and display a very high level of aggressiveness between individuals of the same species. Some species have a vast range, others are, on the contrary, very localised: the Rwenzori Shrew *Ruwenzorisorex suncoides*, for example, is endemic to the Albertine Rift; the Hero Shrew *Scutisorex somereni* inhabits only inundatable forests of the north-eastern part of the Congo basin, the Rift region and a part of Uganda; genus *Paracrocidura* is known only from the Cameroon region, the eastern part of the Democratic Republic of the Congo and Uganda.

Very different are the Giant Otter Shrew *Potamogale velox* (FIGURE 6.94) and the Rwenzori Water Shrew *Mesopotamogale ruwenzorii*. These two relict species belong to the family Tenrecidae, limited to Africa and Madagascar. The first is found from Nigeria to Uganda. The second, distinctly smaller, inhabits mountains of the Albertine Rift. Both are nocturnal, hunting somewhat like otters in forest streams and feeding on insects, crustaceans, mollusks, amphibians and fish. The larger species has a vertically flattened tail with which it can swim somewhat like a fish.

The Macroscelidae – elephant shrews or sengis

These small diurnal, terrestrial mammals, the size of a rat or a mouse, with long legs and a long tapering muzzle, feed on arthropods of the litter. They represent a strictly African family whose genealogy can be traced back over several tens of millions of years. Their main diversification took place in the eastern and southern parts of the continent. They apparently did not penetrate Central Africa except very marginally. The Four-toed Elephant Shrew *Petrodromus tetradactylus*, a consumer of ants and termites, can, nevertheless, be encountered in Caesalpiniaceae forests south of the Congo River. The much larger Chequered Elephant Shrew *Rhynchocyon cirnei*, of the Rhynchocyoninae subfamily, is found in forests east of the Congo River and between the Congo and the Ubangi rivers. Its diet also includes caterpillars and grasshoppers.

Studies completed within the framework of the ECOFAC programme in Ngotto forest by Marc Colyn's team from the University of Rennes show that shrews are distinctly more diversified in old forests, called "primary", than in young or disturbed forests. Some species are also more frequent along watercourses. In this way these studies add an element to the value of old and riverine forests for conservation. Some species of shrews can therefore be good indicator species in the evaluation of the conservation value of a forest.

6.94. The Giant Water Shrew Potamogale velox *has a vast range and occurs even in forest galleries far outside of the main forest massif. It is fairly common in Gabon.*

6 THE FAUNA

6.95. The Fire-footed Rope Squirrel Funisciurus pyrropus *occurs in all the forests of Central Africa. It likes palm thickets under a closed canopy. Three-quarters of its food is composed of fruits often gathered on the ground.*

6.96. The African Giant Squirrel Protoxerus stangeri *is also mainly frugivorous. It is recognisable by its large size and long, barred tail.*

6.97. Lady Burton's Rope Squirrel Funisciurus isabella *is recognisable by the four dark stripes that run from head to tail. This rope squirrel inhabits Lower Guinean forests, from Nigeria to the Congo River, and feeds essentially on fruits gleaned in the canopy.*

Rodents

The 17 species of squirrels or Sciuridae that haunt the forests of Central Africa nowadays are all apparently derived from terrestrial squirrels coming from Asia some ten million years ago. Today, they are divided into three groups.

The *Funisciurus* or rope squirrels are small squirrels with a rather unbushy tail often held curved forward over the back. They live in trees and frequent dark places, but they do not hesitate to move about on the ground. Seven species are known from Central Africa. Some, including the Fire-footed Rope Squirrel *Funisciurus pyrrhopus* (FIGURE 6.95), are arboreal and frugivorous. Others, like the Ribboned Rope Squirrel *Funisciurus lemniscatus,* are strongly terrestrial and largely insectivorous. Still others, like Lady Burton's Rope Squirrel *Funisciurus isabella* (FIGURE 6.97), are almost exclusively arboreal and eat nothing but leaves and fruits. Carruther's Mountain Squirrel *Funisciurus carruthersi* is endemic to the forests of the Albertine Rift where it is found between 1,500 and 2,800 metres altitude. The African Pygmy Squirrel *Myosciurus pumilio* is a form near to the species of genus *Funisciurus*. It is very small, with a thin tail, and is limited to the vicinity of the Gulf of Guinea, from eastern Nigeria to Gabon. In the east, it does not cross the limit of the Ogooué basin.

The *Paraxerus,* or bush squirrels, are fairly variable in size as much as in their pelage. Like the *Funisciurus* species, they have a liking for the lower levels of the vegetation and for dark places. They feed on fruits, seeds, nuts, mushrooms, gums and invertebrates. When danger threatens, they hide behind a branch and remain motionless. Two species inhabit Central African forests north of the Congo River: the Green Squirrel *Paraxerus poensis* occurs from the Ubangi River to Sierra Leone; Boehm's Squirrel *Paraxerus boehmi* has its range from the Ubangi to the Albertine Rift. All other species of this group occur in the forests and grasslands of East Africa.

The giant squirrels and sun squirrels of genus *Heliosciurus* have powerful teeth and jaws that they use to partake of the hardest fruits. Strictly arboreal, they live essentially at the top of the canopy. When danger threatens, they flee with great agility. Three species are known from Central Africa. The Red-legged Sun Squirrel *Heliosciurus rufobrachium* is the most common. The Gambian Sun Squirrel *Heliosciurus gambianus* is more of a grassland species, but in some areas it is also found in dense forests. As for the Rwenzori Sun Squirrel *Heliosciurus ruwenzori*, it is an endemic of the forests of the Albertine Rift where it lives between 1,500 and 2,800 metres altitude. Near to this group are the Biafran Bight Palm Squirrel *Epixerus wilsoni*, inhabiting the forests lying between the Sanaga and the

Congo rivers, as well as the African Giant Squirrel *Protoxerus stangeri* (FIGURE 6.96). This species inhabits most of the dense forests of Central Africa and can be found up to 2,000 metres altitude, and even a little higher in the Nyungwe Forest in Rwanda.

Whichever group they belong to, the squirrels, being diurnal, are fairly visible animals and are identifiable with the help of good binoculars. They often advertise their presence with their strong, piercing vocalisations. Some species build nests of branches. Others occupy natural cavities. Because all of them consume some amount of fruits and seeds from trees, squirrels certainly play a role in their dispersal.

Very different are the Anomaluridae, or flying mice. These are largely nocturnal or crepuscular rodents endemic to sub-Saharan Africa. They are much older than the squirrels, and were probably already in existence 30 million years ago. Today, six species inhabit the forests of Central Africa. All except one, the Cameroon Scaly-tail *Zenkerella insignis*, are distinguished by the presence of a large membrane stretched between the forelegs, hind legs and base of the tail. The anomalurids are thus capable of gliding from one tree to another. They differ from each other mainly in size, habitat and food. With their somewhat beaver-like teeth they are capable of eating bark, as well as "pruning" away branches that get in their path. By day, they generally take shelter in natural cavities, often in groups of five to ten individuals. The smallest species can also simply slide behind detached bark. Lord Derby's Anomalure *Anomalurus derbianus* is the largest and probably the commonest species. It occurs even in montane forests. Beecroft's Anomalure *Anomalurus beecrofti* is a little smaller and also has a vast range, but seems attracted by the presence of *Elaeis* palms. Besides bark, it also feeds on palm nuts, leaves, fruits and sometimes insects. The other species are much smaller and less well known.

There are a good many other rodents that occur in the forests of Central Africa. In the canopy one can find several species of dormice of genus *Graphiurus*. The Brush-tailed Porcupine *Atherurus africanus* (FIGURE 6.99) is a large nocturnal rodent and a close relative of the porcupines of genus *Hystrix*, that inhabit grasslands. Smaller than the latter, the Bush-tailed Porcupine readily climbs trees.

Lastly, there are the innumerable species of rats and mice. They are less diversified in the forest than in grasslands, but the forests of Central Africa have a few dozen species mainly of genera *Prionomys*, *Deomys* and *Cricetomys* (FIGURE 6.98), containing fairly old species, and *Colomys*, *Grammomys*, *Heimyscus*, *Hybomys*, *Hylomyscus*, *Malacomys*, *Oenomys*, *Praomys*, *Stochomys* and *Thamnomys*, which represent a more recent ensemble – the Muridae strictly speaking – whose presence in Africa probably does not date back more than six million years at most. From the mantles can be added species of *Lemniscomys*, *Mastomys* and *Mus*. Some species are omnivorous. Others, on the contrary, are very specialised. Their systematics are very difficult, however, and some species cannot be identified without an osteological study of the cranium or a genetic analysis. In general, 10 to 15 species can be found in the same site, but they vary according to type of forest and mostly according to its state of conservation.

The team led by Annie Gautier-Hion and Marc Colyn demonstrated that the richness and the density of populations of Muridae at Ngotto is greater in disturbed forests than in "primary" forests and that forest-grassland zones of contact are particularly rich. Rodent faunas varied markedly from one site to another and from one season to another, however. They are therefore not suitable for use in biological evaluations, notably for evaluations of forest exploitation.

6.98. *Emin's Pouched Rat* Cricetomys eminii *may reach a length of 70 centimetres from head to tail. This nocturnal animal feeds on fruits, seeds, leaves and roots that it amasses in its cheek pouches before stowing them away in a subterranean cache.*

6.99. *The Brush-tailed Porcupine* Atherurus africanus, *also strictly nocturnal, feeds mostly on roots, tubers and fallen fruits. It is a slow reproducer that unfortunately suffers very heavy hunting pressure. Like Emin's Pouched Rat, it has a liking for man-modified habitats and sometimes causes serious damage to crops.*

6 THE FAUNA

Carnivores

A fauna rich in rodents and herbivores inevitably attracts a fair quantity of predators. They are found in the canopy as much as on the ground and, apart from some Canidae, all the families are represented.

The Mustelidae, originating from Eurasia, have probably been present in Africa for some five million years. On the margins of Central African forests can be found the Striped Weasel *Poecilogale albinucha*. In rivers, swamps and small lakes of the forest also lives the Swamp Otter *Aonyx congica*, while the Spot-necked Otter *Lutra maculicollis* can be fairly abundant in the Great Lakes region, notably Lake Victoria. As for the Ratel *Mellivora capensis*, it can be encountered everywhere, including in dense forest and in Afro-alpine heaths. All these animals are solitary. Males and females live apart, but have territories that partially overlap.

The mongooses, or Herpestidae, represent, on the contrary, an authentic African family, at least 30 million years old, a single genus of which – *Herpestes* – was able to colonise Asia. Many of the species inhabit arid or sub-arid habitats and grasslands. In this way the Ichneumon Mongoose *Herpestes ichneumon* penetrates the forest massif by means of clearings. Some species live in the forest, however. The Long-snouted Mongoose *Herpestes naso* inhabits forests north of the Congo River. It has a liking for the vicinity of watercourses. The Slender Mongoose *Herpestes sanguinea*, smaller and very variable in colour, can be encountered everywhere and is partially arboreal. The three or four cuisimanses of genus *Crossarchus* occupy all of the forest massif, but are separated by large watercourses; *obscurus* and *platycephalus* – perhaps conspecific – live west of the Congo and Ubangi rivers, *ansorgei* south of the Congo and *alexandri* north of the river, between the Ubangi and Lake Victoria. They are diurnal terrestrial species that feed on worms, beetle larvae, snails and small vertebrates in the litter and in decomposing stumps. They live in family groups of 10 to 20 individuals. The Marsh Mongoose *Attilax paludinosus* has a vast range and can be encountered in all of the humid habitats. It has been found up to 2,300 metres in Nyungwe Forest in Rwanda. It is mostly nocturnal, but can be found by day, alone, in pairs or in small family groups. Its food is quite varied and includes fish, crabs, reptiles, birds, arthropods and fallen fruits. The Black-legged Mongoose *Bdeogale nigripes* (FIGURE 6.100) inhabits dense forests from the Cross River to the Semliki River, but is not recorded south of the Congo River. It is a species difficult to see, whose diet is essentially arthropods (mostly ants), molluscs, crabs, rodents and amphibians, occasionally also fallen fruits.

The genets and civets, or Viverridae, represent a group of "primitive" carnivores. They are probably of Eurasian origin and may have entered Africa 25 million years ago. As Kingdon points out, they might be to the felines what the galagos are to the monkeys. They are all nocturnal animals.

The genets, essentially arboreal, occupy all the habitats of the continent, but in the forests of Central Africa they are represented by the Servaline Genet *Genetta servalina*, the Blotched Genet *Genetta tigrina* and the Giant Servaline Genet *Genetta victoriae*. The first is linked to the Congolese forests, from Nigeria to Kenya. The second has a vast range, in forests as well as in grassland regions. The latter is, on the contrary, limited to the areas north and east of the Congo River, from the Ubangi River to western Uganda. It is known from Maiko and Kahuzi-Biega national parks, as well as the Okapi Reserve. All are solitary and more or less omnivorous. In other words, they, and especially the Blotched Genet, complete their diet of vertebrates and invertebrates with fruits. The Central African Linsang *Poiana richardsoni* (FIGURE 6.101) resembles genets, but is more elongated, with a proportionally longer tail and a shorter muzzle. It is a strictly arboreal species.

The Common Civet *Civetta civetta* is nocturnal and terrestrial. Strikingly larger and standing taller than the genets, it has something of the look of a dog. It inhabits most of sub-Saharan Africa, lives alone and marks its territory with piles of droppings and with very strong smelling anal secretions. It is omnivorous.

Known in English either as Aquatic Civet or as Aquatic Genet, *Osbornictis piscivora* most resembles a genet. It is reddish brown with a thick black tail and white on the muzzle. Its range of distribution nearly coincides with that of the Giant Servaline Genet, but apparently does not reach Uganda. It has a liking for the banks of streams in the forest, especially in *Gilbert-*

6.100. The Black-legged Mongoose Bdeogale nigripes *feeds largely on insects – especially ants – snails, crabs, rodents and amphibians, and occasionally also on fallen fruits.*

6.101. *The Central African Linsang* Poiana richardsoni *is related both to the Asian linsangs and to the genets. It is an essentially arboreal species, nocturnal and very poorly known. It feeds on arboreal vertebrates and invertebrates, as well as on fruits. It is found from Nigeria to the Albertine Rift, but its range is still incompletely known.*

6.102. *The Golden Cat* Felis aurata *is an animal that may weigh as much as 18 kilos. It can be russet or grey, spotted or not. It feeds on forest duikers, monkeys, rodents and birds, but its habits and status are not well known*

iodendron forests. Its food consists probably only of fish.

As for the African Palm Civet *Nandinia binotata*, it inhabits all of the forested region and the forest-grassland mosaic regions. It is an elongated animal with a long grey-brown black-spotted tail, omnivorous and very opportunistic.

The Felidae are essentially represented in the forest by the Leopard *Panthera pardus* and the Golden Cat *Felis aurata* (FIGURE 6.102). The first has become very rare in most of the forests of Central Africa and it has even disappeared from the forests of Rwanda and Burundi. In Uganda, it no longer occurs in the Semliki and Kalinzu forests. Not only is this large spectacular feline heavily hunted, notably for its skin, but it has the greatest difficulties in feeding because of the increasing rarity of its prey. On the fringe of the forest massif and in degraded areas, the Serval Cat *Felis serval* is also found, as well as the African Wild Cat *Felis lybica*. A small population of Lions *Panthera leo* still survives in the enclaved grasslands of Odzala National Park.

The Hyaenidae, another family of carnivores of African origin, are animals of open habitats. In Odzala National Park, however, there is a small population of Spotted Hyaenas *Crocuta crocuta*. They live mainly around the large clearings, but also haunt the interior of the forest. Essentially nocturnal, the Spotted Hyaena is seen only rarely, but its numerous tracks along muddy paths of the forest betray its presence. This is not a completely isolated phenomenon since, in Rwanda, it also used to happen that a Spotted Hyaena crossed the forest of the Volcanoes National Park.

Pangolins

Pangolins or Manidae, of the order Pholidota, are very specialised insectivores, covered with large scales. In spite of some resemblance to South American armadillos, they represent a group of typically Old World animals, whose origin can be traced back 70 million years. All species have strong claws on their hind legs with which they tear open termite mounds. Their food consists almost exclusively of ants and termites captured with their long sticky tongue.

The three species that inhabit Central Africa heavily depend on the presence of water. The Tree Pangolin *Phataginus tricuspis* (FIGURE 6.104) is the commonest of the pangolins. It is found just about everywhere in the forest, including in strongly secondarised formations and in cultivation. It is a nearly exclusively nocturnal termite eater that finds most of its food in the litter. The Long-tailed Pangolin *Uromanis tetradactyla* (FIGURE 6.105) is noticeably larger: it measures nearly one metre in length. It is arboreal, partly active during the day and extremely cautious. It feeds mainly on ants found in the canopy of gallery forests and swamp forests. As for the Giant Pangolin *Smutsia gigantea*, it can reach 1.70 metres in length and weigh 35 kilos. Strictly terrestrial and nocturnal, it feeds on termites and ants, but also takes other insects. Much hunted, it is becoming more and more rare and has probably already disappeared from a large part of its range.

6.103. The tree hyrax Dendrohyrax dorsalis.

6.104. The Tree Pangolin Phataginus tricuspis *can be recognised by its narrow scales and pale face.*

6.105. The Long-tailed Pangolin Uromanis tetradactyla, *is a little larger. It has a black face and wide scales.*

Ungulates

The Procavidae are archaic ungulates, limited to Africa. They were very diversified 25 to 30 million years ago, but number no more than 11 species nowadays. Most of them inhabit rocky habitats, but the Western Tree Hyrax *Dendrohyrax dorsalis* lives in the lowland forests of Central Africa (FIGURE 6.104). It is an eater of leaves, fruits and young branches, inhabiting very diverse forests, solitary, strictly arboreal and nocturnal. It can weigh as much as 4.5 kilos. At higher altitude, in the Virunga and Rwenzori mountains, is found a closely related species: the Eastern Tree Hyrax *Dendrohyrax arboreus*. This one is partly diurnal, often terrestrial and tends to form loose-knit colonies. Hyraxes are not often seen, but their long, heartrending cries, fairly variable from one region to another, constitute one of the most characteristic elements of the acoustic ambience of the African forest.

Of the Elephantidae family, there remains today in Africa only the African Elephant *Loxodonta africana* (FIGURE 6.106). This group, whose origin goes back 65 million years, differentiated on the African continent during the course of the Tertiary. The two present-day species would have separated five million years ago and the family had its hours of glory around the middle of the Pleistocene, when there were at least some ten species around the world.

The African Elephant apparently did not extend its range until after the disappearance, some 20,000 years ago, of *Elephas reckii*, a near relative of the Asian Elephant *Elephas maxima*. Before that, it seems to have been limited to dense forests. Forest elephants of the form *cyclotis*, smaller than savanna elephants, would in a way represent the "original" form of the species and the savanna elephants would only be a recent version. The differences observed today between the forest and the grassland populations would then be too recent to justify the recognition of two separate species, as some scientists attempt to do. In many places, the two forms used to enter in contact and to mix together.

The historical range of the African Elephant probably covered most of the continent except for the most arid regions. Today, its range is more and more fragmented and the species has completely disappeared from many regions. Elephants still exist just about everywhere in the forests of Central Africa, but it is difficult to accurately evaluate the number of individuals, especially since herds move about a lot and the fact that some regions are difficult of access. The largest remaining populations, of the order of several tens of thousands of individuals, are probably found in Gabon, in northern Congo and south-eastern Cameroon. For Gabon alone, Allard Blom and Marcel Alers estimated that there were about 70,000 elephants in 1985-1986. From the Democratic Republic of the Congo, we have only very little recent information but we know that the species has been virtually eliminated over vast areas, including in the heart of some of the most remote forests. In the Albertine Rift region, small populations survive in of Uganda, but they number in the hundreds. Even in the Nyungwe Forest, in Rwanda, a few individuals survive in the Kamiranzovu marshes (PAGE 55). In Burundi, the species became extinct during the 1980s.

6.106. Three African Elephants Loxodonta africana *in a clearing in Odzala-Kokoua National Park. They belong to the form* cyclotis, *well adapted to the forest, smaller than the grasslands form, with finer, straighter tusks and more rounded ears. A form* pumilio, *even smaller, has been described, but its existence has never been demonstrated.*

According to Quérouil and his collaborators, 629 individuals were observed in the Maya-Nord clearing in eight months and the total number of individuals visiting this site in 102 days was estimated at 1,900, which represented the entire population of a region of more than 200,000 hectares. Very often, the trails linking one clearing to another cover several tens of kilometres.

Between 1979 and 1988, 2,825 tons of ivory left Central Africa. These figures corresponded to a population of 120,000 to 200,000 elephants. Such an amount of killed animals could obviously not have been possible except through the setting up of well organised networks, benefiting from the complicity of the armies, customs and highest authorities. As for the hunters, they often formed veritable gangs equipped with weapons of war and not afraid of anything. In places villagers abandoned their traditional activities in order to devote themselves entirely to hunting elephants. Lots of meat was wasted. With the ban on legal trade set up by the CITES in 1989, elephant hunting slowed down and the remaining elephants had some rest.

Forest elephants live in small groups. At Maya-Nord in Odzala National Park, the size of groups varies from two to 16 individuals but, on average, it is no more than 3.5. In addition, around 30% of individuals live alone. These are mainly males who do not join the groups of females and young except during brief periods.

These forest elephants feed essentially on the foliage and fruits that they find inside the forest. They visit clearings, but only to bathe and drink – water is generally not lacking in tropical forests – and to take up the mud. Perhaps they are in search of minerals, which they find more easily in these clearings than inside the forest. But also perhaps the frequentation of these clearings has a social role and is just a way to encounter other groups. Studies at Maya-Nord show that elephants that visit this clearing come from a territory of more than 200,000 hectares. Groups move about considerably and cover great distances. Their movements are in part seasonal and related to the fruiting of some plants. In the coastal forests of Gabon, for example, Lee White was able to demonstrate movements related to the fruiting of *Sacoglottis gabonensis*. In general, males migrate over greater distances than females.

In spite of their size, elephants easily go unnoticed. Often, all that is found of them are their tracks, their dung and the trees they have knocked down. Since they like secondary vegetation and do not shun cultivated plants, they easily enter into conflict with Man. The main cause of their decline remains poaching for their ivory, to which has been added in the past few years the commercialisation of their meat. Ivory has always been the object of international trade and, already in the second half of the 19th century, the Arabs of Zanzibar reported its growing rarity. But this historical trade did not compare with the veritable rush that was unleashed around the middle of the 1970s. Within 20 years, the populations of all of Africa were decimated, including those of the most remote forests of Central Africa. In places losses were of the order of 90%. This massacre was slowed during the 1990s following the dispositions taken by CITES, but for the past two or three years there has been a resumption of activity and hunting for ivory is again a major problem (PAGE 319).

The fact that for hundreds of thousands of years, elephants have lived in the forest signifies that very complex interactions have developed between them and the forest. Just by their size, these animals have a profound impact on their habitat, that they model in their way. They tip over trees, create clearings, dig "salines" and transport seeds. In the tall primary formations, they maintain the open understorey and put in paths connecting watering places, clearings and the main feeding areas. Because of this, many other species, as much animal as plant, depend on their presence. In addition, many seeds from forest trees do not germinate except after having passed through their digestive tract. Elephants have therefore played an important role in the selection of species and, when after the last glaciation the forests were reconstituted, they certainly played a considerable role in the "migration" of those whose heavy fruits could not propagate otherwise.

For some millions of years, the African forest therefore not only bore the increasing impact of climate cycles. It also was "gardened" by the Elephant. Today, this fact is too easily forgotten. At the beginning of the 1970s, the *Hagenia* forests of the Virunga and the *Ficalhoa* forests of some ridges in the Nyungwe Forest, for example, were criss-crossed by veritable "autoroutes" created by herds of elephants and strewn with felled trees. Today, 30 years later, some still remain visible and the tree cover remains more or less open. But the elephants having disappeared, young forestry workers, formed in western schools and who have never seen an elephant in nature, systematically attribute them to ancient human activity. To speak to them of elephants is like talking about dinosaurs.

This volatility of the human memory would not matter if it were not that the interactions between the forest and the elephants really are important and that the disappearance of elephants will not be without an impact. It will inevitably provoke an impoverishment of the forest cover. Considering the slowness of reaction of forest formations and the complexity of the relationships involved, these transformations will be difficult to predict and they will appear only very late, when nothing can be done to counteract them.

The Hippopotamidae represent a lineage of large amphibious animals, probably of African origin, nearly 40 millions years old. Towards the middle of the Pleistocene, around a million years ago, Africa had no less than eight species. Today, there remain only two, of which one is limited to the forests of western Africa. As for the Hippopotamus *Hippopotamus amphibius*, it occupied, only a short while ago, nearly all of sub-Saharan Africa and could be encountered everywhere where expanses of open water and pastures coexisted. It was therefore never very numerous in the interior of large forests because of the absence or rarity of grasses. A small population occurs in Odzala-Kokoua National Park, but the species is not known from the Dja faunal reserve. It exists in all the countries of Central Africa, nevertheless, even though its populations are rapidly reducing as a result of unrestrained hunting.

In Gabon, for example, it has disappeared from the area around the Lopé National Park and survives sporadically along the Ivindo, notably just below Koungou Falls, on the Lower Ogooué, from the outskirts of Lambaréné to Port-Gentil, and in the coastal lagoons of the Gamba complex of protected sites. Without mentioning the immense population of Virunga National Park, that numbered more than 20,000 individuals but which has been greatly reduced as a result of the war and the presence of Rwandan refugees, the populations of the Democratic Republic of the Congo are themselves also in a bad situation. Because of its size and its abundance, the Hippopotamus can have a strong impact on grassland vegetation. Its absence can, on the contrary, favour reforestation.

The Suidae represent another Old World lineage, having appeared around 10 million years ago and having experienced its apogee five million years ago. Species of the genus *Potamochoerus* are the African equivalents of Eurasian boars. Mainly occurring in Central Africa is the Forest Hog *Potamochoerus porcus*, a "red" animal (FIGURE 6.108), but south of the Congo basin and in the Albertine Rift region it is replaced by the Bush Pig *Potamochoerus larvatus*, a greyer animal. These pigs inhabit forests, mainly the humid formations along watercourses, but they readily come out in the open into clearings. They are omnivorous, but especially seek out roots and bulbs that they can extract by digging in the soil with their strong snout. Very often, they betray their presence only by their foot prints and patches of dug-up earth. They live in small groups of five to ten individuals, but sometimes form temporary groups of more than 40. In spite of hunting pressures, these wild pigs manage to hold their own fairly well and occur in all the protected areas of forested Central Africa. They are inevitable visitors to the clearings of Odzala-Kokoua National Park.

The Giant Hog *Hylochoerus meinertzageni* (FIGURE 6.107), the largest of the African Suidae, does not have, or no longer has, near relatives outside of Africa. In Central Africa, it is found from eastern Nigeria to Uganda, but it is absent south of the Congo River. Its populations are, however, very fragmented and in many places it has become very rare. In the east, it is still fairly abundant in the Maramagambo Forest and is known from the forests of Bwindi, Kasyoha-Kitomi, Rwenzori and Kibale in Uganda, as well as the Kahuzi-Biega Forest in the Democratic Republic of the Congo. It can easily be observed in Queen Elizabeth National Park where it willingly leaves the dense thickets to venture into open grasslands. It has disappeared from Rwanda and Burundi, however. In the west, Giant Pigs can be seen fairly easily in a few clearings in Odzala-Kokoua National Park in Congo, but in Gabon it is not known except in the north-east, or more exactly, in the Min-

6.107. *The Giant Hog* Hylochoerus meinertzhageni *is seen only in some clearings in Odzala-Kokoua National Park, notably the Maya clearing. It can also be seen in the Queen Elizabeth National Park.*

6.108. *A group of Red River Hogs* Potamochoerus porcus *looks for "abandoned" seeds in a small pool of the Maya-Nord clearing in Odzala-Kokoua National Park.*

6.109. The Okapi Okapia johnstoni *is rarely seen in the natural state and remained for a long time unknown. It has been totally protected since 1933, but with the conditions reigning in the eastern part of the Democratic Republic of the Congo, one can imagine that this measure does not mean much.*

6.110. The Water Chevrotain Hyemochus aquaticus.

kébé National Park. It mostly has a liking for humid places in the forest-grassland mosaic, but, especially in the west, it can also be encountered in old dense forests. It feeds on herbaceous plants, especially species rich in minerals, and seeks out roots much less than the bush and river pigs. Occasionally, however, it eats eggs, droppings or earth. It lives in groups of five to 15 individuals, made up essentially of females with their young. The males are more independent and associate only temporarily with these groups, but they defend their territory against other males and protect the females and young from predators. Because of its size and habits, the Giant Pig makes an easy prey and is much sought after by hunters. This is why it has now disappeared from many of the regions where it was still abundant hardly 50 years ago.

The Tragulidae are the most primitive of the present-day ruminants. They were abundant throughout the Old World 40 to 25 million years ago. A single species survives in sub-Saharan Africa: the Water Chevrotain *Hyemochus aquaticus* (FIGURE 6.110). This small reddish-brown animal, marked with whitish spots and lines, and without horns, superficially resembles the forest duikers but also has some resemblance to the Suidae. Well adapted to the dark understorey of the great lowland forests, it haunts streams and backwaters where it swims with ease. It lives alone and feeds on fruits, mostly fruits fallen to the ground, insects, crabs and occasionally other animal food. It occurs across the entire Guineo-Congolese forest massif, but has never crossed the escarpment of the Albertine Rift. In Uganda, the Water Chevrotain is unknown except in the forests of Semliki National Park. Being essentially nocturnal, it is difficult to observe, but its flesh is much appreciated by hunters.

The Giraffidae, a family of African origin, are no longer represented by more than two species, one of which is endemic to the forests of Central Africa: the Okapi *Okapia johnstoni* (FIGURE 6.109). This animal is presently known only from the north-eastern part of the Congo basin, mainly from the forests located between the Uele and Congo rivers. Isolated observations outside of this area indicate that the species must have had a much greater range in the recent past. It seems to have occurred in all of the eastern part of the central basin and in the forests of the Semliki in extreme western Uganda. The Okapi avoids humid places and preferentially inhabits terra firme forests above 500 metres altitude, where it seeks out places with a dense understorey. It mainly eats foliage – *Rinorea* and *Drypetes* seem to be favourite species – but also consumes herbaceous plants and mushrooms. It lives alone and pairs do not form except during the receptive period of the females. Otherwise, females occupy a territory of the order of 500 hectares, males hold a much vaster territory. The intensification and commercialisation of hunting, as well as socio-political troubles and the war, weigh heavily on the future of this unique species. One can only hope therefore that the reserve created for it in the Ituri Forest – the Okapi Faunal Reserve – can assume its role.

The Bovidae represent the ruminant family having developed the most complex digestive system, capable of getting some good out of cellulose thanks to a symbiosis with bacteria. Of Eurasian origin, bovids probably colonised Africa in several waves and particularly engendered grassland species, more or less well adapted to dry, arid habitats.

The sub-family Neotraginae is probably descended from the very first wave of bovids to reach Africa at least 25 million years ago. Among the species able to colonise Guineo-Congolese forests, there is first of all the Dwarf Antelope *Neotragus batesi*, a small, very slender species weighing no more than 5.5 kilos (FIGURE 6.111). It is distributed in a fragmented way mainly between the Sanaga and Ogooué rivers in the west and between the Congo River and the Albertine Rift in the east. It is also known from some forests in Uganda where it remains rare, however. It inhabits the dense understorey of lowland forests and has a liking mostly for areas around watercourses, tree-fall gaps and young secondary forests. Locally it adapts to plantations, notably of cocoa, and densities can reach 75 individuals per 100 hectares. The diet of this solitary animal is fairly eclectic and comprises essentially leaves and young shoots.

This first wave of bovids also gave rise to more "modern" species, represented in the forest by the duikers, an African lineage of 16 or 17 species, of which eight or nine inhabit the forests of Central Africa. They are small antelopes, common in places but most of the time difficult to observe. Very often they do not betray their presence except by the tracks of their hoofs or when they are caught in hunters' snares. If by luck a duiker can be observed, it is possible to see its vaulted silhouette, small straight pointed horns and the crest of long hairs decorating the top of its head. They share the forest according to differences in size (going from three to 80 kilos), habitat, food and rhythm of activity. All are territorial and exclude from their territory congeners of the same sex.

The smallest and probably the most common is the Blue Duiker *Cephalophus monticola* (FIGURE 6.112). It is known from all the countries of Central Africa except Rwanda, and inhabits all sorts of forests, including submontane forests. Usually diurnal, it feeds essentially on fruits, but partly on leaves, also, and occasionally animal matter. It sometimes follows groups of monkeys in order to gather the fruits that they make fall. In Gabon, Blue Duikers can reach a density of 60 to 70 individuals per square kilometre, but generally populations have been greatly reduced by hunting. In spite of that, Blue Duikers manage to survive in galleries or in small, very isolated forests.

The group of "red" duikers comprises three middle-sized species inhabiting the great forest, from the Gulf of Guinea to the Albertine Rift. All three are diurnal. The Black-fronted Duiker *Cephalophus nigrifrons* (FIGURE 6.115) mainly inhabits humid and swamp forests, but also high altitude thickets and heaths, up to 3,500 metres in the Virunga. On the Rwenzori, above 3,000 metres, a particular form occurs. Jonathan King-

6.111. *The Dwarf Antelope* Neotragus batesi *in a forest of north-eastern Gabon. This species is not well known, but has been studied in considerable detail at the Makokou research station in Gabon by F. Feer.*

6.112. *The Blue Duiker* Cephalophus monticola *is the most common of the forest duikers. It has been studied at Makokou station by G. Dubost.*

don considers it as a full species: the Rwenzori Duiker *Cephalophus rubidus*. Peter's Duiker *Cephalophus callipygus* seems to be the most abundant in forests regenerating after exploitation, but it, too, frequents montane forests. In Nyungwe Forest in Rwanda, it is represented by the form *lestradei* with a nearly all black coat. The White-bellied Duiker *Cephalophus leucogaster* occurs only in forests north of the Congo River. Its range is very fragmented and it avoids inundatable forests. Lastly, north of the forest massif, from Senegal to the Nile, lives the Red-flanked Duiker *Cephalophus rufilatus*. At times diurnal, at times nocturnal, it inhabits the isolated remains of forest and thickets bordering watercourses.

A third group of duikers is made up of the large species. In Africa Central, the group is represented by the Yellow-backed Duiker *Cephalophus silvicultor*, an animal weighing 45 to 80 kilos, a mainly nocturnal animal. It inhabits montane forests and lowland forests, and likes large intact massifs as well as forests along watercourses, degraded forests and forest-grassland mosaics. It has become very rare in the Albertine Rift region.

The fourth and last group of duikers is comprised of middle-sized nocturnal species that feed on fibrous fruits. The Bay Duiker *Cephalophus dorsalis* is known from all the forests of Central Africa but it is now extinct in Uganda. Ogilby's Duiker *Cephalophus ogilbyi* occurs only in forests bordering the Gulf of Guinea. In Central Africa, it is known from Cameroon, Equatorial Guinea, the isle of Bioko, and Gabon. In this latter country, it is represented by the race *crusalbum*, recognisable by its white legs. These two duikers have powerful hindquarters and are capable of impressive leaps.

6.113. A male Sitatunga Tragelaphus spekei *searching for fruits in a pool of Odzala-Kokoua National Park.*

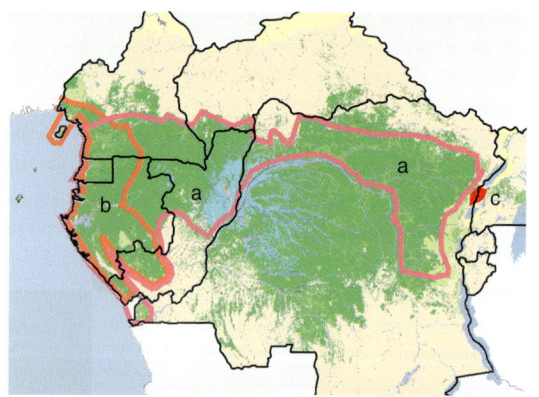

6.114. Most of the seven or eight species of forest duikers of Central Africa can be found all over the forest massif. The White-bellied Duiker Cephalophus leucogaster *(a) is restricted to the forests north of the Congo River and south of the Sanaga, however. Ogilby's Duiker* Cephalophus ogilbyi *(b), known from West Africa, is limited to the coastal regions of Lower Guinea, from Cameroon to the Congo Estuary. As for the Rwenzori Duiker* Cephalopus rubidus *(c), considered by most scientists as a subspecies of the Black-fronted Duiker* Cephalopus nigrifrons, *it is restricted to the high altitude regions of the Rwenzori Mountains.*

6.115. The Black-fronted Duiker Cephalophus nigrifrons.

The Tragelaphinae or spiral-horned antelopes represent a second wave of bovids to reach Africa. It gave mainly rise to large grassland species, but three species can be found in the forests of Central Africa.

The Bushbuck *Tragelaphus scriptus* is a solitary animal that haunts the dense thickets of mantles and gallery forests along watercourses. It is not found in the middle of the forest, except in degraded areas. It is thus commonly observed in the Lopé National Park and in the southern part of Odzala-Kokoua National Park at the forest-grassland contact zone. It feeds on foliage, herbaceous plants, seed pods of leguminous plants and on fruit.

The Sitatunga *Tragelaphus spekei* (FIGURES 6.113-116) is clearly heavier. It inhabits the surroundings of watercourses in the forest where it prefers dense thickets, clearings with grasses and sedges, notably the bais, and swamp forests with palms. Outside the forest massif, it also inhabits large open swamps. It feeds on foliage, herbaceous plants and grasses. In the clearings of Odzala-Kokoua National Park, it has a liking especially for *Rhynchospora corymbosa* meadows, streams and pools where it searches for seeds on the bottom, and *Enydra fluctuans* formations. According to Florence Magliocca, the plants consumed are significantly richer in minerals than the surrounding vegetation. This species is usually solitary, but it can form loose-knit groups. In the Maya-Nord clearing, concentrations of 15 to 25 individuals are observed. Groups comprise one to two adult males, some females and young.

The Bongo *Tragelaphus euryceros* (FIGURE 6.1) is a splendid antelope with a brownish-red coat, marked by long vertical white stripes and beautiful slightly spiralled horns. The male may reach up to 400 kilos. Its distribution is very fragmented. In Central Africa it is found in the northern part of the Democratic Republic of the Congo, in the southern Central African Republic, in northern Congo, in north-eastern Gabon and south-eastern Cameroon. Curiously, its range coincides very closely with that of the Giant Hog. It is an eater of leaves that mainly inhabits the forest, but prefers areas with dense thickets or mosaics of forest and more open spaces. In the Minkébé National Park in Gabon, it is found principally around inselberges. At Odzala and Nouabalé-Ndoki, it can be encountered in the clearings. In dense forests, it is a solitary animal. Each individual occupies a fairly vast vital space that can, however, overlap with that of congeners. As in the Sitatunga, gatherings may form in clearings, notably at Dzanga-Sangha. Altogether, the Bongo is an extremely shy animal that has suffered enormously from hunting and whose survival is threatened in many places. It is fairly frequently observed only in the Central African Republic.

6.116. *A female Sitatunga with young in a pool of Odzala-Kokoua National Park.*

6.117. A small group of buffaloes Syncerus caffer *in the "saline" at Lango. In the back-ground, Dwarf Date Palms* Phoenix reclinata.

6.118. A solitary buffalo in a clearing at Odzala-Kokoua National Park.

There remains the African Buffalo *Syncerus caffer* (FIGURE 6.117 and 6.118). It probably belongs to a last wave of bovids coming from Asia that did not have the time to adapt to the driest habitats. In spite of its size, it can be very unobtrusive and it survives in small numbers in most of the forest massifs, especially in those that touch the grasslands national parks. It has disappeared, however, from the forests of Rwanda and Burundi and has become very rare in most of the forests of the Democratic Republic of the Congo.

Buffaloes of the dense humid forests are brown or rusty brown. Their horns are strongly back-swept and less large than those of the grasslands. Their weight does not exceed 300 to 320 kilos. In the Albertine Rift region, notably in Queen Elizabeth National Park, intermediate animals are found, however, and it is obvious that in places there is a gene flow between grassland and forest populations.

Unlike the grassland buffaloes, those of dense forests live in small groups. At Odzala-Kokoua National Park, in the Maya-Nord clearings, groups are of 12 to 19 individuals. In the enclaved grasslands of Lopé National Park or in the Wonga-Wongué Reserve, it is not rare, however, to see 30 or 40 together in very open spaces. These groups are very territorial, which probably helps to avoid overgrazing in areas where pastures are not very abundant. Competition between males is very strong and combats are frequent. Females reproduce only once every two years and there is apparently no particular season for calving. Young males leave the group as soon as they reach maturity. The food of Buffaloes consists in large part of grasses and sedges. In the clearings of Odzala they also have a marked preference for *Killinga erecta*, *Rhynchospora corymbosa* and *Paspalum conjugatum*.

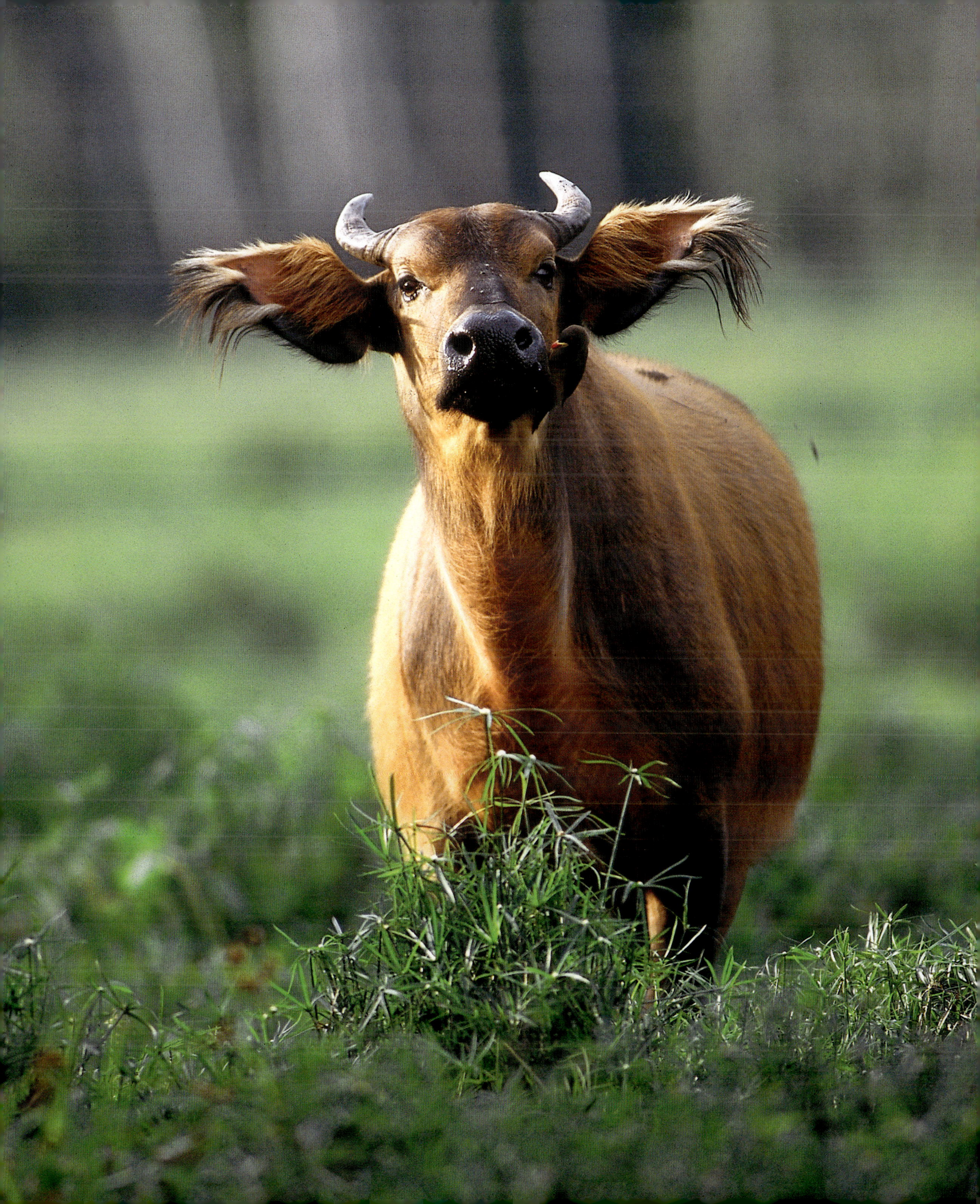

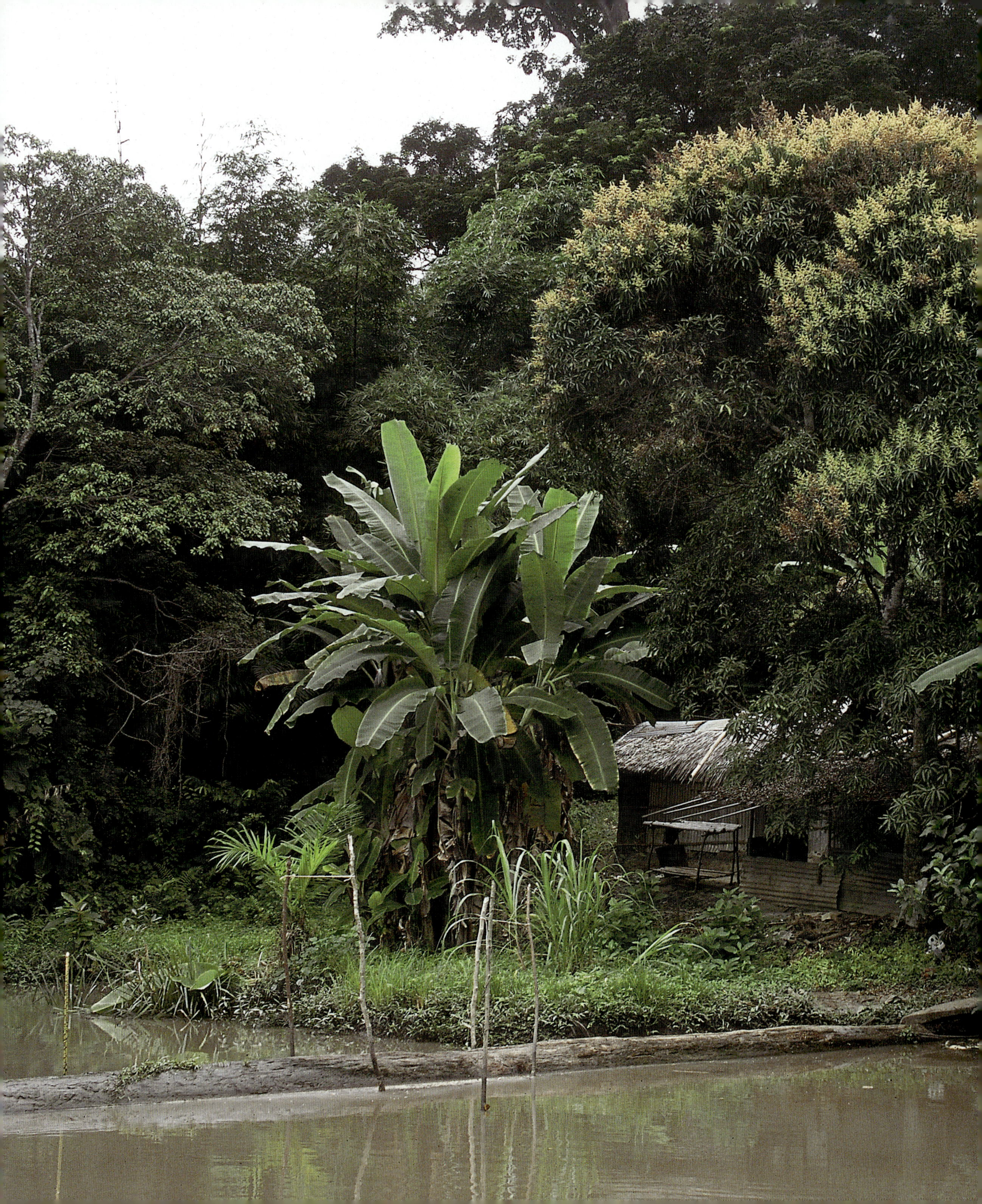

7 | Man and the forest

When, soon after 1850, the first Europeans penetrated the heart of Central Africa, they discovered fabulous landscapes, a natural environment that was hostile but rich and exuberant and peoples who seemed to be emerging from prehistory. Their disorientation, their feeling of strangeness, was of a magnitude comparable to the efforts they had just accomplished and they believed they had found the Garden of Eden. This romantic vision would take form, consolidate, spread, attract adventurers, inspire artists and even profoundly influence a great many scientists. Still to this day, it haunts the spirit of the most nostalgic.

In fact, by 1850 already, immense areas of forest had been replaced by grasslands, wooded grasslands and shrub grasslands had given way to grassy or semi-desert landscapes, and the large mammals, although still very spectacular, represented even then no more than a pale reflection of what the first modern men knew 50,000 years earlier. In a few places, mostly in highland regions of the Albertine Rift or of western Cameroon, the high densities of human populations had even reduced the natural habitats to a few poor relics. These spaces, which the explorers perceived as veritable oases of humanity lost in the middle of a wild and hostile world, did not, however, cover more than very small areas. Everywhere else, very scattered populations inhabited grandiose landscapes where the natural and the human confronted each other so intimately that they gave the illusion of a nature having escaped all human intervention.

They made one forget that the present-day populations had colonised these forests for at least 2,000 or 3,000 years, that the different cultural groups had mixed together and integrated into new societies, that since the 16th century the very centre of the Congo basin had been plundered by the slave trade, that the entire region had after that been profoundly influenced by the introduction of American plants, that in the 19th century Arab slave traders had reached the most remote regions. In short, with the exception of the highest summits, swept by icy winds, man was omnipresent and his history was already long. He had even learned to survive in the "cauldron of fevers" of the immense swamps of the Congo, the Ubangi and the Sangha, in the very heart of the central basin. In spite of everything, the Central Africa of 1850 still harboured landscapes, flora and fauna that the other continents had long since lost. Perhaps because this human history had allowed it?

7.1. A family has built its house by a small river that empties into the Gabon estuary, in the shade of a few remaining old forest trees, some secondary growth species and some fruit trees – here a mango in bloom – and banana trees. An abandoned log serves as a mooring pontoon for pirogues.

7.2. The Ancient Stone Age. The only traces for Central Africa of Homo habilis, *the very first species of the genus, 2.2 to 2.3 million years old, come from Lusso on the north shore of Lake Edward. They are worked stones of the Oldowayan type. Of his successor,* Homo erectus, *who inhabited all of eastern and southern Africa until 500,000 years ago, there has never been a finding that could attest to his presence, neither in the Congo basin nor even in the rift region. Some 1.8 to 1.9 million years ago, he had already colonised the Caucasus and south-eastern Asia, however.*

Traces reappeared with archaic forms of Homo sapiens, *in the form of tools of the Acheulean type, 400,000 to 500,000 years old. They are scattered around the edge of the forest massif, but there is a finding for the centre of Gabon.*

7.3. Of the Middle Stone Age, tools of the Sangoan type were the first to be found in all the countries of the region and, for the first time, also in the heart of the Congo basin, followed by those of the Lupembian type. The climate, then fairly cool and dry, allowed Man to continue his penetration of the Congo basin where the humid forests had largely given way to grasslands.

Stone and fire

Unlike eastern Africa, Central Africa does not yield its archaeological secrets except by chance. The immense majority of findings then consist of worked stones that were gleaned from the surface after having been freed by natural erosion or excavation works. Their exact dating is often difficult if not impossible, and systematic archaeological digging has not been undertaken except in very few places. The acid soils of the forest and the hot, humid climate prevent the fossilisation of organic matter. The forest cover hides everything and erosion has often profoundly reworked the most ancient deposits. The work of archaeologists is therefore not easy in forested areas of Central Africa. A multidisciplinary approach, and the implementation of new techniques arising from physics, chemistry and biology have, however, enabled the main points of this region's prehistory to be outlined.

Although the common ancestors of Man, Gorilla, Chimpanzee and Bonobo were frugivorous primates of the forest, the archaeology of Central Africa does not teach us anything about the very beginnings of the human species. The appearance of genus *Homo*, 2.5 million years ago, is indeed an East African story and, even if he plunges his distant roots into the dense humid forests, Man is above all an animal of the grasslands.

Of *Homo habilis* and *Homo erectus* practically nothing is known in Central Africa (FIGURE 7.2) and, until proof to the contrary, the history of Man in forested Central Africa does not commence until around 500,000 years ago with the appearance of archaic forms of *Homo sapiens*. Some researchers call them *Homo heidelbergensis*. Others include them in *Homo neanderthalensis* in the broadest sense. Whatever name is given them, their Acheulean tools have been found in several places in the Great Lakes region and along almost the entire periphery of the Congo basin, with the exception of Cameroon (FIGURE 7.2). These artefacts are generally massive, with a simple structure, but already testifying to a great mastery of tool-making. In Gabon, not far from the Lopé National Park, Richard Oslisly has found artefacts that are at least 400,000 years old.

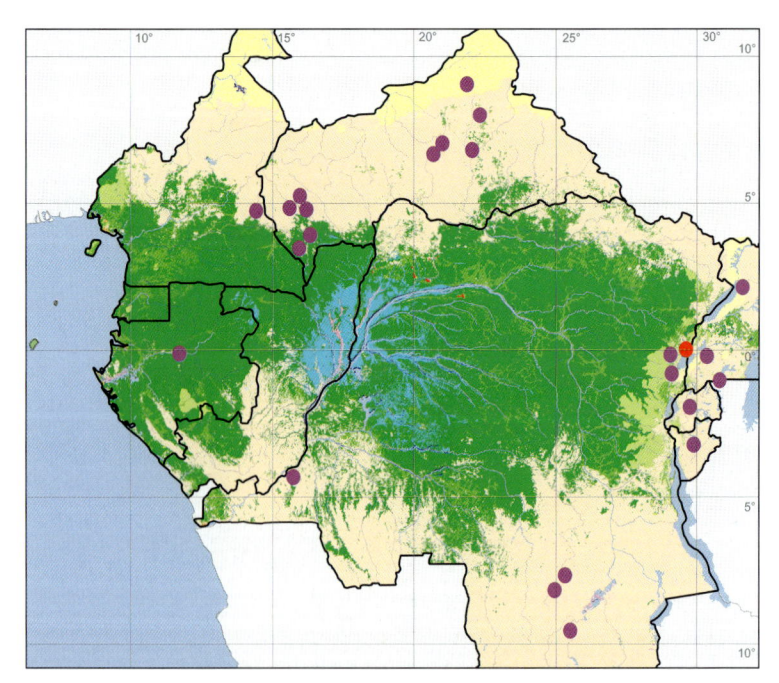

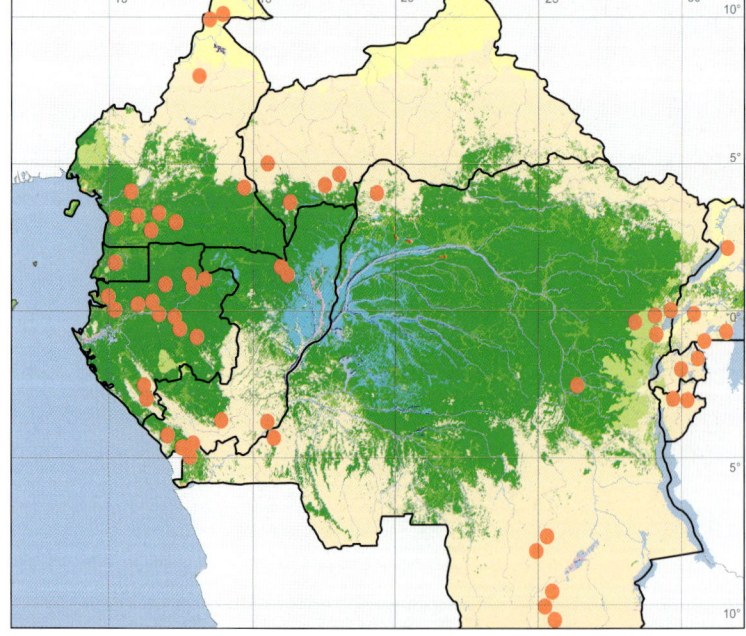

What might have been the relationships of these distant ancestors with the nature that surrounded them? The Australopithecines were probably mainly frugivorous and consumed meat only occasionally, probably a little like Chimpanzees do today. They apparently did not hunt but were themselves hunted by large carnivores. With the advent of genus *Homo*, the consumption of meat became more important. For a long time, however, Man remained dependent on large carnivores whose prey he stole. It is only with the increase of his cerebral capacity and the development of more effective weapons that he himself became a hunter. Exactly when that took place is not known, but many researchers think that this could have happened around a million years ago. The density of humans was so low during this period, however, that they must have had very little impact on the fauna. Man was still no more than one primate among so many others. He suffered nature, all the more strongly so because it was just during this same long period that cyclic variations of the climate became more and more pronounced. The ebb and flow of plant formations as well as the appearance and disappearance of lakes and marshes must have continually influenced his range of distribution and perhaps even accelerated his evolution.

With the appearance and dispersal of modern Man between 100,000 and 60,000 BP this situation had to change. Populations increased and their tools became diversified. From 70,000 BP, the Sangoan culture developed in this way, somewhat more developed, but still near the Acheulean. It was replaced by the distinctly more advanced Lupembian (FIGURE 7.3), which would in its turn give way to the first industries of the recent Stone Age, around 40,000 BP (FIGURE 7.4). These activities would not, however, fully develop until 12,000 BP, when the climate had again become warmer and more humid. Alongside the tools that perpetuated previous traditions, these new cultures also used much smaller stones, microliths, that would serve as scrapers, chisels, punches and, for the first time, arrow heads.

Also for the first time, well individualised regional cultures saw the light of day (FIGURE 7.4). The one from the east and the south of the Congo basin and the region of the Great Lakes belonged to the Wiltonian, a culture that spread across all of eastern and southern Africa and was the work of palaeonegritic populations close to the present-day Bushmen of the Kalahari. The culture of the Congo, Gabon and the west of the Democratic Republic of the Congo represented the Tshitolian, a continuation of the Lupembian, which was characterised by a progressively finer working of tools and, at the very end, by the appearance of polishing. A third culture, which never knew polishing, occurred in the coastal regions of Gabon, Equatorial Guinea and Cameroon. Finally, a fourth culture, related to those of western Africa, developed in Cameroon.

The hunter-gatherers of the Middle Stone Age were not very numerous. According to Bernard Clist, the density of their population was probably never more than 0.02 inhabitants per square kilometre, which represented a population of only 5,000 inhabitants for all of Gabon. Those of the Recent Stone Age were apparently more numerous, but – still according to Clist – numbered no more than 0.1 inhabitant per square kilometre, which still represented only 26,000 inhabitants, at most, for all of Gabon. They were distinctly "rarer" than Gorillas and Chimpanzees are today

In spite of their minute densities of population, humans of these periods probably began to have a significant effect on nature, first of all by the use of fire. In Europe and Asia, fire was probably universally employed from the mid-Pleistocene and, even though its earliest traces are from Kenya – dating back 1.4 million years – it is very difficult to know when use of fire became generalised in Africa. We know, however, that grassland fires intensified at four different occasions over a period of some 400,000 years – at least according to the study of carbon deposits in marine sediments off the coast of Sierra Leone. Each one of these intensifications coincides with a period of transition towards a glacial phase. Their causes are obviously unknown, but it is curious that they begin at around the same period as the use of fire in Europe and Asia. A last phase of clear-cut intensification of bush fires took place – at least in eastern Africa – around 50,000 BP. Starting at this period, layers of ashes enclosed in the glaciers of eastern Africa point to a much increased frequency of grassland fires. Very violent but rare fires, leaving thick, widely spaced, layers of ashes, gave way to less violent fires leaving fine, almost annual layers of ash.

In regions situated on the margin of the large dense forests, this change in the regime of fires profoundly affected the vegetation. The vast dry forests of olive *Olea europaea* and *Euphorbia dawei*, that formed the transition between dense humid forests and more open grasslands on the eastern edge of Central Africa were fragmented and reduced to relicts (PAGE 82). Herbivorous antelopes and zebras profited from this opening of a passage, set about proliferating and managed to extend their range, but these changes sounded the knell for other species.

Already during the epoch of *Homo neanderthalensis*, some large mammals had disappeared, but there is

Archaeology at Monte Alén

Julio Mercader and Raquel Marti discovered recently important assemblages of stone tools in some caves and under overhanging rocks of the Monte Alén National Park in Equatorial Guinea. These tools belong to the Recent Stone Age or to the Middle Stone Age. They suggest that the dense forests of Monte Alén are inhabited since at least 40,000 years by human populations living from hunting, gathering, fishing and even the cultivation of some trees. The stones in which they have been made come from nearby rivers and quartz layers.

7.4. The Recent Stone Age, with the Central African culture (1), Wiltonian culture (2), Tshitolian culture (3), Gabonese coastal culture (4), culture intermediate between the two preceding ones (5) and Cameroonian culture (6).

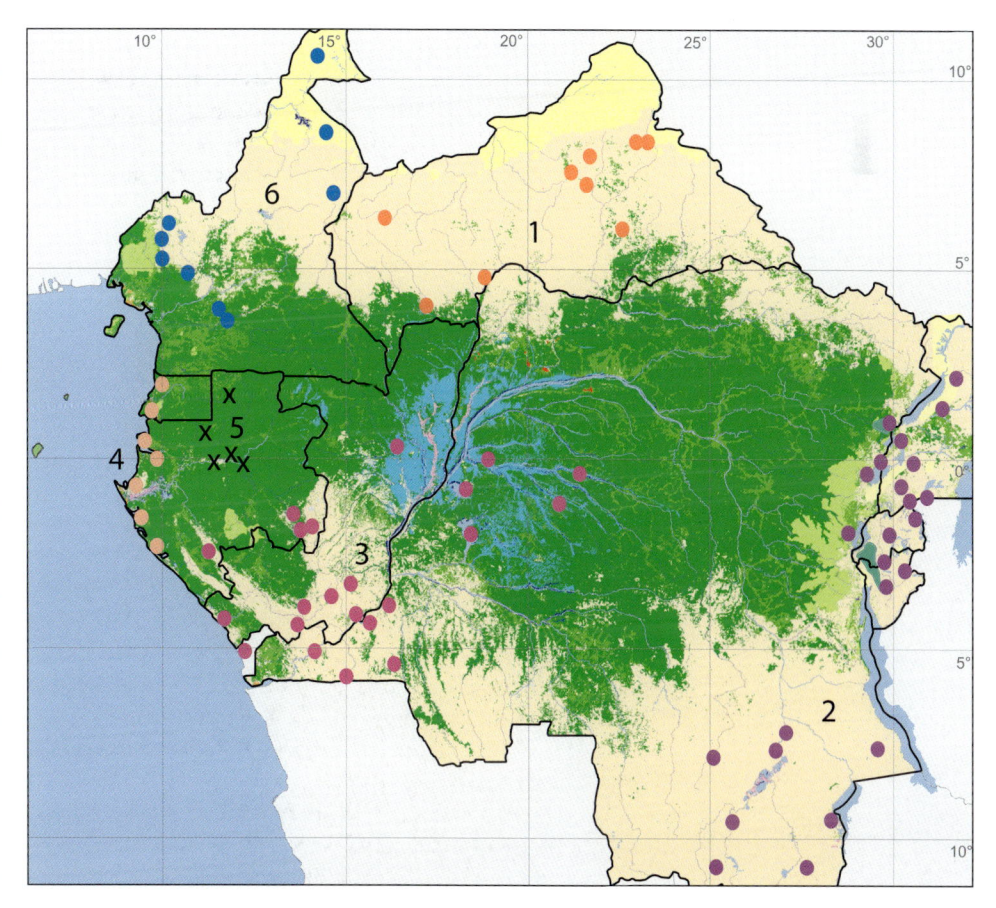

nothing to suggest that humanity was responsible. With the appearance of modern Man, the situation changed. In the past 130,000 years, seven genera of megamammals – mammals weighing more than 44 kilos – disappeared from the grasslands of eastern and southern Africa. Among these can be mentioned *Elephas reckii*, an elephant related to the Asian Elephant *Elephas maxima*, and larger than the present-day African Elephant *Loxodonta africana*, a giant buffalo *Pelorovis antiquus*, several species of Suidae, Giraffidae and at least two Hippopotamidae. Most of these animals were large,

7.5. The different groups of Pygmies are totally dispersed and have no contact with each other, yet they share most of their culture, notably in the domain of music.

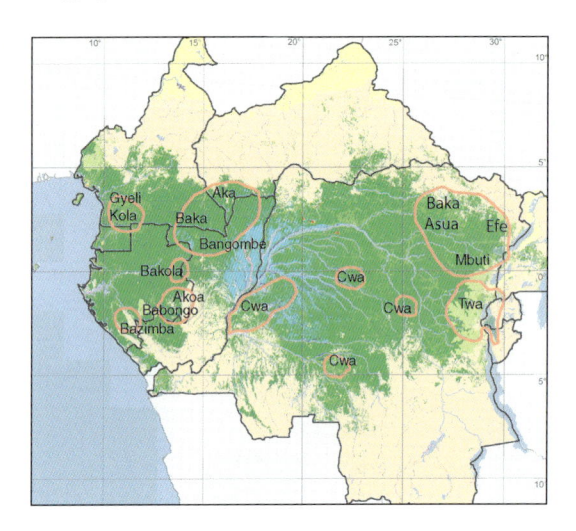

even very large. Their skeletons did not seem to be adapted to running and they probably were too slow, maybe also too unafraid, to escape the more and more perfected weapons of Man.

These extinctions at the end of the Pleistocene or of the Holocene were not a speciality of Africa and P. Martin attempts to show how the appearance of modern Man is accompanied just about everywhere by massive extinction. In Australia, in America, in Madagascar, in New Zealand and across Oceania, all regions colonised late, extinctions were often dramatic. In Madagascar, for example, colonisation did not commence until around the beginning of our era, but the extinctions were completed 900 years ago and all the large and medium-sized mammals wiped out. In Africa, in Europe and in Asia, where the history of Man began much earlier and where some co-evolution had been possible between Man and the large fauna, extinctions were relatively moderate and mostly much more spaced out in time. In this way, Africa lost 21 genera between 1.8 million and 700,000 BP, nine between 700,000 and 130,000 BP and 7 in the last 130,000 years.

For some researchers, these extinctions were linked to changes in the climate and the augmentation of seasonal contrasts. This hypothesis cannot be excluded for the oldest extinctions, but, for more recent ones, it does not hold. How could it explain that they did not happen synchronously across all the continents, that they affected only the very large species, that these species had not already disappeared during preceding climatic variations and that in the Americas they followed the advance of Man very closely? Without being able to totally reject the influence of climate in some cases and without being able to prove formally the implication of Man, everything seems to indicate that our "responsibility" was great.

These extinctions were limited to the grassland fauna, perhaps because animals of the humid forests were more difficult to hunt, perhaps because Man would not penetrate the forests until a good deal later, perhaps because the lost forest species did not leave any traces. Therefore, at first glance they do not concern forested Central Africa. From 50,000 to 10,000 BP, however, grasslands and their fauna penetrated to the very heart of the Congo basin. In fact, large herbivores certainly had an influence on habitats and the disappearance of some large species could not remain without some effect. Besides the direct impact that he had through fire, Man, for at least 50,000 years, therefore also had an indirect impact on the vegetation through hunting. Without even inhabiting the heart of the forest, he would have strongly influenced the forest recolonisation that had begun after the last glaciation. The frequent fires must have operated a strong selection on colonising species, as nowadays, and should have profoundly affected the nature of the forest-grassland transition. The fact that in many places this transition is today so abrupt could be the expression of it.

The arrival of the Pygmies

Man has inhabited the Congo Basin for at least 70,000 years, but he did it mainly when the grasslands were expanding. We have few indications showing that in those ancient times he ventured deep into the forests, except perhaps along large rivers. In the very heart of the forests, gorillas, chimpanzees and bonobos remained the undisputed masters.

About 20,000 to 25,000 years ago this situation began to change. At that time a first group of Humans started to "invade" the forest. Its offspring are the Pygmies: small sized people, physically and culturally well adapted to the life in the dense forest interior (FIGURE 7.6). Today they are found from Western Cameroon to Uganda, Rwanda and Burundi, but their populations are heavily fragmented and the different groups have no contact with each. As they leave no archaeological remnants, their history is very difficult to unravel. Working for the programme "*Avenir des peuples de forêt tropicale*" (APFT), funded by the European Union, Serge Bahuchet studied their culture and history. Comparing the few available data in the fields of genetics and linguistics, he was able to formulate some hypotheses, but much work remains to be done.

The main problem with the Pygmies is that they have no original language of their own. Each group speaks a different tongue, taken from the surrounding agriculturists (FIGURE 7.5). The Kola, Bongo, Twa, Efe and Mbuti speak a dialect, only slightly different from that of their neighbours. The Baka, Aka and Asua speak respectively a Ubangian, a Bantu and a Central-Sudanic language taken from the "Tall Blacks", but these languages have developed a great deal, and are no longer understandable for the people from whom they were taken. However, in many Pygmy languages spoken today words are found which are not taken from the agriculturists. These words are probably the last remnants of an ancient or several ancient, forgotten Pygmy languages.

The studies of Cavalli-Sforza in the domain of genetics show that Pygmies and Tall Blacks have a common origin but diverged about 20,000 years ago. They show also that eastern and western groups of Pygmies have some rare genes in common, but became separated from each other about 15,000 years ago. The Mbuti are also the Pygmies who are most different from the Tall Blacks. In summary, Bahuchet concludes that the Pygmies most probably colonised the interior of the dense forests 20,000 to 25,000 years ago. During the driest period of the last glaciation their different groups became isolated by the strong fragmentation of the forests. Later, when the forests recovered, they never got in touch again with each other. According to the linguists, the Baka from South Cameroon and the Aka from South-West Central African Republic have a recent common origin. Their separation is probably not older than a few centuries. Otherwise these two groups have no relationship with the other groups of Cameroon and Gabon. They seem to come from the East and probably have a common origin with the Mbuti living today in the north-eastern parts of the Democratic Republic of the Congo.

7.6. *For thousands of years the Pygmies have lived in basic huts made of a few branches and leaves, like the one of this woman from South Cameroon.*

Agriculture, metallurgy and pastoralism

Around 5,000 years ago, when the forests of Central Africa were still at their apogee, when the interior of the massifs were inhabited by Pygmies and when the descendants of the Tshitolians populated grassland enclaves or the banks of large watercourses, strangers coming from the north began to infiltrate equatorial regions. Armed with the first rudiments of new technologies – at first agriculture, later metallurgy and cattle-raising – they came to replace, absorb or marginalise the indigenous populations within a few millennia. By producing their own food, they were also going to very profoundly disrupt the relationships between Man and nature.

Their irruption into Central Africa therefore generated an enormous upheaval, of a magnitude comparable to the effects of the climatic variations of the last 15,000 years. Very probably, it was therefore not an effect of chance, but a direct consequence of these changes and of their environmental effects. During the long cool dry period that the first humans had known (PAGE 54), the Sahara was more extensive and the humid forests more reduced and more fragmented than nowadays. Very low density and isolation had engendered a differentia-

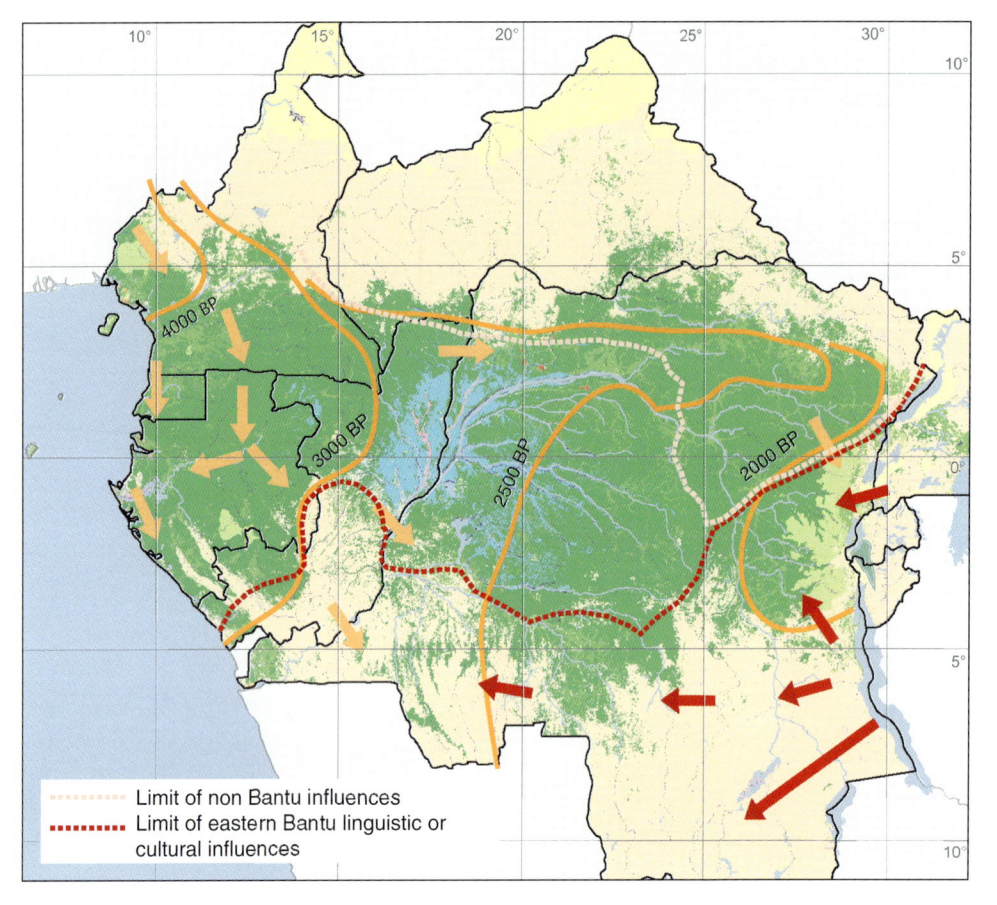

7.7. The migration of the Bantus began around 5,000 years ago. Four thousand years ago, the western Bantus arrived at Sanaga. Three thousand years ago, they occupied most of the forests situated between the Atlantic and the major swamps of the centre of the Congolese basin. Two thousand five hundred years ago, they had gone around these great swamps and reached the north-eastern part of the forest massif. By 1,500 years ago, they had conquered all of the Congo basin. Since then, however, their movements have never stopped. The eastern Bantus migrated to the north of the great forest to reach, 2,500 years ago, the region of the Great Lakes. From there, some colonised the eastern part of the Congo basin and the grasslands situated just to the south of the forest massif (black arrows), most of eastern Africa and southern Africa. Within the forest massif, the Bantus were, however, joined by Ubangian and Central Sudanian peoples whom they absorbed. South of the great forest, western and oriental Bantus finally mingled.

tion among the human populations as much physically as culturally and linguistically. The Tall Blacks inhabited the grasslands of western Africa. The Nilo-Saharans hunted in the vast plains of the Upper Nile. The Paleonegritics or Bushmen haunted the grasslands of eastern and southern Africa, from Somalia to the Cape. The Cushites occupied the shores of the Red Sea. As for the forests of Central Africa, from 20,000 or perhaps 25,000 BP, they were inhabited by the Pygmies (PAGE 242). During the few thousand years that the last glacial maximum lasted, all these peoples must have been driven back into refuges, which must have increased their isolation and accentuated their specialisation.

From 12,000 BP, however, deserts and semi-deserts were rapidly invaded by grasslands, while the most humid grasslands were colonised by forests. Animals followed the vegetation and hunters followed the animals. A southern branch of the Cushites made their way southwards to colonise a large part of the area around the Eastern Rift in Kenya and in northern Tanzania, driving back or amalgamating with the Bushmen they encountered. In the Congo basin, the Pygmies had probably profited from the extension of forest, but were not up to occupying or reoccupying the entire forest massif. The hunter-gatherers of grassland enclaves gathered along the major watercourses and lakes and became fishers. Farther north, hunter-gather-

ers of very diverse origins colonised immense parts of the Sahara, then covered with grasslands interspersed with lakes and wide rivers. Their innumerable rock paintings indicate that during the first millennia, they did not hunt desert animals but species of humid habitats such as the Elephant, the Buffalo, the Hippopotamus, the Giraffe, zebras and various antelopes. Around 7,000 BP, pastoralism made its appearance and pastoral civilisations very rapidly developed. Having reached their apogee around 6,000 BP, these civilisations had certainly engendered a net increase in the density of human populations.

From 5,000 BP, a new drought set in, however. The desert and antelopes of the sands slowly regained the lead. Some human populations remained on the spot and adapted for better or worse to harsher and harder conditions, but most humans began a retreat, some towards the north, others towards the Nile Valley, still others towards the south. The immensity of the territories affected signifies that these movements must have involved a great many people and that finally all of sub-Saharan Africa would be affected. As Europe's history was to be written by the cyclic surges of peoples coming from the east, that of tropical Africa came to be profoundly affected by people coming from the north. Many peoples of tropical and southern Africa still keep the memory of ancestors from the north.

The movements that were to involve all of Central Africa had their origin in the uplands situated on the borders of Nigeria and Cameroon. They began around 5,000 years ago. They reached their apogee 2,500 to 2,000 years ago, but in fact never again ceased completely. Ubangians from the Bénoué Valley in this way set into movement towards the east to colonise a large part of Cameroon, the south of Chad, most of the Central African Republic. Between 2,150 and 1,550 BP, they would reach the Ubangi and the Uele regions in the north-east of the Democratic Republic of the Congo. A little farther south, Bantus, peoples of the forest-grassland margin, set out towards the south and south-east. In 4,000 years, they would colonise all of Central Africa and most of eastern and southern Africa. From their very first movements, 5,000 years ago, they split, however, into two branches (FIGURE 7.8).

The eastern Bantus chose the route of the grasslands and headed eastwards and south-eastwards. Around 2,500 BP, they irrupted in this way into the Great Lakes region where they settled all along the eastern edge of the forests. From the Lake Albert region and from the shores of the Nile-Victoria in Uganda, they continued to push further south and, during the course of the first centuries of our era, they managed to colonise a large portion of eastern and southern Africa. They remained peoples of the grasslands and forest edges. Wherever they encountered forest formations they set about clearing them. Only the forests of the high mountains of the rift would initially resist their advance.

The western Bantus headed southward, crossing the Sanaga River and proceeding towards the estuary of the Congo. They were to colonise all of Central Africa and managed to reach the Kunene River on the border between Angola and Namibia, the middle Zambezi and the present-day border between Zambia and Malawi. Thanks to studies of their languages and to archaeological findings, it has been possible to retrace their migrations. Jan Vansina, in his captivating work *Paths in the Rain Forest*, retraces them in detail. In this way we know that the very first waves were Neolithic. In following the coast, starting in 3,950 BP, they reached the left bank of the Gabon estuary where one of their encampments was discovered near the Denis river, in front of Libreville. Until around 3,500 BP, they were confined to the coastal grasslands. From 2,750 BP, they set about colonising the Ogooué basin. Other groups pursued their route, following along the immense relict beaches and large lagoons in southern Gabon. They reached the Conkouati and Pointe-Noire regions, as well as that of the Bas-Congo around 2,150 BP. Around 2,000 BP, some groups also reached the island of Bioko where they would conserve their Neolithic culture intact until the arrival of the Portugese in 1472. The subsequent waves knew iron metallurgy. They followed the easternmost routes and in this way reached the Middle Ogooué well before appearing on the coasts of the Gulf of Guinea (FIGURE 7.8).

The western Bantus therefore in the first place populated the immense region situated between the Atlantic and the Sangha River. With the Ubangi and the Congo, this river formed, and still forms, a vast swamp difficult to cross for whoever does not know it. After some hesitation, a first group, however, found a passage to the north, crossed all of the northern part of the Congo basin and reached the Uele around 2,400 BP. A second group discovered a passage just to the east of the Batéké Plateau and reached the shores of Lakes Ntomba and Mai Ndombe. A third group crossed the Congo River a little downstream from Malebo Pool and disappeared into present-day Angola. The rest of the basin was occupied between 2,500 and 2,000 BP. However, this colonisation did not end before the first three or four centuries of our era, with the populating of the Kivu and Maniema, enclaved between the upper course of the Congo River and the rift. As these zones represented in fact the last free spaces of the epoch, they were reached almost simultaneously by western Bantus coming from the north and the south, eastern Bantus coming back from the east and some Sudanese populations coming from the north.

The conquest of the forested region of Central Africa by the Bantus therefore took 2,500 years. In part, it was done somewhat by chance, given that villages changed place every five to ten years. As for its general orientation, it was probably in large part conditioned by the environment. The vast swamps of the Congolese basin and the sandy plateaux, where water was sometimes difficult to find, represented obvious obstacles. The vast dense forests also. Because of this, the very first movements followed, coasts, coastal grasslands and watercourses. Ridges, clearings and enclaved savannas were not used until a good deal later. The most important movements, between 2,500 and 2,000 BP, coincided in effect with a phase of climatic worsening that opened wide breaches in the forest massif. Everything leads therefore to the belief that the Bantu migrations were greatly facilitated, if not rendered possible or perhaps even launched, by climatic conditions harmful to the forests that had begun in a fairly abrupt way around 3,800-3,700 BP and culminated between 2,500 and 2,000 BP.

By colonising the Congo basin, the Bantus obviously entered into contact with those who had preceded them. The Pygmies of the great forest interior were, however, so very few that they could not put up any resistance to the newcomers. Having had the time to learn well and thoroughly know their habitat, they became rather their guides and taught them how to find

7.8. Rocks bearing engraved designs are found in the grasslands bordering the Ogooué River in the Lopé National Park region. According to Richard Oslisly, they are the work of early Iron Age populations and are probably about 2,000 years old. They thus belong to the second wave of Western Bantu migrants who invaded Central Africa.

7.9. Elaeis guineensis *is a palm of African origin that is found in a natural or semi-natural state in all the forested regions of western and Central Africa. Some researchers think that its presence in Central Africa is mostly of human origin. It is true that Man considerably helped the dispersal of this species, but it must have existed naturally in this region, given that 7,000-year old nuts from this palm have been found, notably in Gabon, therefore well before the Neolithic migrants arrived. In the natural state this palm would have been a forest-grassland transition species, but in many places it also grows spontaneously in humid habitats.*

7.10. The fruit of Elaeis guineensis, *which become orange when ripe, are sought after not only by Man but by a number of birds, as well, notably the Palm-nut Vulture* Gypohierax angolensis *(PAGE 197).*

their way and how to survive in the dense forest. Relations were probably very different with the fishers of the Congo River. Because of the fact that the Bantus also fished, they very certainly entered into competition and conflicts were inevitable. For a long time, the two groups must have nevertheless lived side by side before the indigenous fishers, submerged by numbers, were absorbed.

Migrations did not stop, however, with the populating of all of Central Africa, From the 9th to the 14th centuries, there was thus a well-known depopulation of the extreme south of Cameroon, Equatorial Guinea, Gabon and the neighbouring regions of the Congo. No one knows the reasons, but, whatever they are, these regions must have been repopulated by peoples coming from peripheral regions, situated further to the north, east, or south. Even in the Congo basin, the history of some populations still relates significant migrations. These were somewhat interrupted during the colonial period and recommenced in the very last few decades. Today, migrant-pioneers have, however, very often become refugees.

The first agriculture

As we have just seen, the first western Bantu migrants were Neolithic. Along with pottery and stone polishing, they also knew the first rudiments of agriculture and lived in semi-permanent villages. As they did not depend on the spontaneous resources of their surroundings, the density of their population could increase. In the Ngounié Valley in Gabon, Clist thinks that this density could have been on the order of 0.2 inhabitants per square kilometre, which represented a population of around 50,000 individuals for the country.

Thanks to studies of their rubbish pits, this people's diet can be reconstituted. We know in this way that they lived mainly from hunting and knew meat smoking. Their menu included Bushbuck *Tragelaphus scriptus*, Mandrill *Mandrillus sphinx*, guenons *Cercopithecus sp.* and duikers *Cephalophus sp.*, but their favourite game seems to have been Buffalo *Syncerus caffer*, Giant Pig *Hylochoerus meinertzhageni* and cane rats *Thryonomys sp.*. Along the rivers and seashore, they also practised fishing but did not yet consume molluscs. They gathered the fruits of *Canarium schweinfurthi*, used oil palms *Elaeis*, that they propagated around their villages, and used many other plants. In the forest, they gathered yams *Dioscorea sp.* and some groups in Cameroon used a cereal. The Neolithic Bantus were therefore probably the first humans of Central Africa to clear the forest in order to cultivate. But their agriculture was mainly based on oil palms and yams. It remained therefore an activity of the forest edge, incapable of becoming established inside the heart of the forest massifs. As the migrants progressed, agriculture evolved, however.

The eastern Bantus, who had set off into the grasslands on the periphery of the forest massif, abandoned oil palms and yams to become, like the Ubangians, cultivators mainly of Sorghum *Sorghum bicolor*. This tall grass, originating from the Sudanese grasslands probably somewhere between the Nile and Chad, appeared from 3,000 BP in association with the squash *Lagenaria siceraria*, the small African aubergine *Solanum aethiopicum*, the Cowpea *Vigna sinensis* and tubers of genus *Coleus*. Having arrived in the region of the Great Lakes, the eastern Bantus also adopted finger-millet *Eleusine coracana*, a small montane grass, known from Ethiopia and the Albertine Rift highlands.

The western Bantus, who had plunged into the heart of the forest, would keep their yams and oil palms. As they progressed they spread this palm. They also learned to cultivate the Bambara ground-nut *Voandzeia (Vigna) subterranea*, the bean *Vigna unguiculata*, several species of squash including *Cucumerops edulis* and green vegetables including various amaranths *Amaranthus sp.* and some *Hibiscus*. Their basic tools were the planting stick and the stone axe, but this latter tool was soon going to be replaced by the iron axe and the machete. Each year a new forested space was cleared – between a half and one hectare per family – and surrounded by a solid fence to keep away wild animals. The women planted yams and other plants there. The next year only a small part of the field was reusable and the whole thing was then abandoned. Unlike grassland agriculture, cultivation in forests involved the association of various crops but ignored crop rotation. Next to the house, each woman had, however, a small personal garden where she cultivated a few vegetables, herbs and spices and a few medicinal plants. In spite of this diversity, productivity remained low and the conservation of food posed a problem. In fact, agriculture did not cover more than 40% of their needs. It offered glucides – rare in the forest – but provided only very few proteins and fats. Forest soils also are very rapidly exhausted, but the main limitation to the extension of this agriculture was the lack of manpower.

Contributions from the forest, by means of hunting and gathering, remained therefore essential. All around the fields, often even in the fences, traps were set out in this way, both to protect the fields and to "produce" some meat. This system of production was completed by some collecting of mushrooms, honey, crabs, molluscs, caterpillars, termites, medicinal plants and fruits, notably those of *Pachylobus edulis*, *Cannarium schweinfurthii*, *Irvingia gabonensis* and various *Cola* and *Ficus*. During the dry season, groups of villagers established temporary camps in the forest to hunt or fish over a greater distance. Collective hunting was practised mostly with nets. The women and children fished by damming up a stream and emptying it, sometimes after having drugged the fish with the help of a plant poison. The men fished rather with the use of lines, fish-traps (FIGURE 7.13) and nets.

Despite the diversification of resources and the development of complementary activities, this way of life did not allow population densities of over four inhabitants per square kilometre. According to Vansina, this figure was probably even a maximum which was never reached. With the rapid degradation of the soil and the growing rarity of game, through hunting as much as by disturbance, it was constantly necessary to

7.11. The polishing of stones is a characteristic activity of Neolithic populations (in red). It is known from Cameroon, Gabon, Central African Republic, Congo and west, west-central and northern Democratic republic of the Congo. It never reached the Great Lakes where human populations changed directly from Late Stone Age to Early Iron Age.

7.12. In high-altitude regions of the Albertine Rift, the western Bantus developed the cultivation of finger-millet Eleusine coracana. *For centuries, this small grass native to eastern Africa and recognisable by its digit-like spikes, constituted their main alimentary resource. Today, this crop is tending to disappear, but it still occurs in Burundi in newly cleared areas (photo opposite) and in some montane regions of Rwanda where the exhausted soils no longer support other crops.*

find new land. This is why, among other reasons, villages were displaced at least once every ten years.

In some marginal regions on the edge of the main forest massif, this first agricultural activity already brought local overpopulation with it, however. Probably in combination with the aridity of the climate, these factors led to ecological crises. Just before 2,000 BP, such a crisis broke out in the Akanyaru Valley in Rwanda. On the basis of charcoal remains gathered in the vestiges of low furnaces, one can guess that dense sub-montane forest galleries, where African mahogany *Entandrophragma excelsa* grew, were completely cleared and that the inhabitants evacuated the region, never to come back again until two centuries later. The forests had by then been replaced by riverine stands of *Acacia* and wooded grasslands. It is, however, very difficult to determine which is the effect of climatic conditions and which is that of Man in the genesis of this crisis. And it is very probable that there was a combination of both factors.

At the north-eastern limit of the forest, in the Ubangi and Uele regions, different peoples developed the Sudanian agricultural system. It consisted of cultivating the same land parcel gained from the forest for several years in a row in a way that completely exhausts it and prevent any forest regeneration. Enormous expanses of forest were in this way irreversibly degraded in the north-eastern part of the Democratic Republic of the Congo, in the Central African Republic and in Cameroon. South of the great forest, on the other hand, more precisely in the forest-grassland mosaic zone of the Kwilu, Kwango and Kasai regions, inhabitants developed – still according to Vansina – an agricultural system based on the cultivation of two different plants during the same year. The first being a cereal planted on a parcel won from the forest, the second was the Bambarra ground-nut *Voandzeia (Vigna) subterranea*. Rotation was known, but the association of crops not very developed. Although less harmful for the soil than Sudanian agriculture, this form of cultivation also led to the degradation of soils.

Iron metallurgy

Iron smelting appeared in Niger around 3,000 BP and in the Great Lakes region around 2,500 or 2,600 BP, or maybe 2,750 BP. The Belgian archaeologist Claire Van Grunderbeek even found a furnace dating back to 3,600 BP in the centre of Burundi. This very ancient date, which poses serious problems of interpretation, has never been confirmed, however. Whatever the case, iron metallurgy was known in Gabon from 2,300, perhaps even 2,400 BP in the Ogooué Valley and from 1,900 BP in the vicinity of the estuary (FIGURE 7.14). In the 2nd century, it appeared in the Lower Congo, in the 3rd century on the Batéké Plateau, in the 4th century in the Malebo Pool area; between the 2nd and the 4th centuries, it reached the woodlands of Katanga and in the 6th century it arrived in the Transvaal. It was necessary to wait until the 13th century, however, for its appearance in the centre of the Congo basin.

According to what is known today, it seems that iron metallurgy was propagated from a single African source, totally independently from the activities of the high Anatolian plateaux. It was situated in south-eastern Niger and was placed therefore well to the north of the starting point of the Bantus, who did not discover it but only disseminated it. Dissemination was, however, extremely rapid and caught up with the first migrants. Those who reached the Great Lakes region were in this way metallurgists right away and this region never knew a Neolithic culture.

In this same region, the culture of the first metallurgists was otherwise replaced from the 6th century by a more recent culture, one which used furnaces of a very different type and produced decorated ceramics turned on the wheel identical to those that potters still make today in Rwanda, Burundi and Kivu. Having no ties of kinship with the preceding one, this culture was therefore not the result of a local evolution but an innovation introduced by a new population. For about two centuries the two cultures co-existed, however, notably in

7.13. For thousands of years, people bordering large watercourses of Central Africa fished with the aid of hoop nets, like these still in use today in the rapids of the Ogooué near Booué.

Rwanda. In Gabon, also, the ancient tradition of iron working was replaced by a more recent culture, but only from the 17th century. This new culture was in every way similar to the one that the Fangs were still using at the time of the Europeans' arrival.

The introduction of iron metallurgy had a considerable impact. Not only did it allow the development of more efficient armies, but it also profoundly transformed social exchanges and cultures. In fact, iron constituted a negotiable material that came to be added to salt and worked stones. since easily exploitable layers of good minerals were not to be found everywhere, iron metallurgy would stimulate commerce and trade.

Copper metallurgy

In the Niari Valley, in present-day Congo, a small centre of copper smelting also developed during the 12th and 13th century (FIGURE 7.14). This industry remained very localised and served only to produce metal for the manufacture of necklaces and bracelets. Its importance was therefore only miner and did not change the course of history nor of relationships between Man and nature.

Pastoralism

The first domestic cattle made their appearance in north-eastern Africa around 7,000 BP. Very rapidly, they were known across the entire Saharan region. The history of these animals was summarised by Hugh Lamprey. They came from two wild species, domesticated in Asia around 10,000 to 11,000 BP. The Aurochs *Bos primigenius* was a large, long-horned bovine of the temperate forests of Europe, Asia and North Africa. The Uru *Bos namadicus* came from India and perhaps other tropical regions of Asia. Compared to these ancestral species, the animals that arrived in Africa were already considerably evolved and belonged to three main races: the long-horned race, the short-horned race and the zebu. Nowadays, the long-horned race is common from Senegal to Uganda while the zebu is found further east. Everywhere, however, where these two types have mixed, they have given rise to sanga cattle characterised by their long horns and a small hump, as can be seen in the *Ankole* cow of the Great Lakes region.

Still according to Lamprey, the history of goats and sheep is more difficult to follow, but it was mostly similar. Goats descended from the Ibex *Capra aegagrus* of the mountains of south-western Asia, while sheep descended from the Arkhar or Argali sheep *Ovis ammon* of Central Asia, the Urial sheep *Ovis vignei* from south-western Asia and the European Mouflon *Ovis musimon* of Asia Minor and Europe.

The domestic ungulates of Africa therefore all come from Asian species, domesticated in Asia. The ancient Egyptians did try to domesticate antelopes, but these efforts were in vain.

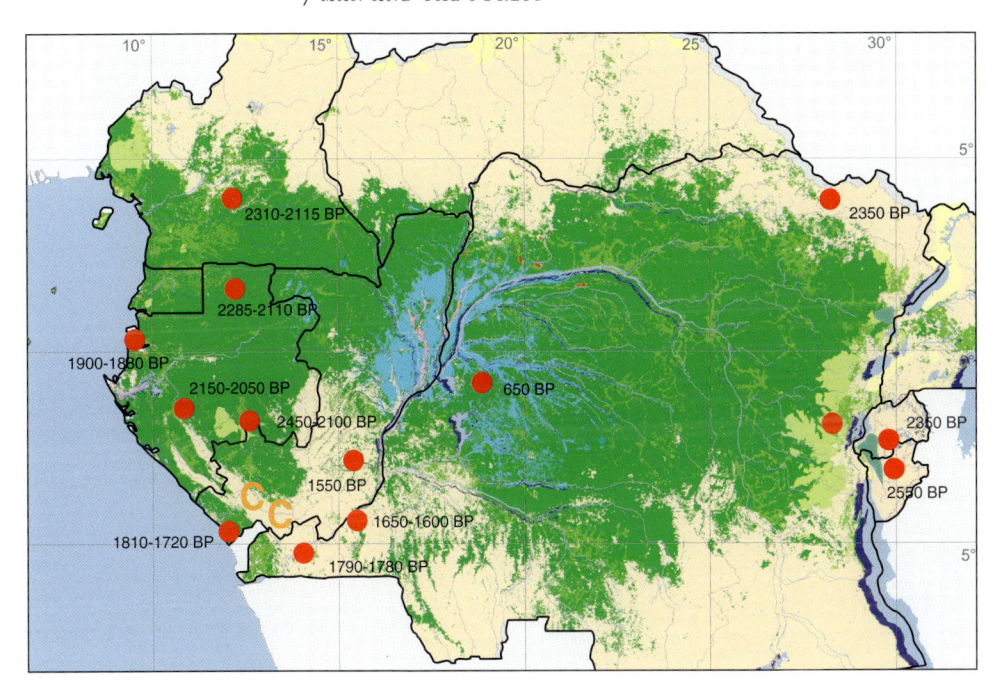

Around 4,000 BP, the first cattle raisers arrived in the Eastern Rift in Kenya. Around 3,000 BP, goat and sheep pastoralists irrupted into the Yaoundé region in Cameroon. The penetration of pastoralism into the humid grasslands and forests of Central Africa, however, was limited by the absence of pastures, and probably mostly by the presence of Tsetse Flies *Glossina sp.*, vectors of various kinds of *Trypanosoma*, unicellular parasites living in the blood of some animals and able to inflict serious illness. Apart from upland regions – we shall return to this – cattle remained on the periphery of the great humid forests where only goats and dogs accompanied the Bantu migrations. Chickens arrived much later, perhaps in the 10th century. The dog became the companion of hunters, while the goat often became an object of ritual, mainly reserved for circumcisions, enthronement of chiefs and marriages. To this day it is a more or less untouchable status symbol in many places, even in regions where game is depleted and protein deficiency and malnutrition are chronic.

It is only in the mountains of the Albertine Rift and Cameroon that pastoralists would have an impact on the forests. However, the mountains of the Rift were protected by a continuous barrier of dense moist forests that extended from the slopes of the Rwenzori to the shores of Lake Victoria, and it is therefore improbable that pastoralists coming from the north would have been able to penetrate southern Uganda, Rwanda or Burundi, or for similar and even more convincing reasons in the east of the Democratic Republic of the Congo, before this barrier was "permeated". This was done perhaps from 3,800 BP, but more probably be-

7.14. *Iron metallurgy appeared around 2,300 BP in southern Cameroon and in Gabon, but it was already known two to three centuries earlier in the Great Lakes region. Iron was known in all the savanna regions on the periphery forest around the 5th century, but it did not appear in the centre of the Congo basin until the 13th century. In forested Central Africa, copper metallurgy (C) remained confined to the Niari Valley in present-day Congo.*

7.15. A herd of cattle of the Ankole type on the high hills of central Rwanda around 1970, before they were massively reafforested with exotic species. These hills were still covered with natural forests 2,500 years ago. After having been cleared by farmers, regeneration of these forests was prevented by pastoralists and their cattle. Independently of all other considerations, Afro-montane forests differ fundamentally in this way from lowland forests in that they offer no resistance to pastoral activities.

tween 2,500 and 2,000 BP, during the brutal climatic worsening which fragmented the dense forests (PAGE 58). Openings very probably appeared in the Mubende region, half way between the Rwenzori and Lake Victoria. In Rwanda, the most ancient traces of cattle date back only to the 3rd century of our era. But this dating comes from the Akanyaru Valley. This region was originally mainly forested, and was not itself open until by the metallurgist-agriculturalists or the worsening of the climate. It is therefore most probable that cattle had been present a good deal earlier in the sub-arid grasslands of north-eastern Rwanda and southern Uganda. In fact it seems that the pastoralists originally settled in dry grasslands that had probably been "freed" by the repeated fires of the hunter-gatherers. The abrupt disappearance of these latter from the region of Akagera National Park around 2,200 BP, as revealed by the diggings of Bernard Lugan, could perhaps coincide with the arrival of the pastoralists. A good deal later, the descendants of these pastoralists would colonise central parts of Rwanda and Burundi from where they would keep pushing westwards, conquer the highlands of the Rift in the middle of the forested region and, indirectly, contribute to the irreversible transformation of the montane forests.

As for the mountains of Cameroon, they are backed up directly against the grasslands that extend northward from the forest massif. However, the pastoralists remained for a long time on the high plateaux of the Adamaoua and it is not until during the 19th century, therefore very recently, that they irrupted onto the Bamiléké plateaux and the Bamenda Highlands.

The introduction of banana plants

The shortcomings of the first agricultural practices were largely compensated by the introduction of banana plants (FIGURE 7.16). Formerly, their numerous varieties were attached to *"Musa sapientium"* and *"Musa paradisea"*. Nowadays, through genetics, we know that they all come from two wild species: *Musa acuminata*, originating from India, Indonesia and south-eastern Asia; *Musa balbisana*, from India, Burma and the Philippines. Some varieties are diploid or triploid clones of *M. acuminata*, notably the "Gros-Michel"; others, notably the plantains, are triploid clones of hybrids of *M. acuminata* and *M. balbisana*.

According to Vansina, their introduction had more impact in Central Africa than that of pastoralism and perhaps even of iron metallurgy. Bananas are indeed well adapted to tropical humid forests and their culture requires less manpower and produces more than yams do. Therefore they constitute an abundant source of food, easy to come by and reliable. The way bananas reached Africa remains a mystery, but it is certain that there was more than one arrival. The first introduction took place before 2,500 BP and by around the year 1000 banana plants were known throughout Central Africa. The plantain was introduced first and became the dominant species in eastern Africa in the region of the Great Lakes. As Uganda seems to have been a centre of dispersal, it is possible that the Nile Valley played an important role in the diffusion of these plants.

In any case, their introduction strongly stimulated the growth of populations of farmers at the expense of hunter-gatherers – the Pygmies – contributing in this way to throwing them into dependence and progressive marginalisation.

Perhaps at the same time as bananas, perhaps independently, other plants of Asiatic origin arrived in Africa: the cocoyam or taro *Colocasia esculenta* (FIGURE 7.16), originating from Afghanistan, the egg plant *Solanum melongena*, from India, and sugar cane *Saccharum officinale*, from New-Guinea. Altogether they were less of a success than the banana and did not succeed in being established everywhere. Nevertheless, sugar cane became the essential crop on the island of Bioko. Taro spread well in the east, especially in the most damp milieus. It became very well established, for example, in marshy valley bottoms of Rwanda and Burundi. In Atlantic regions, on the other hand, it only arrived with the coming of the Europeans.

The integration of cultures

Around the year 1000, all of Central Africa was occupied by Man. Ubangians and Central-Sudanians practised grassland agriculture and pastoralism, being installed all along the northern edge of the forest massif. The eastern Bantus settled all along the eastern edge of the forest east of the Albertine Rift. They were grassland peoples who cleared the forest without ever inhabiting it. Only a few groups had traversed the montane forests of the Rift towards the west. In the 3rd century, the ancestors of the Tembo, who nowadays live around the lower part of the Kahuzi-Biega National Park, had in this way poured into the Congolese basin on the eastern side. There were also some cattle herders having obscure affinities as much with the Nilo-Saharans as with the Cushites installed in the sub-arid grasslands immediately to the east of the forest massif. In the south, the grasslands and gallery forests of Katanga, Kasai and Kwilu had been occupied by western Bantus, but, starting in the 1st century, eastern Bantus, who were cereal farmers, infiltrated the region, bringing to it their cultural and linguistic influences – better adapted to grasslands – and finally became integrated with the western Bantus.

As for the interior of the forest massif, it was essentially inhabited by western Bantus. Trapped by the rapid and generalised forest regrowth begun in the 1st century, they were transformed into true forest people. Here and there, however, isolated groups of Pygmies survived and populations of Ubangians and Central-Sudanians coming from the north or north-east penetrated the forest throughout the first millennium and finally amalgamated with the Bantus. Some Central-Sudanians even arrived in the stretch of land between the Lomami and the Lualaba rivers, nearly in the heart of the basin, and left there a profound influence on the local Bantus.

From the year 1000, all new increases in population inevitably brought with them an increase in population density. Groups that had lived until now side by side, or who perhaps even avoided each other, could no longer prevent contacts. Conflicts in this way became inevitable and groups sharing similar life styles inevitably entered into competition. In the absence of human-rights associations, there was no other way than elimination or integration. For groups who used different resources of the milieu instead, co-existence was possible through the development of complementarity.

In the lowland forests, history was mostly dominated by rivalries between more or less similar Bantu groups in this way. Simplifying somewhat, Vansina tells us how they avoided conflict by integrating and gathering together. Instead of killing each other, the competing groups created larger and larger entities. In the

7.16. A banana plantation with cocoyam or taro Colocasia esculenta *in the foreground, a typically Asiatic association of crops.*

7.17. Unlike most eastern Bantus, western Bantus do not hesitate to settle inside of the forest.

beginning, western Bantus lived in small villages numbering hardly more than 100 persons and gathered in associations of five or six "friend" villages, linked by paths in the forest (FIGURE 7.17). Over the centuries, this grouping together extended and annexed a growing number of villages and inhabitants. Finally entities grouping together several tens of thousands of individuals, even some hundreds of thousands, came into existence. However, at the same time, there was a strong resistance to the establishment of states in the western sense of the term. The Kingdom of the Congo was perhaps an exception – although of short duration – but altogether these forest peoples would be torn between two strong but opposite tendencies: one egalitarian, the other centralist. In spite of the collapse of tradition, these tendencies are still alive to this day.

In the regions inhabited by Pygmies, history was a little different. The Pygmies, very few in number, were not able to oppose the advances of the Tall Blacks – Bantus and Ubangians – but they knew their surroundings very well. Hunting procured proteins for them and, with the gathering of fruits and seeds, they obtained enough fatty material. Their source of glucides, on the other hand, was limited to just a few yam roots *Dioscorea sp.*. In fact, yam production in the forest is generally low and on the average around 90 hectares of forest is needed to nourish a human for one year. Therefore, the density of hunter-gatherers could not exceed one inhabitant per square kilometre. As for the Tall Blacks, they possessed agriculture and produced starches the Pygmies needed. A system of exchanges based on complementarity was very naturally established. However, the Pygmies, being the demanders and very few in number, were subjugated to the Tall Blacks everywhere. They also adopted the languages of the Tall Blacks and kept no more than a few roots of their own languages (PAGE 247).

Studies by John Hart in the forests of the Ituri permit us make out that in most of these forests, consisting essentially of stands of *Gilbertiodendron dewevrei*, the Pygmies could never have survived alone. Their presence in these massifs would not have been possible except in coexistence with farmers. In other words, in this region of the Central African forest massif, the Pygmies would never have been able to settle except at the same time as, or after the Tall Blacks. Curiously, it is in this very same region that Pygmies appear to have suffered least from foreign influences. The subject is therefore not exhausted and much effort must still be deployed before we understand exactly what took place

In highland regions on the edge of the Albertine Rift, relations were more complex because of the presence, not of two, but of three groups of humans. Only the Pygmies, who lived from hunting and gathering, were true forest people but, as everywhere else, they were very few in number. The others were all grasslands people. The eastern Bantus lived from permanent agriculture, based essentially on sorghum and finger-millet or bananas, notably in Uganda. As for the pastoralists, they covered the semi-arid grasslands under a very unpredictable climate and made do essentially with milk and blood from their cattle.

At first glance it should not be necessary to dwell on these people within the framework of this book, but history shows us that their role was not going to be negligible.

These three groups initially inhabited very different surroundings and possessed very different cultures. Because of the irregular topography of the region and its very abrupt eco-climatic gradients, they lived very near to each other, however, and in many places divisions even frankly overlapped. As in the lowland region, they likewise established relationships of complementarity and the different groups adopted the language of the farmers. The eastern Bantus' agriculture, intensive and with brief periods of leaving land fallow, rapidly led to the development of population densities greatly exceeding four persons per square kilometre. The spontaneous resources of the forest, even if they might have interested them in the beginning, were no longer sufficient and the forest finally represented no more than a simple piece of land to clear. As they progressed, they destroyed it. Unlike the western Bantus of lowland forests, they did not open up small, more or less scattered clearings, but advanced on a broad front which progressed inexorably westward.

In regions with a rugged relief, however, they could establish their permanent fields only on hilltops or in small valleys, the only places where soil could accumulate. In other locations, they were very rapidly forced to abandon the shallow, soon-depleted soils carried off by erosion. This process was still visible around 1970-1980 on the Congo-Nile divide in Rwanda and Burundi where lands newly gained from the forest were impoverished and abandoned after 20 to 25 years. While advancing, these farmers left behind them in this way immense desolated areas, which were covered with poor fern heath, *Eragrostis* grassland or heather heath (FIGURE 7.18).

East of the Rift, these degraded lands did not remain unoccupied. Nomadic pastoralists wandering the grasslands just to the east and always in search of new pastures, quickly learned to put them to good use. These uplands were cooler and healthier for livestock than the natural grasslands and represented a real windfall. However, the putting into pasture of these lands ended any chance that forest reinstallation would see the light of day. After having lived apart – each in his own milieu – farmers and herders came closer together and their coexistence overlapped increasingly. Since the farmers had eliminated most of the forest and game, they suffered from a deficiency in proteins and fatty materials which only the pastoralists were able to produce. Animal-raising offered, in fact, an unvaried but well balanced food rich in proteins and fats. It allowed great mobility and rendered the herders less sensitive to the caprices of the climate than the farmers. Agriculture, on the other hand, was still not very diversified and of low profitability. Conservation of provisions was extremely difficult and the unpredictability of the climate did not help matters. Sometimes it rained too much and harvests rotted, sometimes not enough and they dried up on the spot. In these highlands, farmers were therefore the demanders. In exchange for milk and a few head of cattle, they placed themselves in the service of the herders.

This relationship found its expression in social structures. Indeed, from the 13th or 14th century onward, political groupings formed. They were to lead to the creation of the interlacustrine kingdoms. In the previously forested regions of high mountains, these mostly kept their agrarian character. In grassland regions located on the border of the ancient forest massif, they took on a much more pastoral aspect. But everything formalised the superiority of the pastoralists over the farmers.

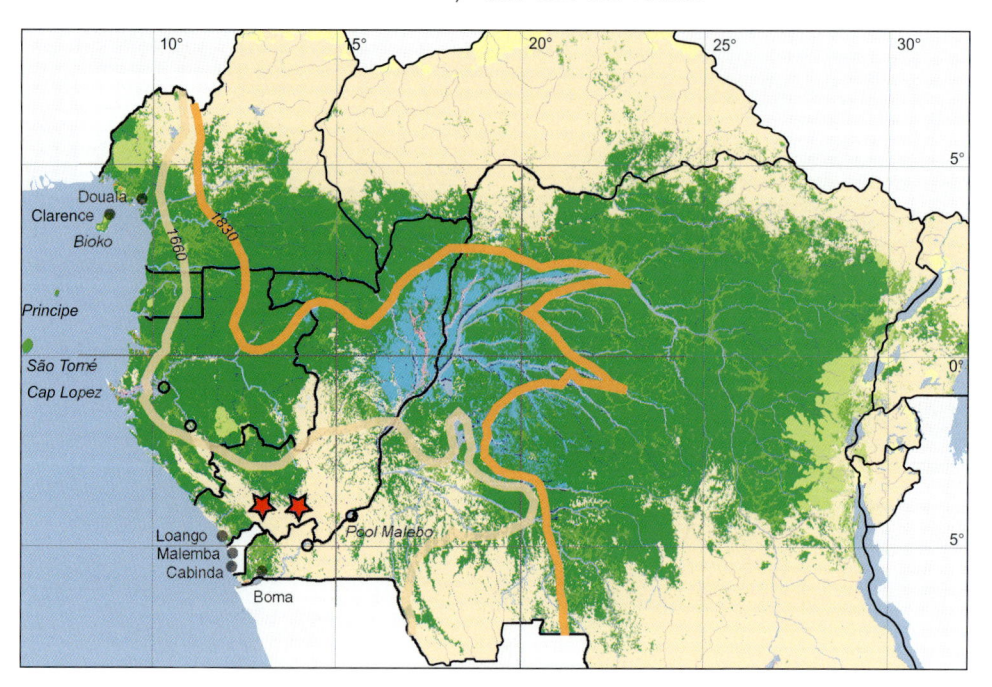

7.18. *Already in 1561, the slave trade front reached Lake Mai Ndombe; in 1568 it arrived at the Malebo Pool. In the north it remained closer to the coast, but in Cameroon it got to the Bamiléké Plateau. The main ports of slave exportation were Loango, Malemba and Cabinda. Boma took over after 1800, while farther to the north, it was Douala's turn to develop (according to Vansina).*

The Atlantic slave trade

In the 16th century, the social and cultural evolution of Central Africa would be profoundly perturbed by the arrival of the Europeans. The Portuguese had been the first to touch the coasts of Central Africa. By 1471, they were established on the island of the Gulf of Guinea that they named São Tomé. In 1472, they discovered the island of Bioko that they baptised Fernando Po. From there, they went on to explore the coasts of the continent and discovered the immense ria which became Rio de Gabão. By 1473, Lopez Gonsalvo had given his name to the cape situated just north of the mouth of the Ogooué and Fernan Vaz had made a reconnaissance of the coasts lying between the Ogooué and Sette-Cama Lagoon. In 1482, Diego Cam finally reached the mouth of the Congo.

Christian missions and merchants' trading posts were very rapidly established. By 1500, however, the slave trade became the main preoccupation. It attained its apogee in the 19th century and did not disappear completely until 1900. The discovery of São Tomé must have played a decisive role in its development. The Portuguese decided in fact to transform this uninhabited island into a vast sugar cane plantation. It was therefore necessary to import manpower. This decision gave rise to the first major slave trading centres on the coast of the continent, particularly in the Kingdom of the Congo.

In the beginning, slaves were obtained on the coast itself. But this caused too much trouble and from 1520 the "hunt" was displaced towards the interior. The Europeans remained on the coast – often even on their ships, for fear of tropical diseases – having the captives delivered by populations from the interior who passed them on to each other all along the trails. In this way an entire commercial system arose whose front stretched constantly eastward. Ahead of it troubles arose. Behind, the destabilised society reorganised in function of a new reality.

From 1660, the English, French and Dutch joined with the Portuguese and the number of slaves exported rose rapidly to reach a peak around 1755-1793. From the end of the 18th century, exports dropped off, however, because of the Napoleonic wars and the official abolition of slavery by Great Britain in 1807. The departure of the English and French merchants was, however, compensated by the appearance of the Spanish and Cubans. At the same time, the slave trade abandoned the too closely watched eastern African coasts to fall back on those of Gabon, the Congo and Angola.

From 1830 to 1880, the slave trade underwent a new transformation. The English and French reinforced their surveillance of the coasts. From 1839, France settled on the banks of the Gabon Estuary. In 1843 Fort d'Aumale was built and Libreville would spring up alongside; in the following years, French control extended to the entire Gabonese coast, from the Komo Estuary to Rio Muni. During this same period, Europe was in full industrial growth and searched both for raw materials for its industry and for outlets for its products. Commercial trading posts therefore appeared all along the coast. Kribi appeared in this way around 1828, opening up the entire south and south-east of Cameron. But with slavery being officially abolished, merchant companies turned to ivory, palm oil, peanuts, ebony and, from 1870, rubber. This switch brought with it notably the creation of specialised groups of elephant hunters, such as the Kélé in the back country of the Ogooué-Ngounié delta. As for slaves, they were no longer exported but used by industrial plantations. On the Ogooué, however, clandestine exportation continued until around 1870 and on the Congo until around 1900.

During the four centuries of slave trading, nearly two-thirds of Central Africa was disrupted socially. Demographically, it was a period of stagnation. Vansina estimates that 1.2 million slaves were exported from the coasts of Central Africa between 1660 and 1840, while the same number of victims had certainly already perished during the troubles and transport to the coast. The total population of the region must have been of the order of four million inhabitants. This loss of 2.4 million individuals was considerable, but, spread over 180 years, it was probably insufficient to reduce the population as a whole. At most, the depletion kept it from increasing. With local slavery, however, whose numbers are impossible to evaluate, there might even have been a reduction of population.

Independently of its demographic and socio-cultural effects, the slave trade also had important effects on the distribution of human populations. In many places, populations fled the hospitable plains and valleys to withdraw into regions difficult of access. It is probably at this epoch that the uplands of the Congolese Mayombe and the highlands of the interior of Bioko were populated. By provoking large-scale displacements of populations, the slave trade therefore considerably modified pressures on the environment.

American plants

In addition to the localised demographic and cultural consequences to regions that it affected, the Atlantic slave trade also had indirect but very profound and somewhat unexpected repercussions on all of sub-Saharan Africa.

From the 16th century, slave traders were confronted with the problem of food for their captives. At this time, the peoples of Central Africa did not cultivate except to satisfy their own needs and disposed of only very little surplus. From 1550, the Portuguese had the idea of importing new plants from America, more productive than those used in Central Africa: cassava *Manihot esculenta* (FIGURES 7.19-7.20), maize *Zea mays*, peanuts *Arachis hypogea*, beans *Phaseolus vulgaris* and sweet potatoes *Ipomoea batatas*. At the same time, they also introduced peppers *Capsicum annuum* and tobacco *Nicotiana tabacum*. Within half a century, these plants were adopted in all the coastal regions. From there they conquered the interior and arrived well beyond the slave trading front. Cassava spread especially in the forest, maize in the grasslands and the peanut in sandy regions, while American beans supplanted several varieties of African lentils in highland regions.

The introduction of American plants deeply influenced the economy of sub-Saharan Africa. Not only did it increase and diversify production but, for the first time, it also provided the possibility of fairly durable food reserves in the form of roots and tubers. In addition, beans were an appreciable source of plant proteins. From this period, agriculture was therefore capable of nourishing more people. In regions where herders and farmers coexisted, such as around the Albertine Rift, the farmers became more independent. The impact of American plants was probably greatest in the highlands. In these cool-climate regions, away from the major tropical illnesses, demography was in fact limited mainly by periodic famines.

The introduction of American plants was of course accompanied by adaptations of cultivation techniques because if these new plants produced more they also consumed more. It is in this way that the use of fire was intensified: after a new field had been cleared, the cuttings were systematically burned and the ashes used as fertiliser.

In regions little or unaffected by slavery, accelerated increase in population set off a heightened quest for new space. The farmers of the high plateaus of Rwanda and Burundi, backed up against the sub-arid grasslands to the east, had no other alternative than to rush to the assault of the high mountains of the Congo-Nile divide to attack the immense forests still intact there. This is how the Bwindi and Virunga forests were separated in

7.19. Like many other tropical plants, cassava Manihot esculenta *defends its roots against aggressors with toxic substances, related to cyanic acid. Before they are fit to eat, it is absolutely necessary to destroy these toxins by retting, usually in a stream, sometimes simply in a vat filled with water. Only sweet varieties can be consumed without particular precautions. In addition to roots, cassava also produces young leaves that can be pounded and eaten like spinach, with or without the addition of dried, salted or smoked fish. Unlike banana plants, cassava grows very well on poor or degrading soils, or soils incompletely regenerated after brief fallow periods. It also resists weeds well. There is presently an increase in the area planted to cassava to the detriment of areas with plantain.*

7.20. *The great success of cassava is largely also linked to the fact that it is easily propagated: sticks put in the ground are all that is needed to start a plantation.*

255

7.21. In a great part of the Great Lakes region, the driest forests were completely cleared to make way for crops. but the shallow soils of the hills were rapidly impoverished or carried away. In a few tens of years, these forests were in this way transformed into heath-lands and poor pastures. It is only on the upper part of the foothills, where colluvia from the slopes accumulate, that an intensive and permanent agriculture could be developed. The more or less swampy valley bottoms, became pastures during the dry season. This process engendered typical landscapes that could still be found around 1975 in Bututsi, a region of southern Burundi (below). More recently, as in all of Rwanda and Burundi, the poorest pastures were planted with eucalyptus or other exotic trees, notably the cypress Callitris calcarata, *while the bottoms of the valleys were invaded by intensive cultivation.*

Uganda, as were the Nyungwe and Gishwati forests in Rwanda. As for the Twa Pygmies, cut off from the forests of the Congo basin by the Rift and having nothing more left to trade after the destruction of their habitat and the decline of game, they had no other choice than to adapt, for better or worse, to the society of pastoralists and farmers. It is perhaps in this way that they became potters.

In socio-political terms, the increase in population density increased tensions, provoked a concentration of power and a reassembly of local state-like entities. In the course of the 18th century, however, the droughts of the "little ice age" occurred. These were unfavourable to the farmers and reaffirmed their dependence on the herders. They therefore had an effect opposite to that of the American plants. At the same time, the pastoralists lost all interest in the most arid grasslands and accentuated their push westwards. It is probably in this way that they appeared on the foothills of the Virunga Mountains, in the high mountains of Itombwe, in the south-west of the Bukavu, in the mountains west of Lake Edward and, much later, in the region of Masisi west of Goma.

The quest for new space also brought with it wars of conquest that ended up with the progressive reassembly of political entities. Six major kingdoms arose. Buganda, a state based mainly on banana growing, arose in the sub-montane forest zones bordering Lake Victoria,

In higher regions along the Albertine Rift, Bunyoro, Toro, Ankole, Rwanda and Burundi were formed. These were kingdoms based on pastoralism and the cultivation of finger-millet, sorghum and beans. Just before the colonial seizure, Rwanda had even started to extend its power to the west of the Rift.

In regard to habitat, the Great Lakes region was characterised by the total absence of villages, each family inhabited its more or less isolated enclosure. In very rugged regions, farmers remained at the top of the foothills, while herders exploited the summits in the rainy season and the swampy valley bottoms during the dry season. (FIGURE 7.21).

The high-altitude regions of central and western Cameroon were also colonised and cleared for a very long time by successive waves of farmers coming from regions situated farther north, mainly the Bénoué Valley. Their forested landscapes gave way to "grass-fields" and only the most inaccessible summits kept some forest remnants. In the 19th century, movements of farmers were precipitated by the expansions of Islamised Peuls throughout northern part of the country. From the high plateaux of the Adamaoua, whose landscapes they had already profoundly remodelled for several centuries, nomadic or semi-nomadic pastoralists pushed towards the south-west. From the 19th century, they thus reached the Bamiléké Plateau in small numbers. Just as the herders of the Albertine Rift, they kept abandoned fields from ever being recolonised by the forest, but they remained very independent and integrated little or not at all with local cultures.

The prelude to colonisation

Even though greatly disturbed by the Atlantic slave trade, the interior of Central Africa continued its evolution out of sight of foreigners. Towards the middle of the 19th century, this isolation would, however, be broken. The Muslim states, set up in the very north of Cameroon by the Peuls or Fulbés, extended their power to the Sanaga River. Ngaoundéré thus became, near the middle of the 19th century, an important centre from where slave raiders took off towards the forested regions of southern Cameroon and northern Congo, notably the region of the upper Sangha River (FIGURE 7.22). These activities destabilised the regions affected and pushed some populations still further into the forests. However the troubles remained localised, and were not as extensive as those caused by the Arabs in the north and east of the Congo basin.

The Arabs

The Egyptians invaded the Sudan in 1820 and by 1840 they had founded Khartoum. They rapidly went up the White Nile to settle at Fashoda in the south of present-day Sudan. From there, they launched their slave caravans towards the present Central African Republic, the Congo basin and northern Uganda. Being part of the Ottoman Empire but having a great autonomy, they dreamed of extending the province of Equatoria, their source of slaves. In this way, around 1865, the Khartoumians reached the Uélé, Rubi and upper Nepoko rivers where they ran up against the Azandes and the Mangbetus (FIGURE 7.22). After some setbacks, they still managed to dominate the country. Between 1872 and 1878 they tried to annex Bunyoro and Buganda also, but in 1885 the Mahdist insurrection and the bankruptcy of the state ended their expansionist dream.

The Zanzibaris – Arabs originating from Oman who inhabited the African Indian Ocean coast since the 10th century – remained on the coast and for a long time were content to commercialise whatever populations from the interior brought them. Beginning in 1820, they themselves set off on the trails. Passing south of Lake Victoria, they reached Buganda and Bunyoro. In 1840, they got to Lake Tanganyika where they founded Ujiji, which rapidly became one of their main bases, and ten or so other trading posts. Further west, on the banks of the Congo River, they created the towns of Kasongo and Nyangwe. In 1876, their leader, Tippo Tib, accompanied Stanley in his descent of the Congo River. Shortly after, he could in this way launch his men towards the heart of the Congolese basin. Finally, in 1887, profiting from the retreat of the Khartoumians, the Zanzibaris reached the Uélé, Ituri, Lopori and upper Tshuapa rivers. In addition to trade in arms, copper, brass, glass jewellery and cloth, the Zanzibaris were especially attracted by ivory. However, slave trading rapidly became their main activity, even more so now that ivory had already become scarce in this period. In the Great Lakes region, they did not hunt themselves but made the local populations deliver slaves. It was usually prisoners of war, but sometimes also individuals sold by their relatives during times of famine. It is only in the east of the present-day Democratic Republic of the Congo that the Zanzibaris themselves engaged in the hunt. In 1892, however, they went to war with the Belgians and, in 1894, they were defeated.

The influence of the Zanzibaris was greater and more profound than that of the Khartoumians. They introduced Swahili – vehicular language of most of eastern Africa – transformed social life, built the first rectangular houses using clay and straw cob, developed shops and commerce. In northern Maniema, they assembled the populations. They spread the cultivation of rice *Oryza sativa* and introduced several new plants, including mangoes *Mangifera indica*. On the sand beaches of Lake Tanganyika, which must have reminded them of beaches on the Indian Ocean, they planted coconut palms *Cocos nucifera*. They also introduced the Kapoc Tree *Ceiba pentandra* in the Great Lakes region.

7.22. *The Arab penetration of Central Africa. The Khartoumians appear in the north-east around 1865, reaching their apogee around 1884 and retreating from 1885. In the south-east, the Zanzibaris crossed Lake Tanganyika soon after 1840 and built the towns of Kabambare, Kasongo and Nyangwe. The latter was founded in 1869. After Stanley's visit in 1876, they set off towards the north: the upper Tschuapa, the upper Lopori, the Aruwimi and the Ituri. Around 1887, they seized the areas abandoned by the Khartoumians. In 1892, however, they went to war with the Belgians of the Independent State of the Congo and were defeated in 1894. Arab domination in the Congo basin lasted only 25 years (according to Vansina).*

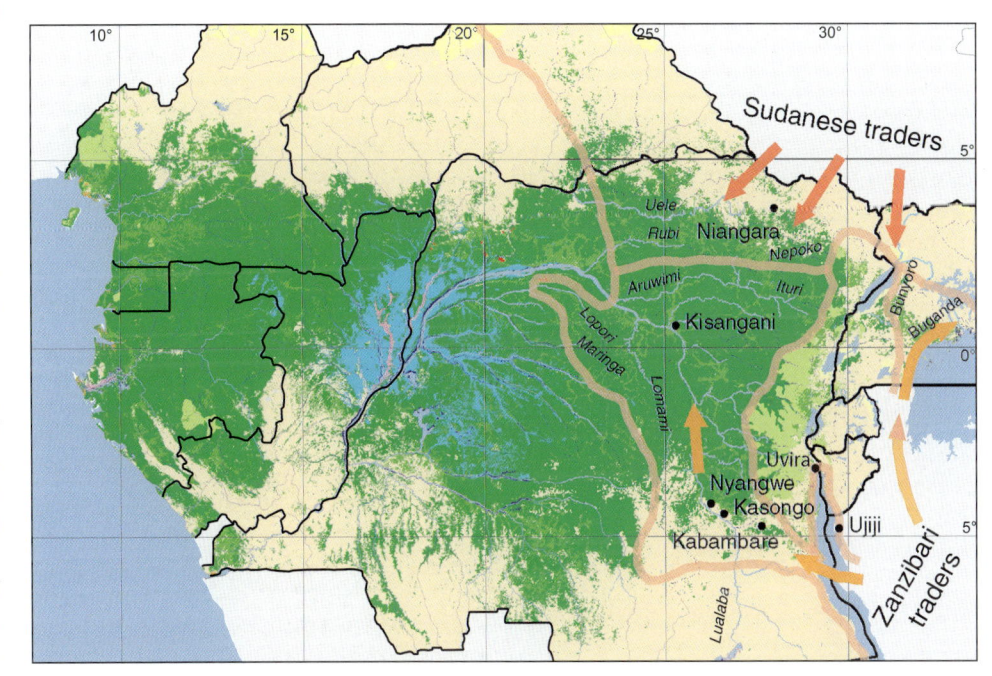

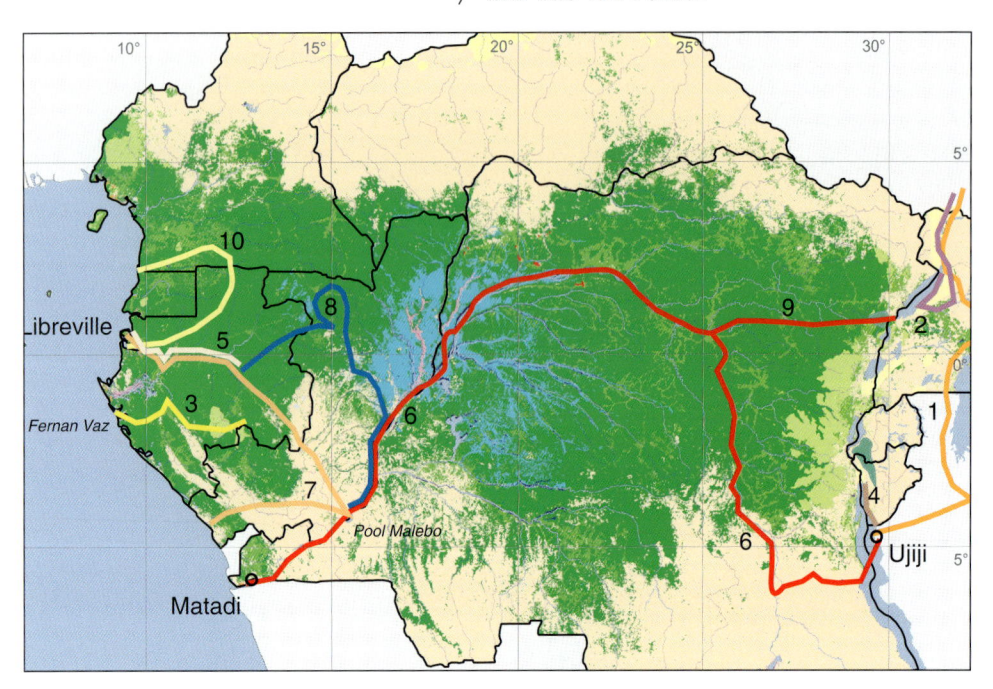

7.23. Peaks of exploration during the second half of the 19th century:
(1) Speke and Burton, 1858; Speke and Grant, 1860-1863;
(2) Baker, 1864;
(3) du Chaillu, 1865;
(4) Stanley and Livingstone, 1871;
(5) Marche and Compiègne, 1874;
(6) Stanley, 1876-1879;
(7) Pierre Savorgan de Brazza, 1877-1880;
(8) Jacques. de Brazza, 1885-1886
(9) Stanley, 1989;
(10) Fourneau, 1889.

The first Europeans

During the entire slave-trade period, Europeans avoided penetrating the interior of Central Africa. In 1816, an English corvette attempted to go up the Congo River, but less than 300 kilometres inland, it was stopped by rapids. Under the leadership of Captain James Tuckey, part of the crew followed the river upstream on foot and got as far as the third cataract, where Stanley would later build a bridge. For lack of means, the expedition had to turn back and Tuckey, weakened by fever, left his life there.

From 1838, the French were settled in the Libreville area and, between 1856 and 1876, a series of explorers – Braouzec, Iradier, Walker and Schultz, Lenz, Aymes, Serval and Griffon du Bellay, Marche and Compiègne – crossed the interior of the country. In 1865, In this way Paul du Chaillu arrived at Lolo, at the foot of the Batéké Plateau. No one went any farther and the immensities of deepest equatorial Africa were still only an enormous white spot on maps. No one in the West knew where the enormous masses of water carried by the Nile and the Congo came from. The location of their sources was one of the last major enigmas to trouble geographers.

Between 1859 and 1894, a series of heroic expeditions would solve the mystery (FIGURE 7.23). In reality, they just fit together the paths of the slave caravans and rediscovered what the Arabs knew for a long time but kept to themselves.

This search for the sources of the Nile has been described in detail by Alan Moorhead. The first expeditions left from Zanzibar. It is in this way that John Hanning Speke and Richard Burton "discovered" lakes Tanganyika and Victoria in 1858. From 1860 to 1863, Speke, returning with James Grant, reached the north shore of the lake where they managed to locate its outlet. The two explorers went upstream along the Nile Valley, reaching Bunyoro, and crossed the river to return to Khartoum. In 1864, two other great English adventurers, Samuel Baker and his wife, discovered the lake they named Albert, in homage to Queen Victoria's dead husband. In 1871, the journalist Henry Morton Stanley – whose real name was John Rowland – and the American missionary David Livingstone visited Lake Tanganyika. From Ujiji, they went in pirogues up the coast of Burundi, landed not far from the present-day town of Bujumbura and reached the mouth of the Rusizi. They thus confirmed that this river did not belong to the Nile basin. In 1875, the English General Gordon went up the White Nile from Khartoum. In 1876, he reached lakes Albert and Kyoga.

In the same year Stanley reappeared in the region, still in search of the sources of the Nile. In carrying out the first circumnavigation of Lake Victoria, he confirmed the observations of Speke and discovered that the main river emptying into the lake was the Akagera. Pushing farther west along the Katonga River, he caught sight of Lake George. Constrained by the local populations, he had to turn round, however, and continued his route southward along the east bank of the Akagera, therefore east of Rwanda and Burundi. From the high hills of the Karagwe, he saw the imposing cones of the Virunga Mountains, but Rwanda remained forbidden to him. He continued his route towards Lake Tanganyika. From there, he followed its outlet – Lukuga River – and reached Lualaba, the upper course of the Congo River. He thus became the first White Man to penetrate the heart of the Congo basin. Finally, in 1879, after three years of travels, he reappeared at Boma near the Atlantic Ocean, where he was the first European to return from the interior, where no one had dared to go before.

The colonial era

These expeditions solved once and for all the enigmas of the Nile and the Congo, but they also marked the beginning of the Europeans' conquest of Central Africa.

In 1876, leaving from the Atlantic coast, Pierre Savorgnan de Brazza reached the Lopé, a small tributary of the Ogooué River, and in 1877 he reached Poubara Falls, upstream from Franceville. In 1880, he came upon the Malebo Pool, where he planted the French flag. In 1885, his brother Jacques de Brazza, at the head of a scientific expedition, discovered the Odzala region. In 1889, Stanley returned to the region, sent to the rescue of Emin Pacha, a German physician named Eduard Schnitzler who had become governor of the province of Equatoria and was cut off from the world by the Mahdist revolution in Sudan. This time Stanley went up the Congo River from the Atlantic to reach the Rift at the level of the southern part of Lake Albert. He could then also visit the region around lakes Edward and George and saw the Rwenzori glacier for the first time. In the same year, the German botanist Stuhlmann crossed the Akagera River at Kanyonza, not far from present-day Kagitumba, and was in this way the first European to walk upon Rwandan soil. In 1892, the Austrian Oscar Baumann crossed Burundi and, in 1894, a German expedition, under the command of Count von Götzen, crossed Rwanda from east to west.

The accounts of all these explorers were best-sellers of the period. Beyond their heroism and scientific revelations, they stirred the emotion of European public opinion, however, in denouncing the slavery of the Khartoumians and the Zanzibaris. From 1877, Christian missions already implanted on the coast since 1840, notably the Sainte Anne Mission in Gabon, launched their missionaries on an assault of the interior of the continent. Until around 1900, however, their impact remained relatively minimal. Before saving souls, it was necessary to create durable settlements. In fact, living conditions in Central Africa were extremely difficult and logistical problems enormous. After a first, often enthusiastic, greeting, local populations sometimes proved to be unfriendly, sometimes frankly hostile. In fact, these missionaries ran up against the same problem as the Portuguese 350 years earlier: the populations of Central Africa disposed of barely enough food for themselves and the passage of foreigners imposed a burden on them they could not cope with.

Nevertheless, each region had its own problems. In the vast deforested landscapes of the mountains of the Great Lakes region, the lack of wood was added to the shortage of food. In order to build their churches, schools and hospitals, the missionaries therefore first of all had to solve their logistical problems. This is why they hunted much game, planted vegetable gardens and introduced eucalyptus trees. On the Congo-Nile Crest, they also cut what was left of the most accessible natural forests (FIGURE 7.25). They did not yet make many converts, but managed locally to take advantage of the sometimes dramatic context into which they had arrived. It is in this way that the White Brothers judged as "providential" the famine that afflicted most of the Great Lakes region between 1887 and 1903. In fact, it very greatly facilitated their hold on the bewildered peasant populations. The missionaries very rapidly took on a political role, also. In order to break the existing systems, they set some social or ethnic groups against others, the young against the old. Their establishment became in this way the prelude to division and colonial seizure.

Division

In the eyes of the Africans, the visits of the first Europeans were no more than local news items without much consequence. In Europe, however, the explorers' writings complemented by accounts of the missionaries immediately enabled people to see the commercial opportunities.

Some companies, pushed by rapid industrial development in Europe, sent representatives and increased the number of trading posts. Like the Churches, however, they at times ran up against slave traders, at times against "recalcitrant" populations and, at still other times, against enormous logistical problems. Together with the Churches they managed to obtain support from their governments, indeed, their military engagement. But each intervention set off another one and without really wanting to, sometimes even against their will, the governments were carried into the colonial adventure.

Inevitably, the continent was divided up and Central Africa fell to France, King Léopold II of the Belgians, Germany, Great Britain and Portugal (FIGURE 7.24). The borders were revised time and again, however. In 1908, the Independent State of the Congo became a Belgian colony and after the First World War, the German territories were divided up between Great Britain, France and Belgium. The present-day borders of the Central African States are thus only 80 to 100 years old.

This dividing-up was done in Europe, without knowledge of the territories involved and without taking into account the local populations. In a few rare places, the European Powers nevertheless agreed to move a boundary to protect the integrity of a community or a kingdom. This was the case for Rwanda, initially divided between Leopold II and the Germans, but reunited under the German authority after von Götzen's visit in 1894. In many regions, however, homoge-

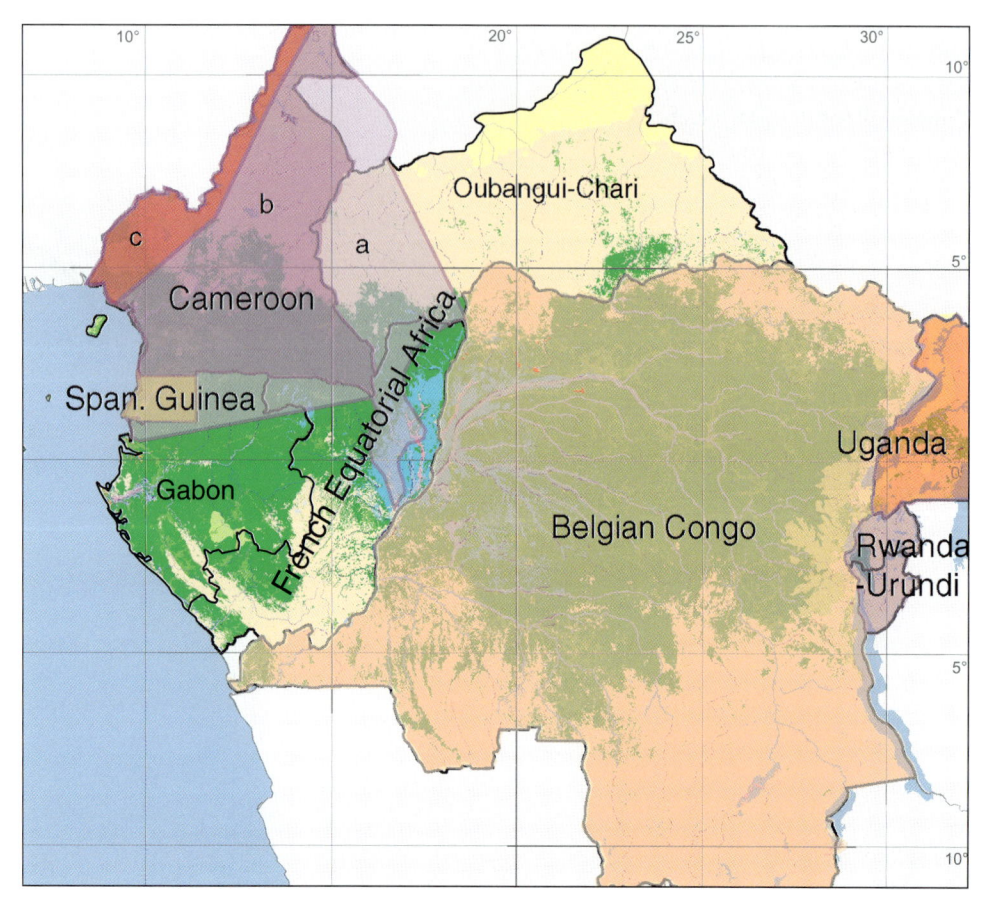

7.24. Colonial divisions were confirmed by the Berlin conference of 1884-1885 that split the Congo basin between France and King Léopold II of the Belgians. The treaty of Heligoland of 1890 terminated the splitting up of the Great Lakes region: the west was included in the Independent State of the Congo; Buganda, Bunyoro, Toro and 'Ankole were given to the English; Rwanda and Burundi went to the Germans. In 1900, a treaty established the border between French Equatorial Africa and Spanish Guinea. The borders between Uganda and the Independent State of the Congo were fixed between 1907 and 1912. However, in 1908, after 24 years in existence, the Independent State of the Congo became the Belgian Congo. In 1911, France ceded part of Gabon and the Oubangui-Chari to Germany in exchange for territories in Morocco (a). After the First World War, these territories were recovered. Cameroon was divided between France (b) and England (c), while Rwanda-Urundi was put under Belgian administration.

nous populations were divided. The Fang world, for example, was scattered between Cameroon, Equatorial Guinea and Gabon. Elsewhere, very different populations, differing in culture and language as well as in religion, were forced to live together and foreign populations were introduced to serve as manpower or to support the administration.

The seizure

The colonial period lasted hardly three-quarters of a century, but it profoundly disrupted the region. The conquest was in fact much harder than most history books make out. It took about 40 years, did not end until around 1920 and wiped out nearly half the human populations of Central Africa because of the combined effects of the wars, destruction, epidemics and famines that it engendered. It was therefore considerably harsher than the Atlantic slave trade and Arab slavery.

First, opening-up to the outside world, development of commerce and population movements were accompanied, from the end of the 19th century, by a series of epidemics. Without being as violent as those that ravaged the Pacific islands, they had very violent social and demographic, as well as environmental, repercussions.

Grassland regions on the immediate periphery of the forest massif were struck by rinderpest. This disease, introduced in 1880 via Djibouti, ravaged the entire continent in a few years, decimated 90% or more of the domestic bovine herd and brought several wild species to the brink of extinction.

In the Great Lakes region, as everywhere, it wreaked tremendous ravages among the livestock and, indirectly, also among the pastoral populations. Its impact hit not only the grassland regions, but also the submontane and montane forests. In Rwanda, for example, a chief had the idea of taking immense herds of cattle into the very centre of Nyungwe Forest in order to remove it for many years from any contact with the exterior. By burning the edges of natural peat bogs, the herders managed to create a vast network of pastures covering thousands of hectares. Regeneration in montane forests being very slow, these spaces are visible to this day in what is now the eastern part of the forest. They are in the form of poor meadows with *Eragrostis*, fern heaths and *Hagenia* or *Cliffortia* thickets. This rinderpest epidemic was followed by the advance of trypanosomiasis or sleeping sickness, unknown in most of the Great Lakes region until this point in time. The momentary collapse of animal populations, wild as well as domestic, probably favoured the regrowth of shrubs and trees in the grasslands and therefore the expansion of tsetse flies, *Glossina sp.*. Human populations of the grasslands, already weakened and decimated by the direct consequences of the rinderpest, were then victims of sleeping sickness in its very virulent eastern form. Arriving in Uganda around 1898, it raged for nearly 20 years and eliminated two-thirds of human populations in affected regions. Faced with the great extent of the disaster, and in the absence of any other means of combating it, the British government decided just to evacuate the infected areas. Between 1907 and 1912, what was left of the human population was evacuated from the lower valley of the Victoria Nile, while farther south, the plains of Lake George and Lake Edward were evacuated. The epidemic was not halted until around 1930.

As if that were not enough, many regions were also hit by smallpox. Adding to the conflicts provoked by colonial seizure, all these microbial catastrophes contributed very largely to the disorganisation of human societies and to the depopulation of vast territories situated on the eastern edge of the Central African forest massif. Nature in this way temporarily took back its rights. Maramagambo Forest expanded northward and managed to recolonise the perimeters of Lake Namusingeri and Lake Kasianduku, now included in Queen Elizabeth National Park. In Kibale Forest, also, vast pastures were rapidly obliterated by the advance of the forest or invaded by dense stands of elephant grass *Pennisetum purpureum*.

Inside the Central African forest massif, epidemics also arose of smallpox, measles, dysentery and trypanosomiasis. The latter disease appeared in places where it had never before been known and some regions had to be evacuated after their populations were decimated.

It is against this background of human tragedy that the first Europeans settled. At the very beginning, however, they had no military forces capable of imposing their power, which was limited to the immediate vicinity of their trading posts and garrisons. They settled for signing treaties with the local chiefs and even made some effort to try to understand the local cultures. Ethnographical notes collected in this way were very disparate in nature, however. Some people were genuinely interested. Others confused observation and interpretation. Often, the information served more to forge arguments justifying the civilising mission of the colonisation process than to really try to understand these peoples.

During the course of the 1880s, armies or local militias were formed, first in the Independent State of the Congo, then in French Africa. From this period, the Europeans imposed their vision of the world. From 1890 to 1897, the Belgians also attacked the Arabs, but in 1897 they had to face up to the revolt of the Batetela. The English had to face the Banyoro and the Germans had to fight in the Yaoundé region and on the slopes of Mount Cameroon. In Rwanda, they had to combat the mutinous elements of the *Force publique congolaise*. Everywhere, these military expeditions, supervised by European officers or in a state of mutiny, pillaged, raped and spread new diseases.

At the same time that these conflicts were resolved, the administration was transformed. In French Equatorial Africa, the Belgian Congo and Cameroon, a direct and authoritarian system was put in place. The traditional authorities lost their power or were replaced by agents appointed by the colonial power. In Uganda, the administration remained more indirect. In Rwanda-Urundi, Belgium initially set up a system of indirect power, following the example of the Germans but, between 1924 and 1929, it completely transformed everything: while keeping the traditional chiefs, it installed a very direct system, somewhat like the one it put in place in the Congo.

The new administration also imposed "customary law", an often very western vision of habits and customs of the people implicated. Above all, the populations were hit by the upheaval of land rights. In the traditional kingdoms of Uganda, a part of the land was attributed to the king, a part to the Crown, therefore the State, and a part – the *mailo* lands – to private parties. The great forests became State property, but many small forests remained in the hands of private parties. In the territories under Belgian or French administration, all land that was not visibly occupied, therefore all the fallow lands, whatever their age, became state property.

Most of the land was thus redistributed in the form of concessions to exploitation companies. In French Equatorial Africa, these numbered about forty and their aim was to extract immediately available natural resources at the lowest cost: rubber and ivory. The inhabitants of the forest would furnish the manpower, or sometimes more directly the desired goods.

This quest for rubber was one of the most sinister pages of colonial history, as much in French Equatorial Africa as in the Independent State of the Congo (FIGURE 7.25). Entire populations were massacred. The Tsayi, who lived south of the Franceville region on the present-day Congo-Gabon border, were practically exterminated: only three villages out of 135 survived. Elsewhere, populations fled into more inaccessible zones. Just as in the Atlantic slave trade period, populations of the Lower Congo thus took refuge in the mountains of the Mayombe.

In 1908, after these exactions, the Belgian State took charge of the colony, but in French Equatorial Africa the "rubber war" raged until 1928, notably in the Odzala region. It drove from their villages most of the populations inhabiting the present border area between Gabon and the Congo and forced these people to take refuge in temporary hunting or fishing camps. Of those who escaped the army, many succumbed to hunger or disease. Villagers inhabiting the periphery of Odzala National Park remember these events to this day.

7.25. *Wild rubber was obtained by collecting the sap of several plants of the Apocynaceae family first from lianas belonging to genera* Landolphia *(as in the old picture below) and* Clitandra, *then the tree* Funtumia elastica. *Exploitation began around 1870 and reached its maximum development around 1910. Following the competition of rubber harvested in the large* Hevea *plantations in Asia and South America, the exportation of wild rubber from Central Africa nearly stopped after the First World War. It experienced a slight reactivation during the Second World War, before being definitively abandoned.*

7 MAN AND THE FOREST

There was no point in harvesting or extracting natural products from the forest unless there was a way to bring them out to the coast. The numerous watercourses already constituted a gigantic natural communication network. However, it was necessary to complete it, not just to reach otherwise inaccessible areas, but mostly to get around the cataracts that separated this network from the sea. The end of the 19th century thus saw the building of the Léopoldville-Matadi and Brazzaville-Pointe Noire railroads. The construction of roads required more time and developed fully only after the Second World War.

The establishment of communication routes had very diverse effects. Beginning with the construction phase, it accentuated population movements by attracting large concentrations of manpower into thinly populated areas. This was notably the case in the Congolese Mayombe. Next, exchanges of products and goods became possible right to the very interior of Central Africa, the supplying of food, notably, was greatly improved and food shortages and famines became rarer. The colonial powers propagated some crops in order to diversify the populations' resources. In highland regions of the Albertine Rift, for example, the Belgians introduced the cultivation of cassava and "Irish" potatoes, another plant from South America, just after the dramatic famine that hit these countries during World War II. Indirectly, the development of communications routes thus contributed to the population growth, perhaps even more than the setting-up of modern health care.

The birth of towns affected the forest world only indirectly, except in subtracting populations from it. A multitude of secondary centres with shops and diverse services sprung up around missions and along routes.

These population movements were in part spontaneous. In part obligatory. Before the First World War, in Cameroon, for example, the Germans displaced some populations during the fight against the trypanosomiasis endemic. Beginning in the 1930s, in French Equatorial Africa and in Cameroon, the French administration obliged the very thinly distributed populations to regroup along roads. Such movements took place in areas of Odzala National Park and the Dja Reserve. This regrouping not only facilitated the organisation of education and the setting up of health care, it also helped the collection of taxes and the requisition of manpower. Moreover, it allowed better surveillance and the prevention of any vague revolutionary impulses. Once more, this policy of regrouping people was used without taking into account the customs and traditional rights of the involved population. It was thus at the origin of a good many conflicts. Different ethnic groups, sometimes even rivals, were brought together. Traditional hunting territories were abandoned or transformed and vast forest massifs too far from roads returned to a "wild" state. In some places, present-day populations have forgotten their ancient villages. Elsewhere, however, the old villages remain their property and still sometimes harbour temporary hunting or fishing camps.

Exploitation of the colonial territories required heavy investment. In French Equatorial Africa and in the Congo these costs were largely covered by the exploitation ivory and rubber, but ivory resources dried up and the rubber market collapsed after the First World War. The colonial powers turned toward industrial crops and mining. The lowland forest regions lent themselves well to large scale exploitation of the oil palm, cocoa *Theobroma cacao* (FIGURE 7.26), *Coffea canephora (robusta)* coffee (FIGURES 7.27-28) and sugar cane *Saccharum sp*. Montane regions were better suited for *Coffea arabica* coffee, tea *Camellia sinensis*, probably originating from the mountains of Tibet or Central Asia, and quinine *Cinchona sp.*, from South America. In many places, coffee and cocoa growing were practised intensively by peasants who allocated part of their land to it. Elsewhere, however, thousands of hectares of forest were cleared to make room for extensive industrial crops.

As for mining, this activity especially concerned regions on the periphery of the central basin. The quest for river gold affected many rivers by drastically dis-

7.26. The cocoa tree Theobroma cacao, *of the* Sterculiaceae *family, is of Amazonian origin. Its flowers, and later its fruits, grow on the trunk or on large branches. Around 1850 this plant was already known on the coast of Central Africa.*

7.27. *Hundreds of hectares of montane and submontane forests have been cleared to make room for industrial crops. The islands of the Gulf of Guinea, including São Tomé, shown here, were profoundly transformed by the establishment of sugar cane fields and later of coffee plantations.*

7.28. *The coffee tree* Coffea arabica, *of the* Rubiaceae *family, is from Ethiopia, but it occurs in various cultivated forms in upland regions of Central Africa. In the forests of lowland regions, numerous other species of the same genus are encountered, notably* Coffea liberica, Coffea congensis, Coffea kivuensis, Coffea mayombensis *and mostly* Coffea canephora. *This latter species is very widespread and has given numerous cultivars of the variety "robusta".*

rupting the bed, by destroying the forest cover and heavily silting the water with the excavated alluvia. The limpid waters were transformed into mud baths. During the Second World War, these mining activities were intensified to support the war effort. It is then that gold mines were opened in Rwanda-Urundi, in the heart of the Nyungwe and Kibira forests, and Cassiterite, or tin oxide mines in the Gishwati forest. In Belgian Congo numerous mines were opened on the western flank of the Albertine Rift mountains. Independently of their direct impact on the environment, these mining activities gave rise to large human concentrations in zones where there had never been any. Not only did large centres arise such as Kamituga, Kalima, Shabunda or Kilo-Moto, but a multitude of smaller ones also came into being.

Finally, there was also wood exploitation. This was not new, since from the second half of the 17th century, Europe has imported precious wood from Western Africa, notably African mahogany – various species of *Entandrophragma* and *Khaya* – and ebony *Diospyros sp.*. Added to this, in the 19th century, was "African oak" *Milicia excelsa*, better known today under the name of iroko or muvula. The seizure of Central Africa at the end of the 19th century caused the discovery of new species, such as Okoumé or Gaboon Mahogany *Aucoumea klaineana* and Makoré or Cherry Mahogany *Tieghemella sp*. The exploitation of these resources, however, ran up against the technical difficulties of tree felling and transport. The least heavy species could be taken down to the sea on rivers by means of simple floating, but the heaviest – mainly "*bois divers*" (miscellaneous woods) according to present-day terminology – require transportation by barge. However, the young profile of the coastal rivers and of the lower Congo River hamper river transport. In the Belgian Congo, timber could be brought fairly easily to the Pool, but from there it had to be transported by train to Matadi or Boma, which was very costly. In practice, lumbering was therefore concentrated essentially on the coastal plains and in the most accessible regions, such as the Mayombe. Nevertheless, wood production increased throughout the colonial period. After the Second World War, it received a new impulse with the mechanisation of tree felling and extraction, along with the improvement of road transportation. It is then that bulldozers and logging trucks appeared. In general, this exploitation was very selective and, as most of the coveted species were rather rare, the impact on the forests fairly small.

Finally, after the implosion of human populations at the beginning of the colonial period, the demographic trend reversed. Beginning in the 1920s, together with the end of the wars, control of epidemics and development of road infrastructures, populations began to increase.

The first foci of overpopulation formed in the Rwanda-Urundi highlands, particularly in the volcanoes region. At first, the Belgian administration organised migrations into the centre of the country: farmers from western and northern parts of the country were settled in humid valleys and swamps in the centre which had served as pastures in the dry season for the pastoralists, a move that inevitably unleashed tensions between pastoralists and farmers. Many Rwandans also emigrated, however, to southern or western Uganda, Kivu and even Kenya. The Belgian administration ended up organising these migrations, which provided manpower to the numerous plantations of the Kivu and the mines of Katanga. In some places, the first clashes took place between local populations and immigrants. Altogether, these population problems were still solvable, however. In Kivu, vast stretches of forest were even granted to colonists for transformation into industrial crops and later into ranches. Montane forests were judged unsuited for wood production and were doomed to conversion

Little by little, the colonial period witnessed the transformation of the landscapes of the area around the Albertine Rift: large pieces of forest still intact were razed, while the mountains that had already been cleared were massively planted with exotic rapid-growth species, mainly *Eucalyptus*, *Grevillea* and cypress *Cupressus* (FIGURE 7.30).

Alongside all these visible manifestations of the colonial seizure, there was also a far-reaching campaign to destroy local cultures and religions. This was mainly the work of missionaries, backed by the colonial powers. In the name of civilisation and "indigenous well-being", everything that was an expression of ancient traditions was prohibited. Not only was the widespread practice of polygamy combated, but the missionaries went to war against dress or the lack of dress, while dances, chants and plastic arts had to obligatorily disappear or become westernised. Tons of statuettes and masks were thus burned on village squares. Initiation cults in particular were fought against. Beginning in the 1930s, they were simply forbidden. But initiation societies probably represented the main vehicle and guaranty of tradition. Finally, what was left of traditions and beliefs was profoundly manipulated in order to "tie in" with Christian principles, to such a point that it is difficult nowadays to separate the original from what has been introduced.

This destruction and manipulation of cultures had serious environmental consequences. The beliefs and rites that had traditionally often tempered the taking of natural resources were replaced by a Christian philosophy according to which Man is the centre of the Universe and can dispose of natural resources without limit. Each fruit of the forest and each piece of game thus becomes a gift of God that no one dares refuse and that is appropriate to gather before someone else does.

7.29. Between the two world wars elephants were domesticated in north-eastern Belgian Congo. These animals were mainly used in agricultural activities, as in the old picture below. Some were also used in the transportation and loading of logs. After World War II bulldozers and tractors put an end to this experience. Some domesticated elephants survive, however, in the Garamba National Park. Rangers use them for patrolling.

1960-2000

Ideas of independence took form in the years that followed the Second World War and succeeded in the independence of the Congo in 1960, followed by the independence of Uganda, Rwanda, Burundi and French Equatorial Africa in 1962. Spanish Guinea, becoming Equatorial Guinea, would have to wait until 1973 and São Tomé and Principe until 1975.

Except for Rwanda and the Belgian Congo, accession to independence was accomplished without too many problems. The new states had very different histories, however, whose details have no place here. Generally speaking, the democratic regimes, put in place by colonial powers in the image of European democracies, were transformed into more authoritarian regimes, however. In 1968, the Congo thus became the Popular Republic of the Congo and sank into twenty years of Marxism.

Nevertheless, whatever the political orientation chosen, galloping demography, poor management by the State, corruption, diversions and an ever more constraining international economic context brought on a slow but inexorable degradation of national economies. At the same time, these Central African countries experienced in a real way the progressive decay of their rural world, bringing with it the degradation, indeed the abandonment of some infrastructures – roads, bridges, ferries, river boats – and the cutting off of rural communities from the rest of the world. In places, notably in the Democratic Republic of the Congo, this situation was exacerbated by insecurity and villagers themselves destroyed bridges giving access to their villages in order to impede pillaging soldiers.

To all this were added the collapse of the exchange of goods, such as tea, coffee, cocoa or palm oil, which precipitated the abandonment of plantations, not only of industrial crops and cash crops, but also of some food producing crops. In Cameroon, for example, the plummeting sales of cocoa and coffee in 1988 precipitated the abandonment of paths, which set in motion a veritable vicious circle.

This economic situation was aggravated by the strong increase in populations. In the lowland regions of Cameroon, Gabon, the Congo and most of the Democratic Republic of the Congo, this augmentation was on the order of 2.1 to 3% per year. Populations thus doubled within 25 or 30 years. This increase was accompanied, however, by heavy migration into the cities to such a point that nowadays urban populations constitute 30 to 80% of the total population. Rural populations have therefore remained relatively low over vast areas: 1.9 inhabitants to the square kilometre in Gabon and in the north of the Congo; 5 to 10 in most of the forested regions of the Democratic Republic of the Congo.

This rural exodus was based in part on the attraction of the city and the hope of finding employment there, in part on the insecurity that reigned in places in the forest. In part, also, in any case in some regions of the Congo and the Democratic Republic of the Congo, it represented a means of escaping the heavy constraints of the traditional world. In the case of the Rega and Tembo peoples in the peripheral areas of Kahuzi-Biega

7.30. In completely deforested regions of Rwanda-Urundi (Pages 250, 256), where the soil has often been completely exhausted and where forest birds have been replaced by birds of the savanna such as bustards, guinea-fowl and francolins, the large-scale introduction of eucalyptus and other exotic trees, such as Grevillea robusta *from Australia and* Pinus patula *originating from Mexico, have profoundly transformed the landscapes. Of course, such reforestations produce timber and fire wood. In places, they relieve pressures on the last vestiges of natural vegetation and on the forests. Unfortunately, in terms of biodiversity, they are really living deserts. The bustards, guinea-fowl and francolins have disappeared, but nothing has come to replace them.*

Demographic and socio-economic indicators in 2001

indicator: country	land area km²	population millions	ann. increase. %	urban popul %	GNP/hab (1999) US$	expenses land educ./health %	popul/ ha cultiv.land cultiv.	life expect. M/F years	illiteracy. >15 y M/F %
Cameroon	465,400	15.2	2.1	49	1,490	– / 1	1.1	49.3 / 50.6	17 / 29
Equatorial Guinea	26,000	0.47	–	48	3,910		1.3	50.4 / 53.6	–
Gabon	265,000	1.3	2.5	81	5,280	3.2 / 2.1	1.0	51.8 / 54	–
CAR	622,980	3.8	1.6	41	1,150	– / 2	4.0	42.7 / 46	39 / 64
Congo	341,500	3.1	3	63	540	8.6 / 2	5.4	49.6 / 53.7	12 / 24
DRC	2,267,050	52.5	3.3	30	–	–	4.0	51.0 / 53.3	26 / 48
Uganda	260,000	24.0	3.2	14	1,160	2.4 / 1.9	2.4	45.3 / 46.8	22 / 42
Rwanda	26,000	7.9	2.1	6	880	– / 2	5.6	40.2 / 41.7	25 / 38
Burundi	27,000	6.5	3	9	570	3.5 / 0.6	5.3	39.8 / 41.4	43 / 58
Central Africa	4,300,930	114.77	2.9	28.18	–	–	2.7	–	–

Source: The state of the world population 2001, United Nations Population Fund (www.unfpa.org/swp/swpmain.htm)

Predictions 2050

	Population in millions	inhab /km²
Cameroon	32.3	69
Equatorial Guinea	1.3	50
Gabon	3.2	12
CAR	8.2	13
Congo	10.7	31
DRC	203.5	90
Uganda	101.5	390
Rwanda	18.5	711
Burundi	20.2	748
Central Africa	399.4	93

Source: The state of the world population 2001, United Nations Population Fund.

National Park, for example, the "progressive" young peasants, who attempted to use selected seeds or to launch into new and more productive farming techniques, systematically ran up against conservative traditional powers jealous of their prerogatives. In many regions, the rural world thus sank into a suicidal immobility and finally lost its most dynamic elements. This phenomenon contributed to the setting-in of a deep cultural rupture between the rural world and the urban world where the elite concentrated.

In Cameroon, the growing shortage of employment in urban centres and the decay of the economy forced many young people, even those with a diploma, to return to the forest, however. This return to the village was accentuated by the general degradation of education. Since, on the other hand, coffee and cocoa were no longer profitable these people resorted to the forest's natural resources. Hunting demanded the least effort.

In the villages around the Dja Faunal Reserve, commercial hunting thus appeared between 1976 and 1988 and game became the main source of monetary revenue from 1993 or 1994. The devaluation of the CFA franc in 1993 and the resulting rise in the price of some basic foodstuffs such as salt, sugar, cooking oil, soap or petrol, amplified the phenomenon. In the Kivu, the development of commercial hunting occurred also in the 1980s with the introduction of the Soviet-made Baïkal rifle. The markets of Bukavu and Goma were literally inundated with monkey meat to the point where the accessible populations of these animals were rapidly destroyed.

In places, the return to the forest and hunting were also accentuated, if not caused by insecurity and the troubles. In the east of the Democratic Republic of the Congo, many rural populations thus moved away from the major trunk roads and plunged into the forest, including into protected areas, in order to escape the exactions of soldiers and armed bands.

Populations in large cities were themselves sometimes reduced to turning to the resources of the forest. Thus, in 1999 and 2000, soon after the end of the war in the Congo, the first meat available in Brazzaville was coming from the north of the country: the regions of Ouesso and Lake Télé.

In many places, pressure on natural ecosystems was also increased by the decline of agriculture resulting from the development of small-scale and totally anarchic mining activities. Instead of cultivating the fields the men went and scraped the earth for gold, diamonds or, more recently, coltan. This phenomenon was favoured by the departure of the large mining companies, the incapacity of the governments to control exploitation, the liberalisation of the trade in mined materials, the organisation of clandestine sales networks, the existence of a large out-of-work population and finally also the decrease in game. Attracted by the possibility or the illusion of being able to earn money fast, men abandoned what remained of the fields. This situation, increased insecurity even more and many mining zones were turned into a "Far West" where gold hunters, bandits of all sorts and soldiers indulged in ruthless competition.

Globally, Central Africa experienced a set of contradictory movements and, even though the rate of urbanisation continued to increase, the general dependence on natural forest resources also increased.

In the highland regions along the Albertine Rift, population increase was much greater than in the lowlands. Supported by intensive agriculture, as in Uganda (FIGURE 7.32), or "generous" external aid, as in Rwanda, it reached 3.5 to 3.7% per annum. These already very dense populations doubled in about twenty years. In many places densities were over 150 inhabitants per square kilometre. In the Ugandan county of Bundibugiyo, at the foot of the Rwenzori Mountains, population density went from 160 to 380 inhabitants per square kilometre between 1948 and 1980. At the foot of the volcanoes in Rwanda, it even reached 400.

This explosive increase had first of all local consequences. Pressure on the last forest massifs of these mountains increased and caused their limits to retreat. Everywhere, there was an intensification of agricultural activity and a generalised putting into cultivation of marginal and fragile land: swamps, pastures and steep slopes. The last spaces still covered with more or less natural vegetation were cleared and fallow land disappeared or was greatly reduced. It was thus not necessary to wait for the arrival of pesticides to see the disappearance of biological diversity in these regions. By encroaching on pastures, this agricultural activity finally left no more space for the pastoralists. The disappearance of the extensive grasslands was otherwise also completed by conversion to extensive plantations of exotic tree species: eucalyptus and Australian acacias, cypresses and Mexican pines. The pastoralists had no other outcome than to plunge into the forests. This was the case in Gishwati Forest in Rwanda and in several regions of Burundi, including protected areas.

In these mountainous regions, the intensification of farming activity also brought a serious increase in erosion. Even if not always very spectacular, it took worrisome proportions and engendered very diverse problems. In spite of erosion control programmes, landslides became very frequent. Each year, entire families were swallowed up in this way in the north of Rwanda and in Kivu. At the same time vast cultivated expanses in the lowlands and valley bottoms were drowned by sudden floods and torrents of mud. Far from the mountains, waters heavily loaded with silt went to pollute the lakes.

In spite of all efforts, this agricultural activity was incapable of keeping up with population increases, all the more so since the suppression of fallow land brought with it a strong decline in production. In the densely populated countryside of the Kivu, on the edge of montane forests, the lack of land took on overwhelming proportions and, in the absence of external aid, malnutrition became omnipresent. In Rwanda also, the capacity of the country was greatly exceeded. Around 1985, it was estimated thus that nearly 10% of the population survived thanks only to external food aid.

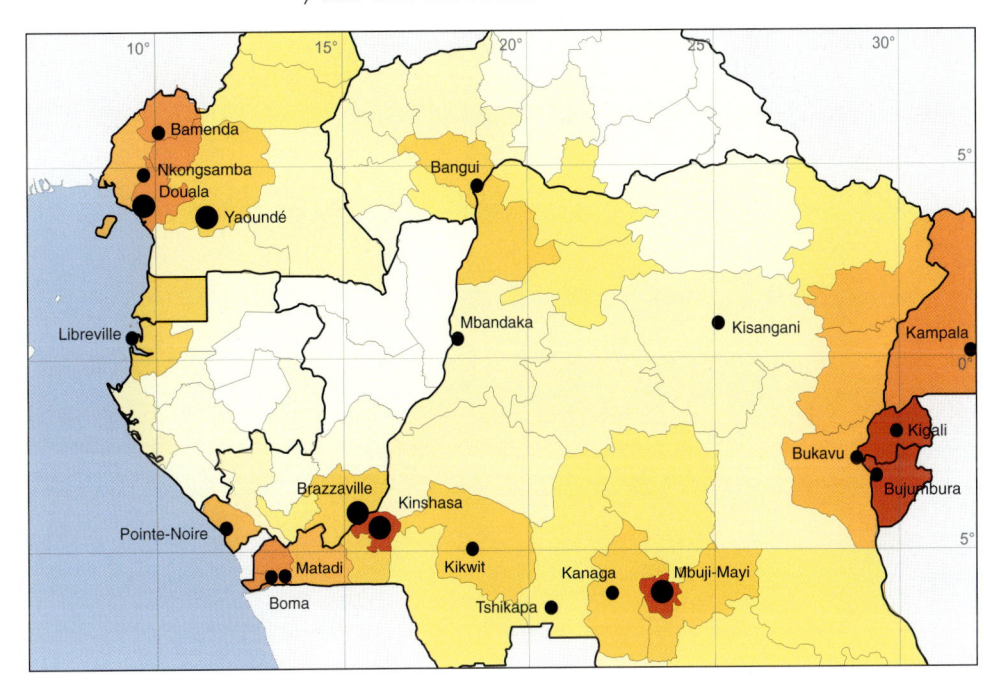

7.31. Rural population density expressed in inhabitants per square kilometre:

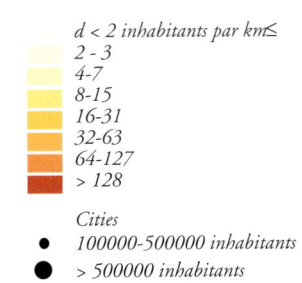

d < 2 inhabitants par km²
2 - 3
4 - 7
8 - 15
16 - 31
32 - 63
64 - 127
> 128

Cities
• 100000-500000 inhabitants
● > 500000 inhabitants

7.32. In formerly forested regions of south-western Uganda, the last traces of natural forest disappeared a long time ago and the landscape today is dominated by intensive terrace farming. The rare woodland tracts are composed exclusively of exotic species.

7.33. The quest for river gold caused enormous damage in the montane forests of the Albertine Rift. During the 1970s-1980s, Nyungwe Forest in Rwanda sometimes harboured more than 6,000 gold hunters whose activities were difficult to control. In the valleys, excavation in the river beds, even though done with pick and shovel, created enormous holes and profoundly upset the watercourses which became loaded with silt. The surrounding forests were inevitably degraded and the gold hunters systematically set traps there to hunt and to eat.

7.34. Small mining centres sprang up spontaneously, like the one at Pindura, becoming the base of all kinds of smuggling, including marijuana Cannabis, *which used to be cultivated virtually industrially in clearings in the nearby forest.*

This process took place on formerly forested highland lands, many of which had been farmed for centuries. It engendered an emigration which affected the lowland forests. The mountainous regions thus became veritable demographic bombs. Initially, the increase in population could be absorbed by migrations towards still uninhabited or little inhabited regions of the same country. In Rwanda and Burundi, populations of the Congo-Nile divide filled nearly all the savannas to the east outside of the protected areas. In Kivu, the Shi, of the lake region, crossed the mountains to go and settle farther west among the Tembo and the Rega. Beginning in the 1980s, numerous Rwandan and Burundian peasants also migrated westward and settled in the lowlands of Walikale, Walungu and Kingulube. Other Rwandans left for Uganda where some even invaded forest reserves.

These continuous migrations caused serious problems between the local people and new arrivals. How, for example, could one make mountain peasants, used to living in isolation, live together with Congolese from the forest used to communal village life? Especially if the newcomers settled just anywhere and any way, without respecting either local traditions or land rights. For these peasants, used to permanent fields, the entire forest, even a recent secondary forest, in fact, represented a free space just waiting to be cleared. They could not imagine that all forests in lowland regions are part of territories belonging to village or clan communities. Moreover, the immigrants' lack of knowledge of lowland forest terrain also resulted in serious ecological damage. The Rega, for example, have long known that some soils within their territory were very fragile and they did not cultivate them. The Rwandans and the Shi, on the contrary, were unaware of this and set off catastrophic erosions.

If mountain and lowland forests are distinct ecosystems at the level of flora and fauna, they have to be considered as one single vast ecosystem when considered at the human level. This situation is not something new, which appeared during the last century. It existed probably already thousands of years ago, when the Bantu people of Western Cameroon and Eastern Nigeria became too numerous, exhausted their environment and started to migrate further south and south-east. Unfortunately this ecological background of history is often very hard to trace, and subsequently easily refuted by historians and social scientists. The archaeologists are the most prone to accept it.

Acute crises

On several occasions and in many places, economic collapse was precipitated by acute crises. Up until around 1990, these were often no more than "palace revolutions" that mostly affected the political class. Sometimes, they were accompanied by massive despoliation of European, Asiatic or even African foreigners. It suffices to recall the expulsion of the Indians from Uganda in 1972, the "Zairianisation" of Congo-Kinshasa in 1973 and the expulsion of the Beninese from Gabon in 1978. Forest populations were little affected by these events.

Over the years, especially from 1990 – the year of the la Baule address and the fall of the Berlin wall – problems deepened. To the struggle for power, was added a still confused aspiration for democracy and a rejection of growing pauperisation, but the result was often no more than interethnic confrontation. After Uganda, which had experienced a long period of troubles and war between 1971 and 1986, Rwanda went to war from 1990 to 1994, Burundi from 1993, Congo in 1993 and 1997, and Zaïre – having become the Democratic Republic of the Congo – from 1996.

Of all these wars, the most catastrophic for the region was the war in Rwanda – the Little Switzerland of Africa. Straightaway, in 1990, it involved Uganda and had negative effects on the economies of neighbouring countries. In 1994, it ended with the genocide of at least one million persons and a massive exodus of refugees, which affected Burundi, Tanzania and the east of the Democratic Republic of the Congo. Apart from the refugee camps in the Kivu, genocide militias and the remains of the defeated army continued, however, their attacks on Rwanda to perfect their "work", as they themselves said. In October 1996, the Rwandan army thus invaded the east of Zaire and the war soon extended to most of the Congo basin. Rwandan refugees, among whom were hidden genocide militiamen and what was left of the Rwandan armies, got to the north of the Congo, the south of the Central African Republic and even Gabon, while the Democratic Republic of the Congo was invaded by Ugandan, Rwandan and Burundian armed forces coming to the aid of Congolese rebel factions, and by the Zimbabwean, Namibian and Angolan forces come to the aid of the Congolese army. For the first time in the history of sub-Saharan Africa seven countries were directly implicated in the same war. In parallel with the human disaster, these events engendered a total environmental disaster in most of the Congolese forests and reduced to nearly nothing more than 70 years of nature conservation and development.

In addition to military operations, destruction, insecurity and cessation of economic activity, massive movements of populations engendered veritable cataclysms and plunged local populations into total privation. Hundreds of thousands of refugees devastated everything for kilometres at a time, razed natural forests and wooded areas, pillaging crops and herds. Western relief organizations brought in medical and food aid, but were unable to furnish to these human masses the space and combustible material that they needed. They were even incapable of disarming them. These refugee movements completed the economic disaster and exacerbated pressures on already greatly weakened environments.

This story, whose ups and downs we have been able to follow in the media, can seem very far removed from the forests of Central Africa. In reality, it plunges its roots into the montane and sub-montane forest regions, transformed, as we have already seen, into veritable demographic bombs. Of course historians, sociologists and anthropologists will find political, economical, social or cultural causes in the phenomenon, but no one can ignore the underlying ecological and demographic factors. These factors finally make of these montane and high plateau regions a constant threat to lowland forest regions.

In fact, the migrations, set off from the uplands of West Cameroon and East Nigeria 4,000 or 5,000 years ago, continue. They are even accelerating. Of course, many things have changed. For populations hopelessly in quest of space, however, the little-populated lowland forests of Central Africa will soon be about the only remaining lands of welcome possible. For how long yet will Gabon and the North-Congo escape invasion by populations coming from the north or the west. Already, numerous Nigerians, Malians, Senegalese and Burkinabes are to be found in villages farther and farther back in the forest where they keep shop or practice their occupation. In the coastal mangrove forests, numerous Nigerians have begun to settle under the nose of the authorities incapable of stopping the onset of this human tide. What will happen when, around 2050, Nigeria reaches 300 million inhabitants?

Through the forced migrations that they unleash, wars have had not only negative effects. Thus Rwandan populations that irrupted into northern Congo, notably in the Ouesso region, introduced market gardening, to such a point that this region now participates significantly in the supplying of Brazzaville. They also introduced the use of baked and adobe bricks. These techniques, known in the colonial period, had been abandoned but are presently in full development.

7.35. In the absence of any control, the mangrove forests are being little by little colonised by foreign populations, mostly Nigerians.

7.36. The establishment of any cultivated land begins with the clear-felling of a parcel of the forest. In most Bantu communities of Central Africa this work is done by the men. All that follows – cleaning of the fields, sowing, planting, tending the crops, harvest and treatment of the harvest – is the work of women. The older a forest is, the more difficult is the felling, therefore men's work, and the lighter is the weeding, women's work. However, the longer a parcel remains under cultivation, the more difficult becomes its upkeep. Therefore, in many cases, the abandonment of a parcel is based more on the increase in the women's work than on the reduction of its fertility.
In many regions of Central Africa, the clearing of old forests, called primary, is the work of young men, while younger forests are left to older men. With the departure of young men to the city, there arises a lack of manpower that is reflected in a shortening of the fallow period and often by the replacement of plantain crops with cassava.

The forest people at the dawn of the 21st century

After conquering the Central African forests, the Bantus progressively adapted their way of life to the constraints of their surroundings. They pushed back or assimilated the peoples that preceded them, or learned to live alongside them. Next, they were affected by the Atlantic slave trade, by Peul and Arab slave raids, by European colonisation, the return to independence and, most recently, the consequences of a ruthless world economy. All these phases of history have profoundly affected communities and ancient traditions have constantly given way to new customs or have amalgamated new elements.

Today, transformations are accelerated by the opening up of the forest massif, building of new roads, penetration by aviation and the introduction of radio and television. Thus, even though inhabitants of the most remote regions spend much time hunting, they can listen to *Radio France International* and some people manage to follow the African Cup of Nations football matches on television. In fact, these changes are so rapid that they create a gulf, not only between rural and urban communities but also between individuals of different generations or education level within these communities. In spite of that, urban communities, which represent about 50% of the population of the sub-region, have kept close ties with the forest world. They are very weakly industrialised and in fact live or survive mostly from the exploitation of natural resources – minerals, petroleum, wood, meat from hunting and medicinal plants – which they increasingly monopolise at the expense of the ever more impoverished rural populations. If the number of persons essentially dependant on forest resources in Central Africa is estimated at three million, then 20 to 30 million humans can probably be estimated to have some connection with the forests.

Not all the peoples of Central Africa's tropical forests live in the same way, but all are torn between the weight of their ancient traditions, of which they are not always conscious, and the attraction of new life styles. To review all the systems that they have worked out would require a book in itself. We will thus settle for a few examples, taken from around some of the ECOFAC sites. With the support of the programme *Avenir des peuples des forêts tropicales* (APFT) (Future of the Tropical Forest Peoples programme) and the Faculty of Agronomy of Gembloux University, the ECOFAC programme has invested much effort in trying to understand how these societies are structured today and how inhabitants on the edge of protected areas live.

The division of space

In spite of a strong increase in national populations, density in most of the forested regions has remained very low. In the area surrounding Dja Reserve, it varies from 0.6 to four inhabitants per square kilometre. In most of Gabon and the North-Congo, it is between one and two. In the Democratic Republic of the Congo, densities reach five to ten inhabitants per square kilometre.

Village communities are on the other hand larger today than they were before the regroupings of the colonial period. The villages studied in the Dja region, together with their satellite hamlets, thus contain 300 to 400 inhabitants, of which 25 to 30% are transient, however. In fact, many people have returned from the cities for lack of employment. In the Odzala-Kokoua National Park region, most of the villages have no more than 100 persons.

Small villages generally contain only a single ethnic group, but in regions where Pygmies live, they often camp near their ancient masters or employers. In large villages of over 500 inhabitants, several ethnic groups may live together. They are then often divided into neighbourhoods.

Whatever the size of the villages, the space that surrounds them is divided into two main zones. (FIGURE 7.38). The area of agricultural exploitation extends over a radius of five to ten kilometres including secondary forests, fields, land fallow for two, three, 10, 20 or even

30 years, collective or individual forests which will one day be returned to cultivation, as well as old sites of villages abandoned notably during the regroupings. In view of the itinerant nature of farming activity, the zone of exploitation does not correspond at all to the area actually under cultivation.

Beyond the agricultural area extends a communal forested territory covering a radius of 10, 20 or 30 kilometres. Hunting, fishing and gathering is practised here. Temporary camps are set up in this zone but it is not, or is no longer, clearly defined and is more or less flexible, probably depending on the displacements of villages and the availability of resources. Following the exhaustion of some of the resources, this communal territory is presently in expansion. In this way, in the Democratic Republic of the Congo, a village of 200 inhabitants used a forest space of 11,000 hectares, or 55 hectares per person. Elsewhere, a hamlet of 34 persons used an area of 1,500 hectares, or 44 hectares per person. According to Cédric Vermeulen and Alain Karsenty, the flexibility of this forest territory is not a result of chance, but is an integral part of a process of conquest of space because the abandoned lands are never totally abandoned. They allow the constitution of a clan or lineage space by successive stages.

In the Kivu Province of the eastern Democratic Republic of the Congo this tradition was and is still a reason for many conflicts. For an immigrant agriculturist from Rwanda or Burundi each parcel of forest which looks "wild" is simply a piece of land awaiting clearing and cultivation. How could he know that all forests

7.37. The space cleared in the forest is sewn with various plants. Depending on the soil and the region, cassava, cocoyam or taro, maize, peanuts, some rice, some Cucumerops squash and some plantain. All of it grows pell-mell in a semblance of disorder, but in reality this type of agriculture is perfectly adapted to the forest environment.

7.38. An imaginary example of the division of space, based on a study conducted in southern Cameroon, shows three villages established along a road that crosses the forest. Around each village extends the agricultural exploitation area, beyond it, the forested land or communal territory with old villages and temporary camps. Finally, beyond the limits of the communal territory are located the limits of the clan or family space. Today, with the exhaustion of faunal resources, communal territory has a tendency to extend to the limits of the family spaces.

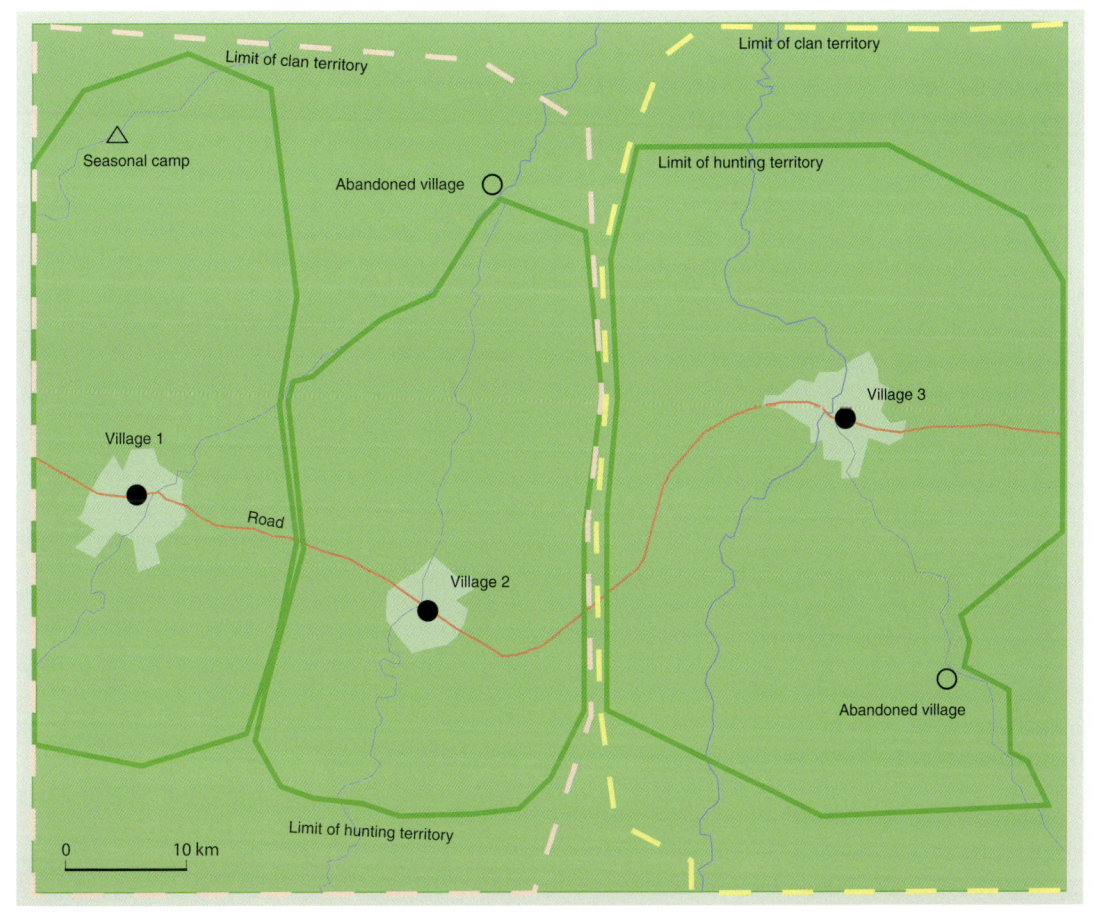

7.39. The seeds of some trees, such as those of Tetrapleura tetraptera *of the Mimosaceae family, are perfectly edible.*

7.40. The roots of the Iboga Tabernanthe iboga, *a small shrub of the Apocynaceae family, contain a drug that is used in initiation ceremonies in Gabon. The root is found in local markets along with innumerable other roots and barks that are used in the making of traditional medicines.*

7.41. Resin from the Okoumé Aucoumea klaineana *is collected for use in making torches.*

7.42. The bark of Bitter Kola Garcinia kola, *a tree of the Clusiaceae family, is sold in small bunches. It is used in the making of palm wine. Only bark from the base of the trunk or the roots is collected.*

belong to a clan, a family group or a village community?

In the case of the Baka Pygmies of south-eastern Cameroon, the agricultural land is much smaller, perhaps because many Pygmies work in the fields of the other people. On the other hand, hunting lands go well beyond the 30 kilometres in use by most of the Bantus and are two or three times more extensive than lands of slash-and-burn farmers. This results largely from the fact that the Bantus chase the Pygmies from their own land or demand a part of the game hunted. The Pygmies thus have no other alternative than to hunt very far from the villages, where Bantus venture only rarely or are unable to exercise control.

Agriculture and the exploitation of natural resources

On the part of the territory reserved for agriculture, itinerant slash-and-burn cultivation remains the main activity of a great many of the forest people (FIGURES 7.36-7.37). It does not truly represent an exploitation of spontaneous natural resources, but constitutes nevertheless a form of use of these resources through the intervention of fire. Seen more broadly, this type of agriculture is an integral part of the forest ecosystems since it constitutes their main mechanism of rejuvenation. Nutritionally, it represents only one strategy among many others.

Cultivated areas vary greatly in size but are generally of the order of a hectare or a little less. The length of time of exploitation is around a year and a half and fallow periods last around ten years, but may be much longer, up to twenty years. After more than eight to ten years, the rate of accumulation of nutrients in the soil slows down, but a too rapid succession of crops, with very short fallow periods, exhausts the soil and eventually requires a very long fallow period to re-establish fertility. After several shortened cycles of cultivation, the forest regenerates with much difficulty: the soil is exhausted, seeds have disappeared and stumps no longer regenerate.

The main problems confronting itinerant farming are the relatively low profitability of the labour involved and the heavy tribute paid to predators. In part, these come from the old forests and comprise small monkeys and ungulates – including elephants – as well as gorillas and even sometimes a chimpanzee. In part, these predators are also linked to man-modified habitats, however. This is the case with Bush-tailed Porcupines, Emin's Rats and Cane Rats. Also, in spite of its diversification with the introduction of American plants (PAGE 254), and in the absence of animal raising, this agriculture must of necessity be complemented by resources from the forest (FIGURES 7.39 and 7.43-46) or by cash crops. Hunting and fishing thus remain activities essential for the acquisition of animal proteins and hunting around villages can be considered as a form of extensive pastoralism. However, men prefer to invest their efforts into more profitable activities than hunting rats since these forest peoples prefer big game.

Independently of any agricultural or ecological considerations, itinerant farming is also, according to Diaw, part of a system of monopolisation of the lineage or clan territory. The fact of regularly displacing fields is based not only on the risk of exhausting the soil, and the risk of increasing the work by women for production (FIGURE 7.36), but is also based on "territorial" considerations.

Apart from the strict domain of food, the forest also offers numerous resources in medicinal plants (FIGURE 7.40), dwelling construction and tool making materials (FIGURE 7.44).

In many regions, population densities still being very low, this itinerant farming does not threaten old forests because most of the clearing involves still relatively young secondary forests. The cash crops do not represent a threat for the forest ecosystem either, because they are generally well integrated into a multi-stage agro-forestry system. For a long time yet, old forests can still coexist with farming.

The gathering of non-ligneous products from the forest can, however, have fairly diverse effects. If the gathering of fruit does not seem to affect the regeneration of the species involved, production of palm wine on the contrary, notably around the Dja Reserve, exerts an exaggerated pressure on populations of *Elaeis* and *Raphia* palms, as well as on Bitter Kola *Garcinia kola* and some other species of the same genus, whose bark is

7.43. Finely chopped, the leaves of Gnetum, *a forest Gymnosperm, are much appreciated and popular replacement for spinach.*

7.44. As for the leaves of the large Marantaceae Megaphrynium macrostachyium, *they are used to wrap food during cooking, notably preparations of "chikwangues" or cassava sticks.*

7.45. Atanga or Safou Dacryodes edulis, *of the Burseraceae family, is rare in the forest nowadays, but it is found around villages where it is planted by the inhabitants. Its fruit resembles those of Ozigo* Dacryodes buettneri, *a near endemic species of Gabon, but is larger. It contains a single large pit enveloped in a very perfumed flesh, and is eaten with a little salt after it has been briefly boiled.*

7.46. The fruit of Ofoss Pseudospondias longifolia, *a medium sized tree of the Anacardiaceae family, comes to the village markets in Gabon in July- August or in November. They taste excellent, but a little bitter.*

The Pygmies

For at least 20,000 years, the Pygmies have inhabited the dense humid forests of Central Africa where for a long time they lived only from hunting and gathering. Today, they are divided into three main ethnic groups (PAGE 247). The Kola and the Bongo live in the west, in Cameroon and in Gabon. The Cwa live south of the Congo River and the Twa farther east, in Kivu, Rwanda, Burundi and south-western Uganda. The Aka, Baka, Asua, Mbuti and Efe live north of the Congo River, from south-eastern Cameroon to the north-east of the Democratic Republic of the Congo and the Semliki Valley (FIGURE 7.7). Since the arrival of the Bantu slash-and-burners, they established relationships of complementarity with the latter, but the Pygmies, being the consumers, have been turned into vassals by the Bantus (PAGE 256-257). With time, their different groups, isolated from each other, have developed particularities on the basis of a probably unique cultural foundation and the contributions of the other Blacks. Today, the Pygmies are divided thus into two distinct cultural groups.

The first is represented by the Aka, the Baka and the Mbuti. The latter comprise in fact very different ethnic groups: the Asua, the Efe, the Sua, the Cwa and the Kango. Altogether, this group is made up of more or less nomadic groups that speak their own language – even if this is derived from Bantu, Ubangian or Central-Sudanic languages spoken by the Tall Blacks – have their own music and a political organisation, concepts of property and beliefs that often differ fundamentally from what is found among the Tall Blacks beside whom they live. These hunters live in small family groups which move about often and take shelter in small crude huts made of a few branches and covered with leaves. They do not have a well-established hierarchy and their chief, often the oldest, is no more than a *primum inter pares* (first among equals). All activities are discussed at length and all decisions are made collectively by the entire community. They know the forest and its inhabitants well and feed themselves on a great diversity of animals, including everything from the termite to the elephant, and a lot of leaves, fruits and roots. The Efe hunt only with bow and poisoned arrows, but most of the groups hunt with nets several hundred metres long placed in the undergrowth. They also use all sorts of traps, especially nooses. Finally, they know the use of glue to capture birds and plant poisons for fishing. It is therefore probable that these populations did not enter into contact with the farmers until fairly recently or in a non-permanent way.

The second group is made up of all the other Pygmies. These live in semi-permanent hamlets, share many cultural traits with the neighbouring farmers, even language and music. These groups therefore have probably been in contact with the farmers for a much longer time and have reached a much more advanced stage of integration. Having nothing more to hunt, most of the Twa in Rwanda and Burundi have become specialised in pottery making.

During the colonial period, however, relationships of complementarity between Tall Blacks and Pygmies were deeply upset, if not broken and, even though efforts were made to settle them, the Pygmies found themselves more and more marginalized. The Baka, Bakola and Medzan of Cameroon, who obtained their starches from their Bantu, Ubangi or Sudanian neighbours suffer a chronic lack of starches now. A few families have turned to farming, but most of the Pygmies have resorted to stealing from their neighbours' plantations, prompting a rise in tensions and conflicts.

Studies show, moreover, that the Pygmy populations of Cameroon could only survive without problems with the resources of the forest available if their population density were reduced by 75% from the present rate. Studies conducted in the Ituri by John Hart show, on the other hand, that in *Gilbertiodendron* forests, where yams are rare, Pygmies have probably never been able to live alone because of a lack of sugars. It is possible to imagine therefore that they did not colonise these forest until after having established relations with the Tall Blacks.

used as a fermenting agent (FIGURE 7.42). This wine is obtained by collecting the sap which runs out from a gash in the terminal bud of a palm tree. "Climbers" thus leave the tree standing and even keep it alive. "Uprooters" on the contrary cut down the palm after having freed its base and kill it outright. A palm tree produces an average of 75 litres of wine. Consumption of palm wine is considerable, however, and it constitutes a daily act whose social and economic impact is appreciable for the forest people.

In many areas, this dependence on the "spontaneous" resources of the forest has increased over the last few tens of years, in places because of the economic debacle, the troubles and the war, elsewhere because forest products have become the object of substantial trade toward the cities and hunting, in particular, has been transformed into a commercial activity. Although the forest peoples traditionally practised a rather prudent exploitation of the natural and semi-natural ecosystems which surrounded them, we now see completely uncontrolled and uncontrollable exploitation under the pressure of the external world, the globalisation of the economy and the introduction of needs unimagined in times past. According to Willy Delvingt, who, with his team from the Faculty of Agronomy of the University Faculty of Gembloux, devoted much time to the study of the inhabitants of the periphery of Dja Faunal Reserve, there is a weakening of basic skills exercised individually or on a family scale and a passage to skills of a public type. Customary basic skills maintain their role as regulators of the exploitation of natural resources in the proximity of villages (agricultural zone). They do not work any more in the forested areas.

This evolution must be placed in its widest context, however. According to Serge Bahuchet, it is, in fact, no less than the expression of the collapse of the social and cultural equilibrium resulting from the weakening of traditional beliefs, monetarisation and the opening up of the market economy, as well as the introduction of new techniques, such as firearms or nylon nets. It finds support, however, in the fact that, for most of its inhabitants, the forest is inhospitable, even frankly hostile, but at the same time inexhaustible. How many times does one hear that the animals do not disappear, but "just go farther away"? In fact, this attitude is perhaps not very different from that of Western Man who believes that science and technology will unfailingly solve all their problems.

Land rights

Among the slash-and-burners of Central Africa, land rights are generally only usufructuary and attachment to the land is of a symbolic type, a priority being given to the descendants of the first one to clear it. This land right is based both on the area of farm exploitation and on the forested land. Vast clan or lineage territories therefore form an entity whose limits are generally defined by rivers and mountains.

Territorialisation of land concerns not only farming but also hunting, fishing and sometimes even mineral extraction and woodcutting. Land is also an inalienable collective good and the right of exploitation is imprescriptible. Within a lineage parcels are allocated concertedly under the surveillance of the chief of the lineage. Once a parcel has been cleared, the ensuing secondary forest returns to the clearer or to his descendants by priority. Fallow land is inherited from one's father, except in some ethnic groups, notably the Fang, where inheritance is from mother to daughter. This difference in customs is today cause for a good many conflicts when mixed couples form.

Problems of property cannot be resolved except by reference to genealogies and to the hierarchy of rights. If, however, the exact ties have been forgotten, the property is subdivided in order to ensure autonomy. Nowadays the dynamics of land rights are influenced by the greater mobility of groups and by the establishment of the market economy. Rents thus appear. The land, or rather its usufruct, becomes a source of financial yield.

Since the cultivation of cash crops is done without fallow land, families are obliged to remain on the spot for several generations. This constraint has reinforced the sense of property and made the land rights evolve from a right based on work to a right based on the land

7.48. *Roads in the forest have various effects. Obviously, they open up forest massifs and make them accessible. They allow the removal of forest products: lumber as much as game. They also allow, however, the removal of agricultural products which from simple subsistence farming can thus evolve into more intensive agriculture with commercial objectives. In facilitating contacts and access to services, roads concentrate people and attract them even outside the forest. In some places, they have even allowed the relocation of populations that inhabited protected areas outside them without conflict. In other words, the problem of roads is complex and consists of positive, as well as negative, aspects.*

itself. So that land is not acquired by strangers, these will be allowed by the chief to farm land but not to plant cash crops on it. The concentration of people along roads and the permanent settlement of villages has also induced the constitution of family plots within the clanic space.

The transmission of land is carried out mainly by inheritance. In childless families, land is handed down to nephews or sometimes to the children of a friend. It can also be the object of a permanent gift or of a loan in case of need, but then without right to fallow land.

In some regions with a high population density, land rights have also undergone a marked evolution because there remain no more virgin lands to clear. An interesting example comes to us from the Bundibugiyo region in the Semuliki River valley in Uganda. Wedged in between the Rwenzori Mountains and the forests of the Semuliki, protected since the 1930s, and a national park since 1994, the density of the Baamba population reached 380 inhabitants per square kilometre during the 1980s. This Bantu people is related to people from the heart of the Ituri forests and is very different from the eastern Bantus of the Great Lakes region. Its society is matrilineal and polygamous. Traditionally the land belongs to no one and cannot be bought or sold. Virgin lands were communal property and each member of a village community could clear a parcel of it and plant it. After the harvest, the land was left fallow, three to four years on average, and the one who had cleared it kept the right to reuse it, except if he waited too long. Then, another could plant it. Today, there is no more virgin land, but since anyone had the right to give up his right to land to someone of his choice, this mechanism of transfer became generalised in exchange for payment. The rental of land for payment also appeared at the same time.

Forest rights

Forest rights regulate hunting, gathering and fishing. In the vicinity of the Dja Reserve, each village has its land, but hunters of the "friend" villages can come and hunt on this land, a situation that introduces problems when it is necessary to control poaching in a reserve. As for rivers, they are no longer divided into fishing "lands", as they were in the past.

In some populations, notably in the Mboko of Odzala-Kokoua National Park, the rightful owners demand rights of usage of the forest space from the other inhabitants, that is to say a portion of the first gathering, fishing or hunting.

Social relationships

The transformation of communities inhabiting the forest remains mostly conditioned by social relationships: inevitably, the men dominate the women and the older dominate the younger. Apart from that, several systems can occur or even coexist.

On the one hand, Pygmy communities are based on the clan, the ensemble of individuals who descend along one line from a single mythical ancestor and their chief is no more than a *primum inter pares*. These communities do not therefore really have chiefs and all decisions are made by consensus. These societies are called acephalous.

On the other hand, some communities of the northeast of the Democratic Republic of the Congo, influenced by Sudanian peoples, and those from the Great Lakes region developed very complex and hierarchical societies, dominated by provincial chiefs who were themselves named by a king.

Most Bantu communities fall between these two extreme types of organisation, like those who live on the border of the Dja Reserve or in the Odzala region. These are lineage communities that descend therefore in a direct line from one or several common masculine ancestors of whom there remains a memory. Real or mythical, these constitute the cohesive factor of the group. The individuals of the same lineage cannot theoretically marry, but are bound by a sentiment of solidarity stronger than towards another lineage. Lineages are also subdivided in segments or even sub-segments which are made up of individuals descending in a direct line from a common relative descending from a common ancestor. While a lineage is generally scattered over several villages, a segment of a lineage is linked to a

7.49. For many of the forest populations of Central Africa, fishing is an important activity. For the immense majority of farmers, who inhabit permanently emerged lands, fishing brings only a supplement to farming. It is practised as much by women as by men. The women often work in teams. After isolating a section of a stream in the forest, they empty it and capture everything they can, fish, amphibians, insects and crabs. This booty is often dried and ground in order to be added to food as a protein supplement.

particular village. As it is strictly forbidden to marry someone of the same segment of the lineage and since men generally remain in the village of their father, women are obliged to go elsewhere, except single women or widows. As for relationships between individuals, the nearer they are in the lineage, the more they are imprinted with solidarity. The head of the family can be the head of a family, strictly speaking, or at the head of an extended family. He can also be the head of a lineage, that is to say the oldest man who still has his faculties.

Since populations have been regrouped along roads and more or less immobilised, a person can occupy a house in a main village and one in a satellite hamlet. The village-hamlet entity then forms an entity, even though hamlets and forest camps are easily abandoned following disease, conflict or isolation. Elsewhere, the village constitutes an entity independent from the neighbouring hamlets. Socio-economic units – lets say families – number five to ten persons, rarely one to five, sometimes as many as 20. Members do not always live together in the same house nor in the same hamlet. They do not work together on the farming land, but share their revenues, and cash crops are transmitted by inheritance within the unit. On the forested land, there is no familial regrouping in the division of hunting areas and men from different units can hunt together. In general, teams of hunters exploit different areas depending on the season and the year.

There are chiefs in most of these Bantu communities but there is a clear disparity between official representatives, notably the village chiefs, and the real leaders. The official chiefs often have only a role of representation, while the real leaders are those who handle internal power. Numerous informal leaders thus intervene in business matters and the official chiefs are subjected to numerous pressures. They have power only when they are listened to and respected. In fact, all these societies remain organised according to the acephalous mode, therefore without a hierarchical structure but with a multiplicity of authorities who manipulate moral and intellectual persuasion with much skill. Leaders emerge and succeed each other which results in a fluidity of the political regime difficult to identify. For the rest, individuals arrange themselves according to their age and sex. Women have no power except through local associations.

In practice, a village has numerous authorities. Heads of family, heads of lineages and committees of wise men constitute the traditional authorities. To these can be added the managers of local associations. The "global" authorities are represented by village chiefs, district and church authorities and the elite. The latter are influential persons living outside of villages and whose wealth and stature, supposed or real, confer on them privileges and duties often used for political and personal purposes. In other words, such a society is comparable to a vast spider web, when one thread is touched, the entire web moves.

Local associations play an important role and are very diverse. First of all there are religious associations whose number has literally exploded in the last ten years (FIGURE 7.50). Next, there are economic associations. They are generally mutual-aid groups composed of a few members who work alternatively for each other according to a pre-established calendar, or groups of subscribers who organise tontines. There are sports associations, generally football, which benefit from the support of the urban elite. Finally, there are associations with a community vocation. These function either at the village level and regroup old women, married women and unmarried women, or youths, or at the administrative unit level. They are then strongly supported by the elite but, often animated by interests that no longer concern the villages, they are not always well understood by the villagers. Or, they are political. They regroup people according to well defined categories (youths, old men, women, etc.) and move into action during celebrations and official visits. They are directed by people who are rich, generous or who have good relations with well-known politicians. They play an important role in civil education.

Some Bantu communities, however, such as the Mboko community bordering Odzala-Kokoua National Park, are organised on a hierarchical model with a chief of the lands who is the guardian of the land holdings and customs. Through contacts with the Kota, Mboko communities have lost a little of their ancestral "rigour", but display a type of organisation that is very widespread in much of the Congo basin.

7.50. The men generally fish the rivers with the aid of cast nets (above), hoop nets (Page 248), fixed nets or lines with fish hooks. In some regions, there are specialised fishermen for whom farming is no more than a back-up. Some live on floating meadows and in inundated forests where they spend four to six months a year. Others live in permanent villages established on the banks of large watercourses like the Sangha, Ubangi or Congo rivers. They are skilled boatmen and in addition to fishing and some supplemental farming, they also. Traditionally, they trade their fishing products for products from hunter-gathering or farming.
Today, as is the case with hunting, they are involved in real commerce. Whatever their economy, and although nylon has made its entry into the Congo basin, these populations still depend largely on the forest for much of the primary material (wood, fibres, lianas) used in the making of their fishing equipment.

7.51. For a good ten years every village and every hamlet has seen the appearance of new churches. Alongside the classical Christian churches, Roman Catholic or Protestant, and already ancient "branches" such as the Seventh Day Adventists, the Jehovah's Witnesses or the Pentecostalists, more obscure formations have sprung up, such as the Church of the True Brothers of Jesus, the Church of the Apostles of the Last Hour and many others. These "alternative" churches find new converts not so much among the animists, but rather among bewildered attendants of established churches. A syncretic religion was even created In the Democratic Republic of the Congo – Kibanguisme – which today numbers several million converts. As for Islam, it is progressing in all of Central Africa, but remains mostly an urban religion. In spite of all this, traditional beliefs are still well alive. The image of this girl participating in an initiation ceremony in the Odzala region testifies to this.

7.52. This ceremony, probably of Kota origin, is practised nowadays as much by the Mongom and the Mboko who live south of the national park. All youths between the ages of 10 and 20 years old are subjected to it.

Beliefs

A century of Christianity and the pullulation of "new" churches over the past ten or fifteen years have deeply transformed beliefs, at least in appearance. The forbidding of most traditional religious customs has nevertheless pushed them into hiding and, even though they are still practised in many places, they have sometimes lost part of their sense and their role in society. Sometimes they are even reduced to the level of mere folkloric manifestations.

The multiplicity of churches, of their leanings and of their interpretation of life, as well as the mixture of Christianity and African beliefs, notably through syncretic beliefs such as Kibanguism, can aggravate the bewilderment of some minds and it is not uncommon to see people stray from one church to another. The attractive power of churches is all the stronger, however, when, in the depths of the forest, they are often the last institutions to guarantee education and health care nowadays, even if it is at times rudimentary. Fundamentalist churches can have a regrettable impact, however, especially when they oppose anti-AIDS or family planning programmes. Instruction in boarding schools accentuates the cleavage between generations.

Very often the impact of churches remains superficial, however, and religion a foreign contribution. Ancestral beliefs, even though repressed, transformed and sometimes even disguised, continue, to orient behaviour and decision-making although often in a muddled way, all the more so since missionaries have deliberately created confusion by inventing correspondences, even amalgams, between Christianity and traditional beliefs.

One of the great differences that exist between traditional beliefs and Christian beliefs resides in the position of Man towards nature.

For forest people it is the equilibrium between supernatural forces – the spirits – and humans that allows the obtaining of natural resources needed by the latter. This equilibrium is not guaranteed, however, and humans do not get what they want except by respecting the rules that dominate relationships within human communities and by communicating with the spirits through offerings and rituals, perhaps helped by "sorcerers", even if it has to be done secretly. Thus, mishaps are never anything but the expression of a break in this equilibrium and no death is natural. Only the Whites believe in bacteria and viruses and dare to ignore all the power of the spirits. For many of the populations who inhabit it, the forest remains in this way the domain of occult forces and represents because of this an irreplaceable property. This does not keep them from emptying it of its fauna, but is it not inexhaustible? In terms of the environment, Christianity has reinforced this notion of unlimited abundance. Above all, it has instructed that nature belongs to Man and that all natural resources are nothing other than a gift from God that it is proper to appropriate. The new churches have often not done more than to reinterpret the old beliefs in the light of the Bible and conduct nights of prayer or Bible readings to the sound of drums serving to ask God or Jesus Christ for what the human communities need.

This muddled religious context deepens the cleavage between generations. Not only are many of the youth from the villages fascinated by the city, which they visit regularly to see their relatives, sell or buy some merchandise or participate in the markets, but for them the city also represents the only means of escaping the pressures of the small traditional chiefs and "sorcerers", in short, everything that tradition has that is negative and opposed to development.

This difference finds its expression in the fact that Christians – also Muslims – pray mainly in churches or temples, while the people of the forest still enter far into the forest to pray and bring offers to their spirits. The protected Mondah Forest just north of Libreville is not only interesting for its biodiversity and for the medicinal plants it harbours. It is also an important site where people from the town come to perpetuate ancient traditions.

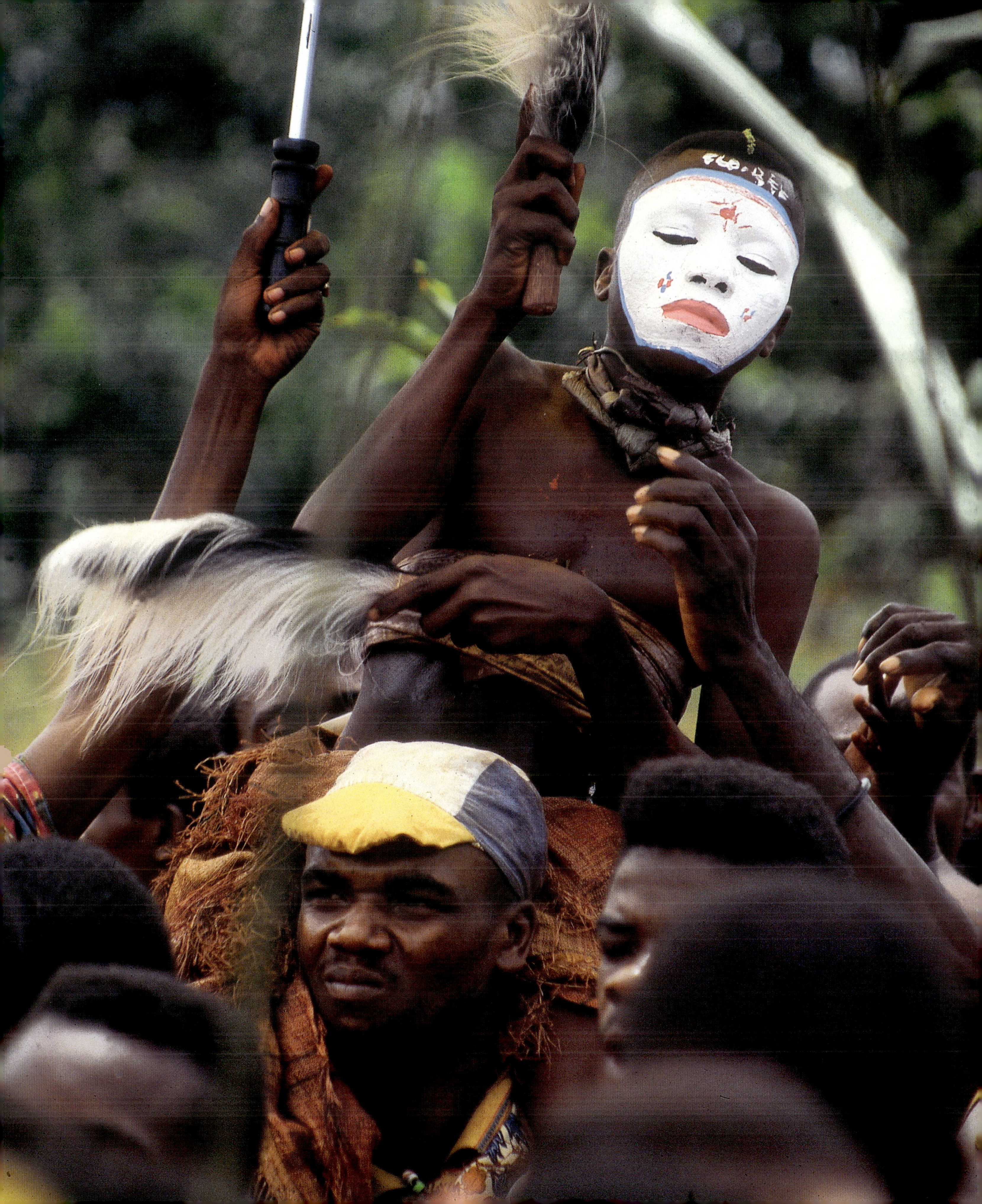

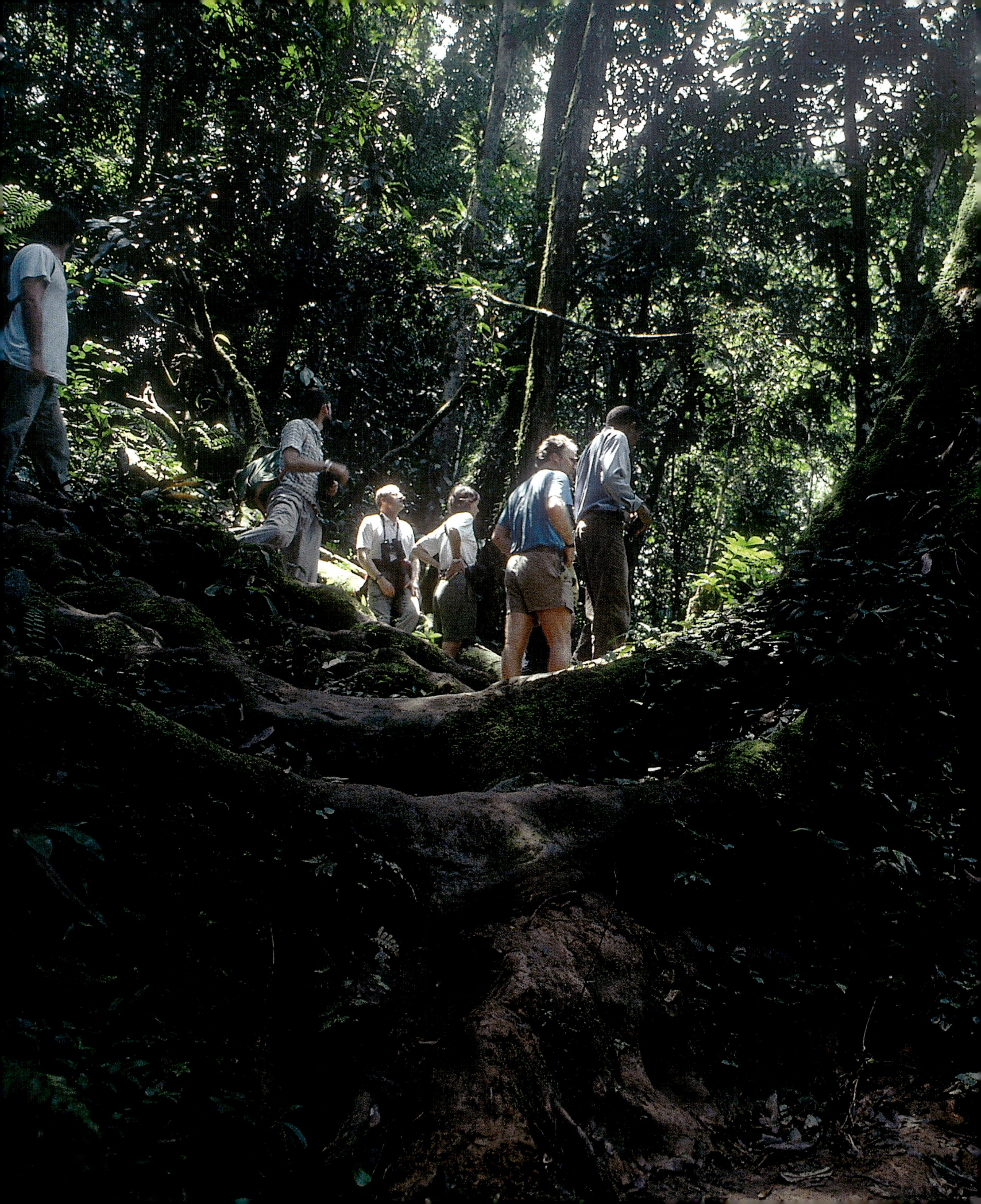

8 | Conservation

Alongside ancient traditions, which prevented unlimited exploitation of the fauna but did not resist colonisation, nature conservation as we think of it today was initiated by the colonial powers, beginning at the end of the 19th century, and particularly from 1925. Curiously, the deep roots of this conservation "*à l'occidentale*" are not to be found in the West, but in the heart of South Africa in the first half of the 19th century. In 100 years, many things have changed. Mostly in terms of context. On several occasions socio-political situations were turned upside down: human populations were multiplied by around eight and in several countries the last 40 years were marked by troubles, war and economic recession.

Then, conservation objectives evolved. At first, only the large grassland fauna was targeted. In the 1970s, they shifted towards ecosystems. During the 1980s and 1990s, they ended up focusing on forests. In 25 years, the surface of protected forested areas was multiplied by ten. In parallel, there was a profusion of actors. In the field, converted hunters and retired soldiers, supported by their all-powerful colonial administrations, were replaced by biologists after 1960 and, more recently, by experts in human sciences and economists. At a higher level in addition to the governments, bilateral and multilateral co-operation as well as international non-governmental organisations and the World Conservation Union (IUCN) appeared. The governments ratified big international conventions and launched several sub-regional initiatives. Unfortunately, financing did not follow the creation of protected areas, or else the programmes were too short in duration. This is even more regrettable since forest conservation was something brand new and required an in-depth revision of the strategies to be considered.

No one had the experience in this field, however, and local capabilities were often very inadequate. In addition, everyone began to meddle in conservation and the process of democratisation which began in Africa around 1990 did not simplify matters. While billions of US dollars are spent on defence without consultation, the least expense in the field of conservation requires broad discussion and acceptance. Meanwhile pressures increase from year to year and, proportionately, there are fewer and fewer conservation agents in the field. A substantial part of the money dedicated to conservation is also monopolised by development. Since the Rio Conference in 1992, however, conservation and the environment have never been so talked about, but the little money that remains serves only to edit strategies and action plans, to organise dialogue or ratification meetings, seminars and workshops. The World Summit in Johannesburg in 2002 could only count the losses and accept that the situation had indeed got worse.

8.1. To conserve is to love, to love is to know. The tropical forest is unfortunately a difficult habitat.

The precolonial period

In the 19th century hunting and fishing were the only sources of animal proteins for the forest populations of Central Africa, who did not have domestic animals. We do not know much about their impact on the fauna, but it is unlikely that it was great or long-lasting. The density of the human populations was too low. Nylon and metallic cables were not yet in existence and, in the absence of firearms, a good part of the forest fauna was out of reach in the canopy.

Many populations also designated species of animals that could not be killed or eaten, like the Bongo *Tragelaphus euryceros* in the Central African Republic. Others established periods of closure to hunting or practised a cyclic rotation of hunting land and everyone hunted on his own land, not outside of it. Some sites were also protected. In many regions, there were funerary places or sacred forests where only a few initiated individuals could enter for rituals. Sometimes these islets were limited to a tree or a group of trees, sometimes they covered several hectares or tens of hectares. In lowland areas, they were drowned in an ocean of more or less intact forests and their importance was more cultural than biological. In the mountainous regions of the Albertine Rift, they often represented the only vestiges of the original vegetation and therefore constituted real natural reserves. Contrary to the lowland forest populations, those of the mountains did not, or no longer, live in harmony with their habitat. By the time the Europeans arrived, the forest cover of Rwanda and Burundi had already been reduced to less than 10% of its initial area and most of the populations did not even have a source of fire wood. In the Cameroon highlands, grass fields had largely replaced the forests for perhaps 2,000 or 3,000 years.

The colonial period

With the arrival of first the Arabs and then the Europeans, the balance between Man and nature was rapidly broken. Beginning around 1850, firearms made their appearance and facilitated the collection of ivory. Ivory was quickly eclipsed by slave trading for the Arabs, but it became a strong currency for the Europeans. Much in demand by the trading posts, this white gold rounded off the salary of more than one colonial functionary.

To ivory hunting was added meat hunting (FIGURES 8.2 et 8.3). It was necessary to eat and to feed the labourers, hired notably during the construction of the first railroads and missions or for the exploration of large concessions. Thus, some companies soon hired professional hunters and meat from hunting completed or replaced the very low or non-existent salaries. At a few missions, it served as bait and attracted even more people than the sermons did. Without any scruples, the first Europeans abandoned themselves to a hunting activity that nothing restrained. Like the Africans, they were convinced that the continent was inexhaustible.

After the end of the fight against slavery, hunting ended up being the only attraction the continent had to offer for many a son of good family. For colonial functionaries lost in the bush and condemned to the uniform and to saluting the flag, it constituted a powerful means of letting off steam and combatting the frustration and monotony of everyday life. In the absence of television, hunting tales, each crazier than the others, filled the long tropical evenings.

In spite of the return of nature which followed the flight of human populations (PAGES 264-265), everything seemed to indicate around 1900 that the large fauna of tropical Africa would be subjected to the same destiny as the southern African fauna, decimated as fast as the advance of the white colonists. The rapid development and urbanisation of the conquered land materialised little by little, however, and resulted in the loss of the utilitarian role of hunting. It thus became a sporting activity (FIGURE 8.3). Most hunters set off to track down trophies, but some turned into naturalists and embarked on the search for new species, perhaps in the

8.2. For Europeans, big game hunting was one of the main attractions of the African continent and the meat obtained from it completed or replaced the salaries of the local labourers.

hope of one day seeing their name attached to it. They considered themselves as the only connoisseurs of nature and firmly believed in their mission. The enormous collections that they helped accumulate in metropolitan museums laid the foundations of our present knowledge of the continent's fauna.

From a subsistence activity hunting became a "noble" sport, reserved for strangers. In the English colonies it was even forbidden to indigenous people and the traditional hunter was no more than a poacher.

The creation of hunting reserves

It is in this ambiguous context that the first ideas of conservation took root. It is therefore natural that they were entirely centred on the large fauna, the grassland fauna, in particular. Just as the great lords of ancient Europe had appropriated forests and large game and withdrew them from the villeins – without which many species would otherwise have disappeared – the colonial powers created hunting reserves to withdraw vast portions of the territory from the hold of the indigenous peoples.

In Central Africa, the movement was initiated by the Belgians: the first elephant reserves of the Congo were set up in 1889 and total protection of the mountain gorilla was promulgated in 1912. Known barely since 1908, this ape was in fact already threatened by hunters and zoos. In Uganda, the first hunting reserves came into being in 1902; in French Equatorial Africa, in 1925.

National parks and faunal reserves

In 1919, just after the First World War, King Albert I of the Belgians made a trip to the United States. His visit to Yellowstone National Park gave him the idea of creating large national parks in the Congo. This dream took form in 1925 when the Belgian colonial administration created the first African national park: Albert National Park, now Virunga National Park.

At first glance, the national parks of Africa were therefore created in the image of those of the United States, though in fact these were the product of ideas that germinated in southern Africa. The first ideas of conservation in the Western tradition came into being in the Cape region just before 1850. They were developed by Ludwig Pappe, an Austrian surgeon and botanist, who, from 1840, was concerned by the deforestation, disappearance of species and degradation of habitats and landscapes in the Cape colony. They led to the promulgation of a first forest law in 1846. Supported by the director of the botanical garden of London and by the governor of the Cape, Pappe became botanist in chief of the colony in 1858. In this same year, the Forest and Herbage Preservation Act was promulgated and the first official faunal reserves were set up. These measures were motivated by the need for protection of the soil, water resources, forests and species and above all by the fear of climate changes.

Pappe's successor, John Brown, pursued these ideas, but was less diplomatic than his predecessor and rapidly came up against the powerful lobbies of the white farmers and lumber merchants. In 1866, he was sent back to Scotland where he continued to write on forest and water conservation. From his writings were born nature conservation. This soon infiltrated the administration of the English territories, especially in India, where the preservation of forests had become a major preoccupation. Brown's writings also made their way to North America and led G.P. Marsh in 1864 to publish his work *Man and Nature*, which was the basis for the creation of the first national park at Yellowstone in 1872.

In this way that nature conservation ideas developed in the West after having sprung up in southern Africa in the minds of a few Whites animated by a naturalist-doctor. The English historian, G. Grove, relates all this in detail and shows us that these ideas were initially much more attached to the conservation of habitats and sustainable development than to the protection of the large spectacular species. They had also been more motivated by the destruction by the white colonists than by the indigenous populations. They were therefore closer to those who today try to reconcile conservation, development and local populations. Unfortunately, they were born too early and ran up against the often hard to reconcile interests of the state, of the colonists and the scientists. And even though the colonial authority was very strong at the time, it could not oppose its colonists. That is why conservation in South Africa finally refocused on the large fauna and marginal habitats and the creation of Albert National Park was followed, in 1926,

8.3. Ivory rounded off the salaries of the colonial functionaries.

8 CONSERVATION

8.4. The end of a mountain gorilla in Rwanda, at the beginning of the 20th century

by Kruger National Park, not in the humanitaro-scientific tradition but in the pure "Nimrodian" tradition as described by Grove. In the rush, while they were at it, Albert National Park was extended to the region of the volcanoes where the mountain gorilla lived. It thus extended from the extreme north-west of Rwanda to the Semliki Valley and included the entire western shore of Lake Edward, as well as the west slope of the Rwenzori Mountains.

From 1925, Belgium persuaded Great Britain to extend the protected area to the east of Lake Edward in Uganda. At first, these steps achieved only the extension of hunting reserves and the first national parks in Uganda did not see the day until 1952. Kazinga National Park, which became Queen Elizabeth National Park after a visit by the Queen of England, was established in the Lake Edward-Lake George region. It completed Albert National Park and englobed Maramagambo Forest. Murchison Falls National Park stretched into the Victoria Nile region.

In 1932 Belgium proposed the creation of three national parks in Rwanda-Urundi: Rusizi and Ruvubu in Urundi; Akagera in Rwanda. However, this project ran up against the Belgians of Usumbura, present-day Bujumbura. The plain of the Rusizi, hardly a few kilometres from the town, constituted their Sunday hunting grounds where they shot birds, while in the Ruvubu Valley they hunted buffalo and the large antelopes.

Only Akagera National Park, far from Usumbura, came into being. This example illustrates how, before the Second World War, only the marginal grassland habitats came into consideration for conservation and how hunting prevailed over conservation.

In French Equatorial Africa, a decree of 1929 provided for the creation, not only of national parks but also faunal reserves. The latter had a statute very close to that of the national parks, but traditional agriculture and pastoralism remained authorised on it subject to control. In the French view, the creation of protected areas could not in any case slow down development. Several large faunal reserves were established in the north of the present Central African Republic in 1925 and in 1939 and the first national parks were established in 1933, just on the eve of the London Convention. Farther west, in the present Congo Republic, Odzala National Park came into existence in 1935, while in Cameroon the Campo and Douala-Edéa faunal reserves were created in 1932, and Santchou Reserve in 1933. Immediately after the Second World War, four faunal reserves were created in the present Congo: Léfini, Lékoli-Pandaka, Mount Fouari and Nyanga-Nord. In Gabon, a national park and a hunting reserve were created in the Lopé-Okanda region in 1946 and, in Cameroon, Dja Faunal Reserve came into being in 1950.

The creation of the first protected areas set off a movement which culminated in the London Convention of 1933 covering the conservation of fauna and flora in the natural state. This convention proposed the creation of protected areas, a list of important species and regulation of the trade in trophies. It also proposed the banning of non-selective hunting methods and encouraged even the preservation deforested regions. Nine countries, including Belgium and the United Kingdom, adhered to this convention, which was in fact not much more than a simple declaration of intent.

The first protected areas were clearly conceived for the protection of the large fauna and ideas of the protection of entire ecosystems, *a fortiori* forests, would not make its way until much later. The protected areas of Central Africa therefore involved practically nothing but grasslands. Even the Lopé-Okanda hunting reserve and Odzala National Park, located in the heart of the forest, owed their existence to their grassland enclaves.

As for the delimiting of the first protected areas, it was done in an opportunistic way, often without cartographical support and generally by colonial functionaries without any notion of biology. It therefore rarely corresponded to biological limits. At the time, that did not have much importance, however for there was still a lot of wild spaces outside the protected areas.

8 CONSERVATION

The forest reserves

Even though the forests harboured a fauna that yielded nothing to that of the grassland species, they were most of all considered as a potential source of wood and, where this resource was not exploitable, they were destined for conversion. The lowland forests of the Congo basin were so vast that no measure of protection would seem reasonable

In Uganda, the situation was very different: wood was a limited resource and ideas of conservation, born around the middle of the 19th century in the Cape, immediately found a favourable terrain. In 1898, the Scientific and Forestry Department of Uganda, which gave rise to the Forest Department in 1917, was created. From 1900 various treaties signed by Great Britain and the Kingdoms of Uganda, designated a great part of the forests as property of the Crown and in 1902 the first graduate foresters arrived. Only the small forests located on private lands escaped the colonial seizure. Clearing continued however. To such a point that a first forest policy was decreed in 1929 and all the large forests were set up as forest reserves of the Crown beginning in 1932. In all, 365 000 hectares were classified in this way. On the scale of Central Africa, this area perhaps does not mean much, but on the scale of Uganda this measure was crucial. Without it, there would probably not be much forest left in the country today.

For the English, the setting aside of reserves did not however, mean total protection. The goal was only to subtract these forests from conversion into farm land with a view to their sustainable exploitation. Budongo Forest, the largest and richest mahogany forest (*Khaya* and *Entandrophragma*) of eastern Africa, whose commercial exploitation began in 1930, was from 1934 given a management plan providing for a rotation of 80 years. Around 1950, most of the large forests of the country were provided with similar plans that provided notably the putting into total reserve of some sectors as reference samples. These did not cover more than very small areas, however, and were not out of reach of illegal exploitation.

Outside the large central forest reserves, the country was also endowed from 1938 with local forest reserves, under the administration of the districts. Smaller, they were destined to meet the local needs of the populations in a sustainable way. Around 1950, the Forest Department was, however, uneasy about poor management of these private forests which covered large areas bordering Lake Victoria. The opportunity was then given to the owners to give their forests in rental to the Department of Forests for a period of 99 years in exchange for 75% of the benefits earned. Between 1930 and 1960, Uganda very strongly developed the sustainable use of its natural forests and its Department of Forests acquired a good reputation in the management of tropical forests.

In Rwanda-Urundi, wood exploitation had only local importance. The small, isolated montane forests had been cleared by missionary builders and some more extensive forests had been degraded by the systematic skimming off, notably, of *Podocarpus*. Because of steep slopes, the large forests on the Congo-Nole crest, like the Bwindi Forest in Uganda, did not lend themselves to extensive exploitation, however. Also, the legislation promulgated in 1930 aimed rather at protecting these forests from conversion to industrial plantations by the colonists while leaving to the local populations the right of taking wood for their traditional uses. The Rwanda-Urundi forest reserves, created in 1934, therefore became real nature reserves. They were delimited by rows of exotic trees – generally cypress – and were consecrated to research on the dynamics of natural forests and the behaviour of indigenous species.

In the Belgian Congo, the first forest law did not come into being until 1949. It defined forest reserves and protected forests, forestry practices at the level of traditional exploitation as well as commercial exploitation, the attribution of logging permits and the payment of a reforestation tax. It also required inventories to be made in order to be able to determine rationally the annual volume of wood that could be cut. Vast expanses

8.5. The forest reserve of Bururi in Burundi covers only 1500 hectares, but it protects a beautiful forest of Entandrophragma excelsum *and harbours several endemic species. The summit of Mount Bururi culminates at around 2,300 metres and the forest occupies all of the west flank of the mountain. Its eastern flank, exposed to the dry winds coming from eastern Africa and the Indian Ocean, is occupied by high-altitude grasslands punctuated with shrubby* Protea madiensis, *and rich in Orchidaceae, Liliaceae and Iridaceae species and therefore probably very old. The mantle between the forest and the grassy habitats has certainly been accentuated by fires, but it nevertheless constitutes a natural transition which merits protection.*

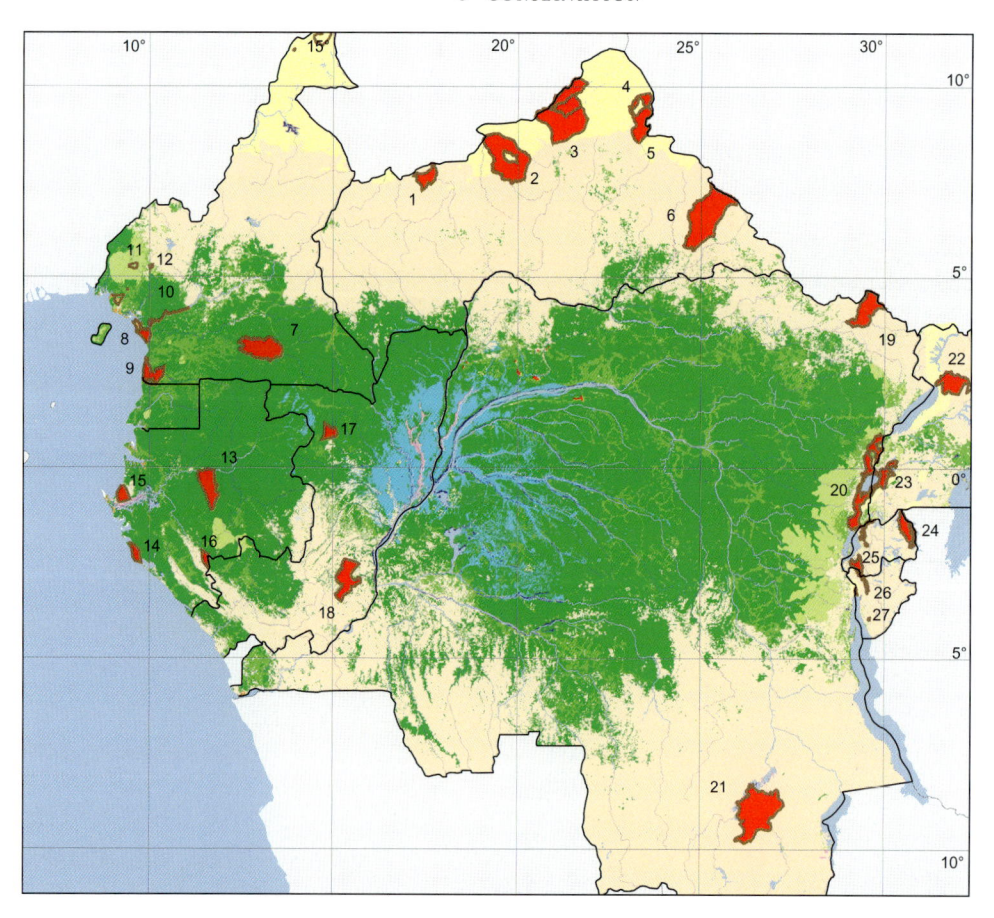

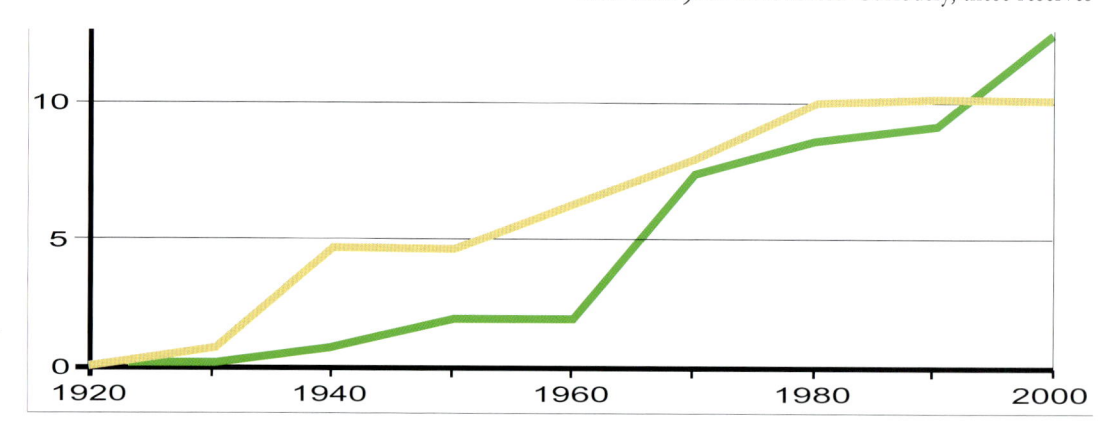

8 CONSERVATION

The last forty years

In spite of all the difficulties that loomed after their independence, the countries of Central Africa who already possessed a network of protected areas extended it and those who did not yet have one created one. On a regional scale, the total area of protected areas went from 9.2 million hectares in 1960 (FIGURE 8.6) to 26.2 in 2001 (FIGURE 8.7). It therefore practically tripled in 40 years. This extension was spectacular mostly in the forest domain where the protected area went from 1.3 to 14.2 million hectares, and was multiplied by ten (FIGURE 8.8). This picture is composed of vast shaded corners, however, linked sometimes to development but more often to the troubles and the war. At times and in places, years of effort were thus reduced to nothing. Strategic approaches and forms of management also underwent profound transformation. In part, these resulted in new relationships engendered by the return of independence. In part, they were the reflection of an irresistible world evolution from which Africa did not remain on the sidelines.

Achievements

Among the countries that already possessed a network of protected areas, Gabon was the very first to modify and extend it. In 1962, the national parks of Lopé-Okanda and of the Sette-Cama region became areas of rational exploitation of the fauna, while the Wonga-Wongué, Ndendé and Moukalaba-Dougoua areas of rational exploitation of the fauna were created. In 1967, the Mount Kouri area of rational exploitation of the fauna were created and, in 1971, the Ipassa integral reserve near Makokou. In 1972, Wonga-Wongué became a presidential reserve. All these sites encompass vast expanses of forest, but, except for Ipassa, they remain centred on islets of grassland. Pure forest protected areas would not see the day until 1998 with the creation of the Mounts Doudou Reserve and in 2000 with the creation of the Minkébé Reserve. In all, 2.6 million hectares were thus put in reserve, of which more than 90% were forest. Curiously, these reserves

8.6. *The network of protected areas in 1960: Nana-Barya Faunal Reserve (1), Bamingui-Bangoran National Park (2), Manovo-Gounda-St Floris National Park (3), . André-Félix National Park (4), Yata-Ngaya Faunal Reserve (5), Zémongo Faunal Reserve (6), Dja Faunal Reserve (7), Douala-Edéa Faunal Reserve (8), Campo Faunal Reserve (9), Sanaga Faunal Reserve (10), Santchou Faunal Reserve (11), Kalfou Faunal Reserve (12), Lopé-Okanda Faunal Reserve (13), Petit Loango National Park (14), Wonga-Wongué Faunal Reserve (15), Ndende Faunal Reserve (16), Odzala National Park (17), Léfini Faunal Reserve (18), Garamba National Park (19), Albert (=Virunga) National Park (20), Upemba National Park (21), Murchison Falls National Park (22), Queen Elizabeth National Park (23), Akagera National Park (24), Nyungwe Forest Reserve (25), Kibira Forest Reserve (26) and Bururi Forest Reserve (27)*

8.7. *Evolution of the total area of grassland protected areas (in yellow) and of forest protected areas (in green) between 1920 and 2000.*

of montane forest in the Kivu region, judged to be without commercial value, were however, granted to colonists to be converted into plantations or pastures

In French Equatorial Africa, the innumerable forest reserves and classified forests came into being from the beginning of the 1930s, but they could also be very rapidly declassified and their importance for conservation was highly variable

protected the fauna but not the flora and many had already been exploited or were in the process of exploitation. In 2000, the Gabonese government decided, however, to devote 10% of its territory to the creation of national parks. The Lopé Reserve then saw its limits redrawn, to exclude from it a concession whose exploitation was imminent. In compensation, new territories were appended in the west. Around the same time, a global evaluation of the critical sites of the country was launched by the ministry in charge of conservation, supported by the Wildlife Conservation Society (WCS) and the WWF. The Gamba complex of protected areas was split up into two national parks : Mount Moukalaba-Doudou in the east, Loango in the west. A stripe of land in between was left to the oil companies. Lopé and Minkébé national parks were based on the existing reserves. All other national parks were totally new. Akanda and Pongara, close to Libreville, are mainly centred on mangrove ecosystems. Monts de Cristal, Waka and Birougou are centred on the rich submontane forests of the Crystal Mountains and the du Chaillu massif. Ivindo National Park includes a large stretch of the river and also the Langoué bai. Mwagné National Park includes swamp forest. This exercise succeed at the end of August 2002 in the creation of 13 national parks.

Cameroon extended its network of protected areas beginning in 1964 by creating the northern national parks. The first national park in the forest came into being in 1986, while the Korup faunal reserve, created in 1937, became a national park. Next, other protected areas were proposed in the forest – the reserves of Lake Lobéké, Nki and Boumba Bek in the south-eastern part of the forest massif, Kilum-Idjim Reserve on the highlands north of Mount Cameroon and Pangar-Djerem Reserve at the limit between forest and grassland – but most are still waiting for the final classification. In 2000, two new national parks were gazetted, however: Campo-Ma'an, in the south-west; Mbam and Djerem in the centre. Cameroon thus brought its network of protected areas to 2.6 million hectares, of which 1.7 million hectares of forest.

In spite of the troubles of the first years of its independence, the ex-Belgian Congo also extended its network of protected areas. In 1969, Albert National Park was renamed Virunga National Park. The next year, two large forest national parks were created: Salonga in the very centre of the basin and Maiko on the western foothills of the mountains of the Albertine Rift. The two of them alone protect 4.7 million hectares of nearly intact forest. Shortly after, Kahuzi-Biega National Park was born. Its initial area of 60,000 hectares of montane forest was extended in 1974 to 600,000 hectares by the addition of a vast block of submontane forests and lowland forests. Lastly, in 1992, Okapi Faunal Reserve was still created, with a area of 1.37 million hectares in the middle of Ituri Forest, and Mangroves Natural Reserve, located just north of the Congo River estuary.

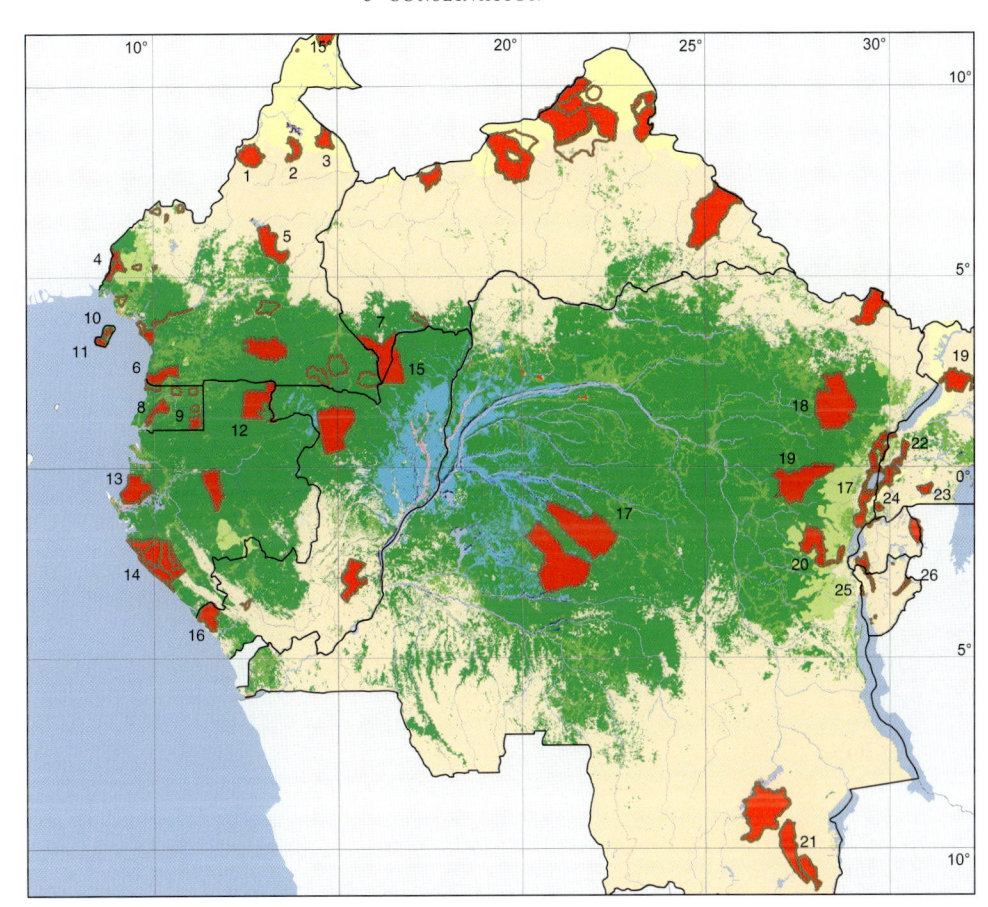

In the Congo, Dimonika Biosphere Reserve was created in the Mayombe in 1983, Tsoulou Faunal Reserve was established in 1984, Nouabalé-Ndoki National Park was founded in 1993 in the extreme north, along the Cameroon border, and Lake Télé and Likouala-aux-Herbes Community Reserve were constituted in 1999. In 2000, Conkouati Faunal Reserve, located in the coastal region, became a national park and, in 2001, Odzala National Park was extended from 126,000 hectares to 1.3 million hectares and renamed Odzala-Kokoua National Park. Altogether, the area of the protected areas, excluding hunting reserves, was thus brought to 3.5 million hectares, of which about 2.9 million hectares were in forests.

In Uganda, extension of the protected areas network came later. Despite the enormous problems of the time, the 1970s witnessed the creation of Lake Mburo National Park in the south-west of the country. Then, in 1994, eight years after the end of the troubles, Kibale National Park, Semuliki National Park, Bwindi Impenetrable National Park, Rwenzori National Park, Mgahinga Gorilla National Park and Mount Elgon National Park were created, all in forest. The protected area, in the form of national parks or forest reserves, was thus brought to 1.3 million hectares, nearly 500,000 hectares of which in forest.

8.8. The network of protected areas in 2000. The new units since 1960 are: Faro National Park(1), Bénoué National Park (2), Bouba-Ndjidah National Park (3), Korup National Park (4), Mbam and Djerem National Park (5), Campo-Ma'an National Park (6), Dzanga-Ndoki and Dzanga-Sangha National Park (7), Monte Alén National Park (8), Altos de Nsork National Park (9), Pico Basilé National Park (10), Caldeira de Luba Scientific Reserve (11), Minkébé Reserve (12), Wonga-Wongué Presential Reserve (13), Gamba Complex (14), Nouabalé-Ndoki National Park (15), Conkouati National Park (16), Salonga National Park (17), Okapis Faunal Reserve (18), Maiko National Park (19), Kahuzi-Biega National Park (20), Kundelungu National Park (21), Kibale National Park (22), Lake Mburo National Park (23), Bwindi-Impenetrable National Park (24), Rusizi National Park (25) and Ruvubu National Park (26). In 2002, Gabon gazetted 13 national parks: some were ancient reserves, some were entirely new sites (Page 352).

8 CONSERVATION

8.9. Destruction of the forest and settling of peasants in the lower part of Volcanoes National Park in Rwanda, around January 1970. While in 1958, just before independence, this forest, which was then part of Albert National Park, had already lost 7,000 hectares to agriculture, a new project financed by the European Economic Community amputated another 10,000 hectares from it in 1969 for the implantation of a pyrethrum programme (Figure 8.10). Ten years later, in 1979, another 1,300 hectares were cleared.

In already well endowed Rwanda no new protected area was created but, among countries that had no protected areas, Burundi moved into conservation in 1980 with the creation of three national parks, including Kibira National Park comprising 40,000 hectares of montane forest and a good population of chimpanzees. Ruvubu National Park covered some 50,000 hectares of wooded grassland and swamps, and also harboured large forest galleries inhabited by a population of the *tephrosceles* form of Red Colobus, estimated to number at least 1,500 to 2,000 individuals. At the same time, Burundi reanimated a few forest reserves, notably Bururi Reserve (FIGURE 8.5).

Equatorial Guinea, which had initially cancelled all arrangements for conservation taken by Spain, created Monte Alén National Park in 1988. In 1997, it proclaimed a forest law, making provision for the establishment of a vast network of protected areas. It came into being in 2000 and comprises two scientific reserves, three national parks, two natural monuments and six natural reserves (Annex). Altogether, Equatorial Guinea thus puts into reserve an area of 514,500 hectares of land – nearly 20% of its terrestrial territory – and 77,000 hectares of marine space. Most of the terrestrial protected areas are occupied by forests.

Lastly, on São Tomé and Principe, a proposition for the creation of Obo National Parks – one on São Tomé, one on Principe – is awaiting signature since 1999.

The reverse of development

While protected areas multiplied, development projects also proliferated, some of which unfortunately ran counter to conservation projects. In large countries, there was no reason to directly attack protected areas, but, in the small countries where available space was much rarer, protected areas were the last "virgin" lands and made the dreams of managers and the developers.

Rwanda was particularly affected and the limits of Volcanoes National Park were constantly eaten away (FIGURES 8.9 and 8.10). This park, harbouring rare habitats, was thus reduced to 15,000 hectares, less than half of its initial area, and its *Prunus africanus* forests were completely cleared to make room for growing pyrethrum. Towards the end of the 1960s, an area of 600 hectares of very rich old forest was also cleared on the basalt plateau of Gisakura, in the western part of Nyungwe Forest, to make room for tea plantations. During this period, the elephant population, which still numbered at least 200 individuals around 1969, was hunted down and reduced to less than ten by 1975. Finally, in the 1980s, Gishwati Forest was reduced from 22,000 to 4,000 hectares to create pastures within the framework of a World Bank project.

Troubles and war

The collapse of the economy and serious political troubles that affected several countries of Central Africa had the same consequences everywhere (PAGES 269-273) and conservation was particularly hit. The protected areas became inaccessible and, where it existed, tourism stopped. Personnel were no longer paid. Maintenance of the sites stopped. Infrastructure and equipment were abandoned, pillaged or destroyed. Insecurity set in and hunting and nature conservation laws no longer made any sense; the employees of the protected areas were probably among the first to ignore them.

In Uganda, the forest reserves were seriously degraded. During the darkest years, some were invaded by farmers, sometimes even with the approval of what remained of the government. For Idi Amin, these reserves were a creation of the English – a vestige of colonisation – and were condemned to disappear.

In western Rwanda, farmers profited from the social troubles at the beginning of the 1960s to invade part of Nyungwe Forest Reserve, the area of which was thus reduced from 120,000 to less than 100,000 hectares. Much more recently, in 1996, two years after the end of the war, Akagera National Park was reduced from 250,000 to 100,000 hectares, while the 4,000 hectares remaining of Gishwati Forest were invaded and cleared by displaced populations.

In the Congo basin, populations fleeing insecurity plunged into the forest. Residents along the Bukavu-

Kisangani road came into Kahuzi-Biega National Park, adding their numbers to the indigenous populations.

When classic or guerrilla wars broke out, the protected areas became the theatre of predilection for confrontations. What remained of infrastructures was destroyed and numerous members of the personnel were killed – in Virunga National Park, more than 100 guards have been killed since 1996 – while others took flight. The soldiers massacred the fauna, sometimes for meat, sometimes for ivory, and at yet other times, for the simple pleasure of killing. Finally, to the ravages of armies, armed bands and militias, were added those of the foreign forces, African as well as European, come to the rescue.

The damage caused by fighting, hunting and insecurity, however, was very often superficial and transient. Damage engendered by the waves of refugees was deeper and more permanent. Although the national parks of Kivu suffered less from the general decline of the economy and the deterioration of infrastructures than other national parks of Zaïre, they suffered severly from the Rwandan war. The situation became dramatic with the flow of 1.2 million refugees into the Bukavu and Goma regions in 1994. With the aid of humanitarian organisations, camps were set up in the immediate proximity, and even inside, Virunga National Park. For many agents of these organisations, especially the young as well as the religious, of good will but often with little education, conservation was nothing but a heresy that should disappear and some went so far as to launch veritable crusades against conservation and conservationists among local populations. In the Kivu, the result was that, over a surface of many square kilometres, trees were cut down and the volcanoes area of the national park lost around 15,000 hectares of forest. Bands of armed refugees hunted tirelessly in order to sell meat in the camps. The High Commissary for Refugees also wanted to set up a camp in Kahuzi-Biega National Park, but thanks to the efforts of the warden Germain Mankoto and the support of UNESCO, this catastrophe was averted.

With the war, which began in 1996 and is still going on, the degradation of the national parks of Kivu took on even more alarming proportions. Much of these

8.10. Around 1975, the forests of Prunus africanus *had effectively given way to neat rows of pyrethrum. From the onset of the programme, the peasants, however, were not interested in pyrethrum and planted potatoes in the middle of the fields or in areas that were not very visible. Some years later, pyrethrum had lost its value and these crops were then replaced by food crops. By this ephemeral project, the EEC thus achieved at great cost what the peasants of the region would have been able to do alone. But the damage to Volcanoes National Park is irreversible.*

8.11. During the troubles of 1993, the infrastructure of Ruvubu National Park in Burundi was reduced to ruins. For an occasional observer, the determined onslaught of these peasants shown by these poor slabs, could indicate how much they hated conservation. During the same period, however, they also destroyed numerous bridges, hospitals and schools.

parks fell into the hands of the Mai-Mai rebels and three-quarters of the gorillas of Kahuzi-Biega National Park were probably exterminated.

In the Congo, the serious troubles of 1993 remained confined to areas near Brazzaville, but the war of 1997 inflamed the entire country. The national parks were relatively spared but were cut off from the capital. Because of its regional structure, the ECOFAC programme could maintain its presence at Odzala National Park. The co-ordination of the programme had to be transferred to Libreville. The trans-border road linking Odzala to Gabon was rebuilt so as to guarantee the provisioning of the park and the surrounding region (PAGE 295).

After the end of hostilities, stricken countries are invariably confronted with the problem of landmines and the proliferation of weapons. Animal populations are greatly reduced, even though forest species resist better than grassland animals. In spite of some losses, the mountain gorillas of Virunga National Park experienced an increase in population of around 10% between 1990 and 2000. The vegetation also suffered damage. To the thousands of hectares of forest on the lava field cleared by the Rwandan refugees can be added vast portions of Kibira National Park in Burundi, degraded by repeated fires, and anarchic wood extraction in forests of the eastern part of the Democratic Republic of the Congo by Ugandans.

More difficult to correct is the human situation. In Uganda, everywhere where it was still worthwhile, populations illegally settled in forest reserves have been relocated since 1986, although some forests have had to be abandoned. Refugees returning to the country after exile are themselves a source of problems. Totally impoverished, these people have no other alternative than to fall back on immediately available natural resources: land, wood and game. In addition, frustrated by their years of exile, they resist all interference in their affairs and refuse any advice. They settle anywhere, including in the reserves, and do only what they think best, generally in ignorance of the habitats they are invading.

Finally, to complete the picture, the new authorities very often have a tendency, in part through ignorance, in part in bad faith, to make a clean sweep of the past or to challenge the achievements of their predecessors. Once more, Rwanda furnishes an example. Before the war, the 90,000-hectare Nyungwe Forest was the object of many zoological inventories. Its flora was well known, described and published. It even possessed a management plan which had taken years to perfect in collaboration with the local, provincial and governmental authorities, and had obtained the support of several large funding agencies including the European Commission, the World Bank, French Cooperation and Swiss Cooperation. After the war, everything was open to question again and the new functionaries, seeming to ignore everything that had already been done, recently requested the financing of inventories and a management plan. Yet this forest has hardly suffered from the war and has certainly not changed fundamentally. This attitude is not only deplorable, it induces great caution, not to mention mistrust, among funding agencies.

The worst effect of war and civil strife is perhaps on the humans. The most competent and well trained people are the first to leave the country and often the last to come back. Moreover, to survive in such difficult times it is necessary to forget a number of rules, and it is extremely difficult and time-consuming to reintroduce these – especially when the government has many other priorities. So once poaching and illegal woodcutting has become well installed it is not easy to convince people to start other activities. Reconstruction of a country takes many years, and the fauna often does not have the time to wait.

Actors

Behind these facts are hidden very profound changes that have intervened in the field of nature conservation since 1960. Firstly, there was a sharp increase in the number and diversity of actors.

The States

The newly created States remained, of course, in the forefront. To convince the world of it, they confirmed, from 1968, their will to conserve nature by signing the African Convention for the Conservation of Nature and Natural Resources, better known as the Algiers Convention. Accessible to all independent African countries, this convention, like the London Convention of 1933, set up a general framework for the protection and management of natural resources by defining concepts and establishing prescriptions to be taken up in national texts. But it went farther. In addition to recommendations on fauna and flora, it contained numerous recommendations relative to the conservation of soil and water, education, the identification and the taking into account of ecological factors in the formulation of development plans. Being equipped with neither a permanent secretariat nor a budget nor even the means to put it into action, this convention ended up receiving only limited attention from the Organisation of African Unity (OAU), which remained an essentially political organisation.

Then, the countries of Central Africa adhered to a series of major international conventions: the Ramsar Convention on wetlands (1971), the Convention on the world's natural and cultural heritage (1972), the Washington Convention on the international commerce of endangered species, or CITES (1973), the Bonn Convention on migratory species (1979) and the Rio Convention on biological diversity (1992). More recently, several countries have shown an interest also for the Kyoto Convention on climate change (1997).

The fact of signing and ratifying these conventions obviously does not mean that all the resolutions will be applied to the letter. The implementation of the Washington Convention, for example, was more difficult than foreseen: no government has been able to stop the trade in ivory and, for the last two years, an increase in this traffic has even been observed in several countries.

Independently, Uganda, Zaïre, Gabon, Cameroon and the Congo also participated in the *Man and Biosphere* programme (MAB) initiated by UNESCO during the 1970s. The objective of this international scientific programme is the creation of a network of biosphere reserves and comprises research, follow-up, formation and education, as well as direct conservation action. The cultural and human aspects are of vital importance. In the region that interests us, nine countries participate in it through 20 sites. Most of them are existing protected areas that benefit from a supplementary label. However, some, such as Dimonika Reserve in the Congo, have no other status.

As for the World Heritage or Biosphere Site labels, they were sometimes no more than honorific distinctions without consequence. At other times, they contributed to the protection of threatened sites. Dja Faunal Reserve, which bears both labels, could have been reduced or declassified without this international recognition. In the Democratic Republic of the Congo, a vast programme in favour of World Heritage sites was recently launched, involving Salonga National Park, Garamba National Park, Virunga National Park and Kahuzi-Biega National Park, as well as the Okapi Reserve.

The Rio Convention

The Convention on Biodiversity is a result of the World Summit or the United Nations Conference for Environment and Development, held in Rio de Janeiro in 1992. Its global objectives are conservation of biodiversity, sustainable use of biological resources and equitable sharing of benefits provided by the exploitation of these resources. The Member States can still use their resources according to their own environmental legislation, but they have the obligation to see that activities undertaken in their own area or under their supervision have no detrimental effects on the environment in other States or in regions not under any national jurisdiction.

Convention	São-Tomé	Cameroon	G-Equat	Gabon	CAR	Congo	DRC	Uganda	Rwanda	Burundi
Alger	–	1978	–	–	1970	1981	1976	–	–	–
Ramsar	–	–	–	1987	–	–	1996	1988	–	–
World Heritage	–	1982	–	1986	1980	1987	1974	1987	2000	1982
CITES	2001	1981	1992	1989	1980	1983	1976	1991	1981	1988
Bonn	2001	1983	–	–	–	2000	1990	2000	–	–
Rio	1999	1994	1994	1997	1995	1996	1995	1993	1996	1997

8 CONSERVATION

On the fringe of these large international conventions, the States of Central Africa also launched some regional initiatives. In 1996, the Conference on the ecosystems of dense humid forests of Central Africa (CEFD-HAC) was held in Brazzaville; it is also known under the name of *Processus de Brazzaville*. The main objective of this forum was a discussion involving all parties concerned about Central African forest ecosystems: governments, economic operators, managers of natural resources, co-operation agencies, environmental NGOs, local populations. At the centre of the discussions were conservation and the sustainable use of natural forest resources. The participants adopted the Brazzaville Declaration, setting up bi-annual meetings of all signatories to perpetuate their co-operation in elaborating solutions to problems of environment, forestry and biological diversity. Due to a lack of means the CEFD-HAC has not brought very much to the sub-region yet. In 1999, at Yaoundé, a Conference of Heads of State of Central Africa took place, which clearly reiterated and formalised the will to conserve the forest ecosystems. Finally, in 2002, the COMIFAC was created, a conference of the ministers in charge of the forests. As a political institution, its objectives are mainly a better coordination of the strategies and actions aiming at forest conservation and sustainable use.

Between theory and practice there is often a considerable discrepancy linked to the weakness of the bodies responsible for the management of protected areas and the application of environmental laws. Until recently, States, in an attempt to keep control, had divided these tasks between various departments – Forests, Environment, Fauna – that were a part of ministries sometimes in charge of Agriculture, sometimes of Water and Forestry, sometimes of Environment. Administrative arrangements were very diverse and in some countries they changed over time. In addition, other ministries influenced the management of forest resources and conservation: Planning, Mines, Interior, Tourism, Rural Development, Economy, Finances, Justice and even Defence. Conflicts of interest were frequent and, in spite of interministerial committees, it was difficult to reach a consensus.

The departments in charge of conservation remained the poor relations. In Gabon for example, in 2000 the Directorate of Fauna disposed of an annual budget of 30 million CFA francs (US $42,000) and some 30 agents, of which only a handful worked in the field to oversee more than two million hectares of protected forests.

To remedy this situation, some countries consigned conservation matters to a parastatal organisation, more efficient and more free to find funds outside the budget of the State. In Uganda, the national parks were managed from their creation by National Parks of Uganda. After independence this organisation was put under the Ministry of Conservation and Tourism. In the ex-Belgian Congo, the *Institut national pour la conservation de la nature* (INCN) was created in 1969, initially under the direct responsibility of the President of the Republic. Later, it became the *Institut zaïrois pour la conservation de la nature* (IZCN), in charge of nature protection, scientific research, tourism and management of protected areas. In 1975, this organisation was put under the Department of Environment, Nature Conservation and Tourism. In 1996, it became the *Institut congolais pour la conservation de la nature* (ICCN). In Rwanda, the *Office rwandais du tourisme et des parcs nationaux* (ORTPN) was created in 1973, was placed directly under the responsibility of the president of the Republic. In Burundi, protected areas created at the beginning of the 1980s were placed under the management of the *Institut national pour la conservation de la nature* (INCN), under the responsibility of the Ministry of Agriculture. Finally in Equatorial Guinea, the *Institut national des aires protégées* (INAP) was created. In all cases, the forests remained the domain of the Department of Waters and Forests, a situation which engendered many conflicts.

8.12. Management of the environment and conservation often come up against a multiplication of involved departments and institutions who at times complete, at times refuse collaboration, and at times are in open conflict. For example, how does one handle cases where the substratum of a forest is managed by the department of mines?

Foreign advisers

The civil servants responsible in the field for the management of protected areas generally lacked adequate education and often felt their assignment to the bush as being put on the sidelines by the administration. In the 1970s and the beginning of the 1980s, the setting up and management of protected areas was consequently carried out mainly by foreign advisers or co-operators. Through their enthusiasm and obstinacy, some were even behind the creation of new protected areas that came into existence during these years. In Zaïre, this was the case for the national parks created under the impetus of the Belgian biologist Jacques Verschuren and for Kahuzi-Biega National Park, created at the instigation of Adrien De Schrijver. In Cameroon, Korup National Park was proposed by Steve Gartlan. In the Congo, Nouabale-Ndoki National Park would probably never have come into existence without Mike Fay's obstinacy.

Heads of state

The creation of these new protected areas would not have been possible without the goodwill of the heads of state. Some understood that conservation could be a powerful political tool, capable of attracting favours from the western world. General Habyarimana, who took power in Rwanda in July 1973, very rapidly understood that some countries were ready to invest considerable sums in conservation or in the name of conservation. Like President Mobutu in Zaïre, he adroitly used conservation and, when war broke out in Rwanda in October 1990, he relied on "ecology" and the mountain gorilla to obtain western financial and military aid.

He received honorary distinctions on several occasions for his merits in conservation. During the war, he again received one for his devotion to the protection of the gorillas, while the *Bagogwe* – transhumant pastoralists who lived on the piedmont of the volcanoes only a few kilometres from the gorillas – had just been massacred. For many Rwandans, this gesture clearly left the impression that in the eyes of some Westerners, gorillas were worth more than humans. It is therefore not surprising that after the change of regime the population and the new leaders attacked conservation and that it took several years to repair the damage.

Funding agencies

The chronic lack of financial means that constrains conservation, like most development, was in part compensated by western countries. Until 1985-1990, this was mostly through bilateral aid, very often an extension of old colonial ties. Belgium intervened substantially in Zaïre and in Rwanda. However, most of the aid came from countries whose colonial past was already far behind or had never existed. Through the GTZ, Germany launched a project centred on Kahuzi-Biega National Park. Begun in 1986, it continues to this day in spite of enormous difficulties engendered by the war. The Netherlands are involved through the *Tropenbos* programme in Cameroon. Switzerland was involved in the pilot forest project centred on Nyungwe Forest, in Rwanda. The United States launched several projects through USAID, notably in the Rwenzori Mountains.

Over the years, bilateral aid was progressively replaced by multilateral aid. Since 1963, the European Economic Community, now the European Union, signed a convention between the European States and 18 African and Malagasy States. This was renewed in 1969, then reworked at Lomé in 1975. It involved 46 States in Africa, the Caribbean and the Pacific (ACP). It was followed by the Lomé Convention II in 1979, Lomé III in 1984 and Lomé IV in 1989. The latter was concluded for a period of ten years and a budget of 12 billion euros, or 8,400 billion CFA francs. Among the agreements, there were parts concerning environment and agriculture, rural development and management of natural resources. This convention was revised in 1995

The European Union and forest conservation

Following the signature of the Lomé III Convention in 1984, the European Union allotted a large part of the European Development Fund (EDF) to the conservation and sustainable management of forest ecosystems in Central Africa. Recognition of the value of forests has been expressed since 1989 and the European Union's policy on forests was formalised in 1995 in a regulation and an agreement added to the Lomé IV Convention. Most of the activities are financed by the EDF, therefore conceived and executed in agreement and in partnership with the governments of the ACP (Africa, the Caribbean, the Pacific) countries. Among the most important actions in Central Africa is the ECOFAC programme (PAGE 294). In addition, the European Commission is endowed with a "Tropical Forests Actions" budget line since 1991. More flexible in its setting up than the EDF, it permits the innovative use of resources. Among the activities launched in Central Africa on this budget line are its contribution to the PRGIE (PAGE 296). Overall, around 100 million euros were engaged for African tropical forests between 1992 and 1996. These funds were mainly used for conservation and protection of protected areas, while in Asia they were oriented more toward sustainable exploitation. The European Commission's strategy is based on the principle that the management of natural resources, including use of the forest, can not be separated from sustainable socio-economic development. In practice, it is materialised by the struggle against uncontrolled deforestation and forest degradation, by the increase of areas of forest under sustainable development, the increase in revenues coming from the forests and their more equitable distribution, the maintaining of genetic resources and biodiversity, the creation of institutional structures capable of meeting the contradictory demands on forests while taking into account the interests of all the protagonists and the growth of knowledge about forests through the development of research.

The ECOFAC programme

In 1992, the European Commission initiated a vast regional programme of tropical forest conservation centred on the management of protected areas. At first, it involved seven countries: São Tomé e Principe, Cameroon, Equatorial Guinea, Gabon, the Central African Republic, the Congo and the Democratic Republic of the Congo. In this latter country, activities were interrupted after the interruption of co-operation, however.

At the origin of this programme was the fact that most of the national parks of western Central Africa existed only on paper. Some did not exist at all. In ten years time, ECOFAC was able to considerably change this situation: 28,000 square kilometres of forests and associated habitats, divided among eight protected areas, are today managed and equipped with infrastructures. They are Odzala National Park in the Congo, Monte Alén National Park in Equatorial Guinea, Zakouma National Park in Chad, the Obo National Park project on São Tomé e Principe, Dja Faunal Reserve in Cameroon, Lopé Faunal Reserve in Gabon, Ngotto Forest and village hunting areas in the Central African Republic. In each of these sites, personnel were put in place and trained. Monitoring systems were elaborated and applied research and conservation programmes were initiated. Prospecting and biological inventories permitted substantiation of the importance of some previously unsuspected sites. This activity led to the extension of Monte Alén National Park in Equatorial Guinea, whose area was extended to 260,000 hectares, and the extension of Odzala National Park, whose area increased from 260,000 to 1,360,000 hectares. The needs and aspirations of local populations, still extremely dependant on forest resources, constituted a major preoccupation of the programme which invested much into the search for strategies and means able to reduce hunting pressure on animal populations. Finally, activities were launched in the field of ecotourism, mainly based on the Gorilla. Tourist facilities were constructed in Monte Alén National Park, in Odzala National Park and in Lopé Faunal Reserve. Altogether, over 40 million euros were engaged.

One of the main assets of the ECOFAC programme is its regional approach. This approach has enabled the continuation of activities at Odzala while war raged in southern Congo. Twice a year, it gathers persons in charge of conservation from the seven countries of Central Africa to participate in evaluation of common problems, the exchange of experiences and the conception of new strategies. This regional approach has resulted in the creation of the Central African Network of Protected Areas (RAPAC) that was endowed, in May 2000, with a judicial identity and whose vocation is to promote benefits to the other protected areas in the subregion from the ECOFAC experience. The Congolese Institute for the Conservation of Nature (ICCN) and Chad have already joined the network.

---

and was followed in 2000 by the Cotonou Convention, planned to run for 20 years. This convention attaches much importance to the struggle against poverty and, for the first time, involves the private sector and NGOs. It relies on a budget of 13.6 billion euros for the period 2000-2005.

Among the numerous projects of the European Community are a support programme for Virunga National Park, infrastructure repairs for the Ugandan national parks and the Programme for the Development of the Region North (PDRN) for the Central African Republic including an important section on conservation part.

The United Nations acts through the United Nations Programme for Development (UNPD) and the United Nations Organisation for Food and Agriculture (FAO). The UNPD is involved in several conservation projects, notably in Kibale National Park and Semuliki National Park, in Uganda.

During the 1980s, the World Bank was also involved in development and conservation projects, notably in Gishwati and Nyungwe forests in Rwanda. Later it was also responsible for the management of part of the Global Environment Facility (GEF), a product of the Rio Summit in 1992.

Beginning at the end of the 1980s, more and more projects sprung up, supported by numerous funding agencies. In Nyungwe Forest, in Rwanda, the forest pilot project was replaced by a much more grandiose project, financed by the French Co-operation, the European Community, the Swiss Co-operation and the World Bank. After criticisms of their Gishwati project, the World Bank became more conservative and financed only research and biological inventories.

The 1990s saw at last the appearance of regional projects. Somewhat put off by the misadventures of projects centred on a single country, the European Commission set up the Programme for the Management of Forest Ecosystems of Central Africa (*Programme pour la gestion des écosystèmes forestiers d'Afrique centrale*, ECOFAC) (see box opposite). The justification of this approach was affirmed in 2002 at the Earth Summit in Johannesburg by the launching of a partnership initiative between the countries of the Congo Basin and the main donors: the United States of America, the European Commission, France, United Kingdom, Belgium, Germany, Japan and the NGOs most active in the subregion.

NGOs

With the independence of the countries of Africa, the 1960s also witnessed the expansion of large non-governmental organisations (NGOs) involved in conservation, most of which were created in the years after the Second World War. Among the most notable are the World Wildlife Fund (WWF), the African Wildlife Foundation (AWF), the Fauna and Flora Preservation Society (FFPS), recently renamed Fauna and Flora International (FFI), the Wildlife Conservation Society (WCS), issued from the New York Zoological Society, and Conservation International (CI). These organisations work partly on their own funds, collected from their members, partly also on funds from large or multilateral funding agencies. Beginning at the end of the 1980s, western governments saw that NGOs had qualities and capacities that their organisations in charge of bilateral co-operation found difficult to match in terms of ability to raise the public awareness of development aid as well as in flexibility of implementation. Thus, governments increasingly subcontracted their projects to NGOs. The headquarters of these NGOs are generally in America or in Europe, but, in view of the magnitude of their involvement and in a perspective of decentralisation, several have established delegations or permanent representations in Central Africa.

The WWF is involved in Cameroon, in Korup National Park, Kilum-Idjim Forest and Lake Lobéké Reserve; in Gabon, in Minkébé Reserve; in the Central African Republic, in Dzanga-Sangha Reserve. WCS acts in Lopé National Park in Gabon, Nouabalé-Ndoki National Park and Conkouati Nouabalé-Ndoki National Park in the Congo and Nyungwe Forest in Rwanda. At times, several NGOs join their efforts. In Uganda, Rwanda and the Democratic Republic of the Congo, the AWF, FFPS, People's Trust for Endangered Species and the WWF launched, in 1978, a Mountain Gorilla Project which, while focusing on the gorilla, remains attentive to other conservation problems in the volcanoes region. Another joint project of the WWF and the AWF runs Mgahinga National Park, Gorilla National Park and Bwindi Impenetrable National Park, in Uganda, also harbouring mountain gorillas. In Rwenzori National Park, the WWF is associated with USAID.

In 1990, along with the onset of the process of democratisation, national NGOs also appeared. Their memberships are not always large, however, and are sometimes reduced to only the founders. The life span of these organisations is consequently often short. In fact, most are no more than a trampoline for politics. In order to conform to the fashion of the day, they often amalgamate social matters – especially women and children – environment, sustainable development and the struggle against poverty. Lacking financial means, motivation or political recognition, their action is generally pathetic. Lacking professionalism, they rarely manage to win the confidence of funding agencies open to helping them. And yet, these national NGOs could bring a useful counterweight to the omnipresence of the States and their monopolies.

The IUCN

Founded in 1948, the International Union for the Conservation of Nature (IUCN), now the World Union for Nature, is essential as the organisation that "thinks" conservation and helps states in the multitude of tasks that conservation imposes. It brings together States, public organisations and a wide range of NGOs. Today it comprises more than 900 members in 138 countries. As a union, its mission is to influence the societies of the whole world, to encourage them and to help them conserve the integrity and diversity of nature. It attempts to promote the sustainable use of resources. To achieve these objectives, it relies on its members, networks and partners. Its headquarters are at Gland, near Geneva in Switzerland, but it has established a regional representation for Central Africa at Yaoundé to better adapt its involvement to the realities of Central Africa. This representation houses notably the secretariat of the CEFDHAC (PAGE 291).

In collaboration with large international NGOs, especially the WWF, the IUCN has given rise to the publication of directives for the elaboration of development plans for protected areas and, in 1980, it published its Strategy for Conservation which has become a reference document essential for the years to come. Until very recently, it was also involved in support programmes in the field, notably in Dja Faunal Reserve, but it has abandoned this form of activity. It has also organised specialist groups into world-wide networks of experts in certain groups of animals or plants who voluntarily pool their information and regularly publish their conclusions.

8.13. Inauguration of a 25 metre bridge within the framework of the ECOFAC programme. Many activities have been carried out that at first glance have no direct relationship with conservation. When war broke out in the Congo in 1996, the road linking the Odzala National Park to Gabon had to be repaired. This was the only way to be able to continue activities. These repairs were undertaken with the collaboration of the human populations settled along the road. Unfortunately, funds spent on this achievement, very useful for the entire region, were subtracted from strictly conservation activities. This sort of example incites authorities and populations to ask for other rural development activities. Incomprehension soon settled in.

The ADIE, the REIMP and the Central Africa Forest Watch (FORAC)

As a result of difficulties in information management encountered in the implementation of the Rio Convention and in the management and circulation of environmental information, six countries of Central Africa – Cameroon, Gabon, Equatorial Guinea, the Central African Republic, Congo and the Democratic Republic of the Congo – created, in 1997, the Association for the Development of Environmental Information (ADIE). This association, under Gabonese law and with its headquarters in Libreville, has a regional vocation and comprises representatives from governments, the private sector and NGOs. Its objectives are to develop existing environmental information, to provide support for decision-making, to incite the collection of missing environmental data and to support formation in the field of environmental management. In a first phase, it put into action the Regional Environmental Information Management Programme (REIMP) co-financed by the Global Environment Facility (GEF) through the World Bank, the tropical forests budget line of the European Commission (PAGE 293), the African Development Bank (BAD), the French Fund for World Environment (FFEM) through the French Ministry of Foreign Affairs, Canada, Belgium, Sweden, the FAO and the International Fund for Agricultural Development (IFAD). In 2002, in partnership with the Global Forest Watch initiative (GFW) of the World Resources Institute (WRI) whose headquarters are in Washington, the Joint Research Centre (JRC) of the European Commission, settled in Ispra in Italy, the International Centre for Forest research (CIFOR), based in Bogor in Indonesia, the Research Institute in Forest Ecology (IRET) based in Makokou in Gabon, and the CIRAD-Forêt, based in Montpellier, the ADIE created the Central Africa Forest Watch (*Observatoire des forêts d'Afrique centrale*, FORAC) which brings together in a single internet site (*www.forac.net*) the essential or most used information in the monitoring of forests, as well as in the perspective of their conservation and the sustainability of their exploitation.

Evolution of context, concepts and strategies

While numerous protected areas were being created, conservation actors multiplied and the region was being agitated by deep socio-political crises, the past forty years were also being characterised by deep changes in the global context of conservation and, as a consequence, changes in its strategies.

During the colonial period, conservation in Africa was limited to the creation of protected areas in grasslands. It sufficed then to set apart more or less vast territories, to assign to them some colonial agents or retired military officers to supervise a few guards (FIGURE 9.14) and to keep away poachers. National parks were somewhat like entrenched camps, all the easier to keep since they had been created in generally arid regions that were little inhabited, or not at all. Despite this militarist approach, colonial authorities were much more respectful of the rights of human populations than some experts in human sciences or some NGOs try nowadays to make us believe. For this reason the United Kingdom delayed the creation of national parks in Uganda for several years, for fear of infringing on the rights of the population. In Queen Elizabeth National Park and Murchison Falls National Park, populations were evacuated long before their creation (PAGE 261) and the remaining fishing villages inhabiting the shores of Lake Edward and Lake George were left in place. Only some very small hamlets were displaced and re-housed.

In the first years after independence, authoritarian drift in most of the Central African States did not change much, in fact, quite the opposite. The creation of protected areas in forests, especially in lowland forests, would pose a new problem for conservation actors. The first forests to benefit from protection measures were montane forests, reduced to more or less wild islets in the middle of a densely inhabited world of farming. It was enough therefore to delimit and organise their monitoring, as had been done for the grasslands. Conflicts with local populations were not important since they depended little on the natural resources of these forests, if at all. In lowland regions, the situation was very different, however. Under a nearly continuous and apparently intact canopy, there was a mosaic of habitats at very different stages of modification by man and inhabited by extremely diverse populations, generally not very dense, but all extremely dependent on forest resources. It was necessary therefore to reinvent conservation, to conceive it on greater surfaces than before and to include the human dimension. For that it was necessary to call upon the human sciences.

The task was even more difficult when, during about the same period, conservation was confronted with a growing dilemma: the destruction of biological diversity was taking alarming proportions and required ever stronger action, but the demographic, social and economical context reduced its effectiveness from year to year.

Many people do not believe that this biodiversity crisis is real and few understand its magnitude. However, since 1600, therefore within 400 years, 88 species of mammals, 107 species of birds and 20 species of reptiles have become extinct. Today, according to a report of the World Conservation Monitoring Centre (UNEP-WCMC), published in 2000, 24% of the mammalian species of our planet, 11% of the species of birds and 3% of the species of reptiles, amphibians and fish are threatened with extinction in the short term. For invertebrates and plants, it is more difficult to give figures, but according to estimates of various biologists, summarised by Nigel Stork in the publication *Biodiversity II*, the world number of species has been reduced by around 4 to 10% during the last ten years. Yet independently from the species that have become extinct, many species have experienced a dramatic reduction in their range of distribution or the size of their populations. It is therefore highly probable that the 25 years to come will see a colossal number of disappearances.

In themselves these extinctions are not surprising. There have been extinctions throughout the history of life and our present knowledge suggests that a species will survive on average no more than four to ten million years. Extinction is normally compensated by the appearance of new species. At present, however, species disappear without being replaced. We are headed for a crisis of mass extinction, of the sort our planet has already experienced five times since life first colonised terrestrial habitats. None of these crises has threatened life itself, but all of them have profoundly upset floras and faunas. In fact, long before we were able to put an objective date on the rocks we studied, we allocated these rocks to distinct periods of the Earth's history just looking at the fossils which were associated with them. Each period had its own array of life forms, often totally different from older or younger rocks.

After a phase of five or ten million years of extreme impoverishment, a new diversification has taken place, but each time the dominant species were replaced. In other words, the crisis, whose onset we are now living through, is liable to remove not only a large portion of the biological diversity we now know, but it will also very probably bring an end to the reign of the human species, at least in its present form.

Biodiversity

Created from the contraction of "biological diversity", this new word appeared in the course of the 1980s. Since the Rio Summit in 1992, it has come into everyday use and nowadays everyone believes they understands its meaning. The various definitions that have been given for it are indeed all fairly similar. All of them emphasise the fact that biodiversity designates the variability of living organisms from all origins, at gene and species level as much as at the ecosystem level. Biodiversity is therefore an incommensurable concept – vague and ambiguous – because it is not possible to put genes, species and ecosystems into a single equation. It is like adding up the forest, the trees and leaves from the trees. Many species and genes are still unknown to us and the notion of ecosystem varies according to the scale on which it is placed. Contrary to the principal of "species richness" used by biologists who study communities of plants or animals, biodiversity thus cannot be expressed simply by figures. It is also very subjective, because it is strongly influenced by the values, intentions and interests that animate conservation biologists, as well as by the methods that they use. At the level of ecosystems, integrity and origin play a preponderant role: a natural ecosystem will have more value than an ecosystem modified by human activity. As for species, not all of them have the same value either: species that are symbolic, relict, rare, endemic, vulnerable, useful, beautiful or spectacular are clearly worth more than others. As for genes, their usefulness plays an essential role.

Biodiversity is often expressed in number of species – one of the rare measurable criteria – but it is only one aspect among many others and the presence of a greater number of species does not necessarily indicate a greater biodiversity. In fact, as D. Perlman and G. Adelson, two Harvard University researchers, explain, the biodiversity of a site or of a region is not measured, it is described.

8.14. Alongside public awareness and co-management, monitoring and suppression of illegal acts remain indispensable tools of conservation, especially in the face of large scale poaching, such as of elephants. The results of this repressive approach are often questioned and it is certain that they rarely come up to expectations. It is just as certain, however, that, in the absence of such surveillance and anti-poaching campaigns, there would not be much left, neither within, nor outside, the limits of the protected areas of Central Africa.

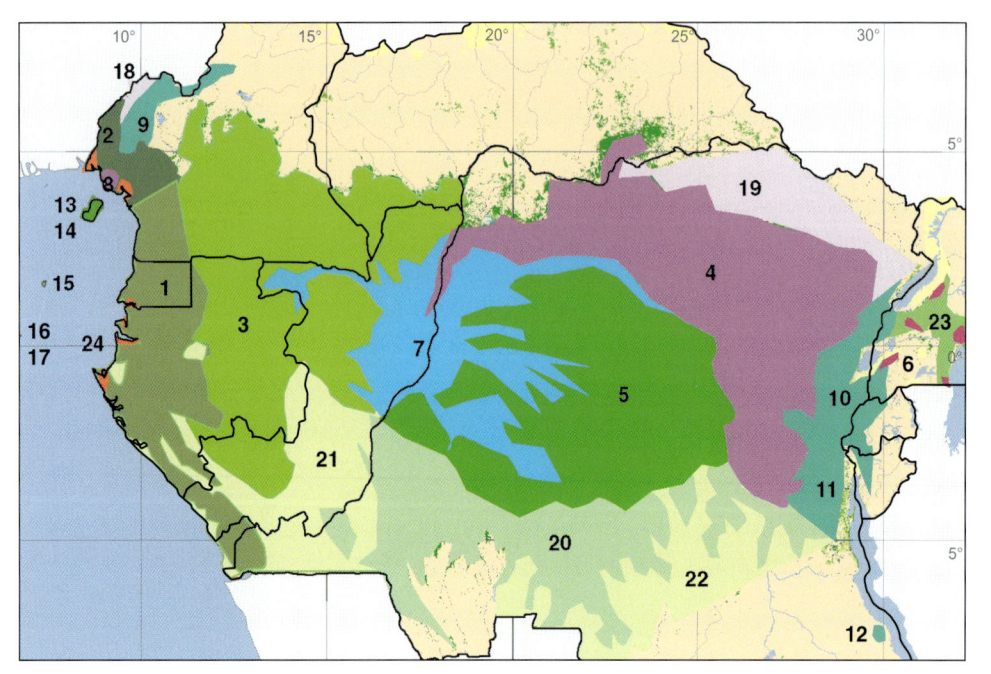

8.15. *According to the* WWF, *Forested Central Africa is divided into 26 ecoregions: (1) Congolian coastal forests, (2) Cross River coastal forests (3), North-western Congolian forests, (4) North-eastern Congolian forests, (5) Central Congolian forests, (6) Lake Victoria moist forests, (7) Congolian swamp forests, (8) Mount Cameroon forests, (9) Western Cameroonian montane forests, (10) Albertine Rift montane forests, (11) Itombwe montane forests, (12) Marungu montane forests, (13) Bioko lowland forests, (14) Bioko montane forests, (15) Principe forests, (16) São Tomé lowland forests, (17) São Tomé e Principe submontane and montane forests, (18) Guinean forest-savanna mosaic (19) North-eastern Congolian forest-savanna mosaic, (20) Southern Congolian riverine forests, (21) South-western Congolian forest-savanna mosaic, (22) Southern Congolian forest-savanna mosaic, (23) Lake Victoria forest-savanna mosaic, (24) Central African mangroves, (25) Afromontane habitats (impossible to map on the scale of the map).*

Edward O. Wilson and Paul Ehrlich think that our cultures and civilisations will not survive the 21st century unless we manage to radically change our behaviour. Considering the difficulties of application of international conventions on the environment as well as the refusal of some nations, like the United States of America, to adhere to these conventions, it is not clear how such a change could ever be possible.

From a "strictly Western fad", conservation has become a problem that concerns all of humanity. Whatever value is attached to biodiversity, it represents the ultimate indicator of the state of our environment and of our perspectives for the future. Unfortunately, Man has never liked birds of ill omen and prefers to take refuge in myths that make him believe that everything is going for the best. The people of the forest continue to believe that the forest is inexhaustible and Westerners have an unshakeable faith in science and technology to solve all their problems.

In the perspective of a grave world crisis, the creation of protected areas covering no more than 10% of the surface of States is a minimal, even an insufficient, measure. Insofar as the forests, in particular, are concerned, it is essential to attempt to save from conversion not the entirety of the forest massifs – that would be utopian – but much more substantial portions of them.

This necessity is based on various reasons.

Some are not easy to understand. They concern the role that tropical forests play in the biosphere. Even though this is difficult to describe with accuracy, and especially to measure objectively, we know that these forests have an important role in climate regulation, principally through the fixing of atmospheric carbon and the regulation of precipitation. If they were reduced to 10% of their present area, it is unlikely that they could still play this role and climatic perturbations could be expected on a world scale as well as on a local scale.

Other reasons are more obvious. Firstly, many species are migrants or transhumants and use different vital spaces in the course of their annual cycle. Elephants can cross vast areas in function of the availability of the natural resources on which they live or the hunting pressures they are subjected to. Such species are hard to confine to a restricted space and therefore the protected areas must not be progressively converted into oases lost in the midst of a biological desert of crops and artificial plantations. In itself, insularisation of natural habitats alone would cause many species to disappear. To remedy this, production forests, although undergoing the inevitable transformations inherent in their status, must be allowed to constitute buffer zones or to serve as corridors between the protected areas.

Secondly, every animal population should have a minimum number of individuals in order to survive in the long term. For large mammals, the work of Michael Soulé in the United States shows that this number is of the order of several thousand. Smaller populations are not automatically doomed to disappearance – the example of the mountain gorilla proves it – but their chances of extinction are great. In tropical forests, many organisms occur only in very low densities. In order to maintain sufficiently large and viable populations, vast areas must be preserved.

Choice of sites to protect

The rapid reduction of biological diversity and the degradation of forest ecosystems constitute an urgent situation that imposes swift reactions. Notably, it will be necessary to decide without delay which sites to protect before the critical ones are irremediably given over to destructive activities. In Gabon, the State has thus decided to designate a quarter of the area of its forests, amounting to 4,000,000 hectares, as Permanent National Forests and to set aside 10% of the area of the country for the creation of national parks. This area must imperatively include the most remarkable sites in order to protect all of the country's biological diversity. How is this to be achieved?

Ideally it should be possible to make the choice by superimposing the distribution of the maximum number of species belonging to very different groups of animals and plants in such a way as to identify the "hot spots", or the zones richest in number of species and the zones richest in number of endemic species. Often these two overlap but not always. This step should allow the definition of an optimal network of protected areas in order to preserve all of the biodiversity. Although fairly laborious, this is possible in Europe and the United States, even in South Africa, but not in Central Africa. The gaps in our knowledge of species and their distribution are too great. Only a few very particular species, rare or endangered, can be taken into consideration. The evaluation of existing networks of protected areas and the definition of priority areas for conservation must therefore be based on other approaches.

The first consists of taking habitats into account. Starting from the principle that each habitat possesses its own cortège of species, it should be enough to protect a representative group of habitats in order to protect a representative group of species. This approach is widely used in Europe, but in Central Africa habitats are still poorly known. In addition, many species use different habitats during different periods and it is obvious that a broader approach is necessary.

The WWF has thus launched the concept of ecoregions. These are "more or less vast regions, independent of the administrative divisions that constitute the countries and represent a geographically distinct assemblage of natural communities sharing a large majority of their species, ecological dynamics and environmental conditions" (FIGURE 8.15). For forested Central Africa taken in the large sense, 25 ecoregions have been identified. With this approach it should be possible to verify the extent to which protected area networks cover the diversity of ecosystems. In practice, ecoregions are variable and at times non-comparable entities. The Central African mangroves ecoregion is a very fragmented but also very homogeneous, ecosystem, while the Lake Victoria Mosaic is made of a variety of habitats whose relationships are not always obvious. It also happens that the divisions are not truly justified. Thus, there are no objective reasons to separate the montane forests of Itombwe from other montane forests of the Albertine Rift, even if they are richer. Either all of these forests should be regrouped into a single entity or else at least nine should be recognised. Lastly, it is sometimes difficult to draw the limit between two entities, as for example between the Congolese coastal forests and the Congolese forests of the north-west.

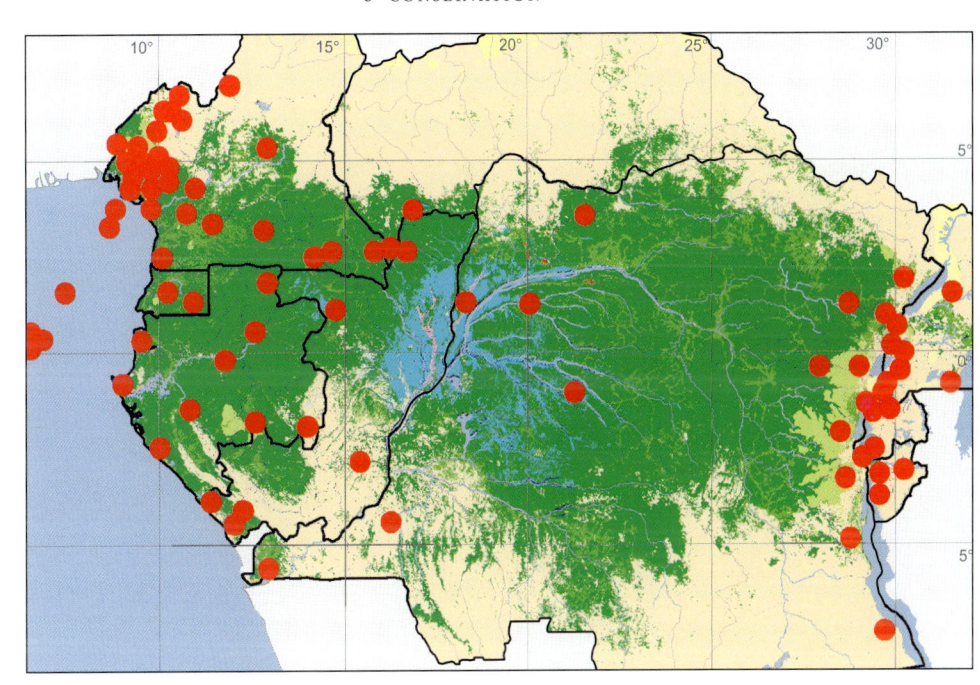

Because of a lack of precise data on many habitats and their distribution, the dividing up of sub-Saharan Africa into ecoregions remains therefore fairly subjective and imperfectly reflects the biogeographical and eco-climatic patterns of distribution. Even in the United States, where knowledge is much more advanced than in Central Africa, the concept of ecoregions is not perfect.

A completely different approach was launched by *BirdLife International*, previously the International Council for the Protection of Birds (ICBP). Starting out with the observation that birds represent one of the best known groups of animals, that they intervene at all levels of the ecosystems and that they consequently constitute one the best indicator-groups for the state and distribution of biodiversity, this organisation, with the help of many local NGOs, defined and identified Important Bird Areas (IBA) (FIGURE 8.16). These represent sites, are very variable in area and can include several habitats. They are chosen according to the species and number of birds that frequent them. Altogether, they form a vast network covering species distribution and are considered to represent the minimum area required to insure their survival. IBAs are an even more

8.16. Among the 125 IBAs recognised by BirdLife International in the ten countries of Central Africa, at least 76 are part of forest ecosystems. Many are centred along the volcanic axis of the islands of the Gulf of Guinea and Cameroon as well as on the Albertine Rift region. The near absence of sites in the central basin of the Congo, notably in the immense swamp zone of the Congo-Oubangui confluence, and along the margin of the forest massif may in part be related to the lack of recent ornithological knowledge of these regions. The centre of the basin and the Kasai region harbour several endemic forms of primates. Do they not also have ornithological value? Or should it then be admitted that birds are not as good indicators of biodiversity as they were thought to be?

8 CONSERVATION

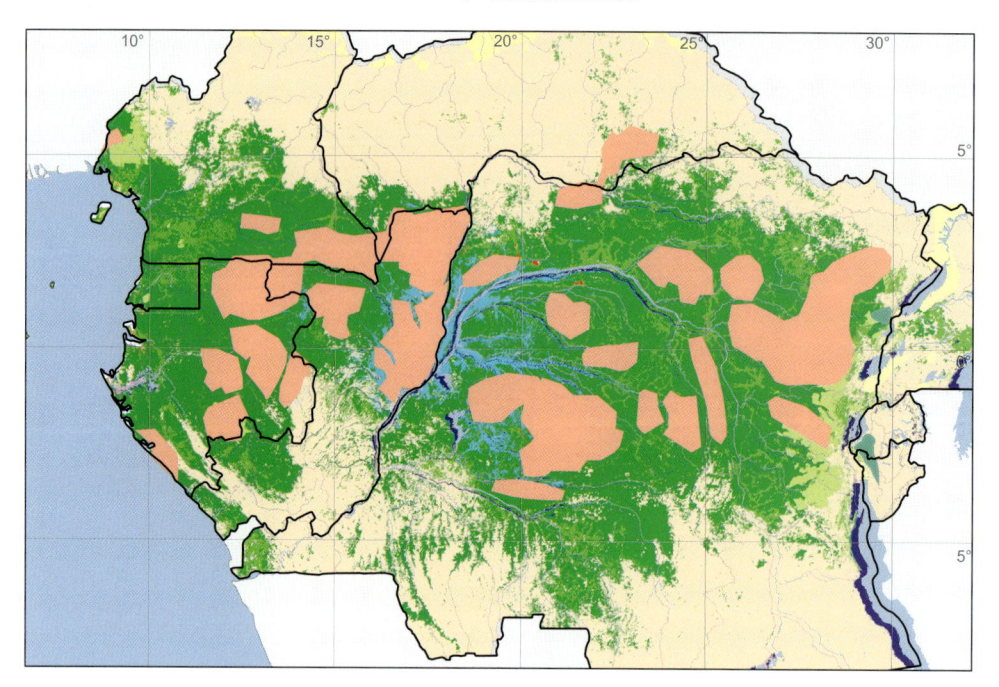

8.17. Frontier Forests constitute blocks of at least 100,000 hectares of primary forest or of non-degraded forests, located away from access routes. They are the last ones able to harbour intact forest ecosystems, without external influence and capable of supporting viable populations of large mammals. The ones represented in red are endangered, those in dark green are not; those in purple have not been evaluated. There is unfortunately no correlation between Frontier Forests and zones of endemism or of biological richness. This map was prepared by the World Resources Institute.

interesting approach in that tens of thousands of passionate and competent amateurs are interested in birds throughout the world and annually furnish an enormous quantity of free information, a much rarer phenomenon in other taxonomic groups. Used in association with data from other taxonomic groups and from socio-economic data, IBAs can thus become a good tool for defining priority zones for conservation. The work of T. Brooks and his colleagues have clearly demonstrated this in East Africa.

Finally, the World Resources Institute (WRI) has launched the Frontier Forest concept (FIGURE 8.17).

These approaches can render great service. Ecoregions, for example, by presenting a simplified vision of a complex problem, are understandable by the public and are a precious tool in the collection of funds. This is also true for the Frontier Forests which, beyond their scientific interest, focus the public's attention on the last large "wild" spaces of our planet. These approaches will never be more than shortcuts whose limitations should be kept in mind, however. The various classes of organisms do not respond in the same way to environmental constraints. Very often they have non-overlapping patterns of distribution. As an example, the distribution pattern of birds (PAGE 200) can be compared with that of mammals (PAGE 207).

As it is not possible to take into account all species and habitats, the only pragmatic and realistic approach is to use the data available, however incomplete. An exercise, costly but very interesting, was organised in March-April 2000 by the WWF. For four days, around 140 natural sciences and human sciences experts, all with field experience in Central Africa, gathered in Libreville to pool, in an informal process, their knowledge, often unwritten, and identify the priority zones for the conservation of forested Central Africa. The results of this exercise have not yet been fully published, and have been incorporated in this book. They have already been used however by six central African countries to establish a regional strategy in order to coordinate and optimize their actions.

Whatever the approaches chosen, between the identification of priority zones for conservation and their eventual protection, the route often remains a difficult one. It is necessary to convince authorities who generally have other more urgent preoccupations to deal with. It is necessary to deal with the local populations who do not understand the problems and are often influenced or manipulated by the political elite. Above all, it is necessary to gain the assent of an ever increasing community of civil servants and authorities whose reaction sometimes amounts to blocking processes. A long route thus lies ahead, all the more so since some fundamental principles of conservation are still not understood. Since the Rio Convention mentioned conservation *ex situ*, this technique is regularly presented as a valid alternative to the setting up of protected areas – goodbye ecosystems and relationships between species. For some, botanical gardens and zoos can replace national parks. In this perspective, the increase in seminars, workshops, conferences, roundtables and congresses are perhaps an inevitable path towards understanding the real problem of conservation, which no teaching in Central Africa deals with seriously.

Taking into account the human context

Along with the disastrous biological condition, whose origin lies in human demography and the incessant increase in the consumption of natural resources, the socio-political condition of Central Africa has considerably evolved and has become much more complex. This phenomenon has amplified mostly since 1990, with advancements in the process of democratisation. Today, conservation should, more than ever, take into account the ever more constraining human context.

This evolution is not particular to Africa: it began in 1962, during the first World Conference on National Parks held in Seattle, in the United States. Progressively, national parks are no longer considered as isolated entities, but as integrated spaces in their human context, at once cultural, historical and economic. They have become a strategic element in an ensemble of measures aimed at setting up sustainable conservation.

In 1980, the experiences of the previous decades led the IUCN to publish its *World Conservation Strategy* aimed at managing natural spaces in the perspective of sustainable development. It was the time of the slogan "Conservation by the people for the people". This orientation was developed during the third conference on national parks, held in 1982 in Bali and, more recently, in the report by the World Commission for Conservation and Development, published in 1987, and the "Caring for the World" strategy, published in 1991.

For years, Central Africa remained on the fringe of this evolution, but following the painful socio-political events of the very end of the 1980s and the 1990s, it was finally taken into account. Central African governments made conservation an integral part of their development policy and drafted national action plans for the conservation of biological diversity. The legal foundations of conservation, embodied in international conventions, became a backdrop to all multilateral or bilateral co-operation negotiations, and socio-economic and cultural considerations became a major preoccupation.

The people of the forest, ever more poor, are paradoxically also increasingly aware of what is happening in the world. With the appearance of parabolic reflectors, even in the most remote areas, some village communities can follow sporting events on television, and, at times, they are better informed of what is happening in Europe than of events in their own country.

Forests, including protected areas, are nowadays subjected to greater and greater pressures. In lowland forest regions where human population densities have remained very low, these repose on a strong dependence on the spontaneous resources of nature. In montane regions, they repose on a an insatiable thirst for land to cultivate.

Quite naturally, the solution envisioned to solve this problem involved the populations, not only in daily management of protected areas, but also in their conception and development. Participative management thus came into existence, concretised in the form of village committees. This strategy begins from the assumption that active involvement of local populations will automatically improve the efficiency of conservation measures by making these populations benefit from conservation. It supposes, however, that these populations possess the necessary knowledge and are prepared to integrate scientific information into it, and are motivated by a genuine will to conserve their environment in its natural state.

Unfortunately, the results of this community approach so far are meagre. The success of the CAMPFIRE programme in Zimbabwe is often cited, but a recent world-wide evaluation shows that results are very limited. This failure probably results from more than one factor. Some populations do not have the knowledge required. Wood exploitation, for example, was never practised by Central African populations, who therefore have no traditional knowledge in this field. Others are not able to integrate scientific information into their traditional knowledge. Still others are simply not disposed to conserve the forests. For populations of the high mountains of the Albertine Rift, whether in Rwanda, Burundi or the Democratic Republic of the Congo, forests only represent land waiting to be cleared and converted. Why involve them in conservation programmes that go counter to their deepest desires? That the disappearance of the last forests could have consequences on their way of life and could endanger their survival in the long term, does not even cross their minds.

The lack of performance of village committees probably mostly results from the fact that the demands of the populations who are the basis of the creation of these committees, are generally incited by members of the village or urban elite with demagogic aims or in the hope of obtaining advantages. These people, among whom are found high civil servants and soldiers, at times have no interest in the solution of the situation. To top it all off, collaboration between biologists and human sciences experts does not always go very well. Too many biologists have no interest for the socio-anthropological aspects and too many sociologists become protectors of the "poor populations despoiled by conservation".

8.18. *These mountains dominating the Bukavu-Walikale road, just west of Kahuzi-Biega National Park, were still covered with dense forest around 1980. Today, the last forest massifs, retreated to the summits and ridges, have also been cleared. How can participative forest management be practised with peasants for whom any forest is no more than wasted land waiting for "valorisation".*

Non-extractive values of the forest

Apart from the recognition of the economic, aesthetic, cultural, emblematic or even recreational value – often through local or regional values – biodiversity is more and more often appreciated for its intrinsic value. Nevertheless, over the course of the last decades, the *"use it or lose it"* concept has gained more and more importance and some have tried by every means to prove that conservation could have or did have an economical justification. This approach is the basis of game ranching in grasslands areas and some sport hunting programmes. It is found in sustainable forest exploitation and, in a more general way, it has aroused an infatuation for the non-extractive values of forests. A great change has come about in the perception of the role of forests. Finally, it has been realised that forests produce not only wood, but also offer many other forest products – mushrooms, fruit, seeds, roots, game, medicinal plants – and that they play an important role in the protection of watersheds, the regulation of watercourses, the stocking of carbon and climate regulation. In the hope of finding economic arguments in support of conservation, various authors have thus tried to assess all these values, especially in America and in Asia. Their estimates are subject to debate, but they show that the only value that greatly exceeds all the others is the value based on the storing of carbon. While a hectare of tropical forest is worth on average 50 US dollars per year through wood exploitation, it is worth 600 to 4,400 dollars through carbon storage. It is not certain that damage caused by the warming of the climate will be as disturbing as is now thought. The value of forests in carbon storage can thus be much less great. Also David Pearce, in an overall synthesis of this problem, estimates that if the conservation of tropical forests should be based only on their non-extractive value, many forests are still at risk of disappearance in the next few years. Other studies show that the non-ligneous resources generated by forests spontaneously do not generally lend themselves well, if at all, to exploitation. Either they are only of limited utility, in which case they interest only a very marginal fringe of the population and their exploitation does not pose a problem. Or they are highly valued, and their exploitation rapidly becomes non-sustainable since natural regeneration is not sufficient for the survival of the species concerned. Generally speaking, these studies conclude that, if a forest product is truly significant, finding ways to farm it is imperative.

8.19. Slash and burn remains the most practised form of farming in forested Central Africa. Fire transforms part of the organic matter of the forest into "fertiliser", but also releases much carbon into the atmosphere. With a fairly low density of human occupation, as in Gabon, the Congo and even in some parts of Cameroon, this technique does not represent a danger to old forests. On the contrary, it allows the creation and maintenance of mosaics of landscapes. It thus contributes to the maintenance of biodiversity. With higher population densities, it is not sustainable.

Apart from reasons directly linked to the deplorable socio-economic context, the low rate of success of participative management perhaps also resides in a strategic error: that of closely linking conservation and development – at times without really wanting to (FIGURE 8.13) – or allowing it to be believed that these two activities are linked. By making promises of development in direct relation with conservation programmes or in the name of these programmes, a sort of blackmail settles in. In the Odzala National Park region, for example, the inhabitants, probably pushed by their elite, ask for a solution to all their problems from the ECOFAC programme. Not only must ECOFAC furnish employment for everyone, but must also provide schools and dispensaries, all without any services in return from the beneficiaries. Forestry companies are subjected to the same pressures. As an example, some villagers demanded from a company located in Gabon the construction of new villages, with new houses for each family, dispensaries, schools, information centres and "factories" to provide work to the men. Unfortunately, some well meaning western NGOs support these sort of unreasonable demands, often completely out of proportion with the local context.

Finally, there remains a fundamental problem: any attempt to make human populations coexist in a sustainable way with natural habitats runs up against the constant increase of these populations which inevitably brings a regression of these habitats. In this perspective, is not the harmonious coexistence between human populations and natural habitats utopian in the long term? It is for this reason that the delimitation of some areas where humans would be excluded, a solution not politically correct at present, seems to be indispensable.

This does not prevent co-operation with the local communities on conservation matters from being sought after in the immediate perspective. A good example is provided by the Lossi Gorilla Sanctuary, just south of Odzala-Kokoua National Park. In this village area, the users of the forest have designated a zone where all hunting activity is banned. The observation of gorillas by foreign visitors and the presence of filming teams has brought them much more than traditional exploitation of this land. In the northern part of the Central African Republic, as well, on the Sangba site, located on the fringe of the Congolese forests, to be exact, local populations have considerably reinforced protection of their area as a result of the development of partnerships with hunting guides. The revenue from hunting safaris are invested in development activities benefiting the community, as well as the development of the conceded areas.

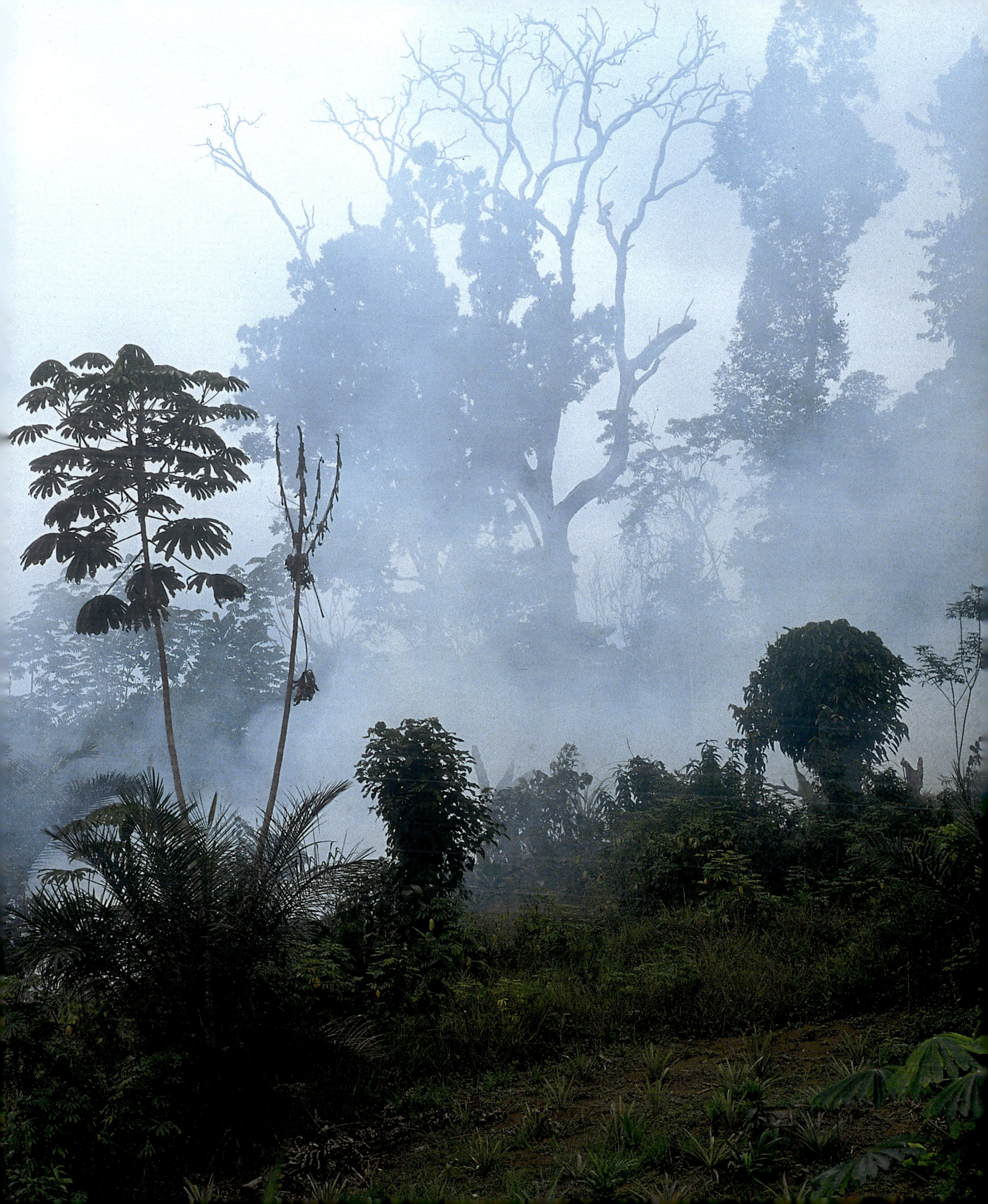

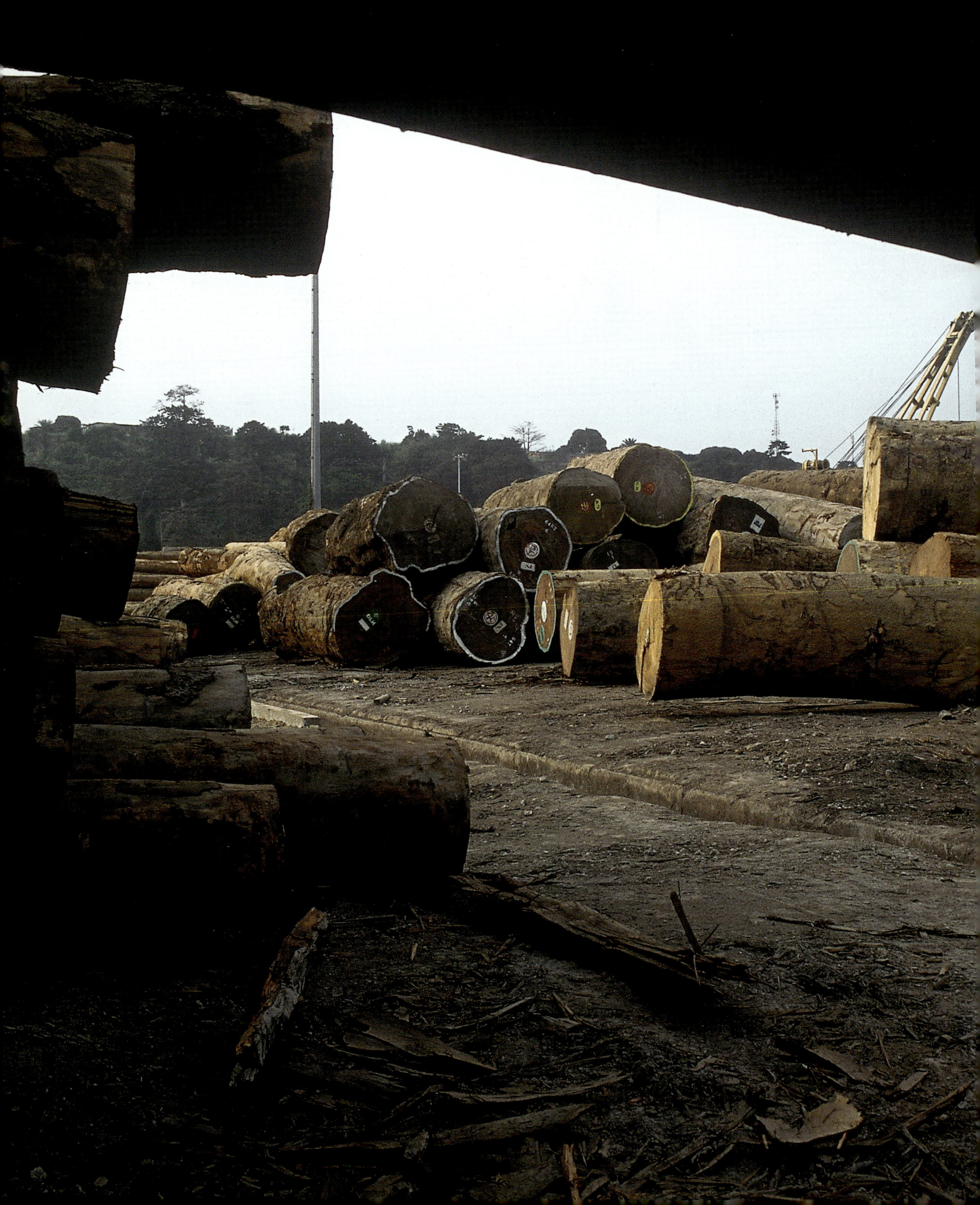

9 Great Challenges
Invasive species, climatic change, logging, hunting, tourism and research

Unlike other tropical forests of our planet, the forests of Central Africa do not yet suffer massive conversion. They are being reduced every year, however, by about 0.5% and, in spite of the effort deployed in their conservation, vast expanses are progressively being transformed into "silent forests" where the trumpeting of elephants and the howling of monkeys no longer resound. In the long term, demographic pressure and the inevitable increase in the needs of the human populations obviously constitute the most disturbing threats. No one doubts it. For the moment, however, the main threats that bear on Central Africa' ecosystems come from forestry exploitation, which is now reaching the most remote forests, with often questionable sustainability, and from hunting, transformed for some decades into a highly commercial activity escaping all attempts at regulation. These two activities irremediably bring with them a degradation and loss of character of the forests, with long-term effects that may be much graver than we realise. Forest ecosystems react with a considerable time-lag to attacks, and when their effects become visible, it is often too late to correct them.

Human in origin, these two activities could be mastered if at least the will existed. Any strategy or action plan going in this direction must, however, consider many constraints. Not only is it necessary to fully envisage the socio-economic and cultural feasibility of each intervention, which seems to be accepted nowadays, but the often ignored or neglected, yet inescapable, biological constraints must also be taken into account. But, more than elsewhere, the means allotted to research, fundamental as well as applied, are ridiculously small in comparison with the stakes, and it could almost be said that for 20 or 30 years science has been asleep in Central Africa. The continent is thus increasingly behind, while hardly a few decades ago it was at the forefront of research in tropical forest biology and witnessed the establishment of institutions and research stations of great repute.

To this picture can be added a few lesser, or still remote, threats such as the wilful or accidental introduction of exotic species which decreases distinctness, the probably inevitable climate changes and the enormous difficulty of developing protected areas, however spectacular, through tourism.

9.1. Today wood is an important resource for the countries of Central Africa. How much longer will it be?

9 GREAT CHALLENGES

In the long term, the monitoring of tropical forests, as of all other natural habitats of our planet, is essentially threatened by the demographic explosion and exponential increase in the needs of humanity. Predictions by the United Nations Population Fund (UNPF) leave no doubt on this subject (PAGE 266). For the isolated villagers of the forests of Central Africa, whose vision of the world stops at the edge of their forests or is reduced to a few televised images, this sort of assertion probably represents no more than yet another of the White Man's follies, all the more so since their forests are still in much better shape than most other tropical forests of the world. Because more than half of the human populations of Central Africa are concentrated in cities and vast stretches of forest support population densities of less than two inhabitants to the square kilometre, the demand for farmland is in effect still far from reaching the scale known in Asia or in West Africa. In Central Africa, it exists, of course, but remains localised at the outskirts of urban centres, along main roads and in a few regions, such as the Mayombe in the Congo and in the Democratic Republic of the Congo or the mountainous areas of Cameroon and the Albertine Rift.

The present state of the forests

According to the IUCN atlas, the total forested area of Cameroon, Equatorial Guinea, Gabon, the Central African Republic, the Congo and the Democratic Republic of the Congo reached 185 million hectares in 1992. According to calculations made within the framework of the TREES programme of the European Commission's Joint Research Centre in Ispra, it was of the order of 183.9 million hectares in 1998. Finally, according to the calculations of the FAO, this same area reached 235 million hectares in 1990 and 227 millions hectares in 2000. The annual rate of destruction was therefore of the order of 0.35% for the entire region, with, however, big differences between countries: 0.85% in Cameroon for only 0.05% in Gabon and 0.08% in the Congo.

However, because of the fact that different organisations have a different concept of what a forest is, these figures should be used with caution. The FAO, for example, includes in its definition of forests wooded grasslands, including a good portion of grasslands with scattered trees, and even exotic plantations. For Rwanda and Burundi, these figures are completely unrealistic. In Rwanda, forest cover was said to have reached 457,000 hectares in 1990, while in reality it was no more than 100,000, unless the innumerable plantations of *Eucalyptus*, *Grevillea* and *Pinus* are included. Thus, when the FAO considers that the Democratic Republic of the Congo, with a loss of 532,000 hectares of forests per year, numbers among the ten countries in the world having suffered the greatest losses, one has a right to ask questions. Especially when one sees that Sudan, Zimbabwe and Zambia are part of the same list!

Other agencies, especially among those working in conservation, also express the retreat of forests by comparing their present area to an "original" area (FIGURE 9.2). According to the WWF and the IUCN, the forests of Central Africa no longer represent more than a fraction of the area they covered before Man began to clear them. This sort of calculation is just as specious, because the result depends both on what is called a forest and what is taken as a starting point. Throughout their history, these forests have experienced continual transgressions and regressions. Between 8,000 and 6,000 years ago, they reached a maximum extension, but hardly 2,500 to 2,000 years ago, they were much more reduced than they are nowadays (PAGE 56). The evaluation of how much this contraction was caused by climate and how much of it was the result of human activity is an exercise that is all the more difficult since Man and the climate have generally operated in synergy. It is therefore probable that vast stretches that could have been entirely covered by forests during the last 10,000 years, such as the Batéké Plateau, in fact never were, and that the reduction of the forest massif is therefore less than some calculations attempt to show.

Whatever the true figures for their loss of area might be, the principal problem of Central African forests does not yet reside in massive conversion as is happening in western Africa and in Asia. This phenomenon will certainly appear one day and the irresistible pressure of people from the north, which has already been the cause of destruction in western African forests, will

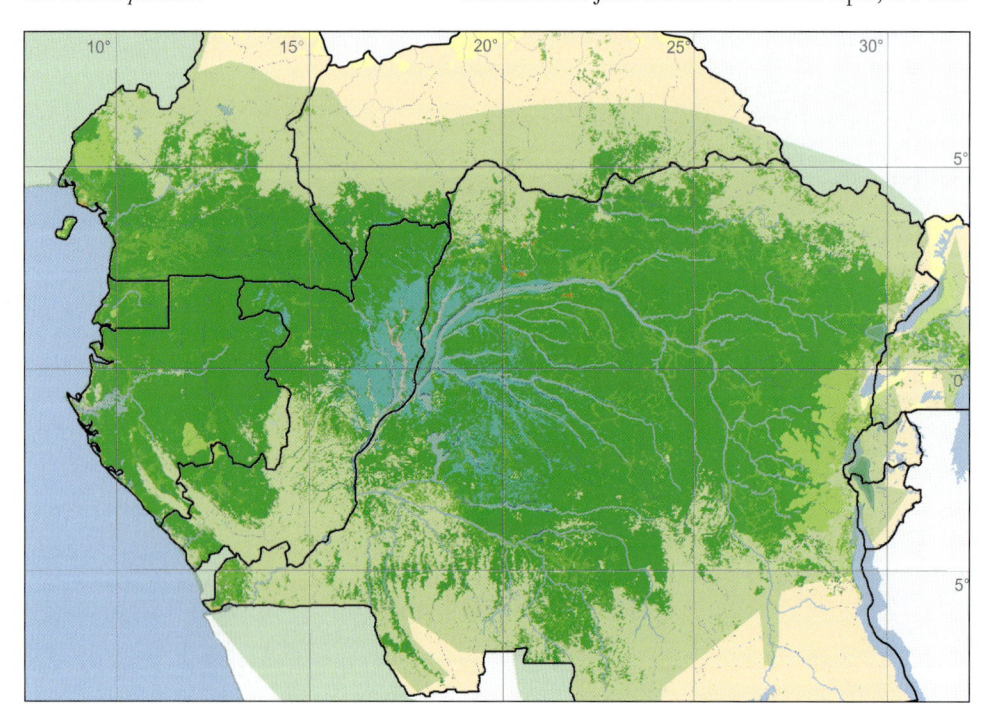

9.2. *According to some estimates, the present dense humid forests of Central Africa represent no more than about two-thirds or three-quarters of their "original" extent, 8,000 years ago. It is certain that at this epoch forests were more extensive than they are today, but there is nothing to prove that all of this hypothetical area was really covered with dense forest formations. Brush fires have occurred for a very long time – they are much older than Man – and vast areas were probably covered with diverse grasslands. Everything goes to show that the grasslands of the Lopé National Park, just as those of the coastal region of Gabon, have never disappeared completely over the course of the last 10,000 years. What then can one say about the vast expanses of the Batéké plateau?*

end up reaching Central Africa. Until further notice these forests will mostly be subjected to considerable degradation and fragmentation, caused above all by forestry exploitation and hunting. The combined effects of these two main threats may be aggravated, however, by more insidious phenomena, such as invasion by exotic species and soon, perhaps, by changes in the climate.

Exotic species

Birds, monkeys, forest duikers and elephants are incontestably good dispersers for trees of the forest, but their efficiency is as nothing compared to that of Man: once a species seems to be useful or beautiful to us, we make it travel around the world. In a while the landscapes of all the continents will hopelessly resemble each other and the principal crop plants will be the same throughout the world's tropical regions. Already, cities are all planted with the same trees – the only one native to Africa is the African Tulip Tree, or Flame Tree, *Spathodea campanulata* with its large vermilion red flowers – and gardens all harbour the same flora, mostly native to South America, Asia and the Pacific. In this domain Man, so inventive in other fields, does not really show much imagination.

As long as foreign species remain in the city there is not much to worry about, but some go farther. Sea coasts in Equatorial Africa are thus progressively invaded by Wild Almond *Terminalia catappa*, Coconut *Cocos nucifera* and Beefwood or Whistling Pine *Casuarina equisetifolia*. The first species comes from India via Madagascar, the two others from the Pacific. Little by little, these plants push back the native vegetation and beaches all end up resembling tourist advertisements (FIGURE 9.3).

Other plants do not remain confined to the coasts and some are true pests. To clearing, exploitation, conversion and development of all kinds, is thus added an undesired but very deep and insidious uniformisation: little by little, primeval and unique floras are replaced by a pan-tropical, undifferentiated, flora. This phenomenon is already very pronounced on oceanic islands and is now increasingly attaining continental habitats. According to the World Conservation Monitoring Centre of the United Nations Environment Programme (UNEP), it contributes very largely to the endangering of over 8,000 of the 100,000 known species of ligneous plants of the world. On the African continent, nearly 300 species of trees are in danger and 1,300 species are vulnerable. In South Africa, the fynbos of the Cape region, a type of vegetation resembling Mediterranean maquis and one of the richest habitats in the world, is seriously threatened by exotic species and costly eradication programmes have had to be undertaken to eliminate these pests.

In Central Africa, this problem may still appear to be insignificant, but it exists and will inevitably increase. The best known of these plagues is the Water Hyacinth *Eichhornia crassipes* (FIGURE 9.4). This free-floating aquatic plant of the Pontederiaceae family originates from the Amazon basin. Because of its beautiful pale lilac inflorescence, which resembles that of a European hyacinth, it was introduced in most of the tropical regions of the world. After the Second World War, it was introduced into the Congo basin where it now carpets many watercourses. Not only does it very seriously hamper navigation, it also profoundly affects the biology of the water where it lives by forming an absolute screen to any light penetration. It probably even has a negative influence on riverine and inundatable forests by removing nutrients that they would normally obtain from the river. It also occurs in the Ogooué, the Akagera, the Lake Victoria and the White Nile basins.

As for terrestrial invaders, they are represented by a whole cortège of very diverse plants introduced by Man for their utility or attractiveness. Among the most well known are *Mimosa pudica* and *Mimosa invisa* (FIGURE 9.5), *Lantana camara* (FIGURE 9.6) and *Tithonia diversifolia* (FIGURE 9.7). More of a problem, however, is the invasion by *Chromolaena odorata*, also known by the name of *Eupatorium odoratum*. This Asteracea from

9.3. *In the vicinity of Cap Estérias, north of Libreville, the sea-front vegetation is progressively invaded by Wild Almond* Terminalia catappa. *This species, recognisable by its large leaves that turn bright red before falling at the end of the dry season, even manages to become established on the narrow dune cordon separating the mangrove forest from the beaches and mudflats.*

9 GREAT CHALLENGES

9.4. *The water hyacinth* Eichhornia crassipes, *of the Pontederiaceae family, is a free-floating aquatic plant of Amazonian origin. By obstructing watercourses and absorbing silt, it can influence swamp forests.*

9.5. Mimosa pudica *is a pan-tropical Mimosacea, probably of South American origin. Today it covers thousands of hectares along roads, on the edge of cities and villages, as well as in humid areas. It is recognised by its small purple inflorescence and its sensitive leaves which close up when they are touched. This very invasive species is used as an indicator plant for climate change.*

9.6. Lantana camara *is a shrubby Verbenacea of South American origin with very aromatic foliage and multicoloured inflorescence. In grasslands its progression is limited by fire and in humid forests it becomes established with difficulty. Just the same, it manages to invade the edge of forest roads, notably forestry roads, the outskirts of villages and fallow land. It also creeps into pioneer formations along beaches and is more frequent in areas of semi-deciduous forest.*

9.7. *Better adapted to the humid forest habitat is the large yellow "daisy"* Tithonia diversifolia, *an Asteracea from South America. In some regions of southern Cameroon, this plant invades abandoned fields and hinders secondary regeneration.*

South and Central America has invaded Asia and subsequently also Africa. It has invaded practically the entire northern edge of the Congolian forest massif – it is present in the Ngotto region, in Central African Republic – and is the cause of interminable controversies. It develops in the forest edge, where it impairs the regeneration processes of the forest, and markedly reduces the biodiversity. In this way it is especially harmful on inselberge where it suffocates the original vegetation. In the Mayombe of South-West Congo it also favours the proliferation of a noxious grasshopper. In plantations it increases seriously the fire risk, and large sums of money are needed to combat it with pesticides, which again cannot be very good for the biodiversity. On the other hand farmers found that *Chromolaena* increases the fertility of soils, and allows a reduction of the fallow period of the land. They also use the stems as firewood, Its compost has a repellent effect on phytoparasitic nematodes, and many people have discovered that the plant harbours diverse medicinal qualities.

In addition to herbaceous or shrubby species, the progression of *Cecropia peltata* has also been noted for some time in south-western Cameroon. This tree of the Moraceae looks much like a native Umbrella Tree, *Musanga cecropioides,* and has a very similar biology. This South American native was imported to West Africa and, in Cameroon, it apparently escaped from the Limbé botanical garden. Everywhere where it becomes established, it replaces the native species.

Finally, there are also invasive species of animals. Well known are the rats that have already exterminated most of the birds and mammals that populated some islands. Less well known to most people is the Fire Ant *Wassmannia auropunctata*. As a native of South America, it was used in biological control of aphids. Unfortunately, it spread at great speed in Atlantic areas of Cameroon and Gabon, where it penetrated the forests along exploitation tracks. Once established, it is impossible to eliminate because its colonies possess nucleus satellites which assure their continuation after the main colony is destroyed. This ant eliminates other ant species, notably the one that protects *Barteria* (PAGE 194). It also eliminates many arthropod species. Butterflies of the family Lycaenidae living in symbiosis with ants seem particularly affected. This is the case notably in the Lopé National Park. So far, *Wassmannia* ants have not penetrated intact forests, but with the extension of forest exploitation and the opening of forest massifs, the question is, where will this ant stop?

Climate change

Another factor that threatens tropical forests is climate change. Its effects are not yet very perceptible, but in the 20 or 30 years to come, they could become catastrophic, according to some scenarios at least. No one any longer doubts the existence of this change, but climatologists are not in agreement on the extent or the exact distribution of the phenomena that it will bring about, or even on its exact causes.

Thus, according to the Intergovernmental Panel on Climate Change (IPCC), created in 1988 at the initiative of the United Nations Environment Programme (UNEP) and the World Meteorological Organisation (WMO), atmospheric concentrations of carbon dioxide will be doubled during the course of the 21st century, world temperatures will rise by 1.5 to 4.5° C on average, world precipitation by 3 to 5% and the level of the seas around 45 centimetres. These changes will not, however, be equally spread over the whole planet.

In temperate regions, where there is a strong latitudinal gradient, it is almost certain that all types of forest will move northward. Thus, the tundra will disappear – it is retreating already – and conifer forests will make way for deciduous forests. These changes could even be so rapid that some types of vegetation will have disappeared before others have had the time to become established.

In tropical regions, there will not be the displacement of vegetation zones towards the north or south. The rise in temperature will be less strong than in regions far from the equator, but it will be more marked in the interior of the continents than on the coasts, and the seasonal variations will be more contrasted. The *El Niño* and *La Niña* phenomena will probably have more marked consequences.

On the basis of palaeo-climatological records and simulations performed for Central America, it is possible to think that sub-montane and montane forests will be displaced upwards and experience a reduction in area, while Afro-alpine habitats will disappear from some mountains and will be greatly reduced on others. As for lowland forests, in spite of an increase of precipitation, they will be subjected to a more pronounced hydric stress because of temperature increase and seasonal contrast. Evergreen forests will thus become dryer. Semi-deciduous forests will be subjected to more frequent and more violent fires which will probably reduce vast areas to ashes, somewhat like what happened 2,000 to 2,500 years ago, or more recently in South-east Asia. Some forests located on the edge of the present forest massif will probably not be able to survive and protected forests will suffer heavy pressure. They will inevitably be transformed and some will even be very difficult to save from total disappearance.

Even though the rise in temperature will be less pronounced on the coasts, it will lift the layer of stratiform clouds that develops in the dry season all along the Gulf of Guinea. The cloud forests of Mount Cameroon, Mount Alén and the Crystal Mountains will therefore also move to higher altitudes, while forests of the coastal plain will be less protected during the dry season. The coastal grasslands of Gabon and the Congo could increase in area again.

As for the mangrove forests, they would retreat before the rising ocean level, but they would not really be able to colonise new ground in the interior because of the human presence. The marine influence, however, would penetrate further up the estuaries and rias. It would thus push brackish and freshwater formations further upstream.

All this remains hypothetical, however, because the more advanced our understanding of climatic phenomena and their simulation becomes, the more apparent it becomes that it is very difficult if not impossible to predict exactly what will happen at a given place.

The more our knowledge of past climates grows, the more we understand that climate never ceased to be modified and often it happened abruptly, contrary to what had been thought a few years ago. Results from the application of new technologies in the study of the arctic ice have thus shown that the last great glaciation was not a gradual phenomenon spread over several thousands of years, but that it was punctuated by brief periods of re-warming that at times reached as much as 10° C in 10 years, as J. W. C. White noted in regard to

9.8. Cassia siamea, *a Caesalpiniacea imported from South America with large golden yellow inflorescence, planted almost everywhere in Central Africa. In places, it invades more or less open or secondary forest formations, as is the case notably in the Semliki Valley in Uganda.*

> ### The Kyoto Protocol: the implications
>
> In the past, forest management was centred on wood production and all other values – biological diversity, non-ligneous forest products, protection of watersheds and cultural aspects – were considered as obstacles to the development of these forests. This mentality is slowly changing and the forest is seen more and more as an ecosystem with multiple functions. Among these functions, the trapping and the accumulation of carbon occupy an increasingly important place and are becoming convertible into cash. Unfortunately, the experts who have conceived the Kyoto agreements have thought mostly of temperate and boreal forests. The problem of tropical forests does not seem to have been at the centre of their worries. Also the Protocol of Kyoto does not anticipate "compensating" those who do not cut down their forest. Only the putting-in of new plantations is taken into account. So that the countries of Central Africa can benefit from these agreements, they must start by razing their forests. For the protection of biodiversity, this is not really what is needed. Like so many other agreements and conventions, the Kyoto Protocol favours a little too much the restoration approach to the detriment of the prevention approach. From the perspective of sustainable wood production and the global renovation of the countries of Central Africa, it will one day be necessary to envisage plantations, because it is unlikely that the exploitation of natural forests will be indefinitely able to yield revenues worthy of notice. Today we are not yet there, however, and, at the moment, the countries of Central Africa – in West Africa it is already too late – must make their voices heard and obtain from the industrialised world compensations for all the forests that for various reasons they do not exploit. The density of carbon accumulated above the ground in old tropical forests is of the order of 120 tons per hectare on average, while in temperate and boreal forests it reaches only half that value.

Greenland. It is therefore difficult to know to what extent the changes we observe today are really the result of the greenhouse effect: perhaps they are still related to the end of the "little ice age" that ended around the middle of the 19th century and which was accompanied, notably between 1760 and 1840, by droughts much more dramatic than those that Equatorial Africa has known during the 20th century.

The only thing we are sure of, finally, is that climatic stability does not exist. In other words, whatever one does, the climate will change and this observation should incite us to review our strategies for the occupation of the land. In the field of forest management, it is essential to be vigilant that massifs that we want to conserve for the future, for conservation of biological diversity or for wood production, are the most resistant possible to these changes. For that, it is necessary to avoid fragmenting them too much, to carefully choose their location – preferably in areas that we know have resisted previous climate changes – and to include the greatest diversity of habitats possible. None of this can be carried out, however, without a detailed knowledge of the individual history of each massif and its reactions to changes in the past.

Carbon management

The fact that climate changes are inevitable does not make us lose interest in the greenhouse effect. If only by simple precaution, it is not useful to add it to the natural variations of the climate. Since this phenomenon is above all the result of the liberation into the terrestrial atmosphere of large quantities of carbon that the living processes have trapped in fossil hydrocarbons, peat, fossil charcoal and the vegetation, we must do everything we can to correct this situation and commence by consuming less fossil carbon. Apart from this essential measure, which rich countries are not, however, ready to put into action, there are only two types of action possible: increase the quantity of carbon fixed in "carbon sinks" or reduce the liberation of carbon already fixed in existing "sinks". This problem thus especially concerns the industrialised world, but the tropical world does not escape. According to some estimates, the tropics will suffer from changes in the climate more than the temperate regions.

In practice, the first strategy could be materialised by the creation of vast plantations or afforestation. It is therefore not or not yet applicable in Central Africa, but it could become so in a fairly near future because of the fact that sooner or later it will be necessary to think of "supporting" wood production by means of artificial plantations. It would also ease pressure on the natural ecosystems.

The second strategy could be materialised by the reduction of forest clearing in order to reduce the liberation of fixed carbon. This would be translated in the field by maintaining as extensive a forested area as possible by the generalisation of reduced-impact exploitation (PAGES 333-334) and control of burning. As for the distribution of the use and production of bio-combustibles, this strategy is not adapted to Central Africa, either. It however become so, if the reduction of the forest massifs increased. On the other hand, the replacement of iron and concrete – as much as possible – by wood is imaginable, notably in construction. The production of iron and cement consume in effect great quantities of energy. The use of wood in a sustainable way in frameworks, among other things, would constitute a way of storing carbon.

Hunting

Hunting has been practised in the forests of Central Africa since Man has inhabited them and it is still an essential means of subsistence for many populations, rural as well as urban. For some decades hunting has weighed heavily on the wild fauna and, if no measures are taken to limit it, many animal species will disappear in the next few years. Consequently, modifying the complex web of ties between species, hunting could then affect entire forest ecosystems. Long minimised by some conservationists, hunting is now at the centre of their preoccupations.

It is in lowland forest regions that hunting and gathering have retained their greatest importance. They are part of the everyday life of rural communities and every able-bodied man is a potential hunter. From a subsistence activity, however, hunting has become transformed into a purely commercial activity over the last decades. It yields not only meat, but also currency, permitting the acquisition of other goods and commodities. This phenomenon is not new. It began when the Pygmies furnished the Bantus with meat in exchange for agricultural goods. The practice has increased since the 1960s, particularly since a general economic decline has become established during the last few decades. Today, the forest feeds not only the rural populations, but also a substantial part of the urban populations. After the war in the Congo, for example, the first meat available in Brazzaville came from forests in the north of the country. In other words, the sustainability of hunting is not just a western preoccupation, it constitutes a major socio-economic problem.

In the high-altitude regions of the Albertine Rift, the situation is different. People there live essentially from agriculture and pastoralism. They thus produce their own food and the natural resources of the forest no longer have more than an accessory importance. Hunting occurs, of course, but it has become a sideline activity, practised by marginal individuals, and often its "sporting" character prevails over its commercial aspects.

Techniques

In many lowland regions of Central Africa, notably in the Democratic Republic of the Congo, very young boys begin hunting birds around villages and along roads with slingshots made from old car tyre inner tubes. Later, they accompany older men in order to learn to hunt with nets, traps or guns.

Hunting with a net is the favourite technique of the Pygmies, but it was also practised by many other populations, notably in eastern Gabon (FIGURE 9.9). In many places it has fallen into disuse, but it is still practised by the Mbuti Pygmies who inhabit the vast forests of Ituri, in the Democratic Republic of the Congo. According to studies by John Hart, a researcher working for the Wildlife Conservation Society (WCS), the technique generally involves a team of ten to 40 men and women, who set up five to 20 nets for a total length of 600 metres to two kilometres. An area of three to 17 hectares is encircled and the animals are driven towards the nets. This hunting is practised at a distance of one to 3.5 or sometimes five kilometres from the temporary camps, themselves located up to 25 kilometres from a permanent village. Most captures (78 to 92%) are of small ungulates, such as forest duikers and water chevrotains.

In the special reserve at Dzangha-Sangha in the Central African Republic, hunting with a net is still practised by the Baaka Pygmies. According to Andrew Noss, it was also practised by other ethnic groups. Hunting with nets is, however, a social activity and cannot be practised by immigrant populations who have lost the indispensable social cohesion. Along with the capture of game, hunting with nets is also the occasion for collecting honey, caterpillars, leaves, fruits, nuts and construction materials. Just as in Ituri, part of the hunt generally involves men as well as women and often even children. Altogether, it requires five to 100 hunters and four to 25 nets from one to 1.5 metres high by five to 40 metres long (on average 18 metres). The nets are arranged in a circle around a densely brushy area, often

9.9. Not long ago, hunting with nets was still practised by the Kota of eastern Gabon. A forest duiker is finished off with a machete in the net in which it was taken.

an old tree-fall gap, attached to branches and fixed to the ground. Some of the hunters enter the encircled area, beating the vegetation in order to chase the animals into the nets, where they are taken by other hunters waiting in ambush nearby. Next, the nets are moved to another place in the forest, usually a few hundred metres from the previous place. The whole operation, from the placement of the nets to their removal, lasts 10 to 15 minutes. On average, a hunt comprises therefore 16 operations spread over four hours of active hunting, without counting the two- to three-hour walk from the camp to the hunting area. A captured animal belongs to the owner of the net in which it was taken, but the person who catches it in the net receives a portion of the meat, like those who help place the nets. At the end of the hunt, those who receive meat often share some of it with those who did not get anything.

Very different is hunting with nooses. The Pygmies of Ituri do not practice it except around the edge of their villages and fields. Elsewhere in Central Africa, even though the use of metal nooses is legally forbidden in most countries, this technique is the most commonly used. It yields more meat per hunter, does not require large communal operations and is much less obvious. It is therefore the favourite method of poachers who work in reserves and national parks. It is practised by a lone hunter, or sometimes a hunter accompanied by a helper or two. The nooses, nowadays almost always wire cables, are usually placed on the ground above a release system carefully hidden and attached to a solid bent branch (FIGURE 9.10). Animals are therefore taken

9.10. A Gabonese hunter from the region of Makokou removes a blue duiker taken in a trap. This hunting method is very economical and therefore much used, but it is not selective and often leads to considerable waste. In a village on the edge of the Dja Faunal Reserve, a quarter of the animals captured were discarded because they were already rotting by the time they were removed from the traps.

9.11. Another Gabonese hunter prepares some bats on his wood fire in the forest.

by a leg. More rarely, nooses are placed vertically across a passage and animals are captured by the neck. Hunters generally place 50 to 100 nooses along a line one to three kilometres long. These lines are left in place for a month at most, but individual nooses may be moved to improve their placement. The hunters camp near their noose lines and these camps are located at a few hours', or one or two days', walk from the permanent village. Meat from the animals taken is smoked in the camps before being brought out to the city or the villages. Game captured with nooses is more varied than that taken in nets. It comprises notably brush-tailed porcupines, large rodents, small carnivores and forest duikers.

Hunting with guns is obviously a fairly recent technique and is not yet generalised. Although guns were already used for hunting in many regions of the Belgian Congo before 1960, they still had not appeared in Ituri in the 1990s. Altogether, the gun has become the principal hunting arm in ex-French Africa and in the Democratic Republic of the Congo. Formerly, locally-made guns were used; they were heavy, inefficient and dangerous – and sometimes even more dangerous for the hunter than for the game. Today, it is often European-made weapons that are used, mostly French or Russian and relatively cheap. These guns are not collectors pieces but they are diabolically efficient. Every furred or feathered animal is potential game. Even a monkey hidden at the top of a tree – as the colobus monkeys are – is no longer out of reach. Only the cost of cartridges and a few rapidly disappearing local taboos still protect some species, notably the bongo in the Central African Republic.

In Uganda, Rwanda and Burundi, carrying a gun has been prohibited for a long time and hunters generally hunt with dogs and lances, bows and arrows and, rarely, with nets. Most often, they practice trapping with nooses, spring traps or pits. Nooses target small antelopes and the main victims are bushbucks, forest duikers and large rodents. Formerly, nooses were made of cord made from plant fibres; today they are made with solid wire cables. Any animal can thus become a victim. As for pits, they were used in the past for hunting river pigs, giant hogs, buffaloes and hippopotamus; nowadays they are rarely used.

Game

The "hunter's basket" varies considerably according to region, method, hunting pressure and status of the fauna. In lowland regions, the most sought-after species are usually even-toed ungulates (antelopes, pigs, forest duikers and chevrotains), which constitute 40 to 70% of captures (FIGURE 9.12). In the region of the Dja reserve, Marc Dethier found, in the course of a study launched by the ECOFAC programme, that these animals may even comprise 85% of the take.

Primates come in second place and most often represent 20 to 25% of the catch, with extreme values of 8 and 40%.

As for rodents, especially brush-tailed porcupines and giant pouched rats, they are in third place, but they are mostly prized in regions where even-toed ungulates and primates have become scarce. On the island of Bioko, they thus represent 32% of captures according to observations made by John Fa. In other words, this percentage is a good indicator of the health of the fauna.

In addition to especially sought-after species, many others appear on the list of species hunted. Probably all mammal species, whatever their status, many bird species and some reptiles are hunted one way or another. Species protected by law do not escape, neither the large carnivores, including panther and golden cat, nor the gorilla, chimpanzee and bonobo.

These large primates mean no more to African populations than do other species of game. Even if some populations do not eat primates, they do not hesitate to kill them just the same, in order to sell them.

In the high-altitude regions of Uganda, Rwanda and Burundi, primates are not eaten – except by the Konjo

9.12. A forest duiker on sale near Odzala-Kokoua National Park in the Congo. According to studies made by the ECOFAC programme, even-toed ungulates number around 750 out of a total of 1,497 pieces of game examined in this region. In most of the villages on the edge of the Dja Faunal Reserve, also, even-toed ungulates constitute the great majority of prey brought back by hunters. In a single village, primates and rodents reached 34%. As for birds, they generally represent less than 3 to 5% of the takes. On the isle of Bioko, ungulates represent 37% of the prey.

9 GREAT CHALLENGES

9.13. A Putty-nosed Monkey Cercopithecus nictitans *on sale along a road in the North-Congo. Around Odzala, according to ECOFAC, primates represent no more than 12% of the take, however, and constituted only 8% of the 15,141 pieces of game brought to the Brazzaville market during a 12-month period. On the isle of Bioko, primates make up 26% of animals killed.*

of the Ruwenzori Mountains and the Baamba of the Semliki – but some species are sought-out for their medicinal qualities. Hunting in Central Africa produces not only meat. The by-products – such as skin, hooves, hairs, teeth – sometimes have even more importance than the meat because they are part of traditional pharmacopoeia. In some regions, black and white colobus monkeys are thus poached for their fur: decorative and used in traditional medicine. Even the fingers, nails and sexual parts of the mountain gorilla enter into traditional preparations, which are politically correct to respect.

The hunting territory

In general, it may be said that formerly most communities had a hunting territory that extended for 20 or 30 kilometres beyond the farming lands (PAGE 271). In places, this territory was strictly reserved to a given community. Elsewhere, groups of "friend" hunters were tolerated. With the concentration of populations along roads and the numerous movements of populations, use of these territories has been somewhat lost and in many places it has even disappeared. For the bands of professional hunters that have appeared with the commercialisation of hunting, which at times operate across borders, the notion of territory has become completely abstract.

Finally, it is often only in protected areas and their periphery that populations remember their former territories and proclaim their rights, especially when the village elite remind them. That is how somewhat peculiar situations arise, such as at Monte Alén National Park, where village land coincides surprisingly with the protected area. The limits of territories have, however, always been adapted to the distribution of the game. Between pure and simple negation of ancestral rights and organised fraud there must be a happy medium. With a population density of about one to five inhabitants per square kilometre, it should be possible to find it.

Impact and sustainability

In many places, the fauna has disappeared and the forest falls silent. Voices are raised everywhere to denounce the massacre and countless testimonies attest that some species are on the brink of extinction. In the 1980s, with the massive introduction of the Soviet-manufactured "Baïkal" gun, the disappearance of primates was observed, notably the Red Colobus in the forests of Kivu and Maniema, but most observations remain anecdotal and poorly documented. Conservationists thus often lack strong arguments to defend their theses.

This lack is all the more serious since experts in the human sciences and biologists very far removed from the realities of the field sometimes have, or have had a tendency to minimise the problem under the pretext that the forest is inexhaustible and that traditional hunting has never put a species in danger. These two illusions die hard. And yet, numerous archaeological data show that, throughout his history and throughout the world, Man has exterminated innumerable animal species and there is no basis for the supposition that the sustainability of hunting has ever figured among his real preoccupations. According to S. J. M. Davis, author of a work on archaeological zoology, only the advent of agriculture and the domestication of animals has saved what remains of the fauna. For many, however, this subject seems taboo and they insist on the need to explain the extinctions by climate changes or by any other phenomenon, even interspecific competition. Man has thus convinced himself that hunting can be sustainable and not profoundly affect wild animal populations. In the Strategy for Conservation published in 1980, the IUCN and the WWF consequently defended the thesis according to which the management of protected areas could be entrusted to local populations. In fact, hunting, even traditional hunting, may seem sustainable in the short or middle term, but it is not if one thinks in terms of centuries.

Today, this debate is completely obsolete, because truly traditional hunting does not even exist any more. Old techniques have been modernised by the use of new materials such as wire cables. In some places, guns, including military weapons, have become the arm of choice. The rules and codes of the elders have given way to the law of supply and demand. Modern means of transport – four-wheel drive vehicles especially – and the increase in the number of roads that penetrate regions previously hardly accessible to hunting and the easy removal of game.

In spite of these realities, there are still people who are not convinced of the ravages of hunting as practised today. Work by John Hart in the forests of Ituri, a vast region in the eastern part of the Democratic Republic of the Congo, is therefore of particular interest. In comparing the density of fauna in areas hunted and areas not hunted, this American researcher, who spent more than 15 years in the region, was able to demonstrate irrefutably that hunting, even traditional hunting, reduced the density of medium-sized forest duikers and water chevrotains by at least 85%. In fact, the hunters of this vast region of the Democratic Republic of the Congo – the Mbuti Pygmies – operate in an area until hunting no longer brings them much. They then move their activities to other sectors, and do not return to the initial land until after at least 15 years. Such a rotation is obviously possible where there is a low human population density in the central parts of these forests (less than 1 inhabitant per square kilometre). Even though their approaches are often very different, studies by Andrew Noss in the region of Dzangha-Ndoki National Park in the Central African Republic, those of Sally Lahm in north-eastern Gabon, those of John Fa on the isle of Bioko and those of Willy Delvingt's team in Cameroon (see box opposite) underscore a significant reduction in the density of most of the hunted species as one approaches villages. Hunting, as it is practised today, thus has a considerable impact on the fauna and all evidence leads to the conclusion that it is no longer sustainable. This observation unfortunately applies to all tropical forest regions of the world.

Elizabeth Bennett and John Robinson, in a recent work, attempt to make a first synthesis of results. Although these come from studies conducted on very different human populations and pertain to several continents, they may be generalised and the conclusions summarised in a few words. Firstly, animal population densities are greatly reduced: light hunting pressure reduces populations of mammals by about 70%; a greater pressure reduces them by 95%. Hunting favours small individuals of a species or those with unremarkable trophies. Hunting also seems to lower the age of first reproduction, increasing female fertility and reducing the number of old individuals. Globally, however, it leads to a considerable reduction of the proportion of reproductive adults and therefore a reduction of future harvest rates. Vulnerable species, generally large species, with a slow rate of reproduction, are susceptible to local extinction. The structure of the biological communities then changes, not only because the proportion of large species diminishes, but also because some trophic groups are more hunted than others.

The Badjoué exploitation of natural resources: a case history

Since 1993, the Sylviculture Unit of the Faculté universitaire des Sciences agronomiques de Gembloux (Belgium) has been involved in a detailed study of the various aspects of the exploitation of the natural resources of the forest by the Badjoué who live just north of the Dja Faunal Reserve in Cameroon. Their studies, conducted in the framework of the programmes ECOFAC, *Avenir des peuples de forêt tropicale* (APFT) and Community Forests (*Forêts communautaires*, PFC), are concentrated on a single population, living in a limited geographical area and have been created by a pluridisciplinary team under the coordination and supervision of Professor Willy Delvingt.

After ten years of work, a first synthesis has come out. Notably, it permits a response to a crucial question: is the present management of spontaneous natural resources sustainable? Slash and burn farming, which can be considered as an indirect form of exploitation of forest resources, does not seem to constitute a threat for old forests, neither in the present nor in the near future, and this in spite of the sedentarisation of the villages. Cash crops (cocoa and coffee) are suffering an obvious decline, as in many other regions of Africa, and many plantations are abandoned. What remains still represents a source of appreciable revenue. They no longer constitute a threat to old forests. Fishing, both collective and individual, still an important activity, remains impregnated with traditional techniques, even if synthetic materials are used. The studies are unable to predict its sustainability, however. As for hunting, it furnishes an indispensable contribution in animal proteins and has become an increasing source of monetary revenue. Because of this it represents an essential activity, but it has been considerably transformed over the past decades: it has changed from a collective to an individual activity, and traditional materials have been abandoned in favour of modern materials. Trapping predominates and everything that can be consumed is hunted. Mammals, especially forest duikers, constitute, however, the main booty. All observations lead to the opinion that it is absolutely not sustainable: in the village of Djaposten, for example, 82% of the biomass hunted comes from the Dja Faunal Reserve, 15% from a forest concession and only 3% from the agroforestery zone. Finally, the collection of non-ligneous forest products, seems to have variable effects. The collection of fruits does not seem to affect the regeneration of species, but the collection of palm wine exercises a very high pressure on natural stands of *Raphia* palms, on plantations of *Elaeis* palms and on natural stands of Bitter Kola *Garcinia kola*. The over-exploitation of the latter species is striking.

Globally, strategies of management of natural resources are based on customary land ownership. These still play their regulating role in the farming zone, located next to the villages, but they no longer work in the forested zone. There, under pressure from the external world, and in particular from the forest exploiters, we are observing, according to Willy Delvingt, a weakening of the land ownership exercised in an individual or familial way and a passage to the public type of land ownership. The taking of natural resources, animal and vegetal, to monetary ends, have become completely uncontrolled. This situation represents a threat at once for the integrity of forest ecosystems and for the life style of the Badjoué populations.

These results are probably true for many of the human populations of Central Africa.

9.14. The Pygmies of the south-west of the Central African Republic still practice hunting with a net, a very social activity that reinforces the cohesion of the group. Increasingly, these Pygmies work for the Bantus who move the game to the centres. In spite of themselves, they are thus become part of commercial hunting. Quite apart from that, hunting with a net is not without effects on the fauna. In the eastern part of the Democratic Republic of the Congo, for example, it is probably responsible for the disappearance of the Congo Peafowl Afropavo congensis *from much of its range. Because of their large spurs, males are very easy to capture in nets.*

Hunters very often have a predilection for granivorous or frugivorous species. Hunting thus represents an unbearable competition for large carnivores, such as the golden cat and the leopard, and provokes a very significant reduction of the global productivity of the forest community.

It is, however, imaginable that after a large initial decrease, animal populations would stabilise. In this case, hunting could still be considered sustainable. But in practice, it appears that, small annual fluctuations aside, populations slowly continue to decrease. At most, it can be observed that once a certain level of reduction has been reached, the rate of decrease becomes very low. At this stage, species are however under high risk of extinction. In any case, they are so reduced that they no longer carry out their functional role in the forest and are no longer of interest for human populations.

These observations draw Willy Delvingt into a more theoretical reflection on what represents or could represent a sustainable management of the fauna. At first glance, it consists of taking the "interest" without touching the "capital". The production of a game population is not, however, proportional to its size, while the complexity and the multiplicity of relationships between species in a tropical forest ecosystem do not allow the definition of a sustainable rate of taking. Even a relatively low harvest rate could bring about the replacement of one species by another, which would lead inexorably to the loss of originality of the system and its modification through the elimination of the seed dispersers. Some trees of great economic value, such as the various species of *Entandrophragma*, would be particularly likely to be affected by this sort of problem.

As for the stabilising effect of species which insure the preservation of biodiversity and influence organisms which appear taxonomically and ecologically very distant, how could it be measured and taken into account? The discrepancy between the speed of reaction of tree assemblages on the one hand, and of populations of animals, on the other, constitutes a major, or even an insurmountable, obstacle to this exercise.

And yet, the sustainability of harvests in a tropical forest ecosystem resides more in the perpetuation of its ecological functions than in the conservation of the species themselves. The two aspects are by definition included in the concept of biodiversity, however, and are probably not dissociable. As Bennett and Robinson have emphasised, by the time an animal species has reached its minimum ecologically sustainable density it is already of no interest for the human populations. In practice, the sustainability of hunting should therefore aim at keeping species above the minimum density that is "socially" sustainable.

However, several species are at present already greatly reduced and some are actually in danger of extinction, notably the Black Colobus Monkey, the Red Colobus Monkey, the drill, Preuss's Monkey, the Owl-faced Monkey, the Giant Pangolin, the Water Chevrotain, Ogilby's Duiker, the Yellow-backed Duiker, the White-bellied Duiker, the Golden Cat and the Leopard. Either these species still have a vast range, like the Red Colobus Monkey, but their population density has become infinitesimal, or their populations are concentrated in a few very scarce places, as is the case of the drill. Even the sun-tailed monkey is vulnerable because of its very restricted range.

The causes of change

Although Man has hunted in the forests for thousands of years, the surviving animal species do not appear to have suffered from this hunting. It therefore must be that some of the conditions have changed.

To understand these changes, it must be remembered that forest ecosystems are much less productive than grassland ecosystems and that in the forest it is easy to exceed the tolerable rate of harvest. John Hart has clearly shown that *Gilbertiodendron* forests are less productive than most more open or semi-deciduous forests. Other studies confirm that the slightest hunting pressure lowers animal population density below optimal levels, in other words, a density that generates a greater rate of reproduction. Apart from species attracted by secondary formations or crops, or those which have a great aptitude for re-colonising lost territory, many tropical forest species are rare and have a low rate of reproduction. Many species also have behaviours that render them extremely vulnerable. Forest duikers, for example, are attracted by the imitation of their calls, while Red Colobus Monkey hide at the top of large trees, freeze and let themselves be shot one after another. Altogether, large-scale changes are not necessary for bringing about an inevitable impact.

Among changes that have taken place over the past decades are first of all the increase in human populations and their density in forest habitats, as much by increased birth rate as by immigration of foreign populations. This factor alone has considerably increased the demand for bush meat. The greatest impact, however, stems from the immigrant populations. Not only do they come and add themselves to local populations, but they often induce profound transformations in local hunting habits and tend more frequently to convert the forests. Often finding themselves in unfamiliar habitats, they do not know the forest and consider some wild species simply as pests to be eradicated. The sustainability of hunting does not concern them. The mingling of diverse populations presently taking place in several of the forested regions of Central Africa accelerates the abandonment of traditions, notably the rotation of hunting areas.

Independently of absolute increase in populations, their increase in relation to the reduction of forest massifs and the productive capacity of remaining forests must be taken into account.

9.15-16. Produce from hunting on display in a market: Emin's Rat, quarters of a cured forest duiker, crabs (above); a Blue Duiker, a Black-fronted Duiker, and a Bushy-tailed Porcupine (left).

9 GREAT CHALLENGES

A major evolution that has intervened over the past few decades is, however, the commercialisation of hunting. Forest populations of Central Africa have embraced a market economy. Their societies have differentiated and some individuals, having rapidly become richer than the others, have acquired new needs that they have often sought to satisfy by an intensification of hunting. It is in part for this reason, as well as because of social changes, that hunting with nets has given way to hunting by means of nooses and guns. But hunting with nooses is accompanied by enormous waste. The method kills indiscriminately and many of the animals killed are lost because they are not retrieved soon enough. Hunting with guns has given access to previously inaccessible species of the canopy. Combined with the use of powerful torch lamps, they enable attacks on nocturnal species. The practice of these new techniques requires money, however, therefore a supplementary hunting effort. Finally, changes in immigration and society have brought with them the disappearance of social, cultural and religious taboos. The dimension of these phenomena was multiplied by the construction of roads and trails, as well as the proliferation of vehicles including four-wheel drive vehicles,

This evolution has been well followed in the villages around the Dja Faunal Reserve in Cameroon, as shown in works summarised by Daou Joiris. Commercial hunting appeared at Mekas in 1976, with the introduction of night hunting, and at Ekom in 1988. Beginning in 1984, it was stimulated by renovations in the President's village of origin, by the creation of commercial plantations and by the installation of a military camp. The fall in sales of commercial produce in 1988

9.17-18. The slaughter of elephants in a swampy clearing in Odzala-Kokoua National Park in the Congo. Formerly this sort of incident was linked solely to hunting for ivory. Today, an elephant also represents a mass of meat that more and more easily finds its way into the trade. The multiplication of light, automatic firearms is a godsend for the poachers.

did the rest. In these villages, game became the main source of income from 1993 or 1994. It was sold fresh or cured. The demand was supported by orders from local functionaries and by hunters coming from outside the region. For the villagers, game in the end represented the only source of money for buying medications, soap, petrol, clothes or for satisfying the expenses of prestige on the occasion of marriages or deaths. The devaluation of the CFA franc in 1994 and the rise in price of some basic staples (salt, sugar, cooking oil, soap, petrol) amplified the phenomenon. The recent rise in coffee and cocoa prices has not changed much.

Elsewhere, commercialisation has been stimulated by the development and implantation of forestry exploitation or mining activity in the centre of the forest massif. Placer mining, notably in the forests of Rwanda, Burundi, Kivu, Cameroon and Gabon, has introduced thousands of gold hunters into the forest ecosystems who, while scraping the earth, set their traps. In Nyungwe Forest, in Rwanda, they have eradicated the buffalo, the red river pig, the giant hog and perhaps even the three species of forest duikers that inhabit it. Indirectly, this rush for gold, or more recently for coltan, has also increased pressures on the biological resources of the forests by removing much of the manpower from farming.

Hunting for ivory

In addition to hunting for meat, there is a more specialised hunt, centred on ivory. The hunters involved are professionals, generally well armed and ready for anything. They are organised into bands and act under the protection of a "pyramid" of political personalities.

This hunting reached its climax in the 1970s. It then slowed down, both because of the ban on ivory trade initiated by CITES in 1987, and because of the considerable reduction in elephant populations. After the war that hit the Congo in 1997, the eco-guards put in place by the ECOFAC programme were almost able to put an end to elephant poaching in the region of Odzala-Kokoua National Park. Elsewhere, poaching continued, of course, but it was not intense enough to put the species in danger. Loss of habitat represented the most serious threat because coexistence between elephants and the farming populations was not easy. Also, as the forests were fragmented, conflicts were more and more frequent and more and more destructive.

Within the last three years, however, ivory hunting has started again in the entire western part of Central Africa. In the Odzala area alone, ECOFAC agents estimate that 300 elephants have been killed each year. This upsurge probably had its origin in the partial liberalisation of the ivory trade by CITES in 1997, but it was supported by a strong demand for ivory in western Africa to supply traditional arts and crafts activity. The elephant populations of this part of Africa are indeed practically extinct. Poaching has also been facilitated by the disorganisation of the country after the war, the uncontrollable circulation of automatic weapons and the complicity of some authorities.

This situation is deplorable because western Central Africa harbours the last large populations of the forest elephant.

However, a solution for ivory poaching is unfortunately difficult to find. Too many aspects and too many different people have to be considered. In Gabon, South-East Cameroon and North Congo elephant populations are still quite impressive, and elephants are everywhere. For many people in these countries cohabitation with elephants is not an easy task: crop damage is frequent, and every year people are badly injured or killed. Protection of elephants is consequently hardly understandable. In southern Africa governments want to be allowed to sell ivory from well managed populations. For Westerners on the other hand elephants are quasi quasi-mythic animals which have to be protected unconditionally.

9.19. Elephant tails, guns, including automatic rifles, and ivory, taken from poachers in the Odzala region of North-Congo.

9 GREAT CHALLENGES

What can be done?

Because of this worrisome situation, many measures have been proposed, but all run up against great difficulties of application. The use of wire-cable nooses is banned in most countries, but no one takes notice. Guns are subject to controls, but these controls are systematically bypassed. Hunting quotas run up against hunters' lack of discipline, corruption and ignorance about game population dynamics. The establishment of reserves in sectors of the forest and the institution of close seasons are just as illusory, since neither protected areas nor protected species are safe from hunters. Supply of alternative sources of proteins by animal raising and fish farming (FIGURE 9.20) is fraught with problems. Game raising – cane rats, river pigs, forest duikers – is hindered by low profit and technical difficulties. Bovine meat is generally easier to supply, but not available except in areas not too far from traditional pastoral regions. Also, these alternative resources are often rejected for cultural reasons. As for the suppression of commercial intermediaries, this irremediably runs up against the social context. Over and above these difficulties, lies an essential factor: the lack of will to change anything and the lack of respect for the law. In Gabon, for example, for 40 years not a single sentence has been pronounced and implemented for serious infractions of the hunting laws. For most magistrates, infractions of hunting laws do not exist.

It is fashionable then to attribute all these problems to a poor understanding and bad planning of conservation or to forest exploitation. Indeed forestry companies favour hunting by constructing trails and roads, by too often closing their eyes to what their personnel do and even by thinking at times that hunting rids them of obligations to this personnel. But to what extent must the companies stand in for the State? The very fact that they are asked to deal ruthlessly with poachers puts them in a very delicate position vis-à-vis their personnel and exposes them to strikes, sabotage and reprisals. On the other hand, large companies, conscious of their image, already take the place of the State in the fields of teaching and health care. Therefore, why not provide help for hunting controls. Besides, the limited experience in this domain show that stricter control of movements on the concession roads not only reduces hunting, it also suppresses a lot of the theft of material and fuel.

As for conservation, it is blamed for not taking the local populations into account enough, for not trying to rely on them more and for being too authoritarian. In reality, the human and financial means put at the disposal of conservation and, more particularly, the means allotted to surveillance, are often ridiculously insufficient. In the Lopé Reserve, for example, there are only five agents to guard 500,000 hectares. In fact, the implementation of hunting regulations runs up against a whole set of fundamental problems.

First of all, too many people do not see the need to limit hunting because they remain convinced that the forest is inexhaustible. This vision is encountered not only among the villagers, but also among foresters, planners, developers, scientists in human sciences and civil service employees at all levels, to whom the power of decision belongs.

Next, there is culture. No villager could report another, stand witness against or *a fortiori* report someone of his village. Any person doing this would be irremediably banned from the community whose cohesion is above any other consideration. A guard must there-

9.20. In the middle of the Gabonese forest, fish-farming lakes have been laid out in the hope of complementing the protein resources of the local populations and of thus easing the pressure on game.

fore protect the hunters from his own community, even help them or inform them, and cannot act except against strangers. On the contrary, when someone does accuse a hunter from his village, it suggests that there is something other than hunting between the two individuals – a family or clan affair – and others doubt the truth of the accusation. To that is added the corruption of Water and Forestry agents. When they intervene and confiscate poached game, it is too often in order to consume it themselves or to resell it for their own profit.

Finally, there is the attitude of government authorities, whose official communications are very often deceiving. Not only do they lack the motivation to support conservation, for which they could easily be forgiven – in the western world it is not very different – but they also very often play a double game. Indeed, hunting in Central Africa is a very popular activity and, as in France, no politician really dares attack it. In towns, moreover, it is often the rich who are the largest consumers of game. Furthermore, much hunting is organised by high functionaries who arm the hunters, protect the commercial networks – from small intermediaries to town restaurants – and pocket the largest portion of the profits. The worst offenders are thus the elite of small provincial towns, but these people could not act without the support of higher authorities.

Thus, good studies can be conducted and the best conceived strategies outlined, but the key to success resides in the awareness not so much of rural populations, but above all of governmental authorities and the magistrature. This job is unfortunately complex and delicate. It takes much time and cannot succeed except by the close collaboration of funding agencies, personalities already convinced – they exist! – and local and international NGOs. It is necessary to succeed in convincing people that the goal is not to forbid hunting but to try to make it sustainable.

In the meantime, it is in any case necessary to try to act on the local scene. First of all, the Water and Forestry agents and the eco-guards – assistants hired for projects but not sworn in – must start by giving the example, and ensure that game seized does not become reintegrated in the commercial circuit but is destroyed and that arms seized are put out of use. In order to achieve these goals, surveillance must be performed by national agents and not by recruits coming directly from the local communities that they will have to penalise, their career must be secure and their profession valorised, not ridiculed. There is an enormous job to be done, probably with the help of sociologists, who could be enlisted not to justify, in the name of culture, immobilism, but to make changes acceptable.

Next, an effort must be made to involve the populations themselves in surveillance. Even though according to the law all game belongs to the State, the fauna must come to be managed as if it belonged to the village communities. It must therefore be possible to limit commercial hunting, bring back the system of hunting territories and put the populations in charge – especially the women – of monitoring their "larder". An experiment in this direction was made by ECOFAC in Ngotto Forest in the Central African Republic where village-operated gates control the tracks leading in and out of the forest. This strategy does not stop hunting but tends to limit it to the real needs of the local populations. It certainly does not constitute a panacea, but is a first step towards an equitable solution.

At the same time, an effort should also be made to try to earn the confidence of the hunters by working with them so as to know their hunting methods better and to better estimate their impact in the ultimate spirit of trying to make hunting more sustainable. This is what the University Faculty of Gembloux has undertaken for the past few years in Gabon and in Cameroon.

In the face of reality, catastrophic in many places, all these measures can seem derisory. Perhaps they are indeed, and the only evolution that may save the African fauna is a profound change which will transform the last hunter-gatherers into producers. Then the fauna will no longer be an essential resource and hunting will become a sport.

9.21. In spite of their difficulties and reticence, the large companies, more and more concerned about their international image, try to take measures to limit hunting.

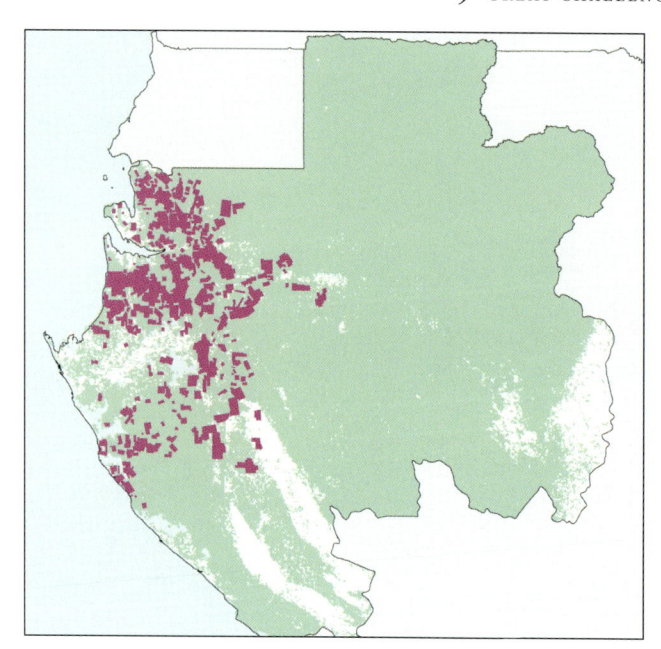

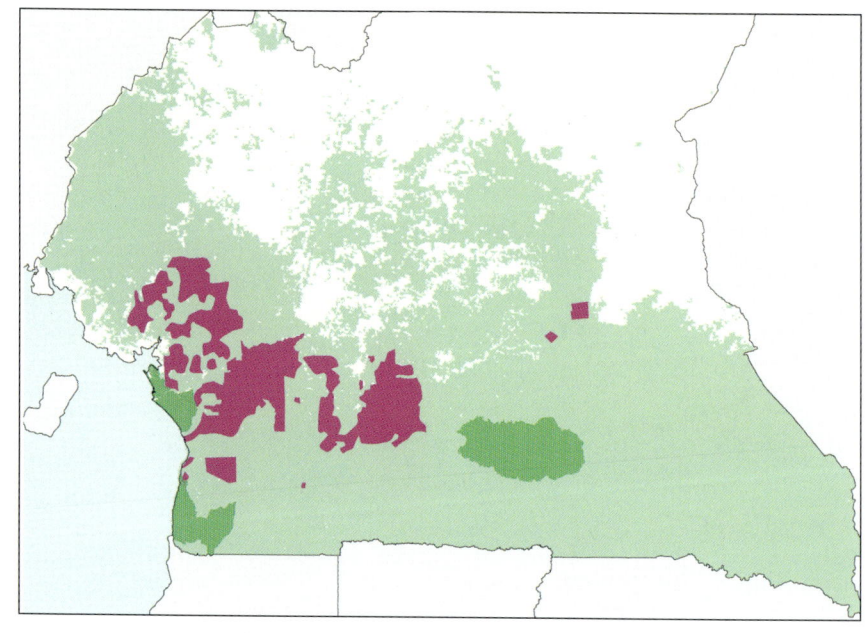

9.22-23. *In 1957, 1.6 million hectares were allocated to forest exploitation in Gabon. In 1959, 1.9 million hectares were allocated in Cameroon.*

In the two countries, exploitation was limited to regions near the coast or situated along large roads.

9.24. *Forest exploitation in Central Africa (in purple) now affects most of the forest block. Only the inundatable forests of the centre of the Congolese basin and those of the south-eastern and eastern parts of the Democratic Republic of the Congo (DRC) still escape industrial exploitation. However, this is not to say that these forests are protected from exploitation by individual long sawyers.*

Logging

For the countries of Central Africa, some of which still have an 80% forest cover, forest exploitation constitutes an essential economic stake. Of course, petroleum or mineral exploitation rank in first place for filling the coffers of the State, but as petroleum and some mineral reserves become exhausted, some governments base their hopes on forest ecosystems.

All the same, logging in the forests of Central Africa remains selective and therefore not very intensive: on average, only one to three trees per hectare are felled, while in South America this number may reach ten and in Asia 20. Exploitation differs greatly, however, from one area to another, depending on the commercial value of the species present and the degree of difficulty related to removal, logistics and exportation. Thus, large-scale industrial logging is essentially limited to the western half of the Central African forest block – therefore mostly west of Kisangani – where timber can be fairly easily transported to the Atlantic. The eastern half is subjected local logging, for local consumption and the exportation of sawn timber to East Africa. This exploitation escapes any regulation, and does not appear in statistics. It will therefore not be discussed here, but its effects on forest ecosystems can be considerable. As for the montane forests of Cameroon and the

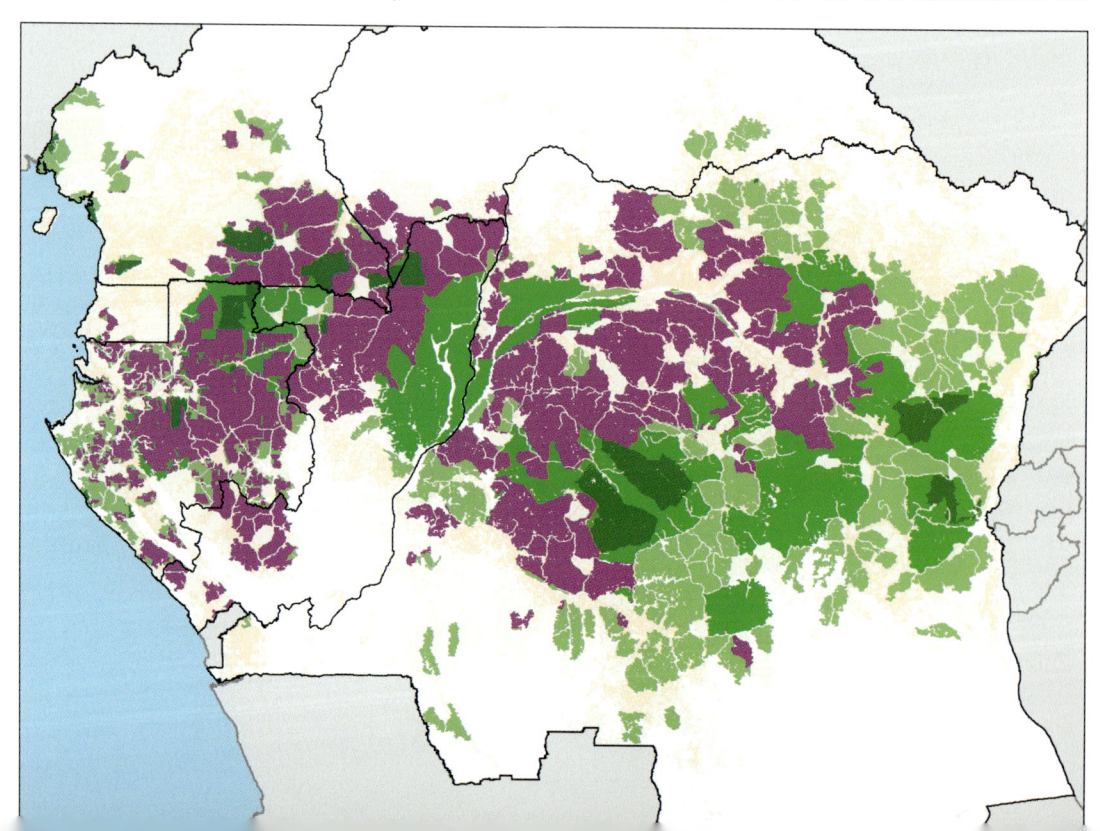

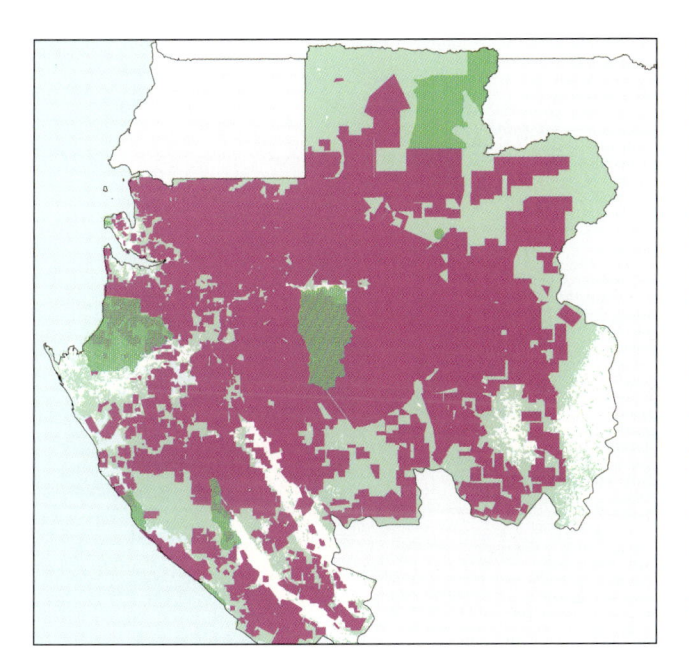

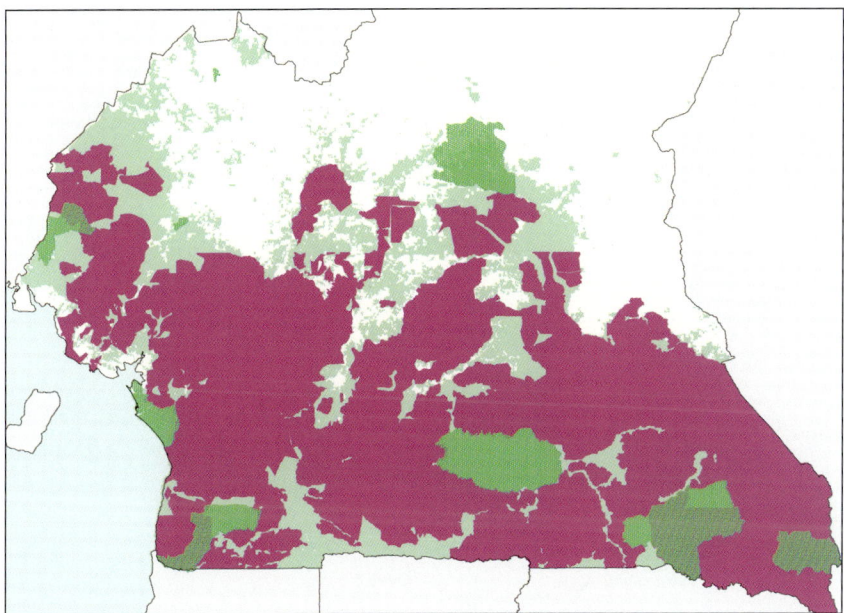

Albertine Rift, they suffer more from conversion for agriculture and pastoralism than from logging. At most, they are the object of very selective and generally very localised exploitation, sometimes for pharmaceutical reasons, such as exploitation of *Prunus africanus* bark on Bioko and in Cameroon.

In spite of the enormous stake forest exploitation represents for the Gulf of Guinea countries, data to document it are few, scattered, not very reliable and sometimes impossible to obtain. Indeed, until very recently, neither the ministries in charge of forestry operations nor the exploiters were willing to divulge their statistics. As for the figures, which would in any case come to be known, they were usually incoherent and contradictory. To cite just one example, extracted from a report by *Forests Monitor:* in 1998, the volume of timber exported from the Congo to Italy was 37,731 cubic metres according to Congolese statistics, 119,102 cubic metres according to Italian statistics. Discounting administrative errors, this discrepancy rests above all on the "haziness" that surrounds the entire timber trade network.

Faced with the retreat and degradation of tropical forests, international environmental NGOs have turned their attention to the problem. Thus the Global Forest Watch initiative of the World Resources Institute was formed, and fairly detailed data were published for Gabon and Cameroon. For Gabon, maps of concessions had already been produced with the aid of the French Co-operation, but they needed considerable up-dating. In Equatorial Guinea, maps had been produced by the CUREF programme (Conservation and Rational Use of Forest Ecosystems) financed by the European Community. Recently, however, the governments of the states involved understood that heightened transparency would enable them to control this crucial economic sector better and to regain the confidence of western countries and funding agencies. One after another they prepared to equip themselves with forestry information systems that could accumulate and compare data from the various departments, in order to have a true "instrument panel" of the forestry sector, based not just on the state of the forests, but also, perhaps especially, focused on the allocation of concessions, taxes and currency revenues.

Areas exploited

Industrial forestry operations were confined to coastal regions until 1960. Progressively, however, they invaded the entire western part of Central Africa, as in Gabon and Cameroon (FIGURES 9.22-23 and 25-26). Today, much of the exploitable forest in Gabon, Cameroon, Equatorial Guinea, the Congo and the Central African Republic was thus allocated to exploitation (FIGURE 9.24). Because of difficult terrain, vast swamps or sometimes prohibitive logging and logistics costs, some parts of the concessions might never be logged. On the other hand, logging operations are being conducted or have already been carried out inside some protected areas. In Cameroon, Lake Lobéké Reserve and Campo Ma'an National Park have been logged. In Gabon, all the reserves have been exploited except Minkébé National Park. Altogether, few of the forests in Central Africa have not yet been, or will not be, modified by logging.

The only country where wood exploitation has declined over the past ten years is the Democratic Republic of the Congo. In 1988, logging still affected an area of 21 million hectares. At present, it does not involve more than eight million hectares, at most. Because

9.25-26. *In Gabon, the area allocated to exploitation reached about 12 million hectares in 2000. More than 80% of non-protected forests, or around 75% of all forests, have thus been allocated. Today, the last forest blocks still intact in the north-east and east have also been allocated. In Cameroon, the area put into exploitation reached 17.3 million hectares in 2002. As in Gabon, this activity has moved to the interior of country and today the last old forests are under attack.*

The maps on this page have been prepared by Global Forest Watch (GFW), an initiative of the World Resources Institute (WRI). The forest cover is that of the TREES programme of the European Commission's Joint Research Centre. The delimiting of concessions in Cameroon is from the Ministry of Water and Forests; in the Central African Republic (CAR), from the Natural Resources Management project; in the Democratic Republic of the Congo (DRC), from the Permanent Service of Forestry Management (SPIAF) (these data are incomplete, however); in Equatorial Guinea, from local partners of Global Forest Watch (GFW); in Gabon, from the Ministry of Water and Forests, from the Official Journal, from the World Conservation Monitoring Centre (WCMC) and the World Wilde Fund for Nature (WWF); in the Congo, from the World Resources Institute (WRI) and the WWF. The protected areas (dark green) are from sources of the Central African Regional Programme for the Environment (CARPE), GFW, WCMC, Wildlife Conservation Society (WCS), WWF and ECOFAC (Conservation and Rational Use of Forestry Ecosystems on Central Africa).

9 GREAT CHALLENGES

of the war, many concessions have had to be abandoned or their activity greatly reduced. In April 2002, however, there was talk of 33 million hectares being granted to Zimbabwe in compensation for its military aid.

Production

In parallel with the expansion of forest exploitation, the volume of timber produced has obviously increased. For the ensemble of the forest massif, it is estimated that production has increased threefold since 1960 (FIGURES 9.27).

According to the FAO, production in 1997 indeed reached just over 12 million cubic metres of industrial timber, only to drop to 11 million by 2000. To that must be added all the timber being traded outside of the industrial circuit, however.

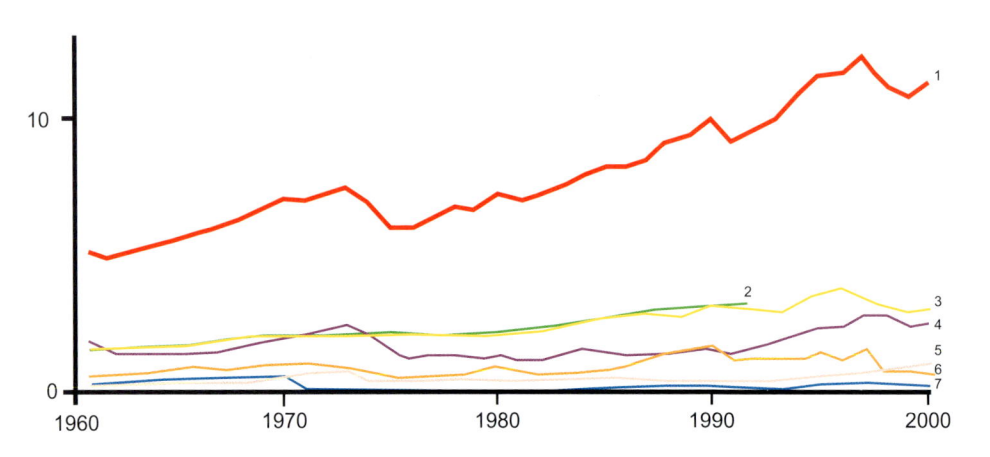

9.27. Evolution of the production of industrial round wood from 1961 to 2000 in six countries of Central Africa, expressed in millions of cubic metres.
(1) total volume.
(2) the Democratic Republic of the Congo
(3) Cameroon
(4) Gabon
(5) Central African Republic
(6) Congo
(7) Equatorial Guinea
(Source: statistics database of the Food and Agriculture Organisation of the United Nations (FAO), ftp://apps.fao.org)

The species of trees logged

The forests of Central Africa have more than 1,000 species of trees – depending on what one calls a tree – but hardly more than 100 are exploited and most of the wood that is produced comes from a small number of species. Thus, in Gabon, Equatorial Guinea and southern Congo, 65 to 75% of the volume of timber produced is Okoumé or Gabon Mahogany *Aucoumea klaineana*, a pale reddish-brown wood, soft and light in weight. Formerly, it was used to make cigar boxes. Today it is mainly used in the manufacture of plywood. The rest of the production comes from "various woods", a collective name for some fifty species among which are Ozigo *Dacryodes buettneri*, Moabi *Baillonella toxisperma*, Padouk or Barwood *Pterocarpus soyauxii* and various Caesalpiniaceae known under the name of andoungs. In southern Congo there is also *Terminalia superba*, known as Limba.

In Cameroon, in the absence of Okoumé, exploitation involves about 80 species. In the south-west, much Azobé *Lophira alata* is logged, but, in the south-east, the two most important species are Ayous or Whitewood *Triplochiton scleroxylon* and Sapelli *Entandrophragma cylindricum*. The former species alone accounts for nearly 50% of production, the latter for about 10%.

In the forests of northern Congo and the Central African Republic, not very different from those of Cameroon, the main species exploited are Sipo *Entandrophragma utile,* Sapelli, Ayous and the bahias *Hallea (=Mytragina) spp.*. In the Democratic Republic of the Congo, it is Sapelli, Sipo, Tola *Prioria balsamiferum*, Iroko *Milicia excelsa*, Afrormosia *Pericopsis elata*, Tiama *Entandrophragma angolense* and Wenge *Millettia laurentii*.

In forests located far from the ocean only wood with a high commercial value is worth logging and exploitation is thus not intensive. Closer to the ocean, distance is not a constraint and logging is intensified. As the best timber disappears or becomes rare and the international demand increases, the number of species taken also increases. It is for this reason that in Gabon Okoumé represents no more than 70% of production while 20 years ago it was as much as 80-90%. Today, species are logged – the kapok tree *Ceiba pentandra*, among others – that no one would have thought of using only five years ago.

Organisation of the forestry domain

While in Uganda the first forestry regulations had already come into being by the beginning of the 20th century, laws regulating forest exploitation in the Congolese region did not exist before the end of the Second World War, or after independence. Because of the frequent irregularities and growing pressure on forest ecosystems, all the laws concerning forestry operations are in need of review or elaboration, however. Those presently in force date from 1990 in the Central African Republic, 1993 in the Congo, 1994 in Cameroon, 1997 in Equatorial Guinea and 2001 in Gabon.

Without going into detail – which would be beyond the scope of this work – most countries have divided their forests into a permanent domain and a non-permanent domain. The first comprises protected forests and production forests that may belong either to the state or to the communes. The second includes forests destined for conversion, including community forests located along roads or around villages. In Gabon, the forest domain is divided into two zones of exploitation: the first covers the coastal area and is reserved for Gabonese exploiters; the second comprises the rest of the country. In addition, these two zones are crossed by the *Zone d'attraction du chemin de fer*, made up of large concessions allotted in 1972 and 1973 for the financing of the Trans-Gabonese Railroad.

Forest exploitation is subject to authorisation by the State and is reserved for residents of the country or for companies registered in the country. In Equatorial Guinea, logging can only be done by private companies, of which, however, the State remains principal stockholder.

Apart from this, logging can be practised in various ways. Industrial logging is conducted on large concessions, limited to 50,000 hectares in Equatorial Guinea, 200,000 hectares in Cameroon and 400,000 hectares in Gabon, but is unlimited in size elsewhere. In the north of the Congo, there are thus concessions of 1.2 million hectares. These concessions are granted for a renewable period of 15 years in Cameroon and in Equatorial Guinea, for 30 years in Gabon and for variable periods in other countries. The granting of a concession imposes the preparation of a management plan that must be approved by the competent services of the ministries in charge of forests. In Cameroon, the Congo and Equatorial Guinea, it also imposes the construction of a processing plant: a sawmill or a rotary peeling unit.

In Cameroon's non-permanent domain, exploitation can be done by means of a *Vente de coupe* (sale of standing volume) on a maximum of 2,500 hectares, *Permis d'exploitation* (logging permit) for a maximum of 500 cubic metres or an *Autorisation de récupération* (logging right, allocated by volume) for a maximum of 30 cubic metres. In Gabon, there is the *Concessions forestières sous aménagement durable* (forestry concession under sustainable management), CFAD, applicable to areas of at least 50,000 hectares, the *Permis forestiers associés* (associated forestry permit), PFA, concerning areas no larger than 15,000 hectares when they are included in a CFAD or 50,000 hectares when they are managed independently, and the *Permis de gré à gré* (mutual agreement permit). The latter two are granted only to nationals and, in all cases, the holders of concessions may not exploit more than 600,000 hectares. There are also community forests planned. Their exploitation is subject to a simplified management plan and one or several contracts to supply local transformation companies.

Whatever they might be, these regulations are constantly bypassed and irregularities in the granting of concessions are innumerable: allocation procedures are often obscure, legal texts are subject to different interpretations; limits are often imprecise and overlap or encroach on protected areas; areas sometimes exceed legal limits; conditions regarding the eligibility of concessionaires are not respected and measures that should accompany the granting of concessions are not fulfilled. To complicate matters, the companies which behave correctly in one country do not always do so in a neighbouring country.

The concessionaires

Knowing who, exactly, exploits the wood of Central Africa is not always simple. Many concessionaires are no more than figureheads and cover influential personalities or companies. In all of the countries of the region, however, a few exploiters hold most of the area allocated.

According to Global Forest Watch, Gabon thus had 221 exploitation permit holders in 1997, but 13 companies held 50% of the area allotted and five companies held 35%. These five exploited areas ranging from 487,000 to 700,000 hectares. These companies, for the most part registered under Gabonese law, were subsidiaries of European companies. In Cameroon, 479 exploitation companies were registered in 1998-1999, but there were no more than 84 valid exploitation permit holders, of which 25 held 75% of the allocated areas. By means of subcontractors, some large companies controlled even more than the official figures show. Many small companies were no more than subsidiaries of large companies. In the Democratic Republic of the Congo, a single company in 2001 held an area of 2.9 million hectares, or more than a quarter of all the concessions in the country.

The fact that much of the forests of Central Africa are in the hands of large European and Asiatic companies worries international NGOs working for tropical forest conservation. The European companies especially are closely watched by these NGOs, but they are surely not the worst. The many Asian, Libyan and Lebanese

9.28. *The felling of a tree, in the past done with an axe, is today done with a chain saw. It is not only dangerous for the cutters, it is also a highly delicate operation. It therefore requires a high degree of professionalism. Before attacking the tree to be felled, it is necessary to eliminate as much as possible the lianas that attach it to neighbouring trees. Next, it must be decided where to make it fall so as to cause the least possible damage to the surrounding trees. Finally, if there are any buttresses, they must be reduced so that the tree will not split when it falls. Only large companies therefore are in a position to organise and finance the necessary training of personnel, as is presently being done by several companies in Gabon and in the Congo. In spite of everything, the profession of tree cutter in tropical forests remains one of the most dangerous in the world according to the International Labour Office (ILO).*

The ways of exploitation

Until a few years ago, most exploitation was of the "mining" type. The great majority still is. The companies invested as little as possible into the planning of their activities. They improvised, relied on an anarchical or unorthodox transportation network and a total lack of knowledge of the resources. Through wastefulness and futile efforts, these logging operations ended up being costly and their effects on the forest and the environment in general were disastrous. This situation went unnoticed because there was no long-term vision and, once the forest was skimmed off and the game consumed, the company moved on, leaving behind a huge mess.

And yet, although we may not know a lot about the population dynamics of a few of the commercial species, sustainable logging techniques have long been known and are beginning to be applied. They unfortunately require a relatively large initial investment of men and of means, which is recouped only in the long term. But the socio-political events which have succeeded each other for forty years in Central Africa, and which persist in places, have not and do not reassure serious investors or real professionals about the timber industry. Also, most companies have pursued a single objective: immediate profit. Their shareholders would not tolerate any other.

Some companies were nevertheless trying to work more correctly by limited operation planning. They first inventoried existing resources and possible obstacles by surveys of the concession conducted along existing paths, elephant tracks and a few machete-cleared trails. They thus recorded exploitable trees in an area that represented at most five percent of the total area of the concession. Next, they determined annual logging areas according to the volume of timber to be produced, decided by the government.

The situation has evolved over the past few years. Under the pressure of governmental demands expressed through new legislation, and especially of international NGOs and funding agencies, more and more companies have understood that in order to sell timber they must launch into sustainable forestry operations and take certification into account. They have also understood that, if they destroy their present concessions, they will not obtain new ones.

One after another, little by little, large European companies are thus launching into the elaboration of more detailed management plans, based on representative inventories of resources and a real knowledge of the habitat. Since such plans will govern logging throughout the life of the concession, they must locate and delimit sensitive areas not to be logged and unexploitable areas, define annual logging zones and conceive the implantation of infrastructures (headquarters,

9.29. The loading of felled trees is now done with the aid of more and more efficient machine-tools which has considerably increased the potential of exploitation and its yield.

Today, some companies still do not hesitate to abandon vast concessions after having skimmed them off, to go and attack still virgin forests. Several, and not among the least, have thus left eastern Cameroon to settle in the north of the Congo. Today, some companies still do not hesitate to abandon vast concessions after having skimmed them off, to go and attack still virgin forests. Several, and not among the least, have thus left eastern Cameroon to settle in northern Congo. Independently from these extreme cases, many "foresters" minimize the forestry laws, and some only pretend to respect them. Written reports and reality are often very different, but only a narrow collaboration between progressive companies and realistic NGOs will promote transparency and confidence in the forestry sector.

companies are much more fear inspiring. They are totally out of control, and even Greenpeace could probably have very little impact on them. The few medium-sized companies held by politically influent local people are also out of reach. The fact that large tracts of forest are managed by a single company brings about abuses and threatens "equitable" sharing of revenues coming from forestry operations. In practice, however, it facilitates the implementation of management plans and the control of logging. Problems encountered by conservation and the rational exploitation of forests thus often come from small obscure companies with no great means in finances or in competence, and no interest in the sustainability of activities. In any case, even if it is true that some large companies behaved or still behave incorrectly, it must also be recognised that it is within these large companies that the first changes were observed in favour of sustainable exploitation. In fact, only the large companies possess the technical and financial means to develop it. Finally, it is easier to negotiate with a small number of powerful actors, as is the case in the northern Congo, than with many people without means, condemned to seeking quick profits.

housing for the personnel, temporary camps, timber yards, processing facilities and permanent roads). They obviously must also take into account the forestry regulations in force. In Equatorial Guinea, for example, operations are prohibited on slopes exceeding 45 degrees, as well as along watercourses and roads; in addition, management plans should include infrastructures for the local population, such as a health centre, a church, a house for a schoolteacher and a school.

The elaboration of such a management plan rests inescapably on a good cartography of the concession, including watercourses, existing infrastructures, villages, possible cultural places to respect (sacred woods, cemeteries), obstacles such as steep slopes, rock outcrops, inselberge and swamps. With the development of new technologies – especially satellite based cartography and positioning systems (GPS) – new possibilities are available, but they are still expensive and require an expertise that most companies do not yet possess. These techniques put off "traditional" foresters and that is why consultancy firms specialised in management are coming into being. They have well trained teams who work according to standardised methods. Resource sampling is performed on plots spread along rectilinear trails, spaced four to five kilometres apart across the entire concession, forming a compass-oriented grid.

Such sampling covers 3 to 5% of the area of the concession. With the superposition of aerial photos or satellite images, the potential of the concession can be estimated with precision. Nowadays there is also computer software available for analysing the forest cover

9.30. *Transport by road done with the aid of powerful logging trucks that can theoretically carry 30 tons of wood, in practice between 25 and 30 tons depending on the specific weight of the wood transported. Many countries limit the load to 20 or 25 tons, however, in order to protect the road network.*

9.31. *For a long time, watercourses played a preponderant role in Central Africa in the removal of wood from the interior to the coasts. The Congo and Ogooué rivers have carried enormous quantities of wood in the form of floating rafts. This means of transport, still used, is not suitable for light wood, such as Okoumé, however. Heavy woods must be transported on barges.*

9.32. Much of the wood – 30 to 70 % of the marketable volume– is abandoned in the forest. This waste constitutes a very serious ecological problem. This waste comes in part from the techniques used, but in good part also from the fact that the European market is interested only in top quality and that, for "scrap", the local market is limited.

and automatically drawing the zonation of the forest into plant formations – what the foresters call stratification. Based on this management plan, the concession is divided into annual logging sectors, each in their turn being the object of a complete logging inventory, with the location of all the trees of a given species larger than a given diameter, together with their dimensions and their quality. This allows the preparation of a detailed map, the verification of sensitive areas not to be exploited, the drawing up of a secondary road network, the location and marking of trees to be logged and finally an optimisation of the layout of the log-removal tracks and the timber loading yards.

The annual logging plan is therefore a very precise document, defining in detail how each sector is to be logged, what will be its production, who will be in charge of operations, exactly when they will take place and what will be the means necessary to implement the work. It is prepared from year to year by a specialised team that does only that.

As for the logging itself, it is done by cutters (FIGURE 9.28), followed by teams who cut the felled trees into sections and load the logs (FIGURE 9.29) to be brought out to timber yards were they will be loaded onto trucks and transported to processing facilities or to the coast to be exported (FIGURE 9.30).

The low density of trees felled – one to three per hectare – makes planning indispensable to logging operations in order to limit unnecessary costs. Despite precautions and good intentions, the waste is enormous. Depending on the species, it is estimated that 30 to 70% of the volume that can be theoretically commercialised – the amount of wood between the buttresses and the first large branches – of trees felled are never used (FIGURE 9.32). Some trees shatter when they fall. Others are not re-found or prove to be – after being felled! – hollow or twisted. Still others are lost during river floating and end up piling up on the beaches.

Just as in the allocation of concessions, irregularities and infractions arising from the way they are exploited are innumerable. Some unscrupulous exploiters do not hesitate to "poach" timber from protected areas or neighbouring concessions, sometimes even across borders. In 2001, a logging track was thus discovered coming from Equatorial Guinea and plunging into the forests in North-Gabon. In Cameroon, over half the concessions not yet allocated have been logged. Apart from these extreme cases, irregularities go from falsification of management plans to fictive or illegal transport of timber by including protected trees and under-dimensioned trees, erroneous labelling of logs and felling of trees in prohibited places, such as on slopes that are too steep or along watercourses. Globally, illegal production is therefore enormous. According to various sources, it represents at least 20 to 30%, if not 50%, of the total production.

Transformation

Until only a few years ago, wood from Central Africa was exported almost entirely in the form of raw logs. For the past 20 years, a portion of the timber harvested is sawn on the spot (FIGURE 9.33), peeled or sliced, in order to increase local revenue and employment opportunities. Also, most governments have taken financial and legal measures to stimulate or require obligatory local transformation. In Equatorial Guinea and the Congo, 60% of timber must therefore be transformed. In Cameroon, exportation of raw logs has been banned altogether since 1999, except for some species, said to be promotional.

Numerous rotary peeling factories and sawmills have thus come into existence. Their anarchic multiplication in itself constitutes a threat to forests and can run completely counter to sustainable exploitation. In the last resort, trees must be felled in order to run the factory. This has already happened in Indonesia, where transformation capacity is four times greater than production, and in Ghana, where it is five times greater. The transformation capacity of factories in Cameroon has soared from 1.2 million cubic metres in 1993 to 2.67 million in 1998, setting off a "race for timber" in which every means was permitted. As for Gabon, the risk of reaching this extreme is imminent.

The transformation of timber on the spot on the very edge of, or even inside concessions, complicates the tracing of products and permits the felling of anything: under-dimensioned trees are hastily peeled or sawn during the early hours of the day, without leaving a trace and before any check can be made.

Despite all efforts, on the scale of Central Africa much timber continues to be exported in the form of raw logs, simply because transformation by rotary peeling or slicing produces better quality and less expensive products when done in Europe where much more sophisticated machines are used and where labour is more skilled and efficient. Factories set up in Africa are very simple. They are usually old factories brought from Europe, taken apart and reassembled at lower cost in Africa, without dryers or industrial carpentry workshops. They permit only mediocre valorisation of the material, which results in a serious disparity between the quality of the transformation and the quality of the raw logs to be transformed.

9.33. *Elements of a sawmill in Cameroon.*

9.34. *The port of Owendo in Gabon has become one of the main shipping points for industrial wood in Central Africa. Because of the absence of a deep water harbour, ocean-going vessels remain anchored in the estuary. Logs are brought to the ships by barge to be loaded. These operations slow down loading and add supplemental costs.*

9 GREAT CHALLENGES

9.35. *The main destinations for wood exported from Central Africa in 1999 (according to Forests Monitors).*

9.36. *In community forests, not only felling, but also sawing on the spot is done with chain saws. This technique is a disaster for the output of wood, because it generates four to five times more sawdust than sawing with a traditional band saw.*

Who buys the wood?

Thirty years ago, most of the wood from Central Africa left for Europe. Subsequently, a growing proportion – up to 30 % in 1996 – was absorbed by the Asian market. With the economic crisis that struck Asia in 1997, the countries of the European Union again became the main buyers. According to Forests Monitor, they thus absorbed 64% of the wood exported in 1999, of which 19% was bought by France, 13% by Spain, 12% by Italy and 7% by Portugal. China comes next with 29% of the market. At scale of the producing countries, these percentages vary considerably, however. China thus bought 69% of Equatorial Guinea's production and 46% of Gabon's production. Portugal bought 46% of the production of the Democratic Republic of the Congo and 30% of the production of the Congo. As for Spain, it bought 64% of the Central African Republic's production. These figures, however, are in constant evolution and it remains to be seen whether the part of the market monopolised by China, Malaysia, Japan and India will continue to increase.

Once more, these statistics are to be taken with caution, however. In Cameroon, for example, many raw logs exploited in this country leave the port of Douala with a certificate attesting that they come from the Central African Republic.

Community exploitation

Forestry laws in force in Cameroon, like laws recently adopted in Gabon, provide that village communities, assembled into associations, may manage forests of a maximum area of 5,000 hectares. These community forests, which are not to be confused with communal forests, began to take shape since mid-2000. The communities involved thus obtain a legal right on all products from the flora and fauna that is in their forest. With the free assistance of the Ministry of the Environment and Forests, however, they must produce a simple management plan for the different resources they propose to exploit.

In order to test the viability of this concept, the European Commission financed a pilot project of community timber exploitation near the Dja Faunal Reserve. Long sawing is indeed a regular, but completely informal activity in Cameroon. It offers products affordable to all layers of the population on the urban markets. Because of this, demand is very high, and so is the damage in consequence. The pilot project therefore attempts to understand and rationalise this activity at the socio-economic level as well as at the environmental level.

Trees are felled and sawn on the spot with the aid of chain saws. Next, the planks are carried to the village to be worked and stored on the men's backs. When the stock reaches a certain number of cubic metres, it is transported to the market at Yaoundé. The proceeds of the sale is split up among the participants according to pre-established allotments. On the basis of a 30 year rotation, 500 cubic metres of timber can be removed each year. In the perspective of the creators of community forests, revenue that they generate could constitute an alternative to revenue from commercial hunting. They amount to 10,000 CFA francs per cubic metre (around 15 euro).

It is probably too early to evaluate the results of this pilot programme, but Willy Delvingt's Gembloux team has taken an interest in it, not because the allocation of communal forests represents an end in itself, but in order to evaluate the extent it might improve the present management of forest resources by village populations. The exploitation of community forests could indeed ensure a revenue supplement to villagers and at the same time even reduce the pressure that they exert on game and palm oil (PAGE 273), since hunting as well as making palm wine have a commercial aim.

Present laws should, however, be reviewed and improved to better integrate the socio-cultural characteristics of the populations involved. In particular, legislation on hunting – it is completely unrealistic and obsolete – must be reviewed, management of natural resources must be redefined and the community forest concept must be inserted in the framework of territory management.

9 GREAT CHALLENGES

In practice, the setting up of community forests confronts considerable obstacles: wood exploitation is not a traditional activity of the populations concerned who consequently have no knowledge in this field; rational management of a community forest does not correspond to traditional political practices and community forests run the risk of becoming a political stake. Also, Willy Delvingt raises the question of the village populations having the required competences in fields as varied as financial management, engineering and forestry technics. The solution is still far away, therefore, but Delvingt estimates that, in spite of everything, community forests could represent an alternative and an interesting supplement to the methods of industrial exploitation that prevail today in the forests of Cameroon.

Before drawing the final conclusions, it is necessary perhaps to see the progress of other community forest initiatives, not only in Cameroon, but also in Gabon.

Impact on the flora and fauna

In spite of low-intensity exploitation, the forests of Central Africa are subjected to degradation often out of proportion with the volume of wood extracted, although their canopy may appear relatively undisturbed when viewed from above (FIGURE 9.37). In reality, degradation affects not only the structure and composition of the plant cover, but also the hydric regime, the soil and the fauna of the forest.

The heavy equipment used, especially bulldozers, for opening up log-removal tracks and truck-loading areas, compact the earth, destroy the thin layer of humus, create quagmires, accentuate erosion and in general destroy the forest much more than necessary. In addition, drivers often haphazardly plough passages through the understorey and execute many unnecessary manoeuvres. In the Lopé National Park in Gabon, Lee White has thus observed that the various tracks occupy 11% of the area of exploited sectors. Besides, nearly 17%, was occupied by the crowns and branches of felled trees. Altogether, nearly 30% of the forest area was therefore disturbed. The creation of tracks in uneven terrain can also provoke erosion (FIGURE 9.38) which may sweep away stripped surfaces, pollute the hydrographic system and transform the black waters of rivers into muddy water. As for loading yards, they are in general so compacted or eroded that it takes much time for vegetation to recolonise them (FIGURE 9.39). To these damages is added the dumping of used oil from equipment during maintenance.

Felling itself creates openings which tend to be larger than natural tree-fall gaps, on average. Many young trees, sometimes even large ones, are also crushed or pulled down by lianas attaching them to the tree being felled. In principle this damage can be limited by cutting away constraining lianas as much as possible, but it is not always feasible, and this operation represents a supplementary cost that practically no company is willing to pay. What is more, good tree cutters, trained professionally, capable of orienting the fall of a tree and avoiding possible shattering, are rare.

Altogether, the understorey therefore suffers much greater degradation than the canopy. Selective extraction changes species composition and the removal of giant trees, 300 to 400, even 800 years old or more, constitutes a skimming process that leaves no hope of reconstituting the heritage removed. The logging of some species also leads to their local extinction. Thus, *Parinari excelsa* of the forests of the Albertine Rift are not regenerating at present. Their extraction cannot be followed by regeneration and this species is condemned to disappear from places where it was logged. Extraction of large emergent trees could also reduce the numbers of some epiphytes, as well as lichens, ferns and orchids. The excessive proliferation of lianas and twining plants can, on the other hand, slow down the sprouting of secondary species, while new openings and tracks constitute excellent means of entry for foreign, often invasive, species (PAGE 307-308). Lastly, exploitation diminishes flower and fruit production while increasing leaf production.

Selective exploitation also brings about fears of a reduction in the genetic diversity of species exploited by eliminating the "best" and the healthiest individuals. This phenomenon is known in temperate forests and has already been observed in some tropical forests. It

9.37. *Seen from above, the felling of a tree makes only a small opening in the canopy.*

9 GREAT CHALLENGES

9.38. Tracks cut into steep slopes on clay soil and rapidly cause erosion liable not only to carry the soil away but also to seriously pollute the hydrographic system.

could affect populations of Okoumé, Moabi and Sapelli, to mention just a few.

Independently of these immediate consequences, there are also effects that do not show up until after some length of time. During the very first years after logging, the mortality of trees not felled is practically doubled. This phenomenon probably affects trees that were injured, often at the root level, during the passage of machines, and hence are more sensitive to fungal attacks. Moreover, along the mantle of the forest block, forests freshly exploited are also more susceptible to fire and more sensitive to tornadoes, but this phenomenon does not seem to be, or not yet be, a threat in Central Africa.

The effects of exploitation on the fauna are very diverse. Some species of large mammals, if they are not already disturbed by hunting, seem attracted by recently exploited areas and can therefore increase locally. It is true of the elephant, and also, in some places, of buffalo, gorilla and some forest duikers. Other, more specialised species, diminish in number, however, or even disappear from exploited areas. The chimpanzee seems very sensitive, as are most of the arboreal monkeys. Very considerable changes are also observed among the small mammal fauna, notably in the Central African Republic by Jay Malcolm and Justina Ray, and in Kibale National Park, in Uganda, by Tom Struhsaker and his team. The latter even considers that the upsetting of the micromammal fauna could be an important factor in alterations of forest composition after logging. Rodents indeed play a considerable role in seed dispersal and seedling predation.

Immediate reactions by birds are variable, but studies conducted in other tropical regions, notably those of Jean-Marc Thiollay in Guyana, show that exploitation reduces the number of species by about 30%. In Central Africa, studies of this type are rare. In Kibale National Park in Uganda, Christine Dranzoa was nevertheless able to show that although the diversity and richness in species of birds was higher in exploited areas of the forest than in non-exploited forests, most of the birds of the exploited areas belonged to generalist species. What is more, seven out of 48 species living in the understorey of old forests did not reappear after 23 years.

Independently of these findings, permanent exploitation roads – wide to allow rapid drying of the surface after rains – fragment the forest and facilitate access. There is therefore an inevitable increase in hunting pressure as much by personnel involved in the exploitation as by hunters, usually commercial, coming from elsewhere. It is indeed very usual to see workers going out in the field armed with hunting rifles and to see trucks loaded with game returning. This practice suits some companies by reducing the quantity of meat that must be carried in to their lumbering sites. Increased hunting pressure thus represents very probably one of the gravest misdeeds of forest exploitation.

Forest exploitation does not only affect vertebrates. It also has an impact on arthropods. This phenomenon has not been thoroughly studied in Africa, but in Southeast Asia, Andreas Floren and Eduard Linsemair found that disturbance engendered by exploitation affected not only the species but also the functional web of the entomological fauna. While communities of insects in old forests are very rich and of largely random composition, those of degraded forests are much poorer, but more predictable, because they are dominated by a few species linked to secondary forest species. After exploitation is terminated, the reconstitution of rich communities is very slow. We do not have comparable data for Central Africa, but preliminary observations in the Lopé National Park in Gabon seem to show that, even 25 years after the end of all exploitation, the butterflies of the understorey, notably the various species of genera *Euphaedra* and *Bebearia* (PAGE 186), are still not the ones that inhabited the old forests. And yet, at first view these forests have taken on the same appearance they had before logging.

Finally, logging also facilitates invasion of the forest by alien species. In very early stages this can be considered as an enrichment of the biodiversity. Later, after several logging campaigns, it becomes a major threat to the local fauna and flora. The Fire Ant *Wassmannia auropunctata* is one of the most dramatic examples (PAGE 308).

Social impact

The onset of a new forest exploitation inevitably attracts human populations, not only because it offers direct employment opportunities, but also because the new villages and camps that spring up in the vicinity represent opportunities for merchants and smugglers. Forest exploitation thus stirs up populations, locally increasing their density and sometimes exacerbating tensions between social or ethnic groups. As for the original populations of the region, they often enter into conflict with the forestry companies.

In some cases, these conflicts are justified because companies do not respect ancestral rights or customs. At other times, they originate from the fact that companies are regarded by the populations as a solution to all their problems. Whatever social actions the companies may undertake and whatever material aid is brought to the populations, they always want more. This type of conflict is often stirred up by local politicians who excite the populations out of pure demagogy, at times in the hope of gaining advantages themselves. The populations are therefore clearly caught between the companies on the one hand and their elite on the other hand and it can be asked by whom they are truly taken hostage.

For some environmental NGOs, forestry companies do not do enough on the social level and it is true that many do nothing at all, remunerate their personnel badly or house them in deplorable conditions. It can be asked, however, to what extent must forestry companies substitute for governments. Of course, they should pay the various taxes linked to their activities, all the more so since they have never invested in the creation of the forests that they exploit and that in Central Africa the charges imposed on them are much less than in Asia or in Amazonia. They can in this way contribute to the development of the countries concerned and to the human communities. But why should they be implicated in sectors of activity that are not their own?

Sustainability

All of these impacts are real, but what is their significance in the long term? This question is all the more difficult to answer because tropical forests are not only very complex ecosystems, they also react very slowly or with a considerable time-lag to some pressures and it is therefore difficult to determine in the long term which damages will be irreversible and which will not.

This incertitude leaves the way open to two completely opposed attitudes. For some, including scientists, tropical forests are very ancient, composed of specialised and therefore very fragile species, intolerant of any form of exploitation. For others they are perfectly "tolerant". Those of the latter opinion support their argument on the fact that forests are subjected to continual regressions and transgressions, that their structure is constantly influenced by minor but frequent events and that the impact of Man, perceptible for the past few thousand years, only adds to a series of other pressures.

According to this opinion, forests are therefore extremely dynamic ecosystems and the many species that constitute them even depend on these disturbances. Forest exploitation thus should be much less catastrophic than is imagined. Most of the impact cited above is transitory and could be reduced by more rigorous planning. It is from this school of thought that the concept of reduced-impact exploitation arises, a condition of sustainability.

This obviously does not mean to say that exploitation has no impact at all. Indeed, whatever is done, an exploited forest is inevitably transformed and it would be dishonest to pretend, as some do, that cutting down or poisoning large emergent trees, cutting away lianas and opening up 30 or 40% of the canopy – the three essential preliminary phases of sustainable exploitation according to some schools – does not alter the biodiversity of a forest. The sustainability of forest exploitation does not rest on the suppression, real or imaginary, of impact. This option belongs to protected areas and it must be acknowledged that only large national parks are capable of conserving vast natural ecosystems and some large-sized animal species, even if some conserva-

A dilemma?

Large forestry companies open the forest to hunters and introduce considerable disturbances, both in biological and in social terms, often totally disproportionate to the volume of wood extracted. It is therefore to be expected that they would do everything possible to minimise their impact. But to what extent should they substitute for the state, whose presence inside the forest massifs is most often virtual, and take charge of education, development, health care? Is this not relapsing into paternalism, so decried in the case of the Belgian Congo?

9.39. *In loading yards, criss-crossed by tracks of elephants and machine-tools, and strewn with left-over wood, the soil, deprived of its surface layer and compacted, requires much time to recover.*

The twelve stages of sustainable management of a forest concession

1. A first inventory covering 1 to 1.25% of the concession area, and obtaining data on 150,000 to 250,000 trees per 100,000 hectares.
2. Drawing of maps showing the spatial distribution of biodiversity and wood resources.
3. Socio-economical studies.
4. Studies on biodiversity, forest dynamics...
5. Drafting of a management plan covering a time period of 20 to 30 years, in function of the production, protection and social promotion objectives.
6. Drafting of 5-year exploitation plan.
7. Implementation of a detailed exploitation inventory.
8. Analysis of the data, and drafting of an annual logging plan.
9. Logging with reduced impact techniques.
10. Organizing the follow-up of the exploitation and tracability of the wood.
11. Implementation of all other activities.
12. Follow-up of the whole process in view of certification.

tion strategists or experts in human science believe that there is no longer a place today for this approach. Sustainable exploitation rather consists in reducing impact to acceptable proportions so that the forest conserves its essential functions, in wood production as well as in non-timber forest products, conservation of biodiversity, carbon storage and hydrothermic regulation.

Thus, to begin with, it is advisable to define what are the functions that should be preserved and what is an acceptable degree of transformation. This exercise is all the more important since, in the long term, a forest never again produces as much wood as during its very first exploitation – many even wonder about the profitability of a second or third passage 30 or 40 years after the first – and which aspects other than wood production play an important role. The introduction of sustainable exploitation encounters many obstacles, however, and, despite efforts to demonstrate that a reduction of environmental impact is compatible with an increase in the profitability of the enterprise, "wood mining" remains the most widespread form of exploitation.

Why? Some exploiters, especially among the small and middle-sized enterprises, do not see the interest of sustainability simply because they believe that their exploitation sites will be converted into fields. Many companies, holding concessions in the permanent forest domain, do not have, on the other hand, any idea of the damage they cause and are not ready to change methods, especially since their present benefits are considerable. They are thus frightened by the supposed high cost of management which they readily contrast with the insecurity of investments and the uncertainty of the real duration of their exploitation. Indeed, in Africa the State can go back on its engagements with impunity and transfer a concession from one exploiter to another. How, then, can an exploiter be convinced to leave a large seed tree standing, to renew inadequate equipment or to invest in the training of his personnel and in his management plan? All he is looking for is a rapid return on his investment.

As for governments, they have asserted since 1990 and on many occasions that they want to exploit their forests in a sustainable way. Their new legislation confirms this intention. Nevertheless, they fear a lack of earnings because of the lack of immediate results and they hesitate to promulgate financial incentives. It is true that the profit margin available to them is already relatively small. In any case, for lack of personnel and motivation, they are mostly completely incapable of enforcing the laws that they proclaim, or else these laws are simply bypassed by the functionaries themselves.

In Central Africa, the setting-up of low-impact exploitation is hindered by the hazards of the market, the large number of species and the increasingly hostile relations with rural populations. Exploiters may thus find themselves confronted with real dilemmas. As for example, secondary trails are normally to be re-closed when exploitation is terminated, but sometimes, on government order, they must be kept open for the local people with the consequences they bring: commercial poaching and clear-felling of forests bordering trails to establish plantations.

In short, as Nick Brown of Oxford University emphasises, foresters must abandon harmful practices and start to see the forest as an ecological ensemble with multiple and complex functions – which is slowly beginning to happen. "Conservationists" must cease to regard foresters as the cause of all misfortune that hovers over tropical forests and acknowledge that a forest, "rejuvenated" but well managed, is still worth much more than wasteland and can play a considerable role in conservation.

In this spirit, the workshop organised in October 2000 at the Lopé National Park by the International Agency for the Development of Environmental Information (ADIE) represented the culmination of a process that had gone on for some years, notably in the framework of the Conference on the Ecosystems of Dense Humid Forests of Central Africa (CEFDHAC). Although it targeted only one very particular aspect of reduced-impact exploitation – the control of hunting in concessions – it brought together conservationists, agents of the ministries, representatives of funding agencies and representatives of the timber industry.

In practice, the application of reduced-impact exploitation techniques requires two types of approach. On the one hand, knowledge of these techniques must be made available to all who are to apply them. On the other hand, persons in charge of exploitation companies must be convinced of the utility of these techniques. Without these two approaches, it is unlikely that the techniques can be applied in the field. This is why ADIE, in partnership with FAO and with financing from the European Commission and the French Fund for World Environment (*Fonds Français pour l'Environnement Mondial*, FFEM), has undertaken a reduced-impact forest exploitation development project (DEFIR) that essentially consists in popularising known techniques by producing a manual adapted to the specificities of Central Africa and by organising training programmes.

Certification

For more than 20 years, biologists have studied the effects of destruction by forest exploitation, conservation NGOS have inundated the world with their apocalyptic figures concerning the regression of tropical forests and methods of sustainable exploitation have been perfected but rarely applied. Also, to force the timber industry to show more comprehension, the process of certification was designed. It guarantees to the consumer that the wood certified comes from a sustainable exploitation.

In reality, this process covers three very different aspects: sustainable management of forests, transformation of wood and traceability of forest products. It therefore affects economy as well as biology and involves exploiters, states, local communities, wood industrialists, biologists, conservationists and the general public. At the international level, the WWF and Greenpeace instigated the creation of the Forest Stewardship Council (FSC) in charge of certifying the sustainable management of forests. The International Standards Organisation (ISO) also perfected a series of environmental standards that could be applicable for foresters. At the regional level, the African Timber Organisation (ATO), in collaboration with the Centre for International Forestry Research (CIFOR), defined and tested some of the principles, criteria and indicators of sustainable management of forests in Central Africa.

In order to evaluate the real impact that eco-certification could have, it is necessary to know that the commerce in tropical woods represents only 1% of the world commerce in wood and that most of this quota is assured by Southeast Asia. Central Africa's part represents only 5%. After the Asian economical crisis at the end of the 1990s, a large proportion of wood from Central Africa was again exported to Europe, but in the United Kingdom, the Netherlands, Germany, Austria and Belgium, the total percentage of wood certified was no more than 20% and, in the United States, it did not even reach 10%. Elsewhere, notably in Latin countries, the public is not yet prepared to pay for certified wood. In other words, certification will not change everything, even if more and more transformation industries are refusing to use wood not produced from sustainable exploitation. The process has its limits: if certified wood costs too much, it runs the risk of being replaced by other materials. Finally, wood certification is confronted with total indifference on the Asian markets and massive competition from timber originating from New Zealand and Chilean plantations or from Siberian

9.40. *Logs are bought in storage yards at the shipping ports by specialised buyers who can judge their real quality. Between the felling of a tree in the forest and its storing at the port, many things can indeed happen. These buyers are usually very independent. They reside in the exporting countries and offer their services to interested companies. They have a long experience with wood and know the habits and demands of these companies.*

forests.

Nevertheless, the costs entailed for enterprises by certification must be recouped through savings made by sustainable management. Christian Fargeot and Alain Penelon, who studied this problem in the framework of the ECOFAC programme, believe that for an enclaved country like the Central African Republic, these costs could constitute a real obstacle, considering the present weakness of the market.

It is therefore advisable to be conscious of the present limits of certification policies and not to overestimate their consequences on forest conservation, all the more so because the different systems of certification hardly take into account the impact of exploitation on flora and fauna. The most complete system, that of the Forest Stewardship Council (FSC), takes into account ten main criteria. Most concern exploitation itself, three concern "indigenous peoples" and manpower, but none really touches on the basic questions that concern the fauna. The principles used are general and target only rare, threatened or endangered species, not the immense commerce that affects the common species and which finally causes most of the damage. Game is seen more as a possession of the indigenous peoples than as something having an intrinsic value. The criteria of the International Tropical Timber Organisation (ITTO) and the African Timber Organisation (ATO) are no better and take into account only protected areas and rare species. They thus disregard 90% of the forested area and 95% of the species.

This situation is explained by the fact that organisations involved in the definition of these criteria are comprised mostly of foresters and sociologists, but very rarely of biologists, not to mention conservation biologists. But in conservation, as in medicine, it is generally better to prevent than to cure. The workshops organised under the auspices of ATO, notably in Cameroon and Gabon in 1999 and 2000, to discuss principles, criteria and indicators of sustainable exploitation, thus brought together masses of civil servants most of whom had never waded about in the mud of logging extraction tracks. Among those gathered, biologists, other than a few teachers, were very rare. Without denying the necessity and value of these exercises, it can be asked – apart from the problem of their application – whether they truly invisaged the forest in all its aspects as their texts assert. According to Markku Simula, a Finnish expert who has looked into the principles, criteria and indicators of ATO, it seems that even these had not been taken fully into account for the certification.

Finally, there remains a last problem. An increasing number of African countries tend to question the principles, criteria and indicators established by northern countries and want to draw up their own standards of certification. This can be understood to the extent that systems established in the "north" do not necessarily take into account all the specificities of tropical forests and of the populations that inhabit them. In reality, this argument is probably no more than a pretext and it raises the question, to what extent do these countries wish to remain judge of their own actions? But if each country develops its own system, the exercise will end up being worthless.

The role of foreign financing

The extraordinary expansion of industrial exploitation in the forests of Central Africa responds obviously to the needs of the international market, but it also reflects the pressures exercised on the states by various multilateral or bilateral creditors, especially the International Monetary Fund (IMF) through structural adjustment. This expansion would never have been possible without the supply of foreign capital. The European Union and some Member States thus furnish, in the name of sustainable development, considerable financial assistance which represented, in 1998, 66% of the world contribution of aid in favour of tropical forests. Given the colonial history of several of the countries in Europe, the largest part of this aid was concentrated on Africa. In 1995, the will to come to the aid of the management of tropical forests was formalised by a protocol on forests which was added to the Lomé Convention. This provides support to the conservation and sustainable management of forests and insists on the necessity of putting wood from managed forests on the market. Among projects financed within this framework, is the ECOFAC regional programme (PAGE 294), as well as several actions responding to geographically more local needs, such as support for the community forests of Cameroon, or thematically, such as support for the creation of a system of forestry certification. Part of the aid contributed by France serves to support private companies in the preparation of their inventories and management plans.

As for the World Bank, it has supported various programmes, notably the Forests and Environment Programme (*Programme Forêts et Environnement*, PFE) in Gabon. These have contributed to reforms in the forestry sector, the revision of forestry laws in Cameroon and Gabon and the elaboration of a system of forestry information in Gabon. More recently, the African Development Bank, through the ADIE, supported development of such systems in Equatorial Guinea, the Central African Republic, the Congo and the Democratic Republic of the Congo.

9 GREAT CHALLENGES

Artificial plantations

In some parts of the world, much of the wood exploited comes from plantations. In Central Africa, there are a few plantations, but their production is largely irrelevant.

Most are based on exotic species. In the Pointe-Noire region of the Congo, there are 70,000 hectares of eucalyptus plantations and on the Batéké Plateau, near Kinshasa, there are 8,000 hectares of it for the supply of firewood. In the Albertine Rift region, also, especially in Rwanda and Burundi, there are tens of thousands of hectares of plantations of eucalyptus, *Acacia mearnsii*, *Acacia melanoxylon*, *Pinus patula*, *Grevillea robusta* and *Callitrys calcarata*. In these severely deforested regions, plantations compensate the disappearance of natural wood resources, but their effect on the soil can be harmful, notably because conifers contribute to their acidification. In terms of biological diversity, these planted forests are veritable deserts. Despite this, they ease pressures on the last natural forest blocks, allowing their protection. Consequently, in Rwanda as well as in Burundi, the natural forests have been surrounded by exotic plantations.

Plantations of local species have not had much success. In Gabon, however, several attempts have been made with Okoumé (FIGURE 9.41). In Rwanda, trials were done on the edge of Nyungwe forest with *Podocarpus*. Despite some promising results, they have not been pursued or extended. The causes of this abandonment lie in several factors. To begin with, the development of tree farming is a very long-term activity and it is probable that no experiment has been let to run long enough. Then, we do not know well enough about the behaviour of local species and their parasites and predators. Also, many trees planted in pure stands finally do not produce anything. Finally, it is also probable that as long as there are natural forests in sufficient quantity, no one will invest in plantations which even in a tropical milieu, require decades to produce.

Some day it must be thought about, however, and, as in the case of the last montane forests of the rift, it is entirely possible to imagine that only artificial plantations will be able to ease pressure on the natural forests, for which sustainable exploitation is not likely to satisfy the need for wood.

9.41. Plantations have been tried in Gabon, notably based on Okoumé. Most of these experiments began during the colonial period and those that have been continued are often carried out in collaboration with research institutions in Europe. Through their colonial history, many European countries have, in fact, acquired and developed a great knowledge of tropical forestry. Some countries were or remain mainly focused on Latin America or Asia, but the United Kingdom, France, Belgium and the Netherlands are interested in Central Africa, and continue to be active there. It is thus that the University Faculty of Gembloux (Belgium) participates in research programmes on forest ecosystems in the Congo basin, as much in the field of their management as in their conservation.

9.42. *A tourist face to face with a group of mountain gorillas in Volcanoes National Park in Rwanda.*

Tourism

Tourism is by definition one of the few activities allowed in national parks. Without being their main raison d'être, tourism is a non-extractive means of developing those immense territories, of generating much needed revenues and, above all, of making nature known. While being the "spearhead" of conservation, national parks are therefore expected to also play an important role in education. The experience of eastern Africa and southern Africa indeed shows that national parks that are regularly visited are better accepted by neighbouring populations.

Tourism in Central Africa

On a world scale, tourism is at present the number one industry, as much in overall turnover as in number of jobs generated, but its distribution and growth remain very unequally distributed between the continents. In spite of Africa's considerable potential, both in beach tourism and in nature tourism, the continent receives only about 3% of world tourism and Central Africa's share is infinitesimal.

This deficit is largely caused by the sensitivity of this economic activity to socio-political events. Large-scale world events, such as the Gulf War in 1991 or the terrorist attacks in New York in 2001, of course provoke an immediate slowdown. In addition, problems arising in one country of the subregion chase tourists from neighbouring countries, in a radius of a thousand kilometres. The Rwandan war thus had repercussions, not only in Uganda but also in Kenya and even in Zimbabwe. More recently, the repeated outbreaks of Ebola fever also became a problem in Gabon and northern Congo.

Before 1960, nature tourism was based on the large savanna parks visited essentially by expatriate residents. In the Belgian Congo, tourists were barely tolerated, subjected to strict regulations and permitted access to only some parts of the national parks, such as the Rwindi Plain and the high slopes of the Rwenzori in Albert National Park (now Virunga National Park). The English developed a more pragmatic approach and their efforts to develop African nature engendered the safari business. This took off in Kenya and in northern Tanzania during the 1960s. It was based essentially on the spectacular fauna of the great plains, in particular lion, leopard, elephant, rhinoceros, buffalo and giraffe. Its success stimulated similar initiatives in northern Cameroon and in the Central African Republic, but the ex-French countries of Africa remained better known in the field of hunting tourism. Until the 1970s, nature tourism in the national parks of Central Africa therefore remained essentially oriented to the savannas and their large mammal fauna. The only exceptions were the high slopes of the Virunga volcanoes, the Rwenzori Mountains and Mount Cameroon which attracted a few hikers more interested in sport than nature. The forests and their fauna remained completely outside the circuits.

This situation changed around the end of the 1970s in the Great Lakes region with the very controversial habituation of a few families of mountain gorillas in Rwanda's Volcanoes National Park. Diane Fossey and her co-workers had indeed found that these "monsters" were in fact peaceful in character and could be habituated to human presence. The gorillas thus very rapidly became a first choice attraction (FIGURE 9.40). The film *Gorillas in the mist*, which retraced the life of Diane Fossey, murdered in 1985, made them a world-wide attraction. Soon, Volcanoes National Park became one of the rare protected areas in the world that truly brought in money.

The annual number of visitors was, however, limited and the demand for visits to the gorillas very rapidly exceeded the supply. In order to increase the number of clients, operators then relocated to Zaire, not only to nearby Virunga National Park, but also to Kahuzi-Biega National Park. In the 1980s, it was thus possible to see two races of Gorilla on a single trip. Around 1990, some ten groups of gorillas were habituated in the Kivu region, which made it possible to show them to nearly a hundred visitors a day.

With the war in Rwanda between 1990 and 1994 and the one begun in Zaire in 1996, this tourism was halted. Around the middle of the 1990s, however, two groups of gorillas were visible in Bwindi Impenetrable National Park in Uganda and, in 1995 and 1996, two other groups were again accessible in Rwanda.

A further and more serious outbreak of troubles in the Volcanoes region in 1996 and the war in the Democratic Republic of the Congo obliged the Rwandan authorities to re-close Volcanoes National Park. Finally, the murder of eight tourists at Bwindi by the Rwandan genocide perpetrators in 1998 put an end once more to tourist activities in the region.

The development of tourism based on the gorillas struck the imagination of those who worked in protected areas. Thus came the idea of habituating groups of chimpanzees. Trials were attempted in Virunga National Park in Zaire and in Kibira National Park in Burundi, but without much success. Finally, the only place outside of Gombe Stream National Park in Tanzania where it is presently possible to observe chimpanzees in a natural state with some certainty is Kibale National Park in Uganda. Chimpanzees are much more mobile than gorillas. Their groups are less stable, they spend most of their time in the canopy and are considerably more aggressive.

Around 1988-1989, a group of 200 to 250 Angolan Pied Colobus monkeys was also about to become an attraction in Nyungwe Forest in Rwanda. It had been habituated to human presence within the framework of a project supported by the New York Zoological Society. Nyungwe Forest was thus included in the regional tourist circuits.

During the same period of time, and parallel to the development of tourism based on primates, ornithological tourism was developed. In Europe and in the United States specialised agencies were formed which sent dozens of visitors in small groups each year. In addition to savanna parks or humid habitats, these people also regularly visited the forests of western Uganda, Nyungwe Forest, Kahuzi-Biega National Park and the Irangi Scientific Reserve 120 kilometres north-west of Bukavu (PAGES 280-281).

From the end of the 1980s, on the eastern side of the Central African forest massif, a promising tourism based essentially on gorillas, but also with a mixture of savannas and forests, was thus developed. However, this tourism depended on operators based in Kenya. Visits to the gorillas were in fact only an extension of Kenyan safaris. For agencies in Nairobi, the gorillas enabled their infrastructures in Kenya to make a profit and it was therefore out of the question to let clients stay too long in the Great Lakes region.

Nowadays, the prospects of re-launching tourism in this region are poor, but the experience acquired could help the western regions of Central Africa to start up similar activities. The protected areas of Gabon, the

For or against the habituation of gorillas?

Some conservation biologists, like Conrad Aveling, Anthony Harcourt, Craig Sholley, Kelly Stuart, Amy Vedder and Bill Weber, to cite just a few, have shown or tried to show that tourism based on the Gorilla could be an enormous asset for the conservation of gorillas and even for conservation in general. Others, including Tom Butyinski, seriously question the wisdom of this tourism, especially if the gorillas visited belong to very small populations as is the case of Mountain Gorillas. Among the negative points that they raise, some are related, however, to all tourism centred on nature and protected areas: the sensitivity to political events, the small percentage of benefits returned to local populations and the possible negative economic repercussions on these populations, such as rises in the price of basic foodstuffs. Other points are specific to gorilla tourism. Thus, habituated groups are more vulnerable to poaching than non-habituated groups. But the greatest risk probably lies in the transmission of human diseases to gorillas or diseases of gorillas to visitors. In the Virunga Mountains, there have been cases of epidemics of bronchial pneumonia or of measles among the habituated gorillas. On the other hand, gorillas are carriers of a herpes virus which could be dangerous for Man.

It is therefore true that tourism based on Gorillas must be submitted to very strict regulation of the number of visitors and the manner in which visits are conducted, as well as strict control of the guards, guides and visitors, notably sanitary controls. For example, gorillas must, at all cost, be prevented from touching visitors and vice versa. Visits must be kept short enough to avoid disturbing the daily activity cycle of these strictly herbivorous primates. These rules exist everywhere, but they are not easy to enforce and numerous infractions have been observed in Zaire as well as in Rwanda and Uganda. As long as Gorilla conservation does not take precedence over economic considerations, as tourism based on gorillas is not subjected to a rigorous scientific control and the share of benefits from this tourism going to conservation is not greater than in the past, Butinskyi proposes to stop all activity in this field. According to him, it is necessary to stop altogether tourism based on the Mountain Gorilla, whose population is too small. From a scientific and rational point of view, these remarks may seem justified. In reality, they should not be exaggerated and some observations show that habituated gorillas reproduce better than others. In the present context of the Great Lakes region, these precautions are even somewhat esoteric and it might have been preferable to run the risks of tourism than those related to war. In any case, gorillas cannot be protected forever in a glass case. The gorillas of the Virunga Mountains must be habituated to more frequent contacts with Man or disappear. But, above all, it must be kept in mind that, if Volcanoes National Park survived the war in Rwanda, it is thanks to the presence of the Mountain Gorilla and, above all, the fact that it had become a "marketable commodity".

9 GREAT CHALLENGES

island of São Tomé, the national parks of Dzanga-Sangha, Odzala-Kokoua and Monte Alén are an ensemble of high quality and varied attractions, extending from the Atlantic coast to the swamp forests of the Congolese basin and offering several opportunities to see gorillas or elusive animals like the bongo. The development of tourism in this part of Africa, however, confronts a series of problems. There are of course good hotels in the large cities and even some infrastructures in the protected areas, notably the hotel at Lopé National Park and the various camps set up by ECOFAC at Odzala (FIGURE 9.41-42), Monte Alén (FIGURE 9.43) and Lopé (FIGURE 9.44). Apart from that, much remains to be done.

In order to remedy these things, governments must be convinced of the interest of tourism. Given that the inception phase is long, that it is necessary to dispose of considerable financial means, possess the professional capacities required and train the necessary personnel, potential developers must feel supported and have a minimum of guarantees. Governmental attitude is therefore very important, but not really perceptible in most of the countries of the subregion, apart from São Tomé, perhaps. However, Gabon seems to be willing to have a try at it and several initiatives are under way. Several countries have also acquired overall tourism

9.43. In French-speaking Africa, the installation of tourist camps inside protected areas was traditionally allowed with much reticence. This attitude was regrettable, particularly in the forested regions, because many things can not be seen except at sunrise or sunset. Happily, Ekanya camp in Odzala-Kokoua National Park is inside the forest and, even allowing for a little rest after a hard day, a visitor can see many things while relaxing.

9.44. The only way to reach this place in the middle of nowhere is by pirogue from the base camp at Mboko. The trip lasts around seven hours, but the journey itself is well worth it. It enables not only the spectacle of the inundatable, and often inundated, forest, but also observation of, notably, the African Slender-snouted Crocodile Crocodylus cataphractus (Page 198). As for Mbomo camp, it is constructed on the edge of a finr forest gallery where it is possible to observe de Brazza's Monkey Cercopithecus neglectus (Page 215).

projects and some funding agencies seem ready to back the sector.

How can private companies be convinced, however, to invest in this sector in Central Africa? Without even considering the socio-political risks, competition with eastern Africa and southern Africa must be taken into account. Having developed their tourist activities over a long period of time, the countries of these subregions offer a whole range of varied and high-quality products at relatively affordable prices. Therefore, in order to carve out a place in this international market, not only must quality products be developed at correct prices, but products not on offer by other countries must be offered. Some attractions may thus be completely valid for residents of the countries concerned, but not for international tourists coming from Europe or the United States.

Besides, a result of this competition is that tourists become more and more difficult to please. Not only must they have their daily dose of impressions, sensations, new species or photos, but lodging and catering must also correspond to their wishes. Someone who buys a Chicago-Libreville ticket has a right to demand something more than the sometimes doubtful inns to be found along the roads in Central Africa.

Tourism in the forest

If infrastructures account for 50% of the success in nature tourism, the remaining 50% depends essentially on the support offered to the visitors, at least in the forest. Developing nature tourism in the forest is indeed not easy: operators must not only know about tourism, they must also have a good knowledge of the forest. In Central Africa, nature is not easy to sell. Biodiversity is not an attraction in itself and the closed forest habitat requires a completely different approach than for the open grasslands of eastern and southern Africa.

Any visitor to the savanna can take the wheel of his vehicle and drive the tracks in hope of seeing the main elements of the fauna, but forested habitats are disconcerting for the non-initiated. The vegetation is opaque and all the trees look alike. The sound ambience is silent or dominated by insects, amphibians and birds that one never sees. The ground is often muddy and, even when the rain has stopped, water continue to drip form the canopy. Invisible insects sting and the unexpected encounter with an elephant or a buffalo can go very badly.

It is therefore unthinkable to let inexperienced tourists venture all alone into the forest. Not only do they risk getting lost or being bothered by some large animal, but as a result of their visit they will swear to never again set foot into the forest. Tourists must therefore imperatively be accompanied. Therein lies another problem. Since usually nothing much is seen, the guide must be able to keep up the interest of the visitors. He must be both a good communicator and have an excellent knowledge of nature. He must be able to speak not only to the ignoramus but also to visitors who perhaps know more than he does. The fact that one never knows in advance what is going to be seen in the forest makes this problem all the more acute.

The training of guides is therefore extremely important. It requires much time and should also include theoretical as well as practical training. In fact, at the moment, the only place where there are well trained and experienced guides is in Kibale National Park in Uganda. The reason is simple: this forest has harboured a research station for over 30 years.

Tour operators

By guiding tourists, tour operators in Europe and America who send clients to Africa play a fundamental role, but they will not be convinced of the interest of some destinations until local operators are able to win their confidence. The development of tourism therefore relies on many factors that are often inter-related.

In Central Africa, there are not yet operators able to offer high quality products in nature tourism. The profession is in its debut. In order to make a company that organises nature tourism in African forest or savanna national parks, prosper, an operator must be good both in business and as a naturalist. These qualities are rarely found together in the same person. "Merchants" therefore must consent to investing in the hiring of good field naturalists. By not doing so, they are condemned to copying the products of their neighbours, who themselves are already copying others. There is therefore rarely room for innovation and discovery, biological diversity, rare species and the real riches of nature.

9.45. At Monte Alén National Park, in Equatorial Guinea, a comfortable "lodge" has been set up in the middle of the forest as part of the ECOFAC programme.

9.46. *Ololo camp was constructed by an* ECOFAC *support programme. Located along the Offoué River, just outside Lopé National Park, it sits on top of a small hill rising between the river and a swamp where elephants often spend the night. At the end of 2001, it welcomed its first customers.*

Ecotourism

The colossal development of tourism since the Second World War has, in many places, entailed distortions and damage. In developed countries, these are mostly environmental. In developing countries, the enormous economic disparity between foreign visitors and their hosts more often leads to socio-cultural or economic problems. Ecotourism was created in reaction to this. Its aim is to establish a near-symbiotic relationship between tourism, peoples living around protected areas and nature. By integrating conservation and development, it tends to link the different actors and nature in a sustainable way while stimulating the wellbeing of the populations concerned. In practice, ecotourism should therefore stimulate the development of indigenous communities by supporting conservation.

The idea is good, but David King and William Stewart, in an evaluation of this problem, think that the possibility of protecting both Men and nature has not yet been demonstrated. It is true that ecotourism, like tourism in general, makes it possible for conservation to generate funds, but the fact that culture and nature become marketable commodities often has negative effects. Ecotourism thus requires a thought-out and experienced approach, a real involvement of the populations concerned, monitoring at all levels and a real determination on the part of the operators and the states.

In Central Africa, ecotourism is much talked about – often without even knowing what it is – but there are not yet any realisations really worthy of the name. On the contrary, relations between tourists and local inhabitants often leave much to be desired. Some Africans from the villages resent tourism as an intrusion in their life. Others, especially in the ministries, see tourism as a form of neo-colonialism, even a means of maintaining underdevelopment. The behaviour of some tourists unfortunately tends to prove them right. Their lack of tact, notably in habits of dress or the obsessional way of photographing everything – whether or not desired by the people – creates conflicts and often even ruins chances for later positive contacts. But the behaviour of local populations often has something to do with it. The extreme poverty in which most of the forest people live means that amiability and spontaneity easily give way to commercial harassment. Cottage crafts and art degenerate. Spontaneity is replaced by theatrics and traditional performances are transformed into masquerades.

In fact tourism, especially nature tourism, does not yet belong to African culture, and most Africans have still no understanding of what tourism really is or should be. They have no idea of the expectations of a tourist. They cannot even believe that somebody comes to Africa just in the hope of seeing some birds or some monkeys. Many believe that behind that kind of tourism there must be something else – a hidden target – and every American tourist is potentially a CIA agent.

If tourism is to help conservation, contacts with the populations must be guided. Man himself must not become merchandise. Besides, many naturalist visitors are bothered by the idea of coming to see not only the gorillas, elephants or birds but also humans.

Apart from socio-cultural problems, tourism can also cause strictly environmental damage. In the great majority of Central Africa's national parks, however, this danger is still far off or non-existent. Only the Afro-alpine habitats are fragile in this perspective. In parks in the forest, the fact of having to walk and the closed nature of the habitat mean that only a restricted category of visitors ventures there. And even then, if tourists trample a few flowers, the damage is nothing compared to what the elephants do.

Research

Throughout this book, we have seen that scientific knowledge of the forests of Central Africa is often insufficient and that this sometimes seriously hinders decision-making in conservation and development.

The distribution of species, of plants as much as of animals, remains poorly known because vast regions have been little explored. The behaviour of many species remains unknown, as well as relationships and interactions between species. Lastly, many species are not yet described or named. In order to understand ecosystems and to protect them, shouldn't one begin by knowing the species that make up part of them? How can we know what we are talking about otherwise? Too often the distribution of species reflects only the distribution of collectors. As for example the number of plant species found in a one-degree square. Obviously there are two richer areas: in the west, Lower Guinea, including Southeast Nigeria, West and South Cameroon and Gabon; in the east, the region of the Albertine Rift. In between the vast forests of the central Congo Basin are much poorer, except just one place : the area around Kisangani, which seems to be nearly as rich as most parts of Gabon or South Cameroon. The only reason for that is that Kisangani had a university, and used to be visited by many researchers who went out in the field.

Therefore much work remains to be done. How much is not easily estimated, but an example from Gabon gives an idea of the immensity of the task. The forests of Gabon are reputed among the richest in all of Central Africa, but are still poorly known. Attaining a good level of knowledge of their flora would require the collection of at least 500,000 specimens more than the 65,000 already preserved, a colossal work which would necessitate considerable means and many years to complete. In addition, the collection of specimens is just the first step. They must next be identified, the new forms have to be described, and the findings have to be published. Considering the increasing lack of taxonomists – taxonomy is no longer in fashion –, these steps are problematic.

Although the forests of Central Africa, including those of the central parts of the Congo Basin, were perhaps better known than those of tropical America before 1960, and although technology today enables advances that were unimagined hardly 20 years ago – including teledetection, satellite imagery, genetics, bio-

9.47. *A team of young Gabonese, on reconnaissance in the Minkébé National Park, rests along an elephant track which follows a ridge.*

9.48. *Long walks in the forest, heavily loaded, across rivers and creeks, are not tourist hikes*

Research?

For many people, research is an obscure activity, highly academic and reserved to an elite of "sorcerers" whose deep motivations and results remain a mystery. In reality, it takes in very different activities including field reconnaissance, observation and collection of samples for very technical analyses and, especially, reflection. Its ultimate goal consists in answering questions and proposing solutions to problems. It therefore implicates actors of very diverse disciplines and levels.

9.49. This male gorilla at Odzala has depigmented patches on his face which are probably caused by a mycosis. These large primates are sensitive to many human diseases. As national parks become isolated, populations of some species reduce and contacts between animals and humans will become more frequent, monitoring of some species by veterinarians will be required.

chemistry and physics –, Africa has now accumulated a considerable lag. And yet, actors in the field of research are not lacking. Norbert Gami and Charles Doumenge, within the framework of the FORAFRI programme, have listed them. We will keep to the main points. At the international level, there is the Centre for International Forestry Research (CIFOR) and the International Centre for Research in Agroforestry (ICRAF). At the national level, several western countries possess research centres and institutes which work in close collaboration with institutions in tropical countries. In France, there is the *Institut de recherche pour le développement* (IRD), the *Centre international de recherche agronomique pour le développement* (CIRAD) and the National Natural History Museum (MNHN). In Belgium, there is the Royal Museum for Central Africa (MRAC) and the National Botanical Gardens at Meise; in the United Kingdom, the Royal Botanic Gardens, Kew; in the Netherlands, the Wageningen Botanical Gardens and, in the United States, the Missouri Botanical Garden.

In Central African countries also, independently from universities, there are institutions in charge of research. In Gabon, there is the *Centre national de la recherche scientifique et technologique* (CENAREST) which supervises more specialised institutes, such as the *Institut de recherche en écologie tropicale* (IRET), the *Institut de recherches agronomiques et forestières* (IRAF), the *Institut de pharmacopée et médecine traditionnelle* (IPHAMETRA) and the *Institut de recherches en sciences humaines* (IRSH). In Cameroon, there is the *Institut de recherche agronomique* (IRAD) and the *Jardin botanique de Limbe*. In the Congo, there is notably the *Délégation générale de la recherche scientifique et technique* (DGRST). In Uganda, the Uganda Institute of Ecology can be mentioned.

Among these institutions, some have stations in the field, including, notably, at Ipassa near Makokou, in Gabon, which depends on the IRET (FIGURE 9.52), at Lopé, which depends on the *Centre international de recherches médicales de Franceville* (CIRMF), at Lwiro, on the shore of Lake Kivu, in the Democratic Republic of the Congo, or at Kibale, in Uganda, which depends on the University of Makerere.

To complete the picture, non-governmental organisations also conduct or finance research, usually in direct relation with conservation programmes. The Wildlife Conservation Society (WCS) is thus active in Gabon, the Congo and Rwanda, while Conservation International (CI) intervenes at Odzala. Some petroleum or forestry companies regulary support research programmes or furnish logistical support to researchers. This is the case of the *Compagnie équatoriale des bois* (CEB) in Gabon and the *Compagnie internationale des bois* (CIB) in the Congo. As for conservation and development programmes, such as ECOFAC, they also have activities in the field of research, which they subcontract to universities, notably the *Université libre de Bruxelles* (ULB) and the *Faculté agronomique de Gembloux* in Belgium and the *Université de Rennes* in France.

In addition, networks have been created which bring together researchers and research institutions : the International Union of Forest Research Organizations (IUFRO), the European Forestry Research Network, the West and Central African Council for Research and Agricultural Development (CORAF), the Forestry Research Network in Sub-Saharan Africa (FORNESSA), and the African Forestry Research Network (AFORNET).

Independently of this "explosion" of institutions, research in Central Africa has also undergone some evolution. Until the 1970s, research in protected areas, including rare studies in the forest, were the domain of biologists whose preoccupations were highly academic. Much of their early work consisted of collecting and describing species. Later they published detailed species studies, mainly of ungulates, carnivores and primates, and some of their papers remain unique. Man was not part of their preoccupations, however, and habitats transformed by human activity merited very little consideration. This attitude came mostly from the fact that in East Africa – where this sort of research began – the natural world and the human world were usually well separated or at least seemed to be. Even in the montane forests of the Albertine Rift it was like that. However, in descending into the lowland forests scientists discovered a completely different and new situation: under a canopy stretching as far as one could see a forest and humans exist in close relation, to such a point that to separate the natural from the man-made often becomes very unrealistic. Researchers thus had to change their approach. Behaviour studies made way for ecology and the human sciences made their entry. From the study of habitats, one even went to a higher scale: that of landscapes.

In spite of everything, and if we make an assessment at this beginning of the 21st century, we must admit that the results do not always correspond to expectations. To say that nothing was done would be exaggerated, and especially unjust to all those who in spite of everything made considerable efforts. We think notably of the achievements of numerous scientists who worked in the station at Makokou in Gabon from 1962 to 1986, in the station at Kibale in Uganda, in the Virunga in Rwanda or even in the various agronomic research institutions. Even so, Africa accumulated a net lag compared to tropical America and some scientists even think that the apparent lesser species richness of Africa is accentuated by the lack of research.

Of course, the often difficult social, political and economic context, as well as the lack of interest of the western world for sub-Saharan Africa, account for much of this. Also, the first hindrance to the development of research is the lack of funds. Several stations have thus been abandoned or survive with difficulty. And yet, the sums spent on research are not nothing, but they are derisory compared to those allocated to development and above all the billions of US dollars devoured by armament and war. Apart from this general problem that affects all of development on the sub-continent, some more specific factors stand out.

First of all, there is the attitude of large funding agencies. Some dispense large sums into various projects but do not want to hear of research. Just the word research chases them away. Others keep to a neo-liberal logic, which systematically replaces the long term by the short term, insist on a rapid, if not immediate, spin-off. Because of this, there is no longer a place except for what they call "applied" research, and then only on condition of producing results and rapid applications. But the history of research, in all these fields, shows that this activity does not easily let itself be guided and that the most marked progress has often been made along unexpected paths, often by "accident". Even if research has in some way to be oriented by priorities and by the availability of means, it cannot respond to urgent questions if the inventive spirit is not in place.

The weakness of field research in Central Africa is also largely due to the less than desirable involvement of Africans. Rare are those who have been able to acquire a real mastery in the sciences and in nature. Why? First of all because research in the field requires a real interest and much motivation (FIGURE 9.45-47), two qualities that are not improvised. But, in Africa, most students are obliged to accept the orientation proposed to them and few are those who can afford the luxury of choosing their own path. Next, because the natural sciences are poorly regarded compared with political or economic sciences, medicine, or even studies in philosophy. Lastly, years must be spent in the field in order to acquire a good knowledge of nature. But the rare students who nevertheless start on the path of biology, and who accept the difficult life in the field for some time – usually the time to produce a thesis – do not have the means or the desire to do that for years, since only political posts offer real possibilities of professional or social promotion. In countries where poverty is a daily reality, one cannot really blame youth who opt for something other than forest research.

The result is that the number of scientists having obtained a doctorate in natural sciences and who work in the field can be counted on the fingers. This situation is all the more deplorable now that there is a need for good national scientists having enough experience of the natural habitats of their own countries to lead teams in the field, to train new generations and to occupy decision-making posts. But at present too many posts, in training as well as in administration, are still occupied

9.50. Examination of a darted gorilla in Odzala-Kokoua National Park. The appearance of epidemics, such as of Ebola-type fever, arouses controversy in which only science can bring rational answers. These repetitive and unpredictable events, whose spinoff can be catastrophic at once to involved populations, to fauna and conservation, show to what point research should remain a fundamental activity. Today we indeed still do not know the reservoir of this fearsome virus and what provokes its epidemic flare-ups. Primates are victims of it just as we can be, but they are certainly not the source of the disease.

9.51. In the clearing at Maya-Nord a well camouflaged mirador was set up where numerous animals, including several groups of gorillas, can be observed in complete security and without disturbing them. The installation of this mirador, and of a good many others, has also enabled researchers to collect much information on forest animals which are often poorly known.

9.52. In 1963, the Biological Mission of CNRS-France (National Centre for Scientific Research) set up a research station in Makokou, north-eastern Gabon. In 1969, this station was moved to Ipassa where an integral reserve of 10,000 hectares was gazetted. This new station was build on the northern bank of the Ivindo River. In 1979, this station became the IRET (Institute for Research in Tropical Ecology). In 1983, the Ipassa reserve became a Biosphere Reserve. Since the beginning of the activities, more than a hundred scientists of different nationalities have visited the station. An enormous amount of observations and publications was accumulated. By the end of the 1980s activities slowed down, but since early 2002, the station is being rehabilitated by a programme of the European Commission, and the Ipassa forest have been included in Ivindo National Park..

by agents with insufficient qualifications whose only strength is in political support. This in turn is another problem : once these people get a good position their only way to survive is to stop all those who are more competent. Outsiders are most of the time not aware of this kind of problem because they are themselves not able to tell the difference between really competent people and those who just play the role.

Some people obviously wonder what good is research and even say that the money "wasted in research" could be more usefully injected into conservation. The magazine *Canopées* devoted an entire issue to this problem. It is nevertheless a non-problem. To say that the money available should be used for conservation and not for research is a little like deciding only to eat and not to drink. The sad reality is that, overall, money is lacking, that research no longer interests funding agencies and that the rare funds available for conservation are in part diverted to research – and to development – whose results are rarely immediately usable. Banning research, however, would be like covering one's eyes, but it is true that one way to resolve a question is to not ask it. The blind do not have problems with glasses!

Thus, the world is decidedly very strange : hardly 25 years ago, many researchers did not see the interest in conservation; today, to many conservationists estimate there is no need for research! Yet, research is the basis of knowledge, it plays an important role in education and conservation cannot get by without knowledge or high-quality experts. The more we refine our knowledge, the more we measure the extent of our ignorance and the more we realise the enormous work that remains to be done. Of course, it is true that some research does not lead to much or not immediately to tangible results, but this inconvenience is part of the game. In order to research truly efficiently, one must already know the answers to most of the questions. It is therefore up to those who lead teams and orient work to have the flair for "feeling" the paths worth following. If not, how can one ever protect what remains of nature in a world where pressures are ever greater and more merciless. Those who preach pure, hard conservation and figure that research is useless risk finding themselves one day in front of empty reserves.

In practice therefore, research should be allowed to advance on several fronts.

On the one hand, poorly known habitats and regions must continue to be explored. Species have to be collected, and "new" forms have to be described – the game of the name has not stopped! Only this will allow a better definition of the distribution of the biodiversity. Real "hotspots" must be determined and the capacity developed to evaluate objectively networks of protected areas.

On the other hand, key biological mechanisms must be better understood, in particular the dynamics and reproduction of forest trees, their ecological preferences, phenology, nutrient cycles, carbon storage – a subject very much in fashion but for which in the end there are few concrete data. The history of the forests must be refined, notably by studying much more sites in a detailed way. The study of animal population dynamics, interactions between species, their movements and displacements must be pursued. In a larger framework, much more general studies are needed to define the role of forests better, notably in climate regulation.

In addition this research, which could be qualified as fundamental, each protected area needs a team capable of monitoring the evolution of target species, indicator species and habitats in order to provide of objective and detailed data for the assessment of the management of the area and for the orientation of more fundamental research.

Finally, research should also include very pragmatic aspects: in the field of forestry, it should concentrate notably on exploitation techniques, increase of output and plantations; in the socio-cultural field, it should expand knowledge of the very diverse relationships between humans and forest habitats, propose and test methods of resolving the inevitable conflicts in a world in full transformation.

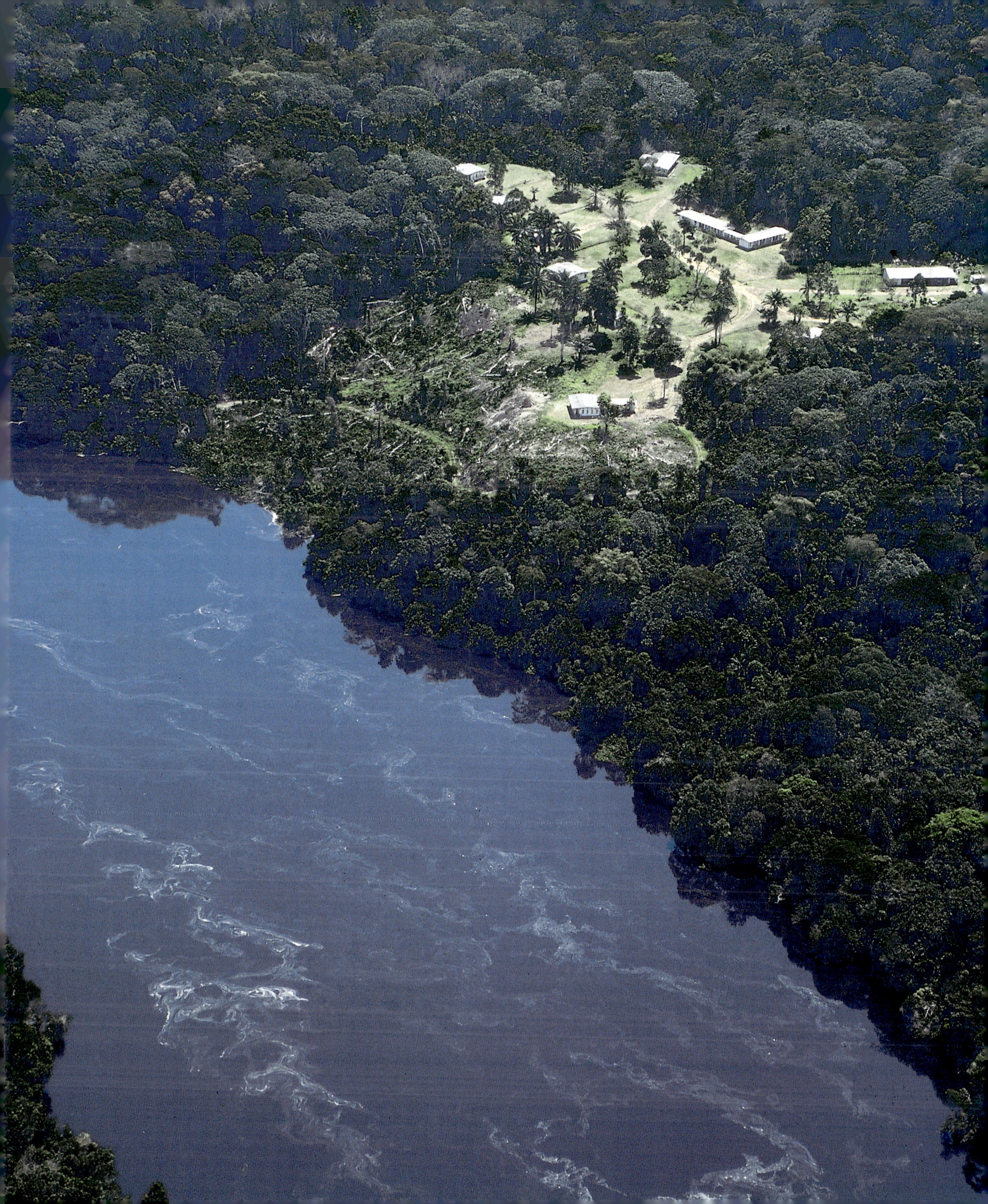

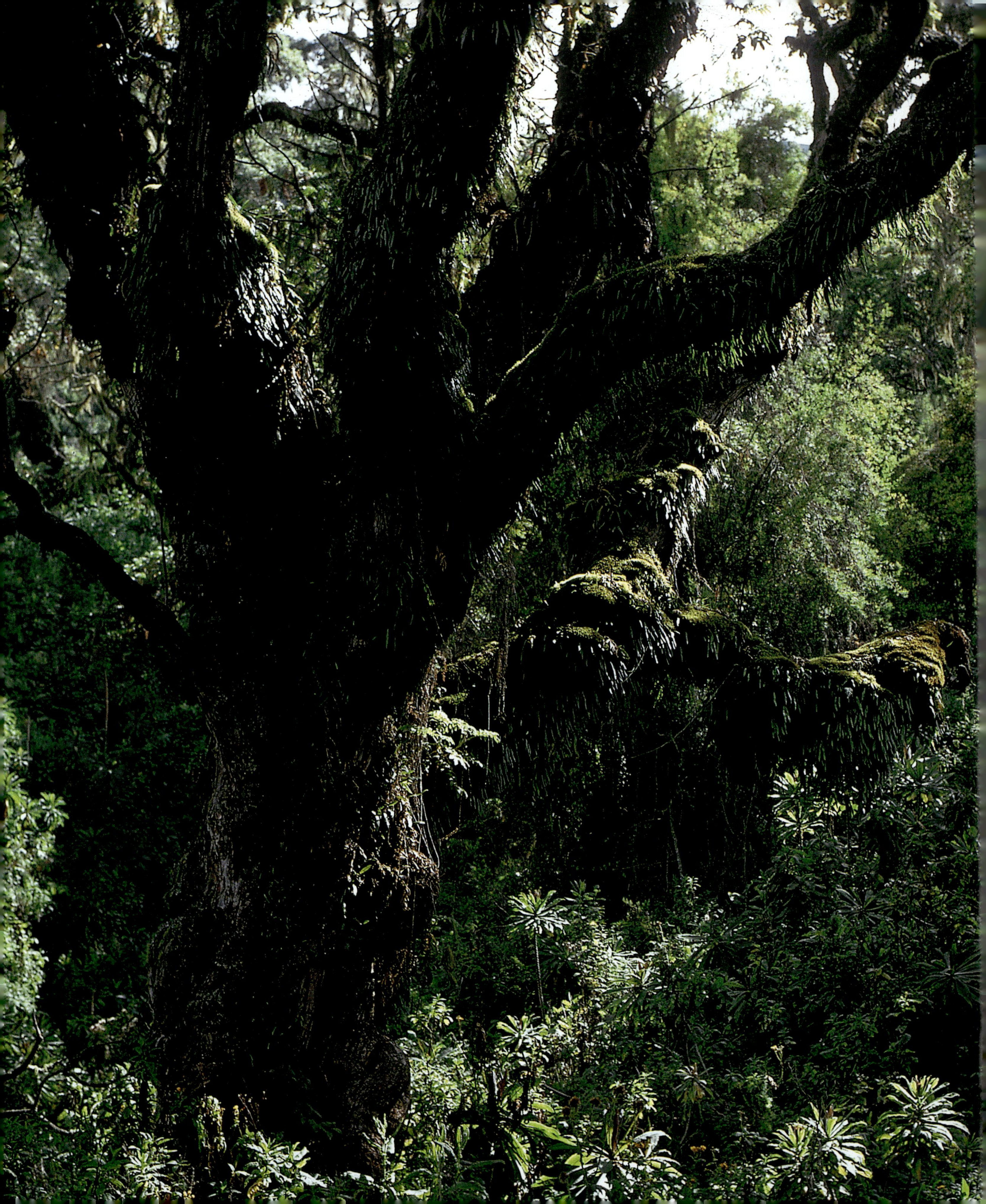

Epilogue

The forests of Central Africa, from the Cross River and the Atlantic to Lake Victoria, constitute the second largest block of tropical forest in the world. Unlike most other tropical forests, they still comprise vast, nearly continuous undisturbed tracts, and the pressures that they are subjected to still seem moderate. This is perhaps only an illusion, however, due to a difference in timing, and it is not excluded that these forests will be destroyed like all other tropcial forests in the world. Human populations continue to grow, indeed, and according to the United Nations Population Fund they will reach nearly 400 million, or more than three times the present population, for the entire region by 2050. How can it be imagined that the vast, sparsely populated expanses of Gabon and the Congo will resist demographic pressures from neighbouring regions, not to mention the 275 million Nigerians.

On the other hand, the world's thirst for the resources of our planet will also continue to increase and the economic context become always more harsh. Inescapably, forest exploitation will intensify, the forest will be fragmented and great expanses will be converted to agriculture, plantations or infrastructures. The silence that is already setting in on many forest tracts is probably no more than the prelude to a much greater degradation. It is indeed to be feared that, with the pressure of politicians and western NGOs, the large forestry concessions granted to foreign companies will be fragmented after the infrastructures have been built and divided between small companies for whom sustainability is no more than talk. Finally, emptied of their resources, the remaining forests will no longer interest anyone and will be converted by a flood of immigrants, as has happened in western Africa.

This scenario, pessimistic, to say the least, will perhaps take some time to be enacted, but why would the forests of Central Africa escape the fate of other tropical forests in the world? A less sombre scenario is nevertheless not necessarily utopian. Numerous examples can be found in the Albertine Rift and the Great Lakes region – regions which at first glance do not have much to do with the deep Central Africa of the Congo basin and the area around the Gulf of Guinea. Forest vestiges in Uganda, Rwanda and Burundi indeed represent no more than 5% of the "original" forest cover of these countries and no tract exceeds 100,000 hectares, but just the same they preserve more than 90% of their original biodiversity. Some animal and plant species have certainly disappeared, but nearly all the large mammals survive – even the elephant – and well over 95% of the birds. Obviously, fragmentation, which frightens many conservationists, is therefore less catastrophic than some imagine – at least for the time being – and most probably only becomes a serious problem in combination with hunting. The only mammals to have disappeared from the forests of these regions are ungulates and larger carnivores, the only species to suffer from hunting pressure. On the other hand, and we have already mentioned it in several places, agricultural development of these regions is such that nature's spontaneous resources represent no more than an infinitesimal proportion of what is used by the populations. Nature conservation is therefore clearly a "by-product" of development.

Montane forests constitute another world, however, and the experience that we gain from them is perhaps not automatically transposable to lowland forests. Much imagination, determination and cooperation will thus be required. Firstly, everyone, including researchers and conservationists, will have to acknowledge that only properly controlled sustainable development will make it possible to save something. The world's history shows that, in the absence of all development, one can observe

10.1. At around 3,300 metres altitude in the Virunga Mountains, forests are extremely fragile and cover only very small areas. Complete protection is the only form of management that is suitable for them.

10.2. The immensity and wildness of some forest landscapes of Central Africa, such as those in the Ipassa-Mingouli rainforest along the Ivindo River in Gabon, contribute to this day to maintaining the myth of the inexhaustible forest. But what will be left of this landscape in 50 years?

an irremediable environmental degradation that, in its turn, maintains and aggravates under-development. Therefore the sentence, often heard in Africa, "if we have to choose between animals and humans, we choose humans" is in fact somewhat meaningless, especially in countries where animals are a basic food resource for most of the people. If the survival of people really depends on the elimination of the last elephants or primates, then the situation is really desperate. In other words, conservation should not be seen as an obstacle to development but as an integral part of it. It will be some time before this will be accepted by many people.

Development planners will have to stop taking away resources from rural populations to feed the urban people and the Western World.

Conservationists will have to accept that a part of the forests be converted and another part be exploited. An increasing part of their work will therefore have to be oriented towards production forests. NGOs, while pursuing their indispensable monitoring efforts and their conservation actions, will have to stop targeting large forestry companies, especially western companies, as the cause of all the evils that befall forests, stop trying to keep forested regions in a state of non-development by opposing, notably, the construction or rehabilitation of roads, stop seeing the salvation of these forests only in the creation of huge national parks that shield them from populations and exploiters. They will, on the contrary, have to exercise more imagination and objectivity, distance themselves from fashions and the politically correct and denounce the dysfunction observed less passionately. Today's technological means allow this.

Forestry companies, including small companies and Asiatic companies, will have to accept a dialogue with conservationists and match their words with their acts.

Scientists will have to learn to work together. Those who work in the human sciences will have to accept that their role is not to keep the peoples of the forests as they are today, but to help them to evolve and find their place in the development process. Biologists will have to understand that conservation biology and fundamental biology are not the same disciplines, and that in conservation human aspects are inescapable.

States will have to realise that forest exploitation has its limits, and that the exploitation of natural forests will never produce all they hope. They should encourage the development of plantations and modified forests to increase wood production, and accept that sustainable forest exploitation is not improvised but, on the contrary, requires a high dose of professionalism. Besides, they will have to ensure that all actors learn to work together. States must show that they have understood all the complexity and gravity of the situation in concretising their declarations of intent through legislation. Finally, at this time of ruthless and irreversible globalisation, they must ensure the world's confidence through transparency and a vigorous fight against corruption and fraud.

Funding agencies, too, will have to understand that conservation is a part, albeit perhaps indirect, of global development strategies. They will have to learn to consider problems on a longer-term basis and acknowledge that research is not an amusement for intellectuals but a preliminary to the development of realistic and sustainable solutions.

As for the peoples of the forest, they will have to accept that conservation is not a milk cow nor a slot machine and that the raising of their standard of living is inevitably connected to production. Otherwise, they will be marginalised, as they themselves have already marginalised the Pygmy hunter-gatherers. A way of life based only on slash-and-burn farming, gathering and hunting is no longer adapted to present conditions and commercial hunting, such as is practised today, is like a scorched earth policy.

This whole process will obviously require much effort and good will, but its achievement will have to fight two main obstacles. Firstly, there is an immense lack of money. Amounts allocated by western countries to forest conservation and sustainable use are given in millions of dollars or euros, while the needs are expressed in hundreds of millions or even billions. These figures seem enormous, but they are just nothing compared to the astronomic amounts spend by Man every year on armament and warfare, even in Africa. The second obstacle is a lack of time. Economic and demographic pressures on the forests rise rapidly, and, in addition, the risks of instability in the region threaten the entire process of establishment and sustainable management of tracts destined to survive as production forests. Sustainable forestry is indeed a long-term activity that requires planning over several human generations, just like the growth of the trees. Many techniques of tropical forest management and exploitation have thus been conceived, but none could be tested over a sufficient time span because of unfavourable socio-economic conditions.

Should we therefore discard every hope? Probably not, and some initiatives, like the partnership launched at Johannesburg in 2002, will perhaps bring some elements of a solution.

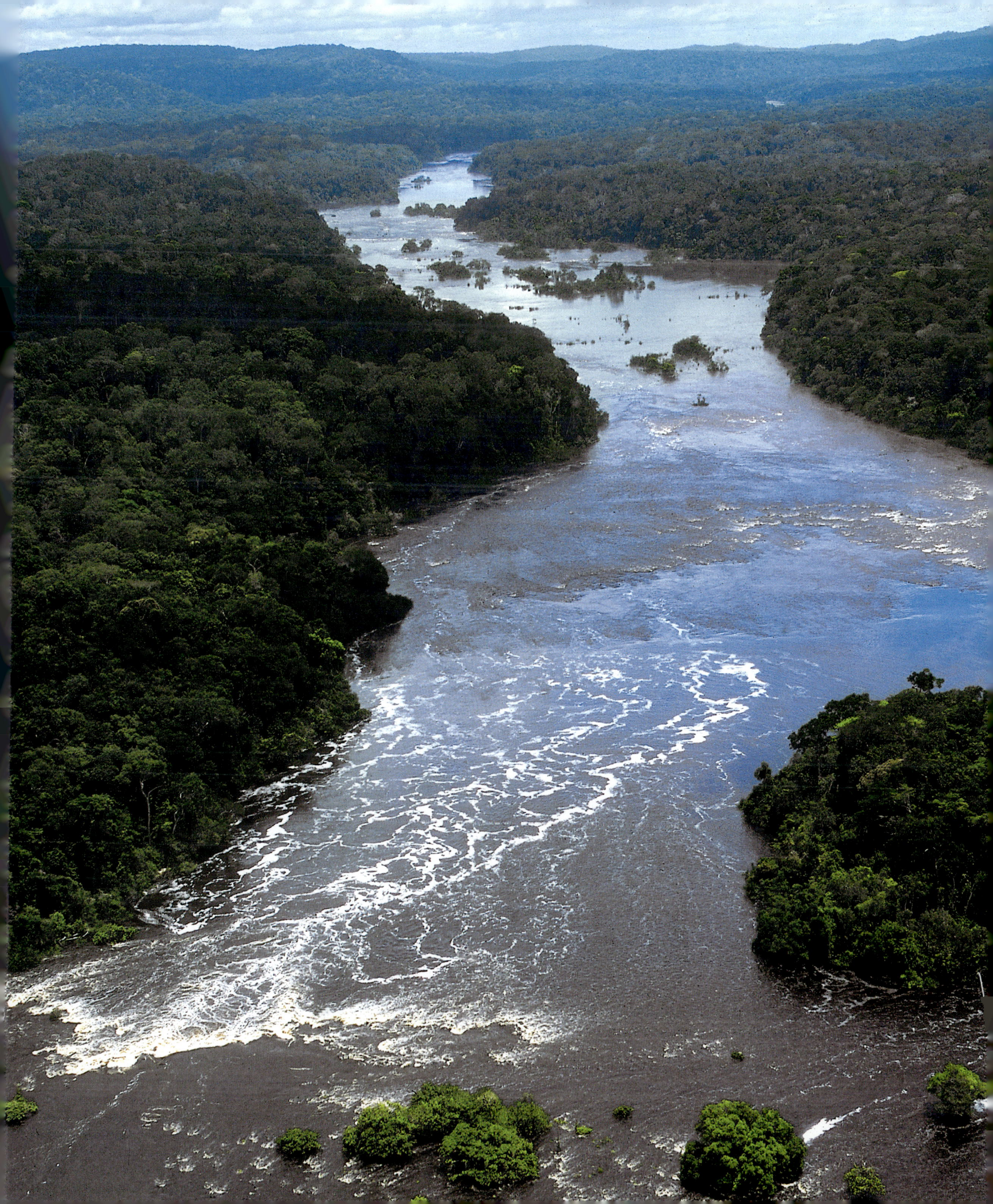

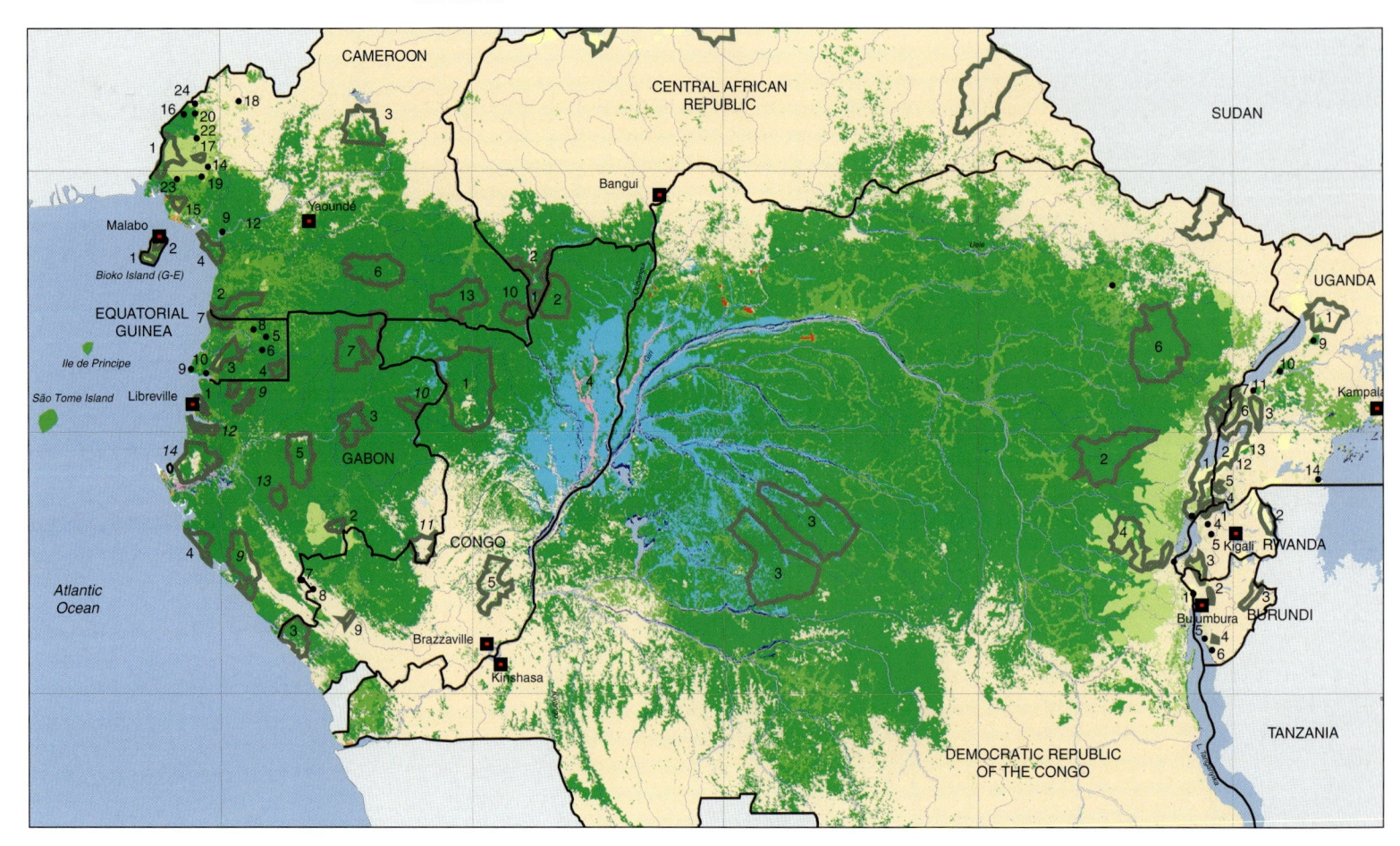

Appendix: the main protected areas of forested Central Africa

Abbreviations

np: national park
far: faunal reserve
for: forest reserve
rs: scientific reserve
nm: natural monument
nr: nature reserve
BR: biosphere reserve
WH: World Heritage site
R: Ramsar site
#: disturbed or destroyed site

UICN categories

I: integral reserve
II: national park
IV: faunal reserve
VI: game reserve, forest reserve

	UICN category	Creation[1]	Area[2]
Cameroon			
1. Korup (np)	II	1986	125,900
2. Campo-Ma'an (np)	II	2000	264,060
3. Mbam et Djerem (np)	II	2000	416,512
4. Douala-Edea (far)	II	1932	160,000
5. Santchou (far, #)	IV	1933	7,000
6. Dja (rfa, BR, WH)	IV	1950	526,000
7. Kimbi (far)	IV	1964	5,625
8. cratère de Mbi (far)	IV	1964	370
9. Lac Ossa (far, #)	IV	1968	4,000
10. Lac Lobéké (far)	IV	1974	43,000
11. Nanga Eboke (far, #)	IV	?	16,000
12. Sanaga (far, #)	IV	?	?
13. Bumba-bek, Nki (for)	VI	proposed	428,000
14. Bakossi (for)	IV	1964	5,520
15. Bambuko (mt Comeroun) (for)	VI	1939	26,680
16. Ejagham (for)	VI	1934	74,850
17. Banyang Mbo (for)	VI	1996	66,220
18. Idjim-Kilum (mt Oku) (for)	VI	1983	11,400
19. Kupe (for)	VI	?	2,300
20. Mawne (for)	VI	?	44,900
21. Mokoko (for)	VI	?	9,070
22. Nta Ali (for)	VI	?	31,500
23. Rumpi Hills (for)	VI	?	44,300
24. Takamanda (for)	VI	1934	67,600

Equatorial Guinea
1. Caldera de Luba (Bioko) (sr)	I	2000	51,000
2. Pico Basilé (Bioko) (np)	II	2000	33,000
3. Monte Alén (np)	II	1988	200,000
4. Altos de Nsork (np)	II	2000	70,000
5. Piedra Bere (nm)	-	2000	20,000
6. Piedra Nzas (nm)	-	2000	19,000
7. Rio Campo (nr)	-	2000	33,000
8. Monte Temelon (nr)	-	2000	23,000
9. Punta Llende (nr)	-	2000	5,500
10. Estuario del Rio Muni (nr)	-	2000	60,000
11. Annobon (nr)	-	2000	23,000

Gabon
1. Akanda (np)	II	2002	54,000
2. Birougou (np)	II	2002	69,000
3. Ivindo (np)	II	(1971) 2002	300,000
4. Loango (pn, R)	II	(1966) 2002	155,000
5. Lopé (np)	II	(1947) 2002	491,000
-. Mayumba (np, marine)	II	2002	8,000
7. Minkébé (np)	II	(1999) 2002	756,000
8. Monts de Cristal (np)	II	2002	120,000
9. Moukalaba-Doudou (np)	II	(1962) 2002	450,000
10. Mwagné (np)	II	2002	116,000
11. Plateau Batéké (np)	II	2002	205,000
12. Pongara (np)	II	2002	87,000
13. Waka (np)	II	2002	107,000
14. Wonga-Wongué (présid. res., R)	-	1972	380,000

Central African Republic
1. Dzanga-Ndoki (np)	II	?	122,000
2. Dzangha-Sangha (special reserve)	VI	1990	335,900
3. Basse-Lobaye (BR)	-	1977	18,200

Congo
1. Odzala-Kokoua (np, BR)	II	1940	126,000
2. Nouabalé-Ndoki (np)	II	1993	386,600
3. Conkouati (np)	II	1999	504,950
4. lac Télé (community reserve)	-	1999	439,960
5. Léfini (far)	IV	1951	630,000
6. Lékoli-Pandaka (far)	IV	1955	68,200
7. Mont Fouari (far, #)	IV	1958	15,600
8. Nyanga Nord (far, #)	IV	1958	7,700
9. Tsoulou (far)	IV	1984	30,000
10. Dimonika (BR)	-	1983	136,000

Democratic Republic of the Congo
1. Virunga (np, WH)	II	1925	780,000
2. Maiko (np)	II	1970	1,083,000
3. Salonga (pn, WH)	II	1970	3,656,000
4. Kahuzi-Biega (pn, WH)	II	1975	600,000
5. Mangroves (?)	-	?	76,850
6. Okapi (rfa, WH)	IV	1992	1,372,625
7. Yangambi (floristic res., BR)	VI	1976	250,000
8. Luki (BR, #)	-	1979	33,000

Uganda
1. Murchison Falls (np)	II	1952	15,000
2. Queen Elizabeth (pn, R)	II	1952	100,000
3. Kibale (np)	II	?	70,000
4. Mgahinga Gorilla (np)	II	1991	5,000
5. Bwindi Impenetrable (np)	II	1991	32,000
6. Rwenzori (np)	II	1991	90,000
7. Semuliki (np)	II	?	22,000
8. Mabira (for)	VI	1900	30,600
9. Budongo (for)	VI	1934	79,000
10. Bugoma (for)	VI	1934	36,000
11. Itwara (for)	VI	1934	8,700
12. Kalinzu (for)	VI	?	?
13. Kashyoha-Kitomi (for)	VI	1934	39,900
14. Sango Bay (for)	VI	?	15,100

Rwanda
1. Volcans (np)	II	1925	15,000
2. Akagera (np)	II	(1934) 1996	100,000
3. Nyungwe (for)	VI	1934	90,000
4. Gishwati (for, #)	VI	1934	4,000
5. Mukura (for, #)	VI	1934	1,500

Burundi
1. Rusizi (np)	II	1980	6,000
2. Kibira (np)	II	(1934) 1980	40,000
3. Ruvubu (np)	II	1980	52,000
4. Bururi (for)	VI	(1951) 1980	1,500
5. Kigwena (for)	VI	(1954) 1980	800
6. Rumonge-Vianda (for)	VI	1980	6,000

[1]) *In case the status of a protected area has changed, the first date gives the year of its creation, the second date the beginning of its current status.*
[2]) *Areas are given in hectares.*

Sources
ADIE (www.adie-prgie.org/forac.net), Doumenge (*et al.*, 2001), Howard (1991), UNEP-WCMC (www.unep-wcmc.org).

Bibliography

General Literature

Archibold, O.W., 1995. *Ecology of world vegetation*. Chapman & Hall, London.

Brosset, A. 1976. *La vie dans la forêt équatoriale*. Nathan, Paris.

Doumenge, Ch., 1990. *La Conservation des Ecosystèmes forestiers du Zaïre*. IUCN, Gland.

Fa, J.E., 1990. *La Conservation des Ecosystèmes forestiers de la Guinée équatoriale*. IUCN, Gland.

Fournier, F. & A. Sasson (eds.), 1983. *Ecosystèmes forestiers tropicaux d'Afrique*. Orstom et Unesco, Paris.

Gartlan, S., 1989. *La Conservation des Ecosystèmes forestiers du Cameroun*. IUCN, Gland.

Golley, F.B. (éd) 1983. Ecosystems of the World, 14a, Tropical Rainforest Ecosystems: structure and function. Elsevier Scientif. Publ. Comp., Amsterdam.

Hecketsweiler, Ph., 1990. *La Conservation des Ecosystèmes forestiers du Congo*. IUCN, Gland.

Howard, P.C., 1991. *Nature Conservation in Uganda's Tropical Forest Reserves*. IUCN, Gland.

Lanfranchi, R & D. Schwartz. (eds.), 1990. *Paysages quaternaires de l'Afrique centrale atlantique*. ORSTOM, Paris.

Larivière, J. (ed.), 1996. *L'atlas pour la conservation des forêts tropicales d'Afrique*. UICN France - Jean-Pierre de Monza, Paris.

Lieth, H. & M.J.A. Werger, 1989. *Ecosystems of the World, 14b. Tropical Rainforest Ecosystems: biogeographical and ecological studies*. Elsevier Scientif. Publ. Comp., Amsterdam.

Richards, P.W., 1952. *The Tropical Rain Forest*. Cambridge University Press, Cambridge.

Stuart, S.N., R.J. Adams, 1990. *Biodiversity in Sub-saharan Africa and its Islands. Conservation, Management, and Sustainable Use.* Occasional Papers of the IUCN Species Survival Commission 6. IUCN, Gland.

Terborgh, J., 1992. *Diversity and the tropical rain forest*. Scientific American Library, New York.

Weber, W., L.J.T. White, A. Vedder & L. Naughton-Treves, 2001. *African Rain Forest Ecology and Conservation. An Interdisciplinary Perspective*. Yale University Press, New Haven and London.

Wilks, C., 1990. *La Conservation des Ecosystèmes forestiers du Gabon*. UICN, Gland.

1. The mineral world

Bagdasaryan, G.P., V.I. Gerasimovskyi, A.I. Polyakov & K.Kh. Gukasyan, 1973. Age of the volcanic rocks in the rift zones of East Africa. Geochemistry International: 66-71.

Beadle, L.C., 1974. *The Inland Waters of Tropical Africa. An Introduction to Tropical Limnology.* Longman, London.

Boulvert, Y., 1982. *Notes de géomorphologie régionale de Centrafrique*. Orstom, Bangui.

Cahen, L., 1954. *La géologie du Congo belge*. Vaillant Carmanne, Liège.

Chernicoff, S. & H. Fox, 1997. *Essentials of Geology*. Worth Publishers, New York.

Chorowicz, J., 1983. Le rift est-africain: Début de l'ouverture d'un océan? *Bull. Centr. Rech. Exploration-Production Elf-Aquitaine* 7: 155-162.

Choubert, G. & A. Faure-Muret (coord.), 1976. Atlas géologique du Monde. Unesco, Paris.

Debelmas, J. & G. Mascle, 1991. *Les grandes structures géographiques*. Masson, Paris.

Ebinger, C.J., 1989. Tectonic development of the western branch of the East African rift system. *Geological Society of America Bulletin* 101: 885-903.

Furon, R., 1968. *Géologie de l'Afrique*. Payot, Paris.

Giresse, P., 1982. La succession des sédimentations dans les bassins marins et continentaux du Congo depuis le début du Mésozoïque. *Sci. Géol. Bull.* 35: 185-206.

Olivry, J.C. & J. Boulegue, 1995. *Grands bassins fluviaux périatlantiques: Congo, Niger, Amazone*. Orstom, Paris.

Palmer, D., 2000. *The Atlas of the Prehistoric World*. Marshall Editions, London.

Petit, M., 1990. Les grands traits morphologiques de l'Afrique centrale atlantique. *In*: R. Lanfranchi & D. Schwartz. *Paysages quaternaires de l'Afrique centrale atlantique*, 20-30. Orstom, Paris.

Peyrot, B., 1991. Hydrologie de l'Afrique centrale. *In*: R. Lanfranchi & B. Clist. *Aux origines de l'Afrique centrale*, 15-17. Centre Cult. Français d'Afrique centrale & Centre Intern. des Civilisations Bantu, Libreville.

Peyrot, B., 1991. La géologie de l'Afrique centrale. *In*: R. Lanfranchi & B. Clist. *Aux origines de l'Afrique centrale*, 7-9. Centre Cult. Français d'Afrique centrale & Centre Intern. des Civilisations Bantu, Libreville.

2. Climate and forests

Amiet, J.L., 1987. Aires disjointes et taxons vicariants chez les Anoures du Cameroun: implications paléoclimatiques. *Alytes* 6: 99-115.

Aubréville, A., 1967. La forêt primaire des montagnes de Bélinga. *Biol. Gabonica* 3: 95-112.

Baillie, J., G. Joffroy & T. Stevart, 2000. Exploration du Pico de Principe. *Canopée* 16: 5-7.

Bengo, M.D. & J. Maley, 1991. Analyse des flux polliniques sur la marge sud du Golfe de Guinée depuis 135 000 ans. *Comptes Rendus de l'Académie des sciences, Pars II* 313, 73-78.

Bigot, S., P. Camberlain, V. Moron & Y. Richard, 1997. Structures spatiales de la variabilité des précipitations en Afrique: une transition climatique à la fin des années 1960. *C.R. Acad. Sci. Paris* (2a) 324: 181-188.

Boltenhagen, E., J. Dejax & M. Salard-Cheboldaeff, 1985. Evolution de la vegetation tropicale africaine du Crétacé à l'Actuel d'après les données de la palynologie. *Bull. Sect. Sc. Paris* 8: 165-194.

Bond, W.J., 1989. The tortoise and the hare: ecology of Angiosperms dominance and Gymnosperm persistance. *Biological Journal of the Linnean Society* 36, 227-249.

Bonnefille, R. & G. Riollet, 1988. The Kashiru Pollen Sequence (Burundi) Plaeoclimatic Implications for the Last 40,000 yr B.P. in Tropical Africa. *Quaternary Research* 30: 19-35.

Bourliere, F. & M. Hadley, 1983. Present-day savannas: an overview. *In*: F. Bourlière (ed.) *Ecosystems of the World 13. Tropical Savannas*, 1-17.

Brenan, J.P.M., 1978. Some aspects of the phytogeography of tropical Africa. *Ann. Missouri Bot. Gard.* 65: 437-478.

Carcasson, R.H., 1964. A preliminary survey of the zoogeography of African butterflies. *E. Africa Wildlife Journ.* 2: 122-157.

Colyn, M., A. Gautier-Hion & W. Verheyen, 1991. A re-appraisal of palaeoenvironmental history in Central Africa: evidence for a major fluvial refuge in the Zaire Basin. *Journ. Biogeogr.* 18: 403-407.

Colyn, M., 1987. Les primates des forêts ombrophiles de la cuvette du Zaïre: interprétations zoogéographiques des modèles de distribution. *Rev. Zool. Afr.* 101, 183-196.

Colyn, M., 1991. L'importance zoogéographique du bassin du fleuve Congo pour la spéciation: le cas des Primates Simiens. *Ann. Mus. Roy. Afr. Centr., Sc. Zool.*, Tervuren.

Colyn, M., 1999. Un nouveau statut biogéographique pour l'Afrique centrale. *Canopée* 14: 3-5.

Conseil scientifique pour L'afrique (CSA), 1956. *Réunion des Spécialistes du CSA en matière de Phytogéographie, Yangambi 1956*. CCTA, London.

Crowe, T.M. & A.A. Crowe, 1982. Patterns of distribution, diversity and endemism in Afrotropical birds. *J. Zool. London* 198: 417-442.

Cruizat, P. 1966. Note sur le microclimat de la strate inférieure de la forêt équatoriale comparé à une clairière. *Biol. Gabonica* 2: 361-402.

Dale, I.R., 1954. Forest spread and climatic change in Uganda during the Christian Era. *The Empire Forestry Review* 33: 23-29.

Diamond, A.W. & A.C. Hamilton, 1980. The distribution of forest passerines birds and Quaternary climatic change in tropical Africa. *J. Zool. Lond.* 191:379-402.

Elenga, H., A. Vincens & D. Schwartz, 1991. Présence d'éléments forestiers montagnards sur les Plateaux Batéké (Congo) au Pléistocène supérieur: nouvelles données palynologiques. *Palaeoecology of Africa* 22: 239-252.

Exell, A.W., 1973. Angiosperms of the islands of the Gulf of Guinea (Fernando Po, Principe, Sao Tome and Annobon). *Bull. Brit. Mus. Nat Hist.* (Botany) 4: 327-411.

Figueiredo, E., 1994. Diversity and endemism of angiosperms in the Gulf of Guinea islands. *Biodiversity and Conservation* 3: 785-793.

Figueiredo, E., 1998. The Pteridophytes of Sao Tomé and Principe (Gulf of Guinea). *Bull. Nat. Hist. Mus. London* (Botany) 28: 41-66.

Fjeldsa, J. & J.C. Lovett, 1997. Geographical pattern of old and young species in African forest biota: the significance of specific montane areas as evolutionary centres. *Biodiversity and Conservation* 6: 325-346.

Gentry, A.H., 1989. Speciation in tropical forests. *In*: L.B. Holm-Nielsen, I.C. Nielsen & H. Balslev (eds.). *Tropical forests: Botanical dynamics, speciation and diversity*, 113-134. Academic Press, London.

Giresse, P., G. Bongo-Passi, G. Delibrias & J.C. Duplessy, 1982. La lithostratigraphie des sédiments hémipélagiques du delta du fleuve Congo et ses indications sur les paléoclimats de la fin du Quaternaire. *Bulletin de la Société Géologique de France* 24 803-815.

Grubb, P., 1992. Refuges and dispersal in the speciation of African forest mammals. In G.T. Prance (éd) *Biological Diversification in the Tropics*, 537-543. Columbia University Press, New York.

Haffer, J., 1974. Avian speciation in tropical South America. *Publ. Nuttal Ornithol. Club* 14, 1-390.

Haffer, J., 1997. Alternative models of vertebrate speciation in Amazonia: an overview. *Biodiversity and Conservation* 6: 451-476.

Hamilton, A.C., 1981. The quaternary history of African forests: its relevance to conservation. *Afr. J. Ecol.* 19, 1-6

Hamilton, A.C., 1982. *Environmental History of East Africa: A Study of the Quaternary*. Academic Press, London.

Hamilton, A.C., 1989. African Forests. *In*: Lieth & Werger (eds.), 155-182 (see general literature).

Hart, T.B., J.A. Hart, R. Dechamps, M. Fournier & M. Ataholo, 1996. Changes in forest composition over the last 4000 years in the Ituri basin, Zaire. *In The Biodiversity of African plants*. L.J.G. Van Der Maesen X.M., Van Der Burgt & J.M. Van Medenbach De Rooy (eds.): 545-563. Kluwer Acad. Publ., Dordrecht.

Jolly, D., D. Taylor, R. Marchant, A. Hamilton, R. Bonnefille, G. Buchet & G. Riollet, 1997. Vegetation dynamics in central Africa since 18,000 yr BP: pollen records from the interlacustrine highlands of Burundi, Rwanda and western Uganda. *Journal of Biogeography* 24: 495-512.

Kingdon, J., 1990. *Island Africa. The Evolution of Africa's Rare Animals and Plants*. Collins, London.

Lauer, W., 1989. Climate and Weather. *In*: Lieth & Werger (ed.), 7-53.

Lawson, D.P. & M.W. Klemens, 2001. Herpetofauna of the African Rain Forest. *In*: W. Weber et al., 291-210 (general literature).

Léonard, J., 1965. Contribution à la subdivision phytogéographique de la Région guinéo-congolaise d'après la répartition géographique d'Euphorbiacées d'Afrique tropicale. *Webbia* 19: 627-649.

Leroux, M., 1975. Climatologie dynamique de l'Afrique. *Travaux et documents de géographie tropicale* 19.

Maley, J., 1990. L'histoire récente de la forêt dense humide africaine: essai sur le dynamisme de quelques formations forestières. *In*: R. Lanfranchi & D. Schwartz (éds) *Paysages quaternaires de l'Afrique centrale atlantique*: 367-389. Orstom, Paris.

Maley, J., 1992. Mise en évidence d'une péjoration climatique entre ca. 2500 et 2000 ans BP en Afrique tropicale humide. *Bull. Soc. Géol. France* 163, 363-365.

Maley, J., 1996. The African rain forest – main characteristics of changes in vegetation and climate from the Upper Cretaceous to the Quaternary. *Proceedings of the Royal Society of Edinburgh*, 104b: 31-73.

Maley, J., 1997. Middle to Late Holocene Changes in Tropical Africa and Other Continents: Paleaomonsoon and Sea Surface Tempertaure Variations. *In*: H. Nüzhet Dalfes, G. Kukla & H. Weiss (éds). *Third Millenium BC Climate Change and Old World Collapse*. NATO ASI Series, vol 149, 611-639.

Maley, J. 2001. The Impact of Arid Phases on the African Rain Forest Through Geological History. In: Weber et al., 68-87 (general literature).
Maley, J. & H. Elenga, 1993. Le rôle des nuages dans l'évolution des paléoenvironnements montagnards de l'Afrique tropicale. *Veille Climatique Satellitaire* 46, 51-63.
Maley, J. & P. Brenac, 1998. Vegetation dynamics, Palaeoenvironments and Climatic changes in the Forests of West Cameroon during the last 28,000 years BP. *Rev. Palaeobot. Palynol.* 99, 157-187.
Maley, J., P. Brenac, S. Bigot & V. Moron, 2000. Variations de la végétation et des paléo-environnements en forêt dense africaine au cours de l'Holocène. Impact de la variation des températures marines. In: M. Servant & S. Servant-Vildary (éd.) *Dynamique à long terme des écosystèmes forestiers intertropicaux*, 205-220. UNESCO, Paris.
Maley, J., G. Caballe & P. Sita, 1990. Etude d'un peuplement résiduel à basse altitude de *Podocarpus latifolia* sur le flanc congolais du Massif du Chaillu: Implications paléoclimatiques et biogéographiques. Etude de la pluie pollenique actuelle. In: R. Lanfranchi & D. Schwartz (eds.). *Paysages quaternaires de l'Afrique Centrale Atlantique*, 336-352. ORSTOM, Paris.
Maloba, J.D. & G. Samba, 1997. Organisation pluviométrique de l'espace Congo-Gabon (1951-1990). *Sécheresse* 8: 39-45.
Malounguila-Nganga, D., 1991. Les environnements marins et littoraux de l'Afrique centrale (du nord Angola jusqu'au Gabon). In: R. Lanfranchi & B. Clist. *Aux origines de l'Afrique centrale*, 35-39. Centre Culturel Français d'Afrique centrale & Centre International des Civilisations Bantu, Libreville.
Mayr, E. & R.J. O'hara, 1986. The biogeographic evidence supporting the Pleistocene forest refuge hypothesis. *Evolution* 40: 55-67.
Midgley, J.J. & W.J. Bond, 1989. Evidence from the southern african Coniferales for the historical decline of the Gymnosperms. *South African Journal of Science* 85: 81-84.
Misonne, X., 1963. Les rongeurs du Ruwenzori et des régions voisines. *Exploration du Parc National Albert* 14, 1-164.
Moreau, R.E., 1966. *The bird faunas of Africa and its islands.* Academic Press, London.
Morley, R.J. & K. Richards, 1993. Gramineae cuticle: A key indicator of Late Cenozoic climatic change in the Niger delta. *Rev. Palaeobot. Palyn.* 77: 119-127.
Mpounza, M. & M.J. Samba-Kimbata, 1990. Aperçu sur le climat de l'Afrique centrale occidentale. In: R. Lanfranchi & D. Schwartz (ed.) *Paysages quaternaires de l'Afrique centrale atlantique*, 31-41. Paris, ORSTOM.
Muloko-Ntoutoume, N., K. Abernethy, L. White, R. Petit & J. Maley, 1999. Utilisation des marqueurs moléculaires dans la reconstruction de l'histoire de la forêt tropicale humide gabonaise: le modèle *Aucoumea klaineana*. In: R. Nasi, I. Ansallem & S. Drouineau (éds). *La gestion des forêts denses afriaines aujourd'hui.* Actes de l'Atelier FORAFRI, CIRAD, Montpellier.

Osmaston, H.A., 1965. *The past and present climate and vegetation of Ruwenzori and its neighbourhood.* D. Phil. thesis, Oxford University.
Peyrot, B., 1991. Climatologie de l'Afrique centrale. in R. Lanfranchi & B. Clist. *Aux origines de l'Afrique centrale*, 19-23. Centre Culturel Français d'Afrique centrale & Centre International des Civilisations Bantu, Libreville.
Philander, G., 1989. El Niño and La Niña. *American Scientist* 77: 451-459.
Poumot, C., 1989. Palynological evidence for eustatic events in the tropical Neogene. *Bulletin du Centre de Recherche, Exploration et Production d'Elf-Aquitaine* 13: 437-453.
Prigogine, A., 1988. Speciation patterns of birds in the Central African forest refugia and their relationships with other refugia. *Acta XIX Congr. Intern. Ornith.* II, 144-157. University of Ottawa Press, Ottawa.
Robbins, C.B., 1978. The Dahomey Gap. A reevaluation of its significance as a faunal barrier to West African high forest mammals. *Bull. Carnegie Mus. Nat. Hist.* 6: 168-174.
Salard-Cheboldaeff, M. & J. Dejax, 1991. Evidence of Cretaceous to rent West African intertropical vegetation from continental sediment spore-pollen analysis. *J. Afr. Earth Sc.* 12: 353-361.
Sosef, M.S.M., 1994. *Refuge Begonias: Taxonomy, phylogeny, and historical biogeography of Begonia sect. Loasibegonia and sect. Scutobegonia in relation to glacial rain forest refuges in Africa.* Studies in Bengoniaceae, Wageningen Agric. Univ. Papers, Wageningen.
Stevart, T., 1999. *Rapport de mission sur les orchidées de Sao Tomé et Principe.* Projet ECOFAC. AGRECO-GEIE, Brussels.
Thompson, B.W., 1965. The Climate of Africa. Oxford University Press, Nairobi - London - New York.
WCMC, 2000. *Global Biodiversity: Earth's living resources in the 21st century.* By B. Groombridge & M.D. Jenkins. World Conservation Press, Cambridge.
White, L.J.T., R. Oslisly, K. Abernethy & J. Maley, 2000. L'Okoumé (*Aucoumea klaineana*): expansion et déclin d'un arbre pionnier en Afrique centrale atlantique (Gabon) au cours de l'Holocène. In: *Actes du Symposium ECOFIT, Dynamique à long terme des Ecosystèmes Forestiers Intertropicaux.* Unesco, Paris.
WWF & IUCN, 1994. *Centres of Plant Diversity. A guide and strategy for their conservation. Vol. I: Europe, Africa, South West Asia and the Middle east.* IUCN Publications Unit, Cambridge.

3. Ecological gradients
Achoundoung, G., 1995. Les formations sub-montagnardes du Nta-Ali au Cameroun. *Bois et Forêts des Tropiques* 243: 51-63.
Achoundoung, G., 1996. Les forêts sommitales au Cameroun. *Bois et Forêts des Tropiques* 247: 37-52.
Aubreville, A., 1965. Principes d'une systématique des formations végétales tropicales. *Adansonia* 5: 153-196.
Aubréville, A., 1966. Les lisières forêt-savane des régions tropicales. *Adansonia* 6: 175-187.

Barthlott, W. & S. Porembski, 1998. Diversity and phytogeographical affinities of inselbergs vegetation in tropical Africa and Madagascar. In: C.R. Huxley, J.M. Lock & D.F. Cutler (eds.) *Chorology, Taxonomy and Ecology of the Floras of Africa and Madagascar.* Royal Botanical Gardens, Kew.
Bouxin, G., 1974. Etude phytogéographique des plantes vasculaires du marais Kamiranzovu (forêt de Rugege, Rwanda). *Bull. Jard. Bot. Nat. Belg.* 44: 41-159.
Caballé, G., 1978. Essai sur la géographie forestière au Gabon. *Adansonia*, série 2, 17: 425-440.
De Foresta, H., 1990. Origine et évolution des savanes intramayombiennes (R.P. du Congo). II. Apport de la botanique forestière. In: R. Lanfranchi & D. Schwartz (Eds.). *Paysages quaternaires de l'Afrique Centrale Atlantique.* Orstom, Pointe-Noire.
Devillers, P. & J.Devillers-Terschuren, 1993. *A classification of Palaearctic habitats.* T-PVS (94) 1. Strasburg, Council of Europe.
Eggeling, W.J., 1947. Observations on the ecology of the Budongo rain forest, Uganda. *Journal of Ecology* 34: 20-87.
Goldsmith, F.B., 1998. Tropical rain forests – what are they like? In: F.B. Goldsmith (éd) *Tropical Rain Forest: A Wider perspective.* Chapman & Hall, London.
Gradstein, G.R. & T. Pocs, 1989. Bryophytes. In: Lieth & Werger, 311-326 (general literature).
Greenway, P.J., 1973. A classification of the vegetation of East Africa. *Kirkia* 9: 1-68.
Hamilton, A.C., 1975. A quantitative analysis of altitudinal zonation in Uganda forests. *Vegetatio* 30: 99-106.
Hart, T.B., 2001. Forest Dynamics in the Ituri Basin (DR Congo). Dominance, Diversity, and Conservation. In: Weber et al. (general literature).
Hart, T.B., J.A. Hart & P.G. Murphy, 1989. Monodominant and species-rich forests of the humid tropics: Causes for their co-occurrence. *American Naturalist* 133: 613-633.
Hedberg, O., 1963. Afroalpine flora elements. *Webbia* 19: 519-529.
Hedberg, O., 1964. Features of Afroalpine plant ecology. *Acta phytogeogr. suec.* 49: 1-144.
Hegarty, E.E., 1983. The climbers - Lianes and Vines. In: F.B. Golley (ed.), 339-354.
Knapp, R., 1973. Die Vegetation von Afrika. Vol. 3. In: H. Walter (ed.). *Vegetationsmonographien der einzelnen Grossräume.* Stuttgart, Fischer.
Lebrun, J. & G. Gilbert, 1954. Une classification écologique des forêts du Congo. *Publ. INEAC, sér. sci.* 63: 1-89.
Lejoly, J., 2000. *Les recherches sur la biodiversité végétale dans les 6 sites du programme ECOFAC entre 1997 et 2000. Rapport final de synthèse.* AGRECO-GEIE, Brussels.
Lejoly, J. & S. Lisowski, 1999. Novitates Guineae Aequatorialis (4) *Polyscias aequatoguineensis*, une Araliaceae nouvelle du Rio Muni. *Bull. Jard. Bot. Nat. Belg.* 67.
Léonard, J., 1952. Les divers types de forêt du Congo. belge. *Lejeunia* 16: 81-93.
Letouzey, R., 1968. *Etude phytogéographique du Cameroun.* Lechevalier, Paris.
Lewalle, J., 1972. Les étages de végétation au Burundi occidental. *Bull. Jard. Bot. Nat. Bot. Belg.* 42: 1-247.

Lind, E.M. & M.E.S. Morrison, 1974. *East African Vegetation.* Longman, London.
Parmentier, I., 1999. La végétation des inselbergs du Rio Muni. *Canopée* 14: 7-9.
Pierlot, R. 1966. Structure et composition de forêts denses d'Afrique centrale, spécialement celles du Kivu. *Mém. Acad. Roy. Sci. Outre-Mer, Cl. Sci. Nat. Med.* 16: 1-367.
Reitsma, J.M., A.M. Louis & J.-J. Floret, 1992. Flore et végétation des inselbergs et dalles rocheuses: première étude au Gabon. *Adansonia* 1: 73-97.
Schnell, R., 1976-1987. *Flore et végétation d'Afrique tropicale.* Vol. 1-3. Paris, Gaulthier-Villars.
Sipman, H.J.M. & R.C. Harris, 1989. Lichens. In: Lieth & Werger, 303-309 (general literature).
Sonke, B., 1998. Les forêts de la Réserve de Faune du Dja. *Canopée* 12: 22-23.
Toelen, P., 1995. *Inventaire, biodiversité, structure des peuplements et biomasse ligneuse dans les forêts d'amltitude de Sao Tomé.* Mémoire de l'Université Libre de Brussels, Section Interfacultaire d'Agronomie. Brussels.
Troupin, G., 1966. *Etude phytocénologique du Parc National de l'Akagera et du Rwanda oriental. Recherche d'une méthode d'analyse appropriée à la végétation d'Afrique intertropicale.* Publication 2: 1-293. INRS, Butare.
Tryon, R. 1989. Ptéridophytes. In: Lieth & Werger (eds.), 327-338 (general literature).
Van Rompaey, R.S.A.R., 1993. *Forest gradients in West Africa. A spatial gradient analysis.* Thesis. Wageningen.
White, F., 1983. *The vegetation of Africa: a descriptive memoir to accompany the UNESCO-AETFAT-UNSO vegetation map of Africa.* Unesco, Paris.
White, L., 1995. *Etude de la végétation de la Réserve de La Lopé.* Rapport Final, ECOFAC, Gabon.
White, L. & R. Oslisly, 1999. Lopé: a window on the history of the central African rain forest. In: R. Nasi, I. Amsallem & S. Drouineau (eds.). *La gestion des forêts denses africaines aujourd'hui.* Actes du séminaire FORAFRI de Libreville. CIRAD, Montpellier.

4. Temporal gradients
Brugière, D., S. Bougras & A. Gautier-Hion, 2000. *Dynamique forestière et processus de colonisation-extinction: relations faune-flore dans les forêts à Marantacées d'Odzala.* AGRECO-GEIE, Brussels.
Dale, I.R., 1940. The forest types of Mount Elgon. *J. E. Africa Uganda nat. hist. Soc.* 45: 74-82.
Dowsett-Lemaire, F., 1996. Composition et évolution de la végétation forestière au Parc National d'Odzala. *Bull. Jard. Bot. Nat. Belg.* 65: 253-292.
Hamilton, A.C., 1989. African Forests. In: Lieth & Werger (eds.): 155-182 (general literature).
Lebigre, J.-M., 1983. Les mangroves des rias du littoral gabonais: essai de cartographie typologique. *Revue Bois et Forêts des Tropiques* 199: 3-28.
Lebigre, J.-M. & C. Marius, 1981. *Etude d'une séquence mangrove-tanne en milieu équatorial, baie de la Mondah (Gabon).* Symp. Internat. Lagunes Côtières, Talence.

Magliocca, F., 2000. Etude d'un peuplement de grands mammifères forestiers tropicaux fréquentant une clairière: structure des populations; utilisation des ressources; coexistence intra- et inter-populationnelle. Thèse, Université de Rennes.

Rollet, B., 1979. Application de diverses méthodes d'analyse de données à des inventaires détaillés levés en forêt tropicale. *Oecol. Plant.* 14: 319-344.

Sonke, B., F. Palla & S. Kuob, 1998. *Etudes écologiques et botaniques de la périphérie de la Réserve de faune du Dja (Cameroun).* ECOFAC, Cameroun.

5. The life in the forest

Bagyaraj, D.J., 1989. Mycorhizas. In: Lieth & Werger (ed.), 537-546.

Baker, H.G., K.S. Bawa, G.W. Frankie & P.A. Opler, 1989. Reproductive biology of plants in tropical forests. In: F.B. Golley (ed.), 183-215 (general literature).

Caballé, G., 1984. Essai sur la dynamique des peuplements de lianes ligneuses d'une forêt du nord-est du Gabon. *Rev. Ecol. (Terre et Vie)* 39: 3-35.

Coley, P.D., 1983. Herbivory and defensive characteristics of tree species in a lowland tropical forest. *Ecol. Monogr.* 53: 209-233.

Cremers, G., 1973. Architecture de quelques lianes d'Afrique tropicale. I. *Candollea* 28: 249-280.

Cremers, G., 1974. Architecture de quelques lianes d'Afrique tropicale. II. *Candollea* 29: 57-110.

Gartlan, J.S., D. McKey, P.G. Waterman, C.N. Mbi & T.T. Struhsaker, 1980. A comparative study of the phytochemistry of two African rain forests. *Biochem. Syst. Ecol.* 8: 401-422.

Gautier-Hion, A., J.M. Duplantier, L. Emmens, F. Feer, Ph. Hecketsweiler, A. Moungazi, R. Quris & C. Sourd, 1985. Coadaptation entre rythmes de fructification et frugivorie en forêt tropicale humide du Gabon: mythe ou réalité. *Rev. Ecol.* (Terre Vie) 40: 405-434.

Golley, F.B., 1983. Decomposition. In: F.B. Golley (ed.), 157-166 (general literature).

Golley, F.B., 1983. Nutrient cycling and nutrient conservation. In: F.B. Golley (ed.), 137-156 (general literature).

Hegarty, E.E., 1983. The Climbers – Lianes and Vines. In: F.G. Golley (ed.), 339-353 (general literature).

Hladik, A., 1974. Importance des lianes dans la production foliaire de la forêt équatoriale du nord-est du Gabon. *C. R. Acad. Sci. Paris, Série D,* 278: 2527-2530.

Hladik, A. & P. Blanc, 1987. Croissance des plantes en sous-bois de forêt dense humide (Makokou, Gabon). *Rev. Ecol. (Terre Vie)* 42: 209-234.

Johnsson, D.R., 1989. Vascular epiphytism in Africa. In: Lieth & Werger (eds.), 183-194 (general literature).

Keay, R.W.J., 1957. Wind dispersed species in a Nigerian forest. *J. Ecol.* 45: 471-478.

Kouka, L.A., 1994. *Etude de la biodiversité des forêts d'Afrique centrale: cas de la forêt de Ngotto (République centrafricaine) et de la Réserve de Faune du Dja (Cameroun).* Mémoire, Université libre de Brussels.

Langenheim, J.H., 1984. The roles of plant secondary chemicals in wet tropical ecosystems. In: E. Medina, H.A. Mooney & C. Velasquez-Yanes (Eds.). *Physiological Ecology of Plants in the Wet Tropics.* Junk, The Hague.

Medina, E., 1989. Adaptations of tropical trees to moisture stress. In: F.B. Golley (ed.), 225-237 (general literature).

Strong Jr, D.R., 1977. Epiphyte loads, treefalls and perenial forest disruption; a mechanism for maintaining higher tree species richness in the tropics without animals. *J. Biogeogr.* 4: 215-218.

Swift, M.J. & J.M. Anderson, 1989. Decomposition. In: Lieth & Werger (eds.), 547-570 (general literature).

Tutin, C.E.G. & M. Fernandez, 1993. Relation between minimum temperature and fruit production in some tropical forest trees in Gabon. *Journal of Tropical Ecology* 9: 241-248.

Waterman, P.G. & D. McKey, 1989. Herbivory and secondary compounds in rain-forest plants. In: Lieth & Werger (eds.), 513-536 (general literature).

White, L.J.T., 1994. Patterns of fruit-fall phenology in the Lopé Reserve, Gabon. *J. Trop. Ecol.* 10: 309-318.

White, L. & K. Abernethy, 1996. *Guide de la Végétation de la Réserve de la Lopé.* ECOFAC-Gabon, Libreville.

Yalibanda, Y. & J. Lejoly, 1998. *Phénologie en forêt dense de Ngotto (RCA). Bilan de 3 années d'observation.* AGRECO-GEIE, Brussels.

6. The fauna

Bernardi, G., 1984. Les théories explicatives du mimétisme batésien chez les insectes lépidoptères, et les mécanismes de la finalité. In: *La fin et les moyens, études sur la finalité biologique et ses mécanismes,* 131-145. Coll. Recherches interdisciplinaires. Maloine, Paris.

Bourlière, F., 1983. Animal species diversity in tropical forests. In: Golley (ed.) 77-92 (general literature).

Brosset, A., 1966. Recherches sur la composition qualitative et quantitative des populations de Vertébrés dans dans la forêt primaire du Gabon. *Biol. Gabonica* 2: 163-177.

Brosset, A. & C. Erard, 1986. *Les oiseaux des régions forestières du Gabon. Vol 1. Ecologie et comportement des espèces.* Société nationale de protection de la nature, Paris.

Brugière, D., D. Sakom & A. Gautier-Hion, 1999. *Structure de la communauté de primates simiens de la forêt de N'Gotto. Rôle des milieux marginaux dans le maintien de la biodiversité.* ECOFAC, Brussels.

Butynski, T.M., C.D. Schaaf & G.W. Hearn, 1997. African Buffalo *Syncerus caffer* extirpated on Bioko Island, Equatorial Guinea. *J. Afr. Zool.* 111: 57-61.

Christy, P. & W.V. Clarke, 1994. *Guide des Oiseaux de la Réserve de la Lopé.* ECOFAC, Libreville.

Christy, P. & W. V. Clarke, 1998. *Guide des oiseaux de São Tomé et Principe.* ECOFAC, Libreville.

Christy, P. & J.-P. Vande weghe, 1999. Liste des oiseaux d'Afrique centrale. *Les dossiers de l'ADIE, Série Biodiversité* 1: 1-32.

Collins, N.M., 1989. Termites. In: Lieth & Werger (ed.), 455-471 (see general literature).

Colyn, M. M., 1987. Les primates des forêts ombrophiles de la cuvette du Zaïre: interprétation zoogéographiques des modèles de distribution. *Rev. Zool. Afr.* 101: 183-196.

Darchen, R., 1965. Ethologie d'une araignée sociale *Agelena consociata. Biol. Gabonica* 1: 117-174.

Darchen, R., 1968. Ethologie d'*Achaearanea disparata* Denis, Aranea Theridiidae, Araignée sociale du Gabon. *Biol. Gabonica* 4: 5-26.

Davies, A.G. & J. Oates (eds.), 1995. *Colobine monkeys: their ecology, behaviour and evolution.* Cambridge University Press, London.

Dejean, A. & B. Bolton, 1995. Fauna sheltered by *Procubitermes niapuensis* termitaries of the African rainforest. *J. Afr. Zool.* 109: 481-487.

Demange, J.M., 1968. Myriapodes Chilopodes du Gabon. *Biol. Gabonica* 2: 281-295.

Dieterlein, F., 1983. Rodents. In: Golley (ed.): 383-400 (see general literature).

Dorst, J. & P. Dandelot, 1972. *Larger Mammals of Africa.* Harper Collins, London.

Dowsett, R.J. & L. Granjon, 1991. Liste préliminaire des mammifères du Congo. *Tauraco Research Report* 4: 297-310.

Dowsett-Lemaire, F. & R.J. Dowsett, 2001. African Forest Birds. Patterns of Endemism and Species Richness. In: W. Weber et al., 233-262 (general literature)

East, R., 1990. *Antelopes. Global survey and regional action plans. Part 3. West and central Africa.* IUCN, Gland.

Eeley, H.A.C. & R.A. Foley, 1999. Species richness, species range size and ecological specialisation among African primates: geographical pattern and conservation implications. *Biodiversity and Conservation* 8: 1033-1056.

Eeley, H.A.C. & R.A. Foley, 1999. Species richness, species range size and ecological specialisation among African primates: geographical patterns and conservation implications. *Biodiversity and Conservation* 8: 1033-1056.

Frétey, T. & C. Blanc, 2000. Liste des Amphibiens d'Afrique centrale. *Les dossiers de l'ADIE, Série Biodiversité* 2: 1-40.

Frétey, T. & C. Blanc, (in press). Liste des Reptiles d'Afrique centrale. *Les dossiers de l'ADIE, Série Biodiversité* 3.

Gautier-Hion, A., F. Bourliere, J.-P. Gautier & J. Kingdon, 1988. *A primate radiation: evolutionary biology of the african guenons.* Cambridge University Press, London.

Gautier-Hion, A., M. Colyn & J.-P. Gautier, 1999. *Histoire naturelle des primates d'Afrique centrale.* ECOFAC, Libreville.

Gautier-Hion, A. & M. Colyn, 2000. Rapport final de synthèse - ECOFAC II: Biodiversité animale et Ecologie. AGRECO-GEIE, Brussels.

Goodall, J., 1996. *The chimpanzees of Gombe: patterns of behaviour.* Harvard University Press, Cambridge, USA.

Grubb, P., 2001. Endemism in African Rain Forest Mammals. In: Weber et al., 88-100.

Guillaumin, M., 1977. Participation des *Euphaedra* et *Bebearia* aux associations mimétiques de la forêt équatoriale africaine. Associations ayant des Hétérocères à moeurs diurnes comme modèles. *Bull. Soc. Zool. Fr.* 101: 603-611.

Haltenorth, T. & H. Diller, 1980. *Mammals of Africa including Madagascar.* Collins & Sons, London.

Holloway, J.D., 1989. Moths. In: Lieth & Werger (eds.), 437-453 (general literature).

Karr, J.R., 1989. Birds. In: Lieth & Werger (éds), 401-416 (general literature).

Kingdon, J., 1997. *The Kingdon field guide to African Mammals.* Academic Press, London.

Knoepffler, L.Ph., 1966. Faune du Gabon (Amphibiens et Reptiles). *Biol. Gabonica* 2: 3-24.

Krafft, B., 1975. Les interactions limitant le cannibalisme chez les araignées solitaires et sociales. *Bull. Soc. Zool. Fr.* 100: 203-221.

Krafft, B., 1979. Organisation et évolution des société d'araignées. *J. Psychol.* 1: 23-52.

Laurent, R.F., 1989. Herpetofauna of tropical America and Africa. In: Lieth & Werger (eds.), 417-427 (general literature).

Lawson, D.P. & M.W. Klemens, 2001. Herpetofauna of the African Rain Forest. In: W. Weber et al., 291-210 (general literature).

Louette, M., 1981. *The birds of Cameroon. An annotated check-list.* Verhandelingen Kon. Acad. v. Wetensch. Lett. Schone Kunsten v. België, Kl. Wetenschappen, 163.

Owen, D.F., 1989. The abundance and biomass of forest animals. In: F.B. Golley (ed.), 93-100 (general literature).

Querouil, S., F. Magliocca & A. Gautier-Hion, 1999. Structure of population, grouping pattern and density of forest elephants in north-west Congo. *Afr. Journ. Ecol.* 37: 161-167.

Richert, S.E., 1985. Why do some spiders cooperate? *Agelena consociata* Denis: a case study. *Fla. Entomol.* 68: 105-113.

Rouland, C. & M. Lepage, 1995. Estimation de l'abondance des nids et des populations de termites de la forêt du Mayombe (République du Congo). *J. Afr. Zool.* 109: 339-347.

Sussman, R.L. (ed.), 1984. *The pygmy chimpanzee: evolutionary biology and behaviour.* Plenum Press, New York.

Tobin, J.E., 1995. Ecology and Diversity of Tropical Forest Canopy Ants. *Forest Canopies:* 129-147.

Wilson, D.E. & D. Reeder, 1993. *Mammal species of the world. A taxonomic and geographic reference.* Smithsonian Institution, Washington.

Wilson, D.E., 1989. Bats. In: Lieth & Werger (éds), 365-382 (general literature).

7. Man and the forest

Atieno, E.S., T.I. Ouso & J.F.M. Williams, 1977. *A History of East Africa.* Longman, London.

Bahuchet, S., 1991. La forêt du Haut-Zaïre: une mosaïque culturelle. In: R. Farris Thompson & S. Bahuchet. *Pygmées?* 115-153. Dapper, Paris.

Bahuchet, S., 1996. Fragments pour une histoire de la forêt africaine et de son peuplement: données linguistiques et culturelles. In: C.M. Hladik, A. Hladik, H. Pagezy, O.F. Linares, G.J.A. Koppert & A. Froment (coordonnateurs). *L'alimentation en forêt tropicale: interactions bioculturelles et perspectives de développement.* L'Homme et la Biosphère, Unesco.

Bahuchet, S., 1997. Un style de vie en voie de mutation: considérations sur les peuples des forêts denses humides. *Civilisations* XLV, 16-30.

Bird, M.I. & J.A. Cali, 1998. A million year record of fire in sub-Saharan Africa. *Nature* 394: 767-769.

Boungou, G., J. Nguembo & J. Sénéchal, 1989. Peuplement et population du Mayombe. *In*: J. Sénéchal, M. Kabala & F. Fournier (éds). *Revue des connaissances sur le Mayombe*. Unesco, Paris.

Cavalli-Sforza, L.L. (ed.), 1986. *African Pygmees*. Academic Press, New York.

Clist, B., 1995. *Gabon: 100000 ans d'Histoire*. Centre Culturel Français Saint-Exupéry, Sépia, Libreville.

Colchester, M., D. Jackson & J. Kenrick, 1998. Forest People of the Congo Basin:past exploitation, present threats and future prospects. *In*: C. Besselink & P. Sips. *The Congo Basin*, 53-63. Netherlands Committee for IUCN, Amsterdam.

Coquery-Vidrovitch, C., 1969. *Brazza et la prise de possession du Congo – la mission de l'ouest Africain 1883-1885*. Mouton, Paris-The Hague.

Davis, S.J.M., 1987. *The Archeology of Animals*. Batsford Ltd, London.

De Wachter, P., 2002. L'agriculture itinérante sur brûlis, base de l'économie Badjoué. *In*: W. Delvingt (ed.), 15-42 (general literature).

Du Toit, J.T. & D.H.M. Cumming, 1999. Functional significance of ungulate diversity in African savannas and the ecological implications of the spread of pastoralism. *Biodiversity and Conservation* 8: 1643-1661.

Gami, N., 1995. *Etude du milieu humain: parc national d'Odzala – Congo*. ECOFAC.

Gowlett, J.A., J.W.K. Harris, D. Walton & B.A. Wood, 1981. Early archeological sites, hominid remains and traces of fire from Chesowanja, Kenya. *Nature* 294: 125-129.

Hamilton, A.C., 1982. *Environmental History of East Africa: A Study of the Quaternary*. Academic Press, London & New York.

Hamilton, A.C., 1984. *Deforestation in Uganda*. Oxford University Press, Nairobi.

Hart, J.A., 1979. *Nomadic hunters and cultivators: A study of subsistence interdependence in the Ituri Forest of Zaire*. Master's Thesis, Michigan State University, Lansing.

Hecketsweiler, P., C. Doumenge & J. Mokoko Ikonga, 1991. *Le Parc national d'Odzala*. IUCN, Gland.

C.M. Hladik, A. Hladik, H. Pagezy, O.F. Linares, G.J.A. Koppert & A. Froment (coord.), 1996. *L'alimentation en forêt tropicale: interactions bioculturelles et perspectives de développement*. L'Homme et la Biosphère, Unesco.

Hladik, A. & E. Dounias, 1996. Les ignames spontanées des forêts denses africaines. *In*: Hladik *et al.*, 275-294.

James, S.R., 1989. Hominid use of fire in the lower and middle Pleistocene – A review of the evidence. *Current Anthropology* 30: 1-26.

Joiris, D.V., 1995. *Systèmes foncier en sociopolitique des populations de la Réserve du Dja. Approche anthropologique pour une gestion en collaboration avec les villages*. ECOFAC.

Lamprey, H.F., 1983. Pastoralism yesterday and today: the over-grazing problem. *In*: F. Bourlière (éd) *Ecosystems of the World 13: Tropical Savannas*, 643-666. Elsevier Scientific Publishing Company, Amsterdam.

Lanfranchi, R. & B. Clist (eds.). 1991. *Aux origines de l'Afrique centrale*. Centres Culturels Français d'Afrique centrale et Centre International des Civilisations Bantu, Libreville.

Lawton, R.M., 1989. The Exploitation of the African Rain Forest and Mans's Impact. *In*: Lieth & Werger: 581-589 (general literature).

Loung, J.-F., 1996. Les Pygmées camerounais face à l'insuffisance des produits alimentaires végétaux de la forêt équatoriale. *In*: Hladik *et al.*, 325-336.

Lugan, B., 1997. *L'Histoire du Rwanda*. Bartillat, Paris.

Maley, J., 2000. L'expansion du palmier à huile (*Elaeis guineensis*) en Afrique Centrale au cours des trois derniers millénaires: nouvelles données et interprétations. *In*: H. PAGEZY *et al.* (ed.), *L'homme et la Forêt Tropicale*. Trav. Soc. Ecologie Humaine, Publ. du Bergier, Paris.

Martelli, G., 1964. *De Léopold à Lumumba (une histoire du Congo belge - 1877-1960)*. France-Empire, Paris.

Martin, P.S., 1966. Africa and Pleistocene overkill. *Nature* 212: 339-342.

Moorhead, A., 1973. *The White Nile*. Penguin Books, London.

Mortehan, M., 1921. L'Agriculture au Ruanda-Urundi. Notes technologiques. *Bulletin agricole du Congo belge* XII: 447-490.

Mworoha, E., J.-P. Chretien, J. Gahama, C. Guillet, F.-X. Nkurunziza, PH. Ntahombaye, Ch. Thibon ET C. Vanacker, 1987. *Histoire du Burundi*. Hatier, Paris.

Ndibi, B.P. & E.J. Kay, 1999. Measuring the local community's participation in the management of community forests in Cameroon. *Biodiversity and Conservation* 8: 255-271.

Oslisly, R., 1994. The middle Ogooué valley: cultural changes and palaeoclimatic implications of the last four millennia. *Azania* 29: 324-331.

Oslisly, R., 1998. Hommes et milieux à l'Holocène dans la moyenne vallée de l'Ogooué (Gabon). *Bulletin de la Société Préhistorique Française* 95: 93-105.

Oslisly, R. 2001. The History of Human Settlement in the Middle Ogooué Valley (Gabon): Implications for the Environment. *In*: W. Weber *et al.* (general literature).

Oslisly, R. & B. Peyrot, 1993. Les gravures rupestres de la vallée de l'Ogooué eu Gabon. Editions Sépia, Paris.

Phillipson, D.W., 1977. *The Spread of Bantu Languages*. Scientific American: 106-114.

Potts, R. 1991., Why the Oldowan? Plio-Pleistocene toolmaking and the transport of resources. *Journal of Anthropological Research* 47: 153-176.

Potts, R., 1993. Archeological interpretations of early hominid behaviour and ecology. in D.T. Rasmussen. *The origin and evolution of Humans and Humaness*. 49-74. Jones and Bartlett Publishers, Boston.

Pourtier, R., 1989. *Le Gabon: Etat et Développement*. L'Harmattan, Paris.

Raemaekers, R.H. (coord.), 2002. *Agriculture en Afrique tropicale*. DGCI, Brussels.

Siegfried, W.R. & R.K. Brooke, 1995. Anthropogenic extinctions in the terrestrial biota of the Afrotropical Region in the last 500,000 years. *Journ. Afr. Zool.* 109: 5-14.

Stringer, B.C., 1993. New views on modern human origins. *In*: D.T. Rasmussen (ed.) *The origin and evolution of Humans and Humaness*. 75-94. Jones and Bartlett Publishers, Boston.

Sutton, J.E.G., 1968. *The Settlement of East Africa*. *In*: B.A. Ogot. *Zamani*. Longman, Nairobi.

Van Den Berghe, L., F.L. Lambrechts & A.R. Christiaensen, 1956. Etude biologique et écologique des glossines dans la région du Mutara (Ruanda). *Acad. roy. Sc. colon., Classe des Sc. nat. med., Mémoire in 8°*, W.S. 4: 1-101.

Van Grunderbeek, M.C., H. Doutrelepont & E. Roche, 1981. *Le premier Age du Fer*.

Van Noten, F., 1983. *Histoire archéologique du Rwanda*. INRS, Butare.

Van Orsdol, K.G., 1986. Agricultural encroachment in Uganda's Kibale Forest. *Oryx* 20: 115-117.

Vansina, J., 1990. *Paths in the rainforests: toward a history of political tradition in equatorial Africa*. University of Wisconsin Press, Madison.

Walker, A., 1993. The origin of the genus *Homo*. *In*: D.T. Rasmussen (ed.). *The origin and evolution of Humans and Humaness*. 29-47. Jones and Bartlett Publishers, Boston.

World Conservation Monitoring Centre, 2000. *Global Biodiversity: Earth's living resources in the 21st century*. By: Groombridge, B & M.D. Jenkins. World Conservation Press, Cambridge.

8. Conservation

Anderson, D. & R. Grove, 1987. The scramble for Eden: past, present and future in African conservation. *In*: D. Anderson & R. Grove (eds.). *Conservation in Africa: people, policies and practice*, 1-12. Cambridge University Press, Cambridge.

Curran, B.K. & R.K. Tshombe, 2001. Integrating Local Communities into the Management of Protected Areas: Lessons from DR Congo and Cameroon. *In*: Weber *et al.* (general literature).

Lahm, S.A., 2001. Hunting and Wildlife in Northeastern Gabon: Why Conservation should extend beyond protected areas. *In*: Weber *et al.* (general literature).

Grove, R., 1987. Early themes in African conservation: the Cape in the nineteenth century. *In*: D. Anderson & R. Grove (eds.). *Conservation in Africa: people, policies and practice*, 21-39. Cambridge University Press, Cambridge.

Mackenzie, J.M., 1987. Chivalry, social Darwinism and ritualised killing: the hunting ethos in Central Africa up to 1914. *In*: D. Anderson & R. Grove (eds.) *Conservation in Africa: people, policies and practice*, 41-61. Cambridge University Press, Cambridge.

Mccracken, J., 1987. Colonialism, capitalism and the ecological crisis in Malawi: a reassessment. *In*: D. Anderson & R. Grove (eds.). *Conservation in Africa: people, policies and practice*, 63-77. Cambridge University Press, Cambridge.

Ntiamoa-Baidu, Y., 2001. Indigenous Versus Introduced Biodiversity Conservation Strategies: The Case of Protected Area Systems in Ghana. *In*: Weber *et al.* (general literature).

ODI, 1998. *Guide des actions de l'UE en faveur des Forêts Tropicales*. Overseas Development Institute, London.

Robyns, W., 1948. Les Parcs nationaux et les Réserves du Congo belge et du Ruanda-Urundi. *In*: *Atlas général du Congo*. Brussels.

Vedder, A., L. Naughton-Treves, A. Plumptre, L. Mubalama, E. Rutagarama & W. Weber, 2001. Epilogue: Conflict and Conservation in the African Rain Forest. *In*: Weber *et al.* (general literature).

9. Great Challenges

Achard, F., H.D. Eva, H-J. Stibig, P. Mayaux, J. Gallego, T. Richards & J.P. Malingreau, 2002. Determinationof Deforestation Rates of the World's Humid Tropical Forests. *Science* 297: 999-1002.

Auzel, Ph. & D.A. Wilkie, 2000. Wildlife Use in Northern Congo: Hunting in a Commercial Logging Concession. *In*: J.G. Robinson & E. L. Bennett, 413-426.

Barnes, R.F.W., 1990. Deforestation trends in tropical Africa. *African Journal of Ecology* 28: 161-173.

Bell, R., 1987. Conservation with a human face: conflict and reconciliation in African land use planning. *In*: D. Anderson & R. Grove (eds.). *Conservation in Africa: people, policies and practice*, 79-101. Cambridge University Press, Cambridge.

Bennett, E. L., 2000. Timber Certification: Where Is the Voice of the Biologist? *Conservation Biology* 14: 921-923.

Bennett, E.L. & J.G. Robinson, 2000. Hunting for Sustainability: The Start of a Synthesis. *In*: Robinson & Bennett, 499-519.

Bennett, E.L. & J.G. Robinson, 2000. Hunting of Wildlife in Tropical Forests. *Environment Department Papers, Biodiversity Series* 76. The World Bank, Washington.

Bermejo, M., 1997. *Expérience pilote de conservation communautaire du sanctuaire de gorilles de Lossi*. ECOFAC-Congo, AGRECO-GEIE.

Bowen-Jones, E., 1998. *The African Bushmeat Trade – A Recipe For Extinction*. Ape Alliance, Cambridge.

Brown, L.H., 1981. The conservation of forest islands in areas of high human density. *Afr. J. Ecol.* 19: 27-32.

Butynski, T.M. & J. Kalina, 1999. Gorilla Tourism: A Critical Look. *In*: E.J. Milner-Gulland & R. Mace. *Conservation of Biological Resources*. Blackwell Science, Oxford.

Caughley, G., H. Dublin & I. Parker, 1990. Projected Decline of the African Elephant. *Biological Conservation* 54: 157-164.

Chapman, C.A. & L.J. Chapman, 1996. Exotic Tree Plantation and Regeneration of natural Forests in Kibale National Park, Uganda. *Biological Conservation* 76: 253-257.

Courouble, M., 1999. Campfire ou l'option du tourisme communautaire. *Canopée* 13: 12-13.

Cuaron, A.D., 2000. A Global Perspective on Habitat Disturbance and Tropical Rainforest Mammals. *Conservation Biology* 14: 1574-1579.

Delvingt, W., 1997. *La Chasse Villageoise. Synthèse régionale des études réalisées durant la première phase du programme ECOFAC au Cameroun, au Congo et en République Centrafricaine*. ECOFAC-AGRECO, Brussels.

Delvingt, W. (ed.) 2002. *La forêt des hommes. Terroirs villageois en forêt tropicale africaine*. Les Presses agronomiques de Gembloux, Gembloux.

BIBLIOGRAPHY

Delvingt, W., M. Dethier, Ph. Auzel et Ph. Jeanmart, 2002. *La chasse villageoise Badjoué, gestion coutumière durable ou pillage de la ressource gibier? In:* W. Delvingt, 2002.

Dethier, M., 1995. *Etude de la chasse dans la réserve du Dja, Cameroun.* ECOFAC-AGRECO, Brussels.

Dethier, M., 2001. Les forêts communautaires, une formule d'avenir pour concilier environnement et développement. *L'Aiélé* 4: 11-12.

Dranzoa, C., 1998. The avifaune 23 years after logging in Kibale National Park, Uganda. *Biodiversity and Conservation* 7: 777-797.

Durieu De Madron, L., 1999. Suivi de l'application d'un plan d'aménagement en Centrafrique. *Canopée* 14: 13-14.

Ehrlich, P.R. & A.H. Ehrlich, 1981. *Extinction: the causes and consequences of the disappearance of species.* Random House, New York.

Eves, H. & R. Ruggiero, 2000. Socioeconomics and the Sustainability of Hunting in the Forests of Northern Congo (Brazzaville). *In*: Robinson & Bennett, 427-454.

Fa, J.E., 2000. Hunted Animals in Bioko Island, West Africa: Sustainability and Future. *In*: Robinson & Bennett, 168-198.

FAO, 2001. *Situation des forêts dans le monde 2001.* Rome.

Fairhead, J. Et M. Leach, 1998. Reframing deforestation. Global analyses and local realities: studies in West Africa. *In: Global Env. Changes Series*, Routledge Publ., London.

Fimbel, Ch., B. Curran & L. Usongo, 2000. Enhancing the Sustainability of Duiker Hunting Through Community Participation and Controlled Access in the Lobéké Region of Southeastern Cameroon. *In*: J.G. Robinson & E.L. Bennett, 356-374.

Floren, A. & K.E. Linsenmair, 2001. Anthropogenic disturbance changes the diversity and the structuring mechanisms of arthropod communities of primary rain forest. *News* 33: 31-33. European Tropical Forest Research Network, Wageningen.

Gami, N., 1999. *Encadrement des ayant-droits des terres de Lossi (sanctuaire de gorilles). Extension du Parc National d'Odzala (cogestion des zones périphériques avec les communautés villageoises).* ECOFAC Congo - AGRECO, Brussels.

Gami, N. & Ch. Doumenge, 2001. Les Acteurs de la gestion Forestière en Afrique Centrale et de l'Ouest. *Document de travail*, 1. FORAFRI, Libreville.

Goodwin, H. & I.R. Swingland, 1996. Ecotourism, biodiversity and local development. *Biodiversity and Conservation* 5: 275-276.

Goodwin, H., 1996. In pursuit of ecotourism. *Biodiversity and Conservation* 5: 277-291.

Grieser Johns, B., 1996. Responses of Chimpanzees to Habituation and Tourism in the Kibale Forest, Uganda. *Biological Conservation* 78: 257-262.

Happold, D.C.D., 1995. The interactions between humans and mammals in Africa in relation to conservation: a review. *Biodiversity and Conservation* 4: 395-414.

Harcourt, A.H., 1986. Gorilla conservation: anatomy of a campaign. *In*: K. Benirschke (ed.). *Primates: The Road to Self-Sustaining Populations*, 31-46. Springer Verlag, New York.

Hart, J.A., 2000. Impact and Sustainability of Indigenous Hunting in the Ituri Forest, Congo-Zaïre: A Comparison of Unhunted and Hunted Duiker Populations. *In*: Robinson & Bennett, 106-153.

Hecketsweiler, P., C. Doumenge & J. Mokoko Ikonga, 1991. *Le Parc national d'Odzala, Congo.* UICN, Gland and Cambridge.

Heuse, E., 2001. Des standards de gestion durable: quel avenir pour les forêts du bassin du Congo. *L'Aiélé* 4: 3-5.

Hutter, C. (ed.), 2000. *An Overview of Logging in Cameroon. A Global Forest Watch Cameroon Report.* World Resources Institute, Washington.

Iovéva, K., 1999. *La commercialisation de viande de brousse et l'impact de l'exploitation forestière sur le gibier.* APFT, Brussels.

Kasenene, J. 2001. Lost Logging: Problems of Tree Regeneration in Forest Gaps in Kibale Forest, Uganda. *In*: Weber et al. (general literature).

King, D.A. & W.P. Stewart, 1996. Ecotourism and commodification: protecting people and places. *Biodiversity and Conservation* 5: 293-305.

Kotze, D.J. & M.J. Samways, 1999. Invertebrate conservation at the interface between the grassland matrix and the natural Afromontane forest fragments. *Biodiversity and Conservation* 8: 1339-1363.

Lanjouw, A., 1999. Tourisme aux gorilles en Afrique centrale. *Canopée* 13: 25-26.

Lahm, S.A., 1996.Utilisation des ressources forestières et variations locales de la densité du gibier dans la forêt du nord-est du Gabon. *In*: C.M. Hladik et al., 1996.

Lejeune, G., 2001. *Les projets de partenariat avec les sociétés d'exploitation forestière en Afrique centrale: quelle contribution durable à l'amélioration de la gestion forestière sur le terrain?* WWF-Belgium, Brussels.

Malcolm, J.R. & J.C. Ray, 2000. Influence of Timber Extraction Routes on Central African Small-Mammal Communities, Forest Structure ans Tree Diversity. *Conservation Biology* 14: 1623-1638.

Malonga, R., 1996. *Circuit commercial de la viande de chasse à Brazzaville.* WCS-GEF, New York.

Mayaux, Ph., F. Achard & J.-P. Malingreau, 1998. Global tropical forest area measurements derived from coarse resolution satellite imagery: a comparison with other approaches. *Environnemental Conservation* 25: 37-52.

Nasi, R., 1998. Aménagement durable des forêts. *Canopée* 11: 3-4.

Noss, A.J., 1998. The impacts of BaAka net hunting on rainforest wildlife. *Biological Conservation* 86: 161-167.

Noss, A.J., 2000. Cable Snares and Nets in the Central African Republic. *In*: J.G. Robinson & E.L. Bennett, 282-304.

Parker, I.S.C., 1979. *The Ivory Trade.* Department of Fisheries and Wildlife, Washington DC.

Plouvier, D. & J.-L. Roux, 1998. La certification: utile ou pas? Deux points de vue. *Canopée* 11: 8-9.

Plumptre, A.J., 2001. The Effects of Habitat Change Due to Selective Logging on the Fauna of Forests in Africa. *In*: Weber et al., 463-479 (general literature).

Putz, F.E., D.P. Dykstra & R. Heinrich, 2000. Why Poor Logging Practices Persist in the Tropics. *Conservation Biology* 14: 951-956.

Robinson, J.G. & E.L. Bennett (eds.), 2000. *Hunting for Sustainability in Tropical Forests.* Columbia University Press, New York.

Sholley, C.R., 1991. Conserving gorillas in the midst of guerillas. *Annual Conference Proceedings, Amer. Assoc. of Zool. Parks and Aquariums*, 30-37.

Soulé, M.E., 1987. *Viable populations for conservation.* Cambridge University Presss, Cambridge.

Stager, J.C., 2001. *Climatic Change and African Rain Forests in the Twenty-First Century. In:* Weber et al. (general literature).

Stewart, K.J., 1992. Gorilla tourism: problems of control. *Gorilla Conservation News* 6: 15-16.

Stewart, K.J., 1993. Gorilla tourism: a reply to Zaire. *Gorilla Conservation News* 7: 12-13.

Struhsaker, T.T., 1981. Forest and primate conservation in East Africa. *Afr. J. Ecol.* 19: 99-114.

Struhsaker, T.T., 1987. Forestry Issues and Conservation in Uganda. *Biological Conservation* 39: 209-234.

Struhsaker, T.T., 1997. *Ecology of an African rain forest: logging in Kibale and the conflict between conservation and exploitation.* University Press of Florida, Gainesville.

Thiollay, J.-M., 1997. Disturbance, selective logging and bird diversity: A Neotropical forest study. *Biodiversity and Conservation* 6: 1155-1173.

Tutin, C.E.G., I.S. Porteus, D.S. Wilkie & R. Nasi, 2001. Comment minimiser l'impact de l'exploi-tation forestière sur la faune dans le Bassin du Congo. *Les Dossiers de l'ADIE*, Sér. Forêt, 1: 1-36.

Vanwijnsberghe, S., 1996. *Etude sur la chasse villageoise aux environs du Parc d'Odzala.* ECOFAC-AGRECO, Brussels.

Vedder, A. & W. Weber, 1990. The Mountain Gorilla Project (Volcanoes National Park). *In*: A. Kiss (ed.). *Living with Wildlife: Wildlife Resource Management with Local Participation in Africa*, 60-83. World Bank, Washington DC.

Verschuren, D., K.R. Laird & B.F. Cummings, 2000. Rainfall in equatorial east Africa during the past 1,100 years. *Nature* 403: 410-414.

Weber, W., 1993. Primate conservation and ecotourism in Africa. *In*: C.S. Potter, J.I. Cohen & D. Janczewsky (eds.). *Perspectives on Biodiversity: Case Studies of Genetic Resource Conservation and Development*, 129-150. AAAS Press, Washington DC.

Wetterer, J.K., P. Walsh & L. White, 1999. *Wasmannia auropunctata*, une fourmi dangereuse pour la faune du Gabon. *Canopée* 14: 10-11.

White, J.W.C., 1993. Don't touch that dial. *Nature* 364: 186.

White, L.J.T., 1994. The effects of commercial mechanized selective logging on a transect in lowland rainforest in the Lopé Reserve, Gabon. *Journal of Tropical Ecology* 10: 313-322.

White, L., 1994. L'impact de l'exploitation forestière dans la réserve de faune de la Lopé (Gabon). *Canopée* 3: 2-3.

White, L.J.T., 2001. The African Rain Forest: Climate and Vegetation. *In*: Weber et al., 3-29 (general literature)

Wilkie, D., E. Shaw, F. Rotberg, G. Morelli & Ph. Auzel, 2000. Roads, Development, and Conservation in the Congo Basin. *Conservation Biology* 14: 1614-1622.

Williamson, L., 1999. Le tourisme de vision de gorilles de montagne et le conflit de la région des Grands Lacs. *Canopée* 13: 21

Index

A

Acacia mearnsii 337
– *melanoxylon* 337
Accipiter erythropus (Red-thighed Sparrowhawk) **205**
– *tachiro* (African Goshawk) **205**
Acheulean 240
Acraea 171
– *encedon* **187**
– *epaea* **187**
Acraeinae 186-187, **187**
Acroceras zizanoides 132
Acrostychum aureum **143**
ADIE 337
Aedes 189
Aerangis kotschyana 82, 83
Aframomum 103
– *sangaris* 103
Africa, emergence of continent 18-19,
–, plate tectonics **18-19**, 50, **50**
African Development Bank (ADB) 296, 337
African Timber Organisation (ATO) 336
African Wildlife Foundation (AWF) 295
Afrixalus dorsalis **195**
Afro-alpine habitats 40, 94-97
Afrocrania 70
Afropavo congensis (African Peacock) 201, 316
Afrotrilepis pilosa **92**, 93
Agaonidae 174
Agapornis (lovebirds) 2002
Agriculture, American plants 254-256
–, beginning 246-248
–, cash crops 315
–, first crisis 253
–, industrial 263
–, introduction of banana 250-251
–, shifting **303**, 315
–, Soudanian tradition 248
–, use of fire 256
Akagera (NP) 82, 284, 288, 353
Akagera (river) 35, 258
Akrum (inselberg) **17**
Albert (lake) 25, 35, 36, 258
Albert (NP) 283, 284
Albertine Rift 25-29, **25**, 44, 71, 252, 256, 255, 262, 266, 282, 301, **301**,
Albizia 78, 84
– *adianthifolia* 108
Alcedo cristata (Malachite Kingfisher) 129
– *quadribrachys* (Shining-bleu Kingfisher) 129, 143
Alchemilla johnstoni 95
Alchornea 56
– *cordifolia* 121, 170
Alchornea floribunda 78
Alethe 205
Allenopithecus nigroviridis (Allen's Swamp Monkey) 211-212
Aloe macrosyphon 82
– *mubendesis* 93
– *volkensi* 82
Alstonia boonei (Emien) 112, **113**
Altos de Nsork (NP) 353
Amaranthus (amaranthe) 247
Amauris vashti **187**
Ammania baccifera 132
Amphibians 194-195
Anadelphia 106, 107, 108
Anchomanes difformis **176**

Andropadus 204
Andropogon 107, 108
Angiosperms, current species 51
–, families 50-51, **51**
–, genera and species 61
–, history 49-51
Angraecum eichlerianum 93
– *podochiloides* **159**
Aningeria (cf. *Pouteria*)
Ankole, volcanism 28
Annobon (island) 24, 66, 353
Annona senegalensis 108
Annonidium manni 172
Anogeissus leiocarpus 82
Anomaluridae 225
Anomalurus beecrofti (Beecroft's Anomalure) 225
– *derbianus* (Lord Derby's Anomalure) 225
Anopheles (malaria mosquito) 189
Ansellia (orchid) 159
Anthocleista 100
– *grandiflora* 91, 101
– *schweinfurthii* **101**, 103
– *vogelii* 101
Anthoscopus flavifrons (Yellow-fronted Penduline Tit) 111
Anthreptes aurantium (Violet-tailed Sunbird) 129
– *gabonicus* (Brown Sunbird) 144
Anthrocaryon klaineanum 81
Anthus pallidiventris (Long-legged Pipit) 107, 112
Antonota vignei 125
Ants 191-193
Aonyx congica (Swamp Otter) 130, 226
Apalis flavida (Yellow-breasted Apalis) 144
– *jacksoni* (Black-thoated Apalis) 64
Apaloderma aequatoriale (Bar-cheeked Trogon) **202**
– *vittatum* (Bar-tailed Trogon) 203
Aphanocalyx 78
Appias 185
Apus apus (Common Swift) 206
Arabo-African Rift 19, 25
Arabs (people) 257, 258, 261
Araceae 158
Arachis hypogea (Ground Nut) 254, 255, 271
Araucariaceae 50
Archaeology 240-242
Archean (geologic period) 15
Arctocebus aureus (Golden Antigwabo) **208**
– *calabarensis* (Calabar Antigwabo) 208
Arecaceae (palms) 62
Aristida 108
Artemisia 55
Arthroleptidae 194
Artiodactyls 230-237, 313
Arundinaria alpina (cf. *Sinarundinaria*)
Aruwimi (river) 33
Asio otus (Long-eared Owl) 206
Association pour le développement de l'information environnementale (ADIE) 296
Ataenidia conferta 113, **114**
Atheris **197**, 198
Atherurus africanus (Brush-tailed Porcupine) 225, **225**, 273, **317**
Atmosphere, terrestrial 18
Attilax paludinosus (Swamp Mongoose) 226
Aukoumea klaineana (Okoumé, Gaboon Mahogany) 56, 63, 77, **77**, **169**, 272
Avenir des peuples de forêt tropicale (programme APFT) 270

Avicennia germinans (Great White Mangrove) 138, 139, 140, 148, **148**

B

Baamba (ethnic group) 276
Bacopa crenata 132, 133
– *egensis* 132
Bacteria 165
Badjoué (ethnic group) 315
Baikaea robynsii 129
Baikal (gun) 266
Baillonella toxisperma 81, 176
Bais (cf. clearings)
Baker, Samuel 258
Bakossi (FR) 352
Balanophoraceae 167
Balthasaria 70
Bamboo thickets 117, **117**
Bambouto (mountains) 22
Bambuko (FoR) 352
Bamenda (mountains) 23
Bantu (people) 244-246, 251, 252-253
Banyang Mbo (FoR) 352
Barombi Mbo (crater lake), 23, 54, 55, 56
Barteria 308
– *fistulosa* 192
– *nigritana* 110
Basse-Lobaye (biosphere reserve) 353
Bateke (plateau) 33, 74
Batis minor (Black-headed Batis) 112
Bats 174, 177
Bdeogale nigripes (Black-legged Mongoose) 226, **226**
Beaches, pioneer vegetation 136-137
Bebearia 173, 186
Bees, social 174
Begonia subaccata **85**
Begoniaceae **85**, 158
Belgium, Belgians 257, 260, 261, 262
Belinga (mountains) 15
Berlinia 77, **78**, 125, 129
– *bracteosa* 126
Bias musicus (Crested Shrike-Flycatcher) 204
Biega (volcano) 28
Bikinia 78
Biodiversity, concept 297
–, crisis 297
Biogeography 58-66
–, ancient and recent species **60-61**
–, Atlantic coastal region 62
–, Cameroon-Gabon region 62
–, centres of endemism 62
–, Congolian region 62
–, regions 61-66
Bioko (island) 24, 66, 80, 253, 315
Biomes 40, **40**
Biophytum zenkeri **129**
BirdLife International 299
Birds 174, 176, 200-206
–, aerial screeners 206
–, biogeography 206
–, endemic families 200
–, multispecific groups 206
–, non-passerines of the canopy 202-204
–, passerines 204-205
–, raptors 205-206
–, terrestrial non-passerines 201-202
Bisoke (volcano) 26, **28**, 91
Bitis 197
– *gabonica* (Gabon Viper) **197**
Boiga pulverulenta (see *Toxicodryas*)
Bokoué (river) 142
Bombax buonopozense (Kapok Tree) **73**, 175

Bonobo 314
Booué, rapids **17**
Bostrychia rara (Spot-breasted Ibis) **201**
Boumba-Bek, Nki (FR) 287, 352
Bovidae 233-237
Bovids, domesticated 249, **250**
Brachystegia 77, 81
– *laurentii* 80
Bradypterus grandis (Dja River Warbler) 133
Brazza (mountain) 16, **16**
Bridelia ferruginea 107, 108
British (people) 254, 258, 260, 261,
Brown, John 283
Bryophytes 87
Bubo leucostictus (Akun Eagle-Owl) **206**
– *poensis* (Fraser's Eagle)Owl) 206
Bubulcus ibis (Cattle Egret) **134**
Buchnerodendron speciosum 103
Budongo (FoR) 156, 285, 353
Bufo superciliaris 194
Bukavu (town) 266
Bulbophyllum (orchid) 143, 159, **159**
Bundibugyio (town) 267
Bungoma (FoR) 353
Burseraceae 78
Burton, Richard 258
Burundi 261, 263, 264, 265, **265**, 267, 268, 269, 284
Bururi (FoR) 86, **285**, 353
Bushmen (people) 241, 244
Buteo auguralis (Red-necked Buzzard) 107
– *oreophilus* (Mountain Buzzard) 205
Butorides striatus (Green-backed Heron) 130
Buttresses 150, **150**
Bwindi Impenetrable (NP) 86, 116, 218, 219, 256, 288, **195**, 353

C

Caesalpiniaceae 73, 81
Calabaria reinhardtii (see *Charina*)
Calamus 119, 128, 155
Caldeira de Luba (scientific reserve) 353
Callitrys calcarata 337
Calpocalyx heitzii 77
Calyptrochilum emarginatum (orchid) **159**
Camellia sinensis (Tea) 263
Cameroon (State) 265, 266, 271
Cameroon (volcanism) 22-23, **22**, **23**, 96, 104
Campethera nivosa (Buff-spotted Woodpecker) **203**
Campo (FaR) 284
Campo-Ma'an (NP) 287, 352
Canarium schweinfurthii (Aiélé or Incense Tree) 103, 113, 246, 247
Canavalia rosea 136, **137**
Canirallus oculeus (Grey-throated Rail) 202
Canopy, composition over time 101
–, structure 150-152, **150**, **151**
Capsicum annum (pepper) 255
Carapa grandiflora 86
Carapa procera 80, 84
Carbon, balance 310
–, isotopes 57
Carboniferous Era 18
Cardamine obliqua 95
Cardioglossa 194
Cardoderma cor 223
Carex monostachya 95
– *runssoroensis* 95
Carvings, rock 242
Cash crops 263, 273
Cassia siamea 309
Cassipourea 74, 75

INDEX

Cassytha filiformis 83, 137, **137**
Casuarina equisetifolia (Whistling Pine) 307
Cathormion altissimum 125
Catuna **186**
Cecropia peltata 308
CEFDHAC 291, 292, 295
Ceiba pentandra (Kapok Tree) 80, 103, **120**, 121, 175, 258
Celtis 56, 78, 81, 85
Centhoteca lappacea **152**
Centipede **182**
Central Africa, geology 14-15, **15**
Central African Republic 269, 284
Central-Soudanic 251, 252
Centre international de recherche agronomique pour le développement (CIRAD) 344
Centre international de recherche en agroforesterie (ICRAF) 343
Centropus anselli (Gabon Coucal) 203
Cephalophus callipygus (Peters' Duiker) 130, 234
– *dorsalis* (Bay Duiker) 234
– *leucogaster* (White-bellied Duiker) 234
– *monticola* (Blue Duiker) 130, **233**, 234, **317**
– *nigrifrons* (Black-fronted Duiker) 234, **234**, **317**
– *ogilbyi* (Ogilby's Duiker) 234
– *rubidus* (Rwenzori Duiker) 234
– *rufilatus* (Red-flanked Duiker) 234
– *silvicultor* (Yellow-backed Duiker) 234
Cerambicidae (longhorn beetles) **184**
Ceratophyllum 119
Cercocebus agilis (Agile Mangabey) 211
– *torquatus* (Collared Mangabey) 211, **211**
Cercopithecidae 209-217
Cercopithecus 212-216
– *ascanius* (Red-tailed Monkey) 214
– *cephus* (Moustached Monkey) 214
– *dryas* (Salongo Monkey) 213
– *erythrotis* (Red-eared Monkey) 214
– *hamlyni* (Owl-faced Monkey) 212, **213**
– *lhoesti* (L'Hoest's Monkey) 213, **214**
– *mitis* (Blue Monkey) 215
– *neglectus* (De Brazza's Monkey) 212, **213**
– *nictitans* (White-nosed Monkey) 215, **215**
– *pogonias* (Crowned Monkey) 215
– *preussi* (Preuss's Monkey) 214
– *solatus* (Sun-tailed Monkey) 214, **214**
– *wolfi* (Wolf's Monkey) 216
Cerestis 155
Ceyx lecontei (African Dwarf Kingfisher) 203
Chaetocarpus africanus 103
Chamaeangis 159
Chamaeleo cristatus **198**
Charaxes (butterfly) 186,187
– *ameliae* **186**
– *bohemanni* 117
– *fournierae* **188**
– *opinatus* 187
Charina reinhardtii (snake) **196**
Cheep 249
Chiromantis rufescens 195
Chiroptera (Bats) 222-223
Chlidonias hybrida (Whiskered Tern) 120
– *leucopterus* (White-winged Tern) 120
Christianism 278
Chromolaena odorata 308
Chrysobalanus icaco 83
Chrysococcyx flavigularis (Yellow-throated Green Cuckoo) 203
– *cupreus* (Esmerald Cuckoo) 203

Chrysophyllum 85
– *africanum* 125
– *gorungosanum* 89, 117
Chrysops (Fly) 189
Churches, alternative 278
Ciconia episcopa (Woolly-necked Stork) 143
CIFOR 335, 343
CIRMF 344
Cisticola anonymus (Chattering Cisticola) **133**
– *brachypterus* (Siffling Cisticola) 112
– *brunescens* (Pectoral-patch Cisticola) 107
– *galactotes* (Winding Cisticola) 120
– *natalensis* (Croaking Cisticola) 107, 112
CITES 230, 291, 319
Civetta civetta (African Civet) 226
Clappertonia ficifolia 107, **107**, 108
Clearings, swampy 130-135
Cleome afrospina 118, **119**, 132, 133
Cliffortia 54, 91
Climate, altitudinal gradient of rainfall **71**
–, austral summer 43
–, boreal summer 43
–, change 309-310
–, current 42-47
–, cyclic glaciations 53
–, dry season 46, **46**
–, during Cretaceous 50-51
–, during Tertiary 51
–, *El Niño* and *La Niña* 47, 57, 171, 309
–, evapotranspiration and evaporation 45
–, insulation 44-45, **44**
–, invisible rainfall 45
–, isohyets **42**, **43**, **44**
–, last 10,000 years 56-57
–, last 200,000 years 54-55, **54**
–, mist 45
–, ocean surface water temperature 47
–, pluvial periods 48
–, rainfall 42-44, **47**, **94**
–, solar radiation 45, **45**
–, stratiform clouds 43
–, temperature 42
–, upwellings 47
–, variations 46-47, **47**, 48
–, worsening 57
Club mosses (see Lycopods)
Cnestis **176**
Cocos nucifera (Coco Palm) 258, 307
Coelocaryon preussi (Ekoune) 78
Coffea (Coffee Tree) 263, **263**
Cogniauxia podolaena 128
Cola 81, 84, **176**, 247
– *flavolutina* **77**
– *griseiflora* 78
– *lizae* 176
Coleus (tubers) 247
Colobus angolensis (Angola Colobus) 216, 339
– *guereza* (Guereza Colobus) 216, **216**
– *satanas* (Black Colobus) 216
Colocasia esculenta (Cocoyam or Taro) 251, **251**, 271
Colonisation 259-265
–, the division 259-260, **260**
–, the first Europeans 258
–, the prelude 257-258
–, the seizure 260-265
–, traditional law 261
Combretaceae 81, 82
Combretum 175
– *platypterum* **155**, 176
– *racemosum* **153**

Commelidinium 152
Congo (Independent State) 258, 260
Congo (river) 30, 33-35, **34**, **122**
Congo Basin, origin 20
–, satellite picture **122-123**
Congo, Belgian 261, 262, 263, 264, 265
Congo, Democratic Republic (Zaire) 265, 266, 268, 269, 271, 290
Congo, Republic 265, 284, 290
Congo-Nile (divide) 26, 259
Conifers 49
Conkouati (NP) **46**, 288, 295, 353
Connaraceae 176
Connocarpus erectus (Grey Mangrove) 138, 140
Conraua goliath (Goliath Frog) 194
Conservation International (CI) 295, 344
Conservation, actors 291-296
–, choice of sites 298-300
–, colonial period 282-286
–, concepts et strategies 296
–, effects of development 288
–, foreign advisors 293
–, from 1960 to 2000 286-302
–, funding agencies 293-295
–, heads of states 293
–, human context 300-302
–, NGO 295
–, pre-colonial 282
–, social strife and war 289-290
Convention, Alger 291
–, biodiversity 291
–, Bonn 291
–, London 284
–, Ramsar 291
–, Rio 291
–, Washington (CITES) 291
–, World Heritage 291, 292
Convolvulaceae 128, **128**
Cooperation, French 290, 296
Copaifera 78
– *religiosa* (Anzem) 81
Cordia 174
Cordylobia anthropophaga 189
Corythaeola cristata (Great Blue Turaco) 202
Costus 103, 171
– *lucanusianus* **103**
– *violaceus* 77
Coula edulis (Coula) 78
Crab **138**
Craton 15, **15**
Cretaceous, (geologic period) 18-19
Cricetomys eminii (Emin's Rat) **225**, **317**
Crinum purpurascens 31
Crocidura 223
Crocodylus cataphractus (Long-snouted Crocodile) **133**, **196**
– *niloticus* (Nile Crocodile) 196
Crocuta crocuta (Spotted Hyena) 130, 227
Crossarchus (mongoose) 226
Crossopteryx febrifuga 107, **108**, 111, 112
Croton mubango 103
Crown (from a tree), form 151, **151**
Crystal (mountains) 15, 21, **21**, 83, **146**
Ctenium 106
Cubitermes 191
Cucumerops (Squash) 247, 271
Culcasia 155
Culex 189
Culture, agricultural area 271
–, diversification 241
–, fishing 273
–, gathering 273

–, hunting 266, 273
–, integration 251-253
–, mountain agriculture 267
–, religion 278
–, shifting agriculture **270**, **271**, 272-273
–, social relations 276-278
–, traditional forest rights 276
–, traditional land rights 275-276
Cupressus (cyprès) 264, 267
CUREF 323
Cushites 244, 251
Cyanobacteria 18, 90
Cyathea manniana (Tree Fern) 87, 88, **89**
Cycads 49, **49**, 50
Cyclosorus tottae 132
Cylicodiscus 81
Cymothoe **186**
Cynometra 78
– *alexandri* 80, 82
– *megalophylla* 125
Cyperus difformis 132
– *papyrus* (Papyrus) 121
Cyrtorchis 159, **175**
Cyrtosperma senegalense **173**

D

Dacryodes buettneri (Ozigo) 77, 78, 113, 273
– *edulis* (Atanga, Safu) **273**
– *pubescens* 81
Dalberghia ecastaphyllum 137, **137**
Danainae 186-187, **187**
Danaus chryssipus **187**
Delonix regia 171
Demography 264, 265-266, 267
Dendroaspis jamesoni (Green Mamba) 198
Dendrohyrax dorsalis (Tree Hyrax) 228, **228**
Desbordesia glaucescens (Alep) 77
Desert, biome 40-41
Detarium macrocarpum 176
Diaphananthe 159
Dichaetanthera corymbosa 89
Dicranopteris linearis 88, **105**
Didelotia 78
– *letouzeyi* 77
Digitaria 108
Dimonika (biosphere reserve) 288, 291, 353
Diodia scandens 132
Dioscorea (jam) 246
Diospyros (Ebony) 264
– *abyssinica* 85
Disa 96, 160
Diseases, epidemic 260-261
Dissotis 52
Distemonanthus 81
Dja (FaR) 15, **46**, 216, 218, 231, 262, 266, 270, 273, 284, 292, 313, 315, 318, 352
Dombeya goetzenii 75
Donella ogowensis 129
Dorylus **192**
Douala-Edéa (FaR) 284, 352
Doudou (mountains) 83
Drosera pilosa 132
Drynaria 158, 159
Dryotriorchis spectabilis (Congo Serpent Eagle) 111
Drypetes afzelii 125
Du Chaillu (mountains) 15, 21, 83
Du Chaillu, Paul 258
Duiker 315, 317
Duparquetia orchidacea 110
Dutch (people) 254
Dyaphorophya concreta (Yellow-bellied Wattle-Eye) **204**

Dzanga-Ndoki (NP) 315, 353
Dzanga-Sangha (special reserve) 236, 311, 353

E

Echinochloa 119, 120, 121
Echuya (forest) 215
ECOFAC (programme) 130, 214, 220, 270, 290, 294, 295, 302, 319, 321, 336, 337, 341, 342, 344
Ecoregions 298, **298**
Edward (lake) 35, 36
Eichhornia crassipes (Water Hyacinth) **34**, 131, 307, **308**
Eidolon helvum 222
Ejagham (FoR) 352
Ekanya (tourist camp) **340**
Elaeis 315
Elaeis guineensis (Oil Palm) 119, 161, 273, 246-247, **246**
Elaeocharis geniculata 132
Elapidae 198
Elephant bath (cf. clearings)
Elephantidae 228-230
Elephas 229, 242
Eleusine coracana (Finger Millet) 247, **247**
Elila (river) 33
Elyonurus 106, 108
Emin Pacha 259
Ensete ventricosa (Wild Banana Tree) 87
Entandrophragma 78, 80, 81, 175, 264
– *angolense* (Tiama) 81
– *candollei* (Kosipo) 81, 113, 172
– *congoense* (Tiama) 81
– *cylindricum* (Sapelli) 81, 113
– *excelsum* 74, 75, 86
– *utile* (Sipo) 81, 113
Enydra fluctuans 132, 134
Epiphytes 156-161, **156-160**
Epitola (butterfly) 185
Epixerus wilsoni (Biafran Bight Palm Squirrel) 225
Equatorial Guinea 265
Eragrostis 89
Eremospatha 119, 155
Erica 95
– *arborea* 95
– *kingaensis* **94**, 95
Erosion circus 20, **20**, **21**, 105
Esobe (see clearings)
Estuario del Rio Muni (NR) 353
Estuary, Gabon 56, **99**, 142, **239**
Eucalyptus 264, 267
Euclea 83
Eugenia 80
Eulophia (orchid) 107
Eulophia oedoplectron 107
Euoticus elegantulus (Elegant Needle-clawed Galago) 208, 209
– *pallidus* (Pallid Needle-clawed Galago) 208
Eupatorium odoratum (cf. *Chromolaena odorata*)
Euphaedra (butterfly) 173, 186,187
– *hewitsoni* **186**
–*hybrida* **187**
Euphorbia dawei 82
– *letestui* **92**, 93
Euphorbiaceae 104, 107
European Commission 290, 293-294, 296
Eurystomus gularis (Blue-throated Roller) 206
Evaro (lake) **32**, **120**, 121
Exploitation, of natural resources 272-275
Extinctions 19, 242
Ezanga (lake) 121

F

Fagara (cf. *Zantoxylum*)
Fang (ethnic group) 275
FAO 306
Fauna and Flora International (FFI) 295
Fauna, diversity and abundance 182-184
Faurea saligna 89, 117
Fegimanra **83**
Felidae (Cats) 227
Felis aurata (Golden Cat) 227, **227**, 314
– *lybica* (African Wild Cat) 227
– *serval* (Serval Cat) 227
Fernan Vaz (cf. Nkomi)
Ferns 88, **158**, 159
Ficalhoa 70
– *laurifolia* 89
Ficus (fig trees) 80, 161, **161**, 174, **175**, 247
Filariosis 189
Fimi (river) 35, 126
Fish ponds **320**
Fishing **248**, **276**, **277**, 315
Flies 174, 189
Flora, inventories 58
–, specificity 62
Fogging, canopy 184
FORAFRI 343
Forest Stewardship Council (FSC) 335, 336
Forest, *Afrocrania volkensi* 91
–, Biafrean coastal 76, **76**, 76-77
–, Congolian lowland 76,
–, Congolian mixed 78, **79**
–, current situation 306-307
–, definition 40-41
–, evergreen 72, 76
–, evolution during lest 10,000 years 56-57
–, frontier 299-300, **300**
–, *Gilbertiodendron* **76**, 252, 317
–, *Hagenia* 91, **91**
–, humid 40
–, hyper humid 72, 77
–, inundatable 74, **124**
–, inundatable, *igapo* type 125-126
–, inundatable, *varzéa* type 124-125, **125**
–, lowland, endemic families 62
–, lowland, floristic composition 62-63
–, Marantaceae 76, 112-115, **113**, **114**
–, mist 309
–, mixed with Caesalpiniaceae 76, 77, **78**
–, mountain 63-65, **64**, 70-71, 76, 86-91, **86**, 175, 309
–, mountain, biogeography 64-65
–, mountain, endemic genera 70
–, mountain, floristic composition 64
–, mountain, in Albertine Rift **65**
–, mountain, in Cameroun **65**
–, mountain, pioneer 116-115, **116**
–, mountain, swamp 91
–, *Musanga* 102
–, non extractive values 302
–, *Olea europaea* **82**, 241-242
–, original area 306
–, *Parinari excelsa* 86-87
–, *Phoenix reclinata* **140**
–, pioneer 111-112
–, *Podocarpus* 90, 91
–, primary, definition 103
–, riparian 75
–, sclerophyl 82-83
–, sclerophyl coastal **83**, 83
–, secondary 175
–, secondary, lowland 103
–, secondary, mountain 104
–, secondary, submontane 104
–, semideciduous 72, **76**, 80-81, **81**, 80-82, 103
–, submontane 71, **76**, 83-85, **84**
–, swamp 64, 76, 126
–, tropical, fragility 102
–, with monodominance of Caesalpiniaceae 78-79, **79**
FORNESSA 344
Fossey, Diana 338
France, French 254, 258, 260
Francevillian (geology) 16
Francolinus finschi (Finsch's Francolin) 109
– *lathami* (Latham's Francolin) **201**
French Equatorial Africa 261, 262, 263, 265, 284
Frugivory 177
Funisciurus carruthersi (Carruther's Mountain Squirrel) 224
– *isabella* (Lady Burton's Rope Squirrel) **224**
– *lemniscatus* (Ribboned Rope Squirrel) 224
– *pyrrhopus* (Fire-Footed Rope Squirrel) 224
Funtumia africana 103, 175
– *elastica* 113
Fynbos 307

G

Gabon 265, 268
Gahinga (see Mgahinga)
Galago matschei (Spectacled Galago) 209
Galagoides demidoff (Demidoff's Galago) 209, **209**
– *thomasi* (Thomas's Galago) 209
Galagonidae 208-209
Gallago alleni (Allen's Squirrel Galago) 209, **209**
Gamba (protected area complex) 353
Garcinia 84, 128
–*kola* (Bitter Kola) **272**, 273, 315
– *lucida* 84
Genetta servalina (Servaline Genet) 226
–*tigrina* (Blotched Genet) 226
– *victoriae* (Giant Genet) 226
Genocide, Rwandan 269
Genyorchis 143, **159**
George (lake) 35, 36
Geotrypetes seraphini **194**
Germans (people) 260, 261,
Gilbertiodendron brachystegioides 126
– *dewevrei* (Mbau, Limbali) 77, 79, 125, **169**
– *klainei* 126
– *ogoouensis* 80
Gilletiodendron 77
Ginkgophytes 49
Giraffidae 232
Giri (River) **122**
Gishwati (forest) 104, 288, 294, 353
Gladiolus **93**
Glareola nuchalis (Rock Pratincole) **31**
– *cinerea* (Grey Pratincole) **32**
Glaucidium albertinum (Prigonine's Owlet) 205
–*tephronotum* (Red-chested Owlet) 206
Gleicheniaceae 88
Global Environment Fund (GEF) 296
Global Forest Watch (GFW) 323
Gloriosa superba 172, **172**
Glossina (Tsetse Fly) 249, 261
Gluema ivorensis 125
Gnetum **273**
Goat 249
Golden Monkey 215
Goliatus goliatus **184**
Gondwana, continent 18-19
Gorilla gorilla (gorille) **6**, 130, 133, 177, 217-220, **217**, **218**, **219**, **284**, 290, 314, **338**, **344**, **345**
– gorilla, tourism 338-339
Gorsachius leuconotus (White-backed Night Heron) 129
Gosswilerodendron balsamiferum (cf. Prioria)
Granites, intrusive 16
–, metamorphic 16
Grant, James 258
Graphium (butterfly) 185
– *gudenusi* **187**
– *latreilianus* **185**
– *policenes* **185**
Graphiurus (African dormice) 225
Grassfields (region) 256
Great Blacks 244, 252, 274
Great Lakes (region) 259
Greenpeace 335
Greenwayodendron suavolens 78
Grevillea robusta 264, 337
GTZ 293
Guaduella 152
Guarea 78, 81, 113
Guibourtia demeusei 126
Gymnosperms 49
Gypohierax angolensis (Palm-Nut Vulture) **205**

H

Habenaria 96, 107, **107**
– *procera* **93**
–, human 256
Hadean (geologic period) 15
Hagenia 70
– *abyssinica* 54, 55, **55**, 56, 91, 94, 117
Halcyon badia (Chocolate-backed Kingfisher) 203
– *malimbica* (Blue-breasted Kingfisher) 143
Haliaeetus vocifer (African Fish-Eagle) 205
Hallea 75, 139, **148**
Haplocoelum galaense 82
Haplormosia monophylla 125, 126
Harma theobene **186**
Harungana madagascariensis 102, 103, 110
Haumania liebrechtsiana 103, 112-113, **113**, 114
Heath, ericaceous **94**,
–, *Helichrysum* 96
Helichrysum 96
Heliosciurus gambianus (Gambian Sun Squirrel) 224
– *rufobrachium* (Red-legged Sun Squirrel) 224
– *ruwenzori* (Rwenzori Sun Squirrel) 224
Hemisotidae 195
Herbivory 169-170
Herpestes ichneumon (Egyptian Mongoose) 226
– *sanguinea* (Slender Mongoose) 226
Herpestidae 226
Heteranthera callifolia 131
Hibiscus 247
– *tiliaceus* 83, **83**, 137
Himantornis haematopus (Nkulengu Rail) 202
Hippopotamidae 230
Hippopotamus amphibius (Hippopotamus) 120, 230-231
Hirundo nigrita (White-throated Blue Swallow) 129

INDEX

Holoptelea grandis 103
Hominidae 217-219
Homo erectus 240
– *habilis* 240
– *neanderthalensis* 240, 242
– *sapiens* (Modern Man) 240
– *sapiens*, environmental impact 242
– *sapiens*, immigration of current populations 244-245
Horsetails 88
Hot springs 29
Humiriaceae 77
Hunting 275, **282**, 311-321
–, alternatives resources 320
–, for ivory 319, **319**, **320**
–, game species 313-314, **313**, **314**
–, impact 314-317
–, laws 319-321
–, pits 313
–, recent evolution 317-319
–, sustainability 314-317
–, techniques 311-313, 315
–, territories 314
–, with gun 313
–, with nets 311-312, **311**
–, with snares 312-313, **312**
Hyaenidae 227
Hydnoraceae 167
Hydrocharis chevalieri 131
Hydrography 30-36, **30**
–, biogeographic importance 36
–, impact on Man 36
Hydrolea glabra 131
Hyemochus aquaticus (Water Chevrotain) 129, 232, **232**
Hylochoerus meinertzhageni (Giant Hog) 231-232, **231**, 246
Hymenocardia acida 108
Hymenostegia 78
Hyparrhenia 106, 107, 108
Hypericum bequaerti 95
– *keniense* 95
– *revolutum* 91
Hyperolius 195
Hypolycaena 185, **185**
Hypsignatus monstruosus (Hammer Bat) 222, **222**

I

ICCN (IZCN) 292
Idi Amin 289
Ilex mitis 54, 55, 74, 89
Impatiens **61**, 64, 95
Important Bird Areas (IBA) 299, **299**
INAP 293
INCN 292
Independence 265
Inselberg 16, 92-93
Insects 184-193
Institut de recherche en écologie tropicale (IRET) 344, **347**
Institut de recherche pour le développement (IRD) 344
International Tropical Timber Organisation (ITTO) 336
Iolaus 185
Ipassa (integral reserve) 287, 353
IPCC 309
Ipomea batata (Sweet Potato) 255
– *pes-caprae* 136, **137**
Irangi (scientific reserve) **8**, **33**, 280
Irvingia gabonensis 77, 247
– *grandifolia* 126

Irvingiaceae 78
Itimbiri (river) 33
Itombwe (forest) 74, 95, 206, 218, 256
Ituri (region) 315
Itwara (FoR) 353
IUCN 295, 301, 306, 315
Ivindo (river) **12**, 30, 31, **31**, 126, **127**, 231
Ivory 254
Ixobrychus minutus (Little Bittern) 120
Ixora brachypoda 132

J

Jasminum dichotomum 132
Julbernardia 78
– *seretii* 78, 79
Juniperus procera 89
Jurassic (geologic period) 18

K

Kabambare 257
Kahuzi (volcano) 28, 95
Kahuzi-Biega (NP) 86, 116, 117, 218, 226, 251, 266, 287, 289, 290, 292, 293, 339, 353
Kalahari (desert), sands 20
Kalinzu (FoR) 353
Kamiranzovu (swamp) 35, **55**
Karisimbi (volcano) 26, 91
Kasai (ancient mountain range) 16
Kasai (river) 33, 34, 35, 251
Kashiru (swamp) 54
Kasianduku (crater lake) 28, 261
Kasongo (town) 257
Kasyoha-Kitomi (FoR) 353
Katanga (mountain range) 16
Kazinga (NP) 284
Kazinga Channel 28
Khartoumians 257, 259
Khaya 80, 81
Kibale (NP) 261, 288, 294, 353
Kibali (ancient mountain range) 16
Kibanguism (religion) 278
Kibara (ancient mountain range) 16
Kibira (NP) **75**, 86, **86**, 104, 116, 263, 290, 353
Kigelia africana 110
Kigwena (FoR) 353
Kilum-Idjim (forest) 86, 287, 295, 352
Kimbi (FaR) 352
Kingdoms, interlacustrine 161, 253, 256
Kinixys erosa **196**, 197
Kisangani (town) 33, 257
Kitagata (crater lake) **29**
Kivu (lake) 35
Kivu (region) 266, 268
Klainedoxa gabonensis (Eveuss) 78
Kobus defassa (Water Buck) 109
Komo (river) 30, 142
Korup (NP) 194, 287, 295, 352
Kota (ethnic group) 278
Kouilou (river) 30
Koungou (waterfalls) **12**
Kribi (town) 254
Kupe (mountain, FoR) 23, 65, 352
Kupeornis (Babblers) 161
– *chapini* (Chapin Babbler) 84
Kwa (river) 33
Kwango (river) 33, 34
Kwilu (river) 33, 251
Kyllinga erecta 132, 133
Kyoga (lake) 35
Kyoto Protocol 310

L

Laccosperma secundiflora (rattan palm) 119, 129, 155, **154**
Lagenaria siceraria (Squash) 247
Laguna, coastal, Gabon 231
Laguncularia racemosa (White Mangrove) 138
Lake Lobeke (NP) 287, 295, 352
Lake Ossa (FaR) 352
Lake Tele (community reserve) 288, 353
Lambaréné (town) 32
Land slides, colonisation 104-105, **105**
Lango (clearing) **130**, 131, 133
Laniarius bicolor (Gabon Boubou) 144
Lantana camara 308, **308**
Lasiodiscus fasciculiflorus 84
Laurembergia tetrandra 131
Lava, colonisation 104, **104**
Leaves 149, **149**
Lebrunia bushaie 84
Léfini (FaR) 284, 353
Leishmannia 189
Lékoli (river) **118**, **119**
Lékoli-Pandaka (FaR) 284
Leopold II, King of Belgium 260
Lepidochrysops 185
Lepidoptera (butterflies) 184-189
Leptaspis 152
Lianas 153-155, **153**, **154**, **155**
Libreville (town) 254, 258
Lichens 90, **90**, **96**, **160**
Likouala (river) 33, **122**
Likouala-aux-Herbes (river) **122**
Lindernia nummulariifolia 132
Liverworts 87, **87**
Livingstone, David 258
Loa Loa (parasitic disease) 189
Loasibegonia 62-**63**
– *bequaerti* 95
– *lanurensis* 95
– *wollastonii* 95, **95**
Logging 263, 322-337
–, areas 322-324
–, certification 335
–, community forests 330-331, **330**
–, concession holders 325-326
–, environmental impacts 331-332, **332**, **333**
–, first transformation 328-329, **329**
–, funding agencies 337
–, in Cameroon 322, 323
–, in Gabon 322, 323
–, market 329-330
–, principles, criteria, indicators 334
–, process 326-328, **325**, **326**, **327**, **328**,
–, production 324
–, social impacts 333
–, spatial organisation 324-325
–, suitable species 324
–, sustainability 333-335
–, tree felling **325**
–, wood market 335, **335**
Lomako (river) **123**
Lomami (river) 33, 251
Lomela (river) **123**
Lopé (FaR, NP) 15, **16**, **38**, **46**, **68**, 74, 88, **105**, 107, **108**, **109**, **129**, 236, 242, 262, 284, 285, 287, 295, 320, 340, 353
Lophira alata (Azobe) 77, 111, **131**, 172
Lophocebus albigena (White-cheeked Mangabey) 211, **211**
Lopori (river) **123**
Loranthaceae 167, **167**
Loridae 207-208

Lossi (sanctuary) 302
Loudetia 106, 107, 108
Lovoa 81
Lowa (river) 33
Lower Guinea 62
Loxodonta africana (African Elephant) 4, 130, **132**, 133, **135**, 177, 178, 228-230, **229**, 288, **318**
Lualaba (river) 33, 35, 251, 258
Luama (river) 33
Ludwigia stolonifera 132
Luhoho (river) 33
Luki (biosphere reserve) 353
Lukenie (river) 35
Lukuga (river) 33, 258
Lulonga (river) 33
Lupembian 240-241
Luvua (river) 33
Lybiidae 167
Lycaenidae 185
Lycopodium cernuum **88**
Lycopods (club mosses) 88
Lygodium smithianum **88**

M

Mabira (FoR) 85, 353
Macaranga capensis 54, 55, 56, 116-117
– *monandra* 103
– *schweinfurthii* 91
– *spinosa* 103
Macroscelidae 223
Maesobotrya 174
Maesopsis eminii 103, 104
Mai Ndombe (lake) 34-35, 126
Maiko (NP) 218, 226, 287, 353
Maiko (river) 33
Malaconotus kupeensis (Mount Kupe Bush-Shrike) 200
Malaria 189
Malebo Pool 30, 33
Malimbus cassini (Cassin's Malimbe) 129
Mambili (river) **122**
Mammals 207-237
–, biogeography 207
–, carnivores 226-227
–, first groups 19
–, intercontinental exchange 19
–, primates 207-221, 313-314
–, rodents 224-225, 313
–, ungulates 228-237
Mammea africana 176
Man and Biosphere (programme) 291
Mandrillus leucophaeus (Drill) 210-211
– *sphinx* (Mandrill) 210-211, **210**, 246
Manenguba (mountain range) 22, 23
Mangifera indica (mango tree) 257
Mangroves 75, 138-145, **138**, **139**, **269**, 309
Mangroves (NR) 287, 353
–, Gabon Estuary **140**, **142**, **145**
–, Mondah Bay **140**, **142**
Manidae 228
Manihot esculenta (Casava) 169, 255, **255**, 262
Manilkara argentea 125
– *lacera* 83
Manniophyton fulvum 155
Mansonia 81
Mantles 110-111, **111**
Maprounea africana 108
Maramagambo (forest) 28, 80, 261
Marantaceae 103, **103**, 152
Marantochloa 103
Marattia fraxinea 88, **89**

362

Marattiaceae 88
Maringa (river) **123**
Markhamia tomentosa 113
Marquesia excelsa 62, 126
Mawne (FoR) 352
Maya-Nord (swampy clearing) 130, 131, **132**, **135**
Mayombe (mountains) 262
Mayombe (ancient mountain range) 15, 16, 21
Mbam et Djerem (NP) 287, 352
Mbi (FaR) 352
Megaceryle maxima (Giant Kingfisher) 203
Megachiroptera (Fruit Bats) 222-223
Megaglossus woermanni 223
Megaphrynium 114, **273**
Melastomataceae **52**
Meliaceae 81
Melichneutes robustus (Lyre-tailed Honey Guide) 204
Mellivora capensis (Ratel) 226
Merops breweri (Black-headed Bee-eater) 111, 206
– *gularis* (Black Bee-eater) 111, 206
– *malimbicus* (Rosy Bee-eater) 106, 107, 143, 206
Mesanthemum radicans 132, 133
Metallurgy, copper 249
–, iron 248-249
Mgahinga (volcano) 26, **27**
Mgahinga Gorilla (NP) 219, 288, 295, 353
Microberlinia 77
Microcalamus 152
Microchiroptera (Insect-eating Bats) 222-223
Microcoelia 159
Microgramma owariensis **158**
Microlepidoptera 188-189
Micropotamogale ruwenzorii (Rwenzori Otter-Shrew) 223
Microsorium punctatum 143, **143**
Mikeno (volcano) 26, **28**
Milankovitsch (cycles of) 57, 61
Milicia excelsa (Iroko) 80, 103, 113, 264
Millettia chrysophylla 125
– *griffoniana* 128, **129**
– *laurentii* (Wenge) 112, 113
Millipede 164
Mimacraea 185, 187
Mimicry 187
Mimosa invisa 308
– *pigra* 119, 132
– *pudica* 308, **308**
Mimosaceae 81
Mimulopsis 86, 116, **117**
– *arborescens* 87, 117, 171
– *elliottii* 95, 117
– *solmsii* 117
Mining 263, **268**
–, gold 319
Minkébé (mountain range) 15
Minkébé (NP) 287, 353
Miocene (geologic period) 19
Miombo (habitat) 41
Miopithecus ogoouensis (Northern Talapoin) 212, **212**
– *talapoin* (Southern Talapoin) 212
Missions, Christian 253, 259, **259**, 260, 264
Mitragyna (cf. *Hallea*)
Mitumba (mountain range) 26
Mochlus fernandi (lizard) **199**
Mokoko (FoR) 353
Mondah (bay) 30, 77, 142
Mongala (river) 33

Mongiro (hot springs) 29
Mongom (ethnic group) 278
Monopetalanthus microphyllus 126
Monté Alén (NP) 83, 93, 172, 241, 288, 340, **341**, 353
Monté Temelon (NR) 93, 353
Monts Doudou (reserve, NP) 287
Moorland, afro-alpine 55, 95, **95**
Morinda 113
Mosquitoes 189
Moss line 157
Mosses 87, **87**, 157, **160**,
–, epiphyls 157, **157**
–, hanging **157**
Mostuea brunonis 129
Motacilla clara (Grey Wagtail) 129
Mouébé (clearing) 131
Mountain Gorilla Project 295
Muchoya (swamp) 54
Muchrooms, Ascomycetes 90
Mudskipper **138**
Muhabura (volcano) 26, **27**, 96
Mukura (FoR) 353
Muni (river) 30
Murchison Falls (NP) 284, 297, 353
Muridae 225
Musa (banana tree) 250-251, **251**
Musanga cecropioides 100, **101**, 103, 104, 308
– *leo-errerae* 84, 100, 104
Muscicapa cassini (Cassin's Flycatcher) 129
Mushrooms 164-166,
–, Ascomycetes 164-166
–, Basidiomycetes 190
Mussaenda 172
– *erythrophylla* **173**
– *tenuiflora* **173**
Mustelidae 226
Mycorhiza 166
Myonycteris brachycephala 222
Myosciurus pumilio (African Pygmy Squirrel) 224
Myotis 223
Myrianthus arboreus 103, 125
Myrmecocichla tholloni (Congo Moorchat) 109

N

Nandinia binotata (African Palm Civet) 227
Nanga Eboké (FaR) 352
Nasutitermes **203**
National Parks, creation 283-285
Nauclea didderichi 129
– *latifolia* 107, **108**, 111, 112
– *pobeguini* 125
Ndiza (forest) 282
Neafrapus cassini (Cassin's Swift) 206
Nectarinia johnstoni (Scarlet-tufted Malachite Sunbird) 96
– *reichenbachii* (Reichenbach's Sunbird) 129
– *rockefelleri* (Rockefeller's Sunbird) 117
Neoboutonia macrocalyx 75, 104
Neolithic period 246-247
Neotragus batesi (Bates's Antilope) 233, **233**
Nephila (spider) 110, **182**
Nephrolepis (fern) 88
– *biserrata* **158**
Nesogordonia 81
Newboldia laevis 110
Newtonia 78
– *buchananii* 74, 75, 85, 86
– *leucocarpa* 81
Ngombé (Light House) **24**
Ngotto (forest) 46, 81, 172, 321

Ngounié (river) 31
Niangara 257
Nicotiana tabacum (tobacco) 255
Niger (river), pollens studies in delta 53
Nile (river) 35, 258
Nilo-Saharians 251
Nki (FoR, NP) 287
Nkomi (laguna) 33
Nlonako (fault) 22
Nlonako (mountain) 65
Nouabalé-Ndoki (NP) 236, 288, 295, 353
Nta Ali (FoR) 353
Nta Ali (mountain) 22, 84
Ntem (river) 30
Ntomba (lake) 34, 35, **122**
Ntumu 275
Nutrients, absorption 166
Nyamulagira (volcano) 27
Nyamusingeri (crater lake) 28, 261
Nyanga (river) 30
Nyanga-Nord (FaR) 353
Nyangwe 257
Nycteridae 223
Nyiragongo (volcano) 27
Nymphaea lotus **118**, 119, 131
– *maculata* 131
– *nouchali* 119
Nymphalidae 186, **186**, 187
Nymphoides forbesiana 131
Nyong (river) 30, 126
Nyungwe (forest) 45, 86, 91, 104, 116, **116**, **117**,) 215, 226, 230, 256, 260, 263, 268, 288, 289, 290, 293, 294, 319, 339, **351**, 353

O

Observatoire des forêts d'Afrique centrale (FORAC) 296
Ocotea gabonensis 84
Ocotea michelsonii 84
Odzala-Kokoua (NP) **115**, 172, 206, 218, 231, **118**, **119**, **130**, 227, 235, 236, 262, 270, 278, 284, 285, 288, 302, 319, **340**, 340, **346**, 353
Odzala-Kokoua (NP), clearings 131, 131-133
– , rainfall 46
Oecophylla **193**
Office rwandais du tourisme et des parcs nationaux (ORTPN) 292
Offoué (river) 2, 31
Ogooué (river) 17, 30-31, **31**
–, delta 20, 32-33, **37**, 120-121, **121**, 126, 231
–, rocky banks **129**
–, sand banks 32
Oguémoué (lake) 32, 121
Oil, fields 20
Oil, Palm 254
Okapis (FaR) 287, 292, 353
Olea capensis 54, 55, 85, 87, 116, 117
– *europaea* 73
Ololo (tourist camp) 342
Olyra 152
Omphalocarpum **178**
Onangué (lake) 32, 121
Onchocerca volvulus 189
Oncoba dentata 103, **176**
– *glauca* 103
– *welwitschii* 103
Oncocalamus 155
Oplismenum 152
Orchids 62, 66, 159
Ordovician (geologic period) 18
Ortygospiza gabonensis (Black-faced Quailfinch) 107
Oryza sativa (rice) 257

Osbornictis piscivora (Aquatic Civet) 227
Osteolaemus tetraspis (Black Crocodile) 196
Otolemur crassicaudatus (Greater Galago) 208
Otus hartlaubi 206
Ouratea **109**
Oxystigma (cf. *Prioria*)
– *buchholzii* 126

P

Pachylobus edulis 247
Pagalu (cf. Annobon)
Palaeonegritic people 244
Palaeopalynology, research sites **48**
Palisota 103, **176**
Pan paniscus (Bonobo) 221, **221**
– *troglodytes* (Chimpanzee) 220-221, **220**
– *troglodytes*, tourism 339
Panda oleosa (Afane) 78, **176**
Pandanus 75, 119, **119**, 128
Pangea (palaeocontinent) 18
Pannicum 108
Panthera leo (lion) 109, 227
– *pardus* (leopard) 227, 314
Papilio antimachus 185
– *cynorta* **187**
– *dardanus* 187, **187**
– *leucotaenia* 187
– *zalmoxis* 1, 185
Papilionidae 185
Papio hamadryas (Common Baboon) 210
Pappe, Ludwig 283
Paracrocidura 223
Paramo 94
Parasites 167, **167**
Paraxerus boehmi (Boehm's Squirrel) 224
– *poensis* (Green Squirrel) 224
Parinari excelsa 75, 84, **331**
– *robusta* 125
Paspalum conjugatum 132
pastoralism 244, 249-250, **250**, 252, 253, 256, **256**
Pelecanus rufescens (Pink-backed Pelican) 143
Pelorovis antiquus 242
Pennisetum 108, 261
Pentaclethra eetveldeana (Engona) 112, 113
– *macrophylla* 103
Pentadesma lebrunii 84
– *reindersii* 65
Pentodon pentandrus 132
Pericopsis elata (Afrormosia) 78
Permian (geologic period) 18
Petrodromus tetradactylus (Four-toed Elephant Shrew) 223
Peucedanum kerstenii 96
Peuls 257
Phaseolus vulgaris (been) 255
Phataginus tricuspis (Tree Pangolin) 228, **228**
Philippia johnstoni 95
– *trimera* 94, 95
Philothamnus 198, **199**
Phlebotomus (Sand Fly) 189
Phodilus prigoginei 206
Phoeniculus (Wood Hoopoes) 161
Phoenix reclinata (False Date Palm) 118, 119, **127**, 128, 140, 143, 144, 148, **148**
Pholidotes (pangolins) 228
Phragmanthera capitata 167
Phyllanthus atripennis (Capucin Babbler) 204
Phyllastrephus 204

INDEX

Phylloscopus umbrovirens (Brown Warbler) 96
Phymatodes scolopendaria **158**
Phyrrochalcia iphis (caterpillar) 170
Picathartes oreas (Grey-necked Bal Crow) 204
Pico Basilé (NP) 353
Piedra Bere (NM) 93, 353
Piedra Nzas (inselberg) 92, 93, 353
Pierrodendron kerstingii 125
Pinus (pine tree) 267
– *patula* 337
Pipidae 194
Piptadeniastrum africanum (Dabema) 81, 84, 112, 113
Pistia stratiotes (Nile Cabbage) 118, 119, 131
Plagiosiphon emarginatus 125
Plain, coastal, origin 20
Plantations 337
Plants, medicinal 273
Platycerium (fern) 88
– *angolense* **158**, 159
Platycorine (orchid) 107
Ploceus alienus (Strange Weaver) 75, 87
– *insignis* (Brown-capped Weaver) 161
– *nigrimentum* (Black-chinned Weaver) 109
– *subpersonatus* (Loango Slender-billed Weaver) 144
Pneumatophores 148, **148**
Pobeguina (cf. *Anadelphia*)
Podica senegalensis (African Finfoot) 129
Podocarpaceae 51, **53**
Podocarpus 53, **55**, 56, 89, 337
– *gracilior* 89
– *latifolius* 74, 89, 117
– *mannii* **53**, 66, 90
Podococcus barteri (palm) 63
Podostemaceae 17
Poecilogale albinucha (Striped Weasel) 226
Poga oleosa 176
Poiana richardsoni (African Linsang) 226, **227**
Poicephalus gulielmi (Jardine's Parrot) 2002
– *robustus* (Cape Parrot) 202
Polyalthia (cf. *Greenwayodendron*)
Polyboroides typus (African Harrier Haw) 205
Polyscias auquatoguineensis 93
– *fulva* 56, 104
Polystachya (orchid) 143, 159
Population movements 262, 264, 268, 269
Portes d'Enfer (rapids) 33, 35
Portes de l'Okanda (rapids) 31
Port-Gentil (town) 33
Portugais 253
Potamochoerus larvatus (Bushpig) 231
– *porcus* (Red River Hog) 231, **231**
Potamogale velox (Otter Shrew) 223, **223**
Pouteria altissima 85
– *superba* 81
Precambrian (geologic period) 16
Principe (island) 24, 66, 80
Prionops alberti (Black Helmet Shrike) 84
Prioria 78, 81
– *balsamifera* (Tola) 78, 81
– *oxyphylla* (Tchitola) 78, 81
Procavidae 228
Procolobus pennanti (Red Colobus) 217, 288
Procubitermes **190**
Proterozoic (geologic period) 16
Protoxerus stangeri (Giant Squirrel) **224**
Prunus africanus 84, 323
Pseudacraea 187

Pseudochelidon eurystomima (African River Martin) 106, 107
Pseudohirundo griseopyga (Grey-rumped Swallow) 107
Pseudospondias (Ofoss) **273**
Psittacus erithacus (Grey Parrot) 127, **202**
Psorospermum febrifugum 108
Pteridium aquilinum (fern) 75, 87, 88, 105
Pteridophytes 49, 88
Pteridosperms 49
Pterocarpus soyauxii (Padouk) 172
Pterodicticus potto (Potto) 208
Pteronetta hartlaubi (Hartlaub's Duck) **133**
Pterygota 81, 85
Pterygota macrocarpa 103
Pterygota mildbraedii 83
Puelia 152
Pycnanthus angolensis (Ilomba) 78, 80, 103, 112, 113, 172
Pycreus mundtii 132, 134
Pygmies 242-243, **243**, 251, 252, 256, 270, 274, 311, 312, 315, **316**
Pygmy Chimpanzee (cf. Bonobo)
Python sebae (Rock Python) 196

Q

Queen Elizabeth (NP) 28, 80, 83, 284, 297, 232, 336, 353

R

Railway 262
Rana lepus **195**
Ranidae 194
Rapanea 54, 55
– *rhododendroides* 91, 95
Raphia 119, **127**, 171, 273, 315
– *regalis* 78
Raphidura sabini (Sabin's Swift) 206
Rauvolfia vomitoria 103, 110
Redunca arundinum (Southern Reedbuck) 109
Refugees 290
Refuges, forest 55, 56,
–, Haffer's theory 58-59
Rega (ethnic group) 265, 268
REIMP (programme) 296
Relief 14, 20-21
Remusatia vivipara 155
Reproduction 171-178
–, cauliflory 174, **174**, 175, **178**
–, flowering 171-172
–, fruit dispersal 175-177
–, germination 177-178
–, pollinisation 172
Reptiles 196-199
Research 343-347
–, actors 343-344
Reserves, faunal, creation 282-285
–, forest, creation 285-286
–, hunting, creation 283
Resources, air 148
–, light 149-161
–, natural, management 315
–, nutrient conservation 168
–, nutrient cycle 162-166, **162**
–, nutrient decomposition 164-165, **179**
–, water 148-149
Rhamnophis aethiopissa 199
Rhaphidophora africana 155
Rhektophyllum mirabile 155
Rhipsalis baccifera 156, 159
Rhizophora (Red Mangrove) 139, 140, 143, **145**

– *harrisonii* 138, 139
– *mangle* 138, 139
– *racemosa* 138, 139, **139**
Rhynchelytrum 106
Rhynchocyon cirnei (Chequered Elephant Shrew) 223-224
Rhynchospora brevirostris 132
Rhynchospora corymbosa 132, 133, 134
Ricinodendron heudelotii (Essesang) 103
Rinderpest 260-261
Rio Campo (NR) 353
River banks 128-129, **128**, **129**
Rivers 30-36
Roads 262, 275
Rocks, origin 14
Roots **166**, 168
–, adventicuous **154**
–, stilt 102, 103
Roussettus aegyptiacus (Egyptian Fruit Bat) 222
Rubber 254, 261-262
Rubiaceae 107
Rubus 95
Rugezi (swamp) 35
Ruki (river) 33, 122
Rumonge-Vianda (FoR) 353
Rumpi (mountains) 22, 65, 353
Rusizi (NP) 284, 353
Rusizi (river) 35, **36**
Ruvubu (NP) 284, 288, **290**, 353
Ruwenzorornis johnstoni (Rwenzori Turaco) 202
Rwanda **259**, 261, 263, 264, 265, 267, 268, 269, 284
Rwenzori (mountains) 26, 86, 91, 95, 96, 259
Rwenzori (NP) 288, 293, 353

S

Sabyinyo (volcano) 26, **27**, 28
Saccharum officinale (Sugar Cane) 251, 263
Sacoglottis gabonensis (Ozouga) 63, 77, 111
Salonga (NP) 221, 287, 292, 353
Salve Trade, Atlantic 253-254
Sanaga (FaR) 352
Sanaga (river) 30
Sangha (river) 30, 33, **122**
Sangoan 240-241
Sango-Bay (FoR) 353
Sanseviera cylindricum 82
– *parva* 82
Santchou (FaR) 284, 352
Santiria trimera (Ebab) 77, 78, 172
Santiriopsis (cf. *Santiria*)
São Tomé (island) 24, **24**, 66, 71, 80, 253
Sao-Tomé and Principe 288
Sarcocephalus esculentus 80
Sarcophrynium schweinfurthianum 113
Sarothrura pulchra (White-spotted Crake) 202
Saturnidae 188, **188**
Satyrium 96, 160
Savanna, Bateke Plateau 107-108
–, biome 40
–, coastal 106
–, colonisation 111-114
–, fire 41
–, Lope 112
–, Niari 105
–, Odzala-Kokoua 108, **109**
–, vegetation 106-109
Savorgnan de Brazza, Henri 258, 259
Scaphopetalum thonneri 78
Schefflera 161

Schizachyrium 107
Sclerosperma mannii (palm) 63
Scolecomorphidae 195
Scorodophloeus zenkeri (Divida) 78
Scotopelia bouvieri 206
– *peli* (Pel's Fishing Owl) 206, **206**
Scrophulariaceae 167
Scutisorex somerini 223
Scutobegonia 62-**63**
Scyphocephalium ochocoa (Sorro) 78
Scytopetalum klaineanum (Odzikuna) 77, 78
Selaginella congoensis 132, 133
– *myosurus* 88
Selaginellas 88
Semliki (ancient lake) 35
Semliki (river) 35
Semuliki (NP) 80, 288, 294, 353
Senecio johnstoni 95, **95**
Sericostachys scandens 75, 87, 155, 171
Shi (ethnic group) 268
Simulium damnosum (small black fly) 189
Sinarundinaria alpina (Mountain Bamboo) 116-117, **117**, 171
Sindora klaineana 126, 155
Slave Trade, Arabic 257
Sleeping sickness 189
Smallpox 260
Snakes 196-199
Soils 162-163, **163**
Solanum (Irish Potato) 262
– *aethiopicum* (African Aubergine) 247
– *melongena* (Aubergine) 251
Sorghum bicolor (Sorghum) 247
Soricidae 223
Spathodea campanulata (African Tulip Tree) 110, 174, 307
Speciation 58-61, **59**
Species, diversity, origin 58
–, exotic 307-308
Speirops melanocephalus (Cameroon White-Eye) 96
Speke, John Hanning 258
Sphagnum 95, 132
Sphingidae 188
Spiders, social **183**
–, Tarantula 182
Spizaetus africanus (Cassin's Hawk Eagle) 205
Stachyothyrsus 78
Stanley Pool (cf. Malebo Pool)
Stanley, Henri Morton 257, 258, 259
Staudtia gabonensis 80
– *stipitata* (Niove) 78, 112
Stephanoaetus coronatus (Crowned Eagle) 205
Steppe 76
Sterculia 81
Sterculiaceae 81
Sterna hirundo (Common Tern) 120
Stoebe 55
Stone Age 240-242
Stone lines 56, **163**
Strangler 161, **161**
Streptogyna 152
Streptonema pseudocola 125
Strombosia pustulata 134
Strychnos camptoneura 134
– *potatorum* 125
Successions, coastal 136-145
–, primary 104-105
–, secondary 102-104
Suidae 231
Swahili 257

Sylvicapra grimmia (Bush Duiker) 109
Sylviidae 204
Symphonia globulifera 62, 74, 148
Syncerus caffer (African Buffalo) 93, 130, 133, 134, 236, **236**, **237**, 246
Syzygium guineensis 54, 83, 89, 116, 117
– *rowlandi* 91, 125
– *staudtii* 84

T

Tabernanthe iboga (Iboga) **272**
Takamanda (FoR) 353
Tanganyika (lake) 25, 35, 36, 257
Tanne 140, **140**
Tauraco macrorhynchus (Crested Turaco) 2002
– *persa* (Green Turaco) 202
Tchabal Mbabo (mountain) 65
Tchad (lake) 56
Telacanthura melanopyga (Mottle-throated Spinetailed Swift) 206
Telipna (butterfly) **187**
Tembo (ethnic group) 251, 265, 268
Tenrecidae 223
Terminalia catappa (Wild Almond) 307, **307**
– *superba* (Limba) 81, 103, 113
Termites 190-191, **190**, **191**
Termitomyces 190
Testulea gabonensis 63, 77
Tetraberlinia 78
Tetrapleura tetrapleura **272**
Tetraponera aethiops (ant) **192**
Theobroma cacao (Cacao Tree) **262**, 263,
Thickets, ericaceous **94**, 95
Thonningia sanguinea **167**
Thryonomys (Cane-Rat) 246
Tigriornis leucolophus (Tiger Bittern) 143
Timaliidae (Babblers) 204
Tippo Tib 257
Tithonia diversifolia 308, **308**
Tockus hartlaubi (Black Dwarf Hornbill) 202
Torenia thouarsi 132
Toro (region), volcanism 28
Tortoises 196
Tourism 338-342

–, ecotourism 342
–, in Central Africa 338-341
–, in forest 341
–, tour-operators 341-342
Towns 262
Toxicodryas pulverulenta **198**
Trachypogon 108
Tragelaphus euryceros (Bongo) 93, 130, 180, 235, 282
– *scriptus* (Bush Buck) 109, 234, 246
– *spekei* (Sitatunga) 129, 130, 235, **235**
Tragulidae 232
Transgressions (theory) 59
Transpiration 149
Treculia africana 129
Tree-fall gap 100-102, **100**, 178
TREES (programme) 306
Trema orientalis 102, 103
Triassic (geologic period) 18
Trichechus senegalensis (African Manatee) 120
Trichilia heudelotii 80
Trichobatrachus robustus 194
Tridactyle (orchid) 159
Triplochiton scleroxylon (Ayous) 81, 103
Tropenbos (programme) 293
Tropicranus albocristatus (African White-crested Hornbill) 202
Trypanosoma gambiense 189
Trypanosomiasis (Sleeping Sickness) 261
Tshitolian 241
Tshuapa (river) 123
Tsoulou (FaR) 288, 353
Turnix sylvatica (Common Button-Quail) 112
Turtur brehmeri (Blue-headed Wood Dove) 201
– *tympanistria* (Tambourine Dove) 2002
Twa (ethnic group) 242, 274

U

Uapaca 75, 120, 125, **125**
– *guineensis* 84
– *heudelotii* 128
Ubangi (river) 33, 34, 35, **122**

Ubangians 245, 251, 252
Uganda 44, 261, 265, **267**, 268, 289, 290
Ujiji (town) 257
Ulindi (river) 33
Ulmaceae 81
Understorey 152, **152**, **157**
UNDP 294
UNEP 309
Unesco 76, 290, 291
Uromanis tetradactyla (Long-tailed Pangolin) 228, **228**
USAID 293
Usnea 82, 90, **160**
Utricularia 119

V

Varanus ornatus (Forest Monitor Lizard) 143, 196
Vegetation, altitude gradient 70-71
–, classification 76
–, climatic gradient 72-73
–, ecologic gradients 69-75, 76
–, edaphic gradient 73-75
–, effects of mist 71, 86
–, effects of rainfall 72
–, temporal gradients 99-145
Vernonia conferta 102, 103
Victoria (lake) 35, 36
Vigna sinensis (Cowpea) 247
Viola eminii 95
Virunga (NP) 86, 95, 96, 231, 267, 287, 289, 339, 353
Virunga (volcanoes) 35, 215, 218, 219, 230, 26-28, **27**, 104, 256, 264
Vitex ciliata 81
– *doniana* 110
– *madiensis* 108
– *pachyphylla* 81
Viverridae 226
Voandzeia subterranea (Bambara Ground Nut) 247
Volcanoes (NP) 104, 117, **117**, 288, **288**, 289, 338, **349**, 353
Vossia cuspidata 119, **120**, 121, 128

W

Warburgia ugandensis 169
Wasmannia auropunctata 193, 308
Wasp 174
Water Hyacinth (cf. *Eichhornia crassipes*)
WCMC 307
Wele (river) 30
Wetlands, colonisation 118-129
Wildlife Conservation Society (WCS) 287, 295, 344
Wiltonian 241
Wine, palm 273
Wonga-Wongué (presidential reserve) 287, 353
World Bank 290, 294, 296, 337
World Resources Institute (WRI) 299
World War, first 260
World War, second 263, 264
Wouri (river) 30
Wuchereria bancrofti 189
WWF 295, 298, 298-299, 306, 315, 335

X

Xylopia 110
Xymalos 70, 74
Xyris 132

Y

Yangambi (conference) 76
Yangambi (floristic reserve) 353
Yaoundé, (conference of Heads of States) 292
Yellow Fever 189

Z

Zadié (river) 126, **126**
Zanthoxylum 103
Zanzibarites 257, 259
Zea mays (corn) 255
Zenkerella insignis (Cameroon Scaly-Tail) 225
Zoothera 204
– *princei* (Grey Ground-Trush) 204

Acknowledgements

Many people have contributed to this book, often without being aware of it. The author wants to acknowledge Bernadette Abela, Kate Abernethy, Gaston Achoundoung, Marcelin Agnagna, Philippe Auzel, Conrad Aveling, Eulalie Bashige, Simon Bearder, Serge Bahuchet, Andrew Balmford, Pierre Dachet, Didier Bastin, Allard Blom, Frans Breteler, Marius Burger, Tom Butyinski, Robert Closen, Jean-Gael Collomb, Marc Colyn, Pierre Devillers, Jean-Louis Doucet, Ernestine Effa, Paul Elkan, Christian Erard, Brian Fisher, Roger Fotso, Yves Gaugris, Alan Hamilton, John Hart, Terese Hart, Philippe Hecketsweiler, Peter Howard, Jean-Pierre d'Huart, Joseph Ipalaka, Philippe Jaenmart, André Kamdem Toham, Kitsidikiti Luhunu, Sally Lahm, Marc Languy, Emile Mamfoumbi-Koumbila, Sammy Mankoto wa Mbaele, Philippe Mayaux, Emmanuel de Mérode, Timothy Moermond, Robert Nasi, Bruno O'Heix, Isabelle Porteous, Ralph Ridder, Ernst Rohrbach, Fabien Sordet, Marc Sosef, Jacques Stebler, Frank Stennmans, Michael Storz, Tom Struhsaker, feu Georges Troupin, Renaat Van Rompaey, Amy Vedder, William Weber, David Wilkie, Roger Wilson et Lee White. Through discussions, generally informal, those people have inevitably influenced his thoughts.

He also want to express his thanks to Hugo Dall'asta, Willy Delvingt, Charles Doumenge, Jean-Marc Froment, Norbert Gami, Steve Gartlan, Annie Gautier-Hion, Jean-Jacques Landrot, Maurice Leponce, Jean Malais, Richard Oslisly, Ingrid Parmentier, Dominiek Plouvier, Filippo Saracco, Tariq Stevart, Dominique Touranchet, Caroline Tutin, Muriel Vives and Chris Wilks, who have read, commented on and enriched chapters or sections of this book. His thanks go also to Noël Ovono who contributed to the updating of the information regarding protected areas. Above all, he wants to thank Patrice Christy who not only commented on large parts of this book but also agreed to make the final proofreading.

As for the illustrations, the author's thanks go to all photographers who have made their pictures available. He acknowledges Jean-Philippe Biteau, who allowed him to make pictures in his private orchid collection in Libreville, and Gael Vande weghe, who helped him to photograph butterflies in the wild as well as in his collections. He acknowledges especially CNRS-Diffusion and Alain R. Devez who allowed the use of often unique pictures taken in the forests of north-eastern Gabon during the long stay of this photographer at the Makokou Research Station. At the same time he wants to express his thanks to the scientists of CNRS-ECOTROP: Jacques Bradburry, André Brosset, Pierre Charles-Dominique, Gérard Dubost, Louise Emmons, Christian Erard, Jean-Pierre Gasc, Pierre-Paul Grassé, Louis-Philippe Knoepffler and Elizabeth Pages. He also thanks the BIOS Agency for its collaboration in the gathering of the necessary illustrations. He thanks WCS, especially Lee White, who facilitated the aerial survey of some sites in Gabon, and the Joint Research Center (JRC), especially Philippe Mayaux, who generously provided the satellite images included in this book.

He thanks Paul Posso, director of IRET, who allowed him to work at the Makokou Station; Michel Mbomoh Upiangu, regional coordinator of l'ADIE, where this book, started in 1995, was finally completed as part of the Regional Environmental Information Management Programme (REIMP).

Last but not least, his thanks go to Enrico Pironio of the European Commission and Filippo Saracco of the Delegation of the European Commission, first in Libreville, later in Kinshasa, who always have supported this project; the staff of agreco, and especially Muriel Vives, in charge of communication at the ECOFAC programme, who found the so much needed funds for the printing. His thanks go to the ECOFAC Programme and the Fonds français pour l'environnement mondial (FFEM) for their financial contributions, without which this book would never have existed.

Photo credits

All pictures by J.-P. Vande weghe, except:

0.3: Roger Leguen; 0.4: Bruce Davidson; 1.6: Ingrid Parmentier; 1.15: Jean Trolez; 1.17-18: Steve Gartlan; 1.19: Michel Günther/BIOS; 1.23-24: Conrad Aveling; 1.30: ®CNRS/Alain R. Devez/ECOTROP (UMR 8571); 2.20-21: Michel Günther/BIOS; 2.42-44: Michel Günther/BIOS; 3.2: Conrad Aveling; 3.16: Michel Günther/BIOS; 3.28: Michel Günther/BIOS; 3.45-46: Ingrid Parmentier; 3.48-50: Ingrid Parmentier; 3.51: M. & C. Denis-Huot/BIOS; 3.56: M. & C. Denis-Huot/BIOS; 4.10: Y. Thonnerieux/BIOS; 4.22: Philippe Dejace; 4.33: Michel Günther/BIOS; 4.59: Roger Leguen; 4.60: Conrad Aveling; 4.61-63: Philippe Dejace; 4.64: Bruce Davidson; 4.65: Philippe Dejace; 5.10: Bruce Davidson; 5.55-56: Charles Doumenge; 5.68: Michel Günther/BIOS; 6.1: Michel Günther/BIOS; 6.2: Bruce Davidson; 6.6: ®CNRS/Alain R. Devez/ECOTROP (UMR 8571); 6.29: Roger Leguen; 6.30-33: ®CNRS/Alain R. Devez/ECOTROP (UMR 8571); 6.33: ®CNRS/Alain R. Devez/ECOTROP (UMR 8571); 6.36: ®CNRS/Alain R. Devez/ECOTROP (UMR 8571); 6.38: Bruce Davidson; 6.39-45: ®CNRS/Alain R. Devez/ECOTROP (UMR 8571); 6.47-49: ®CNRS/Alain R. Devez/ECOTROP (UMR 8571); 6.50: Bruce Davidson; 6.51-60: ®CNRS/Alain R. Devez/ECOTROP (UMR 8571); 6.61: Bruce Davidson; 6.62-63: ®CNRS/Alain R. Devez/ECOTROP (UMR 8571); 6.65-66: ®CNRS/Alain R. Devez/ECOTROP (UMR 8571); 6.68-70: ®CNRS/Alain R. Devez/ECOTROP (UMR 8571); 6.71: Cyril Ruoso/Cirme/BIOS; 6.73: Bruce Davidson; 6.74: T. Crocetta/BIOS; 6.75: ®CNRS/Alain R. Devez/ECOTROP (UMR 8571); 6.76: D. Heuclin/BIOS; 6.77: J.L. Ziegler/BIOS; 6.79: Roland Seitre/BIOS; 6.80: Cyril Ruoso/Cirme/BIOS; 6.81-85: Bruce Davidson; 6.87: Michel Günther/BIOS; 6.88: M. & C. Denis-Huot/BIOS; 6.89: ®CNRS/Alain R. Devez/ECOTROP (UMR 8571); 6.90: Cyril Ruoso/Cirme/BIOS; 6.92: ®CNRS/Alain R. Devez/ECOTROP (UMR 8571); 6.93: Bruce Davidson; 6.94-105: ®CNRS/Alain R. Devez/ECOTROP (UMR 8571); 6.106: Bruce Davidson; 6.107: Philippe Dejace; 6.108: Bruce Davidson; 6.109: G. Renson/BIOS; 6.110-112: ®CNRS/Alain R. Devez/ECOTROP (UMR 8571); 6.114: ®CNRS/Alain R. Devez/ECOTROP (UMR 8571); 6.115-116: Bruce Davidson; 6.118: Roger Leguen; 7.6: Michel Günther/BIOS; 7.8: Richard Oslisly; 7.25: Musée royal de l'Afrique centrale, Tervuren, photo Michel; 7.26-28: Michel Günther/BIOS; 7.29: Musée royal de l'Afrique centrale, Tervuren, photo Deheyn; 7.47: Conrad Aveling; 7.49: Michel Günther/BIOS; 7.51-52: Conrad Aveling; 8.2: Musée royal de l'Afrique centrale, Tervuren, photo Grauwet; 8.3: Musée royal de l'Afrique centrale, Tervuren; 8.4: Musée royal de l'Afrique centrale, Tervuren; collection Gatti; 8.14: Conrad Aveling; 9.9-11: ®CNRS/Alain R. Devez/ECOTROP (UMR 8671); 9.12-13: Bruce Davidson; 9.14: Michel Gunther/BIOS; 9.15-17: Conrad Aveling; 9.18: Roger Leguen; 9.19: Bruce Davidson; 9.21: Michel Günther/BIOS; 9.29-31: Michel Günther/BIOS; 9.34: Michel Günther/BIOS; 9.37: Michel Günther/BIOS; 9.43: Y. Lefèvre/BIOS; 9.45: Muriel Vives; 9.46: Conrad Aveling; 9.48-50: Bas Huijbregts/WWF; 9.51-52: Bruce Davidson.

The *Agence internationale pour le développment de l'information environnementale* (ADIE, international agency for the development of environmental information) is headquartered in Libreville and has offices in Cameroon, Congo, Gabon, Equatorial Guinea, the Central African Republic, the Democratic Republic of Congo and Chad. This organisation is made up of administrations, public companies, private institutions, associations as well as nongovernmental organisations. Its goal is to circulate and capitalize on environmental information, offer support for decisions, outline and implement reliable methods for the production and handling of environmental information, and to enhance expertise in Central Africa in the environmental area.

The **ECOFAC** programme is funded by the European Commission and has been active in Central Africa since 1992 for the conservation of natural resources through the planning of protected areas. Eight such areas currently make up the first network of the sub-region: the Dja sanctuary in Cameroon, the Odzala-Kokoua national park in Congo, the Lope national park in Gabon, the Monte Alén national park in Equatorial Guinea, the Obo nature preserve in São Tome and Principe, the future Mbaere Bodingue national park and the Manovo-Gounda-Saint Floris national park in the Democratic Republic of Congo, and the Zakouma national park in Chad.

The *Fonds français pour l'environnement mondial* (FFEM, French fund for a global environment) was set up in 1994 to foster preservation and enhance the status of biodiversity in developing countries and in countries in transition. To achieve this, the Fund provides resources in the form of donations to projects that have a positive impact on the protection of biodiversity.

Jean-Pierre Vande weghe
Born in Ghent (Belgium) in 1940, he became fascinated by nature very early on. While in medical school he spent a great deal of time studying birds in various locations (Zealand, the shores of the Channel, the forests of Argonne, the shores of the Danube and Tisza, in Macedonia and in Greece). In 1969, as part of the Belgian Medical Cooperation, he went to Rwanda where he learnt more about African nature (the savannah and then the forests) and studied birds in Rwanda and Burundi. Since 1986 he has devoted himself entirely to the preservation of nature. Until late 1990 he coordinated a WWF-Belgium programme in the Akagera national park; his book, "Akagera: Water, Grass and Fire" shows us what this magnificent site looked like before the tragic events that affected the country between 1990 and 1994. After spending a few years in Burundi and Uganda he has been working in Gabon since 1999 as part of ADIE.

DESIGN: Geert Verstaen
TRANSLATION: Jean Devillers-Terschuren
PRINTED AND BOUND by Lannoo, Tielt, Belgium

© 2004 Ecofac
First English and French edition Lannoo Publishers, Tielt - Belgium 2004
Second English edition, Protea Book House, Pretoria - South Africa 2004

Protea Book House
PO Box 35110, Menlopark, 0102
1067 Burnett Street, Hatfield, 0083
protea@intekom.co.za

D/2004/45/97
ISBN 1 86919 073 4
NUR 653/940

All rights reserved. No part of this edition may be reproduced and/or made public by means of printing, photocopy, microfilm, electronic support and in any form whatsoever without prior written authorisation of the publisher.

The ideas expressed in this book are those of the author and do not necessarily reflect the ideas of the organisations or people who allowed its publication.
This book was made possible thanks to European Commission funds as well as the French fund for worldwide environment (*Fonds Français pour l'environnement mondial*, FFEM).